PRESENTING FUTURES PAST

Science Fiction and the History of Science

EDITED BY

Amanda Rees and Iwan Rhys Morus

OSIRIS | **34**

A Research Journal Devoted to the
History of Science and Its Cultural Influences

Osiris

Series Editors, 2018–2022

W. PATRICK McCRAY, *University of California Santa Barbara*

SUMAN SETH, *Cornell University*

Volumes 33 to 37 aim to connect the history of science with other areas of historical scholarship. Volumes of the journal are designed to explore how, where, and why science draws upon and contributes to society, culture, and politics. The journal's editors and board members strongly encourage proposals that engage with and examine broad themes while aiming for diversity across time and space. The journal is also very interested in receiving proposals that assess the state of the history of science as a field, broadly construed, in both established and emerging areas of scholarship.

33   LUKAS RIEPPEL, EUGENIA LEAN, & WILLIAM DERINGER, EDS., *Science and Capitalism: Entangled Histories*

34   AMANDA REES & IWAN RHYS MORUS, EDS., *Presenting Futures Past: Science Fiction and the History of Science*

Series editor, 2013–2017

ANDREA RUSNOCK, *University of Rhode Island*

Volumes 28 to 32 in this series are designed to connect the history of science to broader cultural developments, and to place scientific ideas, institutions, practices, and practitioners within international and global contexts. Some volumes address new themes in the history of science and explore new categories of analysis, while others assess the "state of the field" in various established and emerging areas of the history of science.

28   ALEXANDRA HUI, JULIA KURSELL, & MYLES W. JACKSON, EDS., *Music, Sound, and the Laboratory from 1750 to 1980*

29   MATTHEW DANIEL EDDY, SEYMOUR H. MAUSKOPF, & WILLIAM R. NEWMAN, EDS., *Chemical Knowledge in the Early Modern World*

30   ERIKA LORRAINE MILAM & ROBERT A. NYE, EDS., *Scientific Masculinities*

31   OTNIEL E. DROR, BETTINA HITZER, ANJA LAUKÖTTER, & PILAR LEÓN-SANZ, EDS., *History of Science and the Emotions*

32   ELENA ARONOVA, CHRISTINE VON OERTZEN, & DAVID SEPKOSKI, EDS., *Data Histories*

Acknowledgments

This volume grew out of a lifelong entanglement with reading and writing science fiction on the part of both editors. We are enormously grateful to the *Osiris* editorial board for allowing us to combine our personal and professional lives in this way, and we would also like to extend our fervent thanks to the general editors, W. Patrick McCray and Suman Seth. Their supportive critical commentary was always productive, often entertaining, and invariably rigorous. We also thank the anonymous reviewers, who provided thorough, generous, and stimulating feedback to each of the essays contained herein, and we offer our gratitude to the contributing authors, whose scholarship has produced a collection of essays with which we are proud to be associated.

While editing this volume, we were also engaged in the Unsettling Scientific Stories Project, which studied ways in which the sciences were used to write the history of the future during the long twentieth century. In March 2018, we held the conference, "Imagining the History of the Future" at the University of York (UK), at which many of the authors were able to meet each other for the first time. This meeting played a crucial role in developing our thinking about the history of the future, the history of science, and science fiction. We would like to thank the participants at that conference in general, the members of the Unsettling Scientific Stories Project team in particular, and we would especially like to thank Sam Robinson, who played a central role in the success of both the conference and the project. Financial support for both was provided by the Arts and Humanities Research Council of the United Kingdom, and we acknowledge this with deep gratitude.

Finally, we would like to remember the late Professor John Forrester. Years ago, he wondered what science fiction could do for the history of science. We wish he was still here to see one set of answers to his question.

Presenting Futures Past:
Science Fiction and the History of Science

by Amanda Rees* and Iwan Rhys Morus[§]

This volume of *Osiris* had, as its inspiration, the question of what science fiction could do for the history of science. Or, to put it another way, to what historiographical, intellectual, and pragmatic uses have historians of science put science fiction, and how might these strategies develop in the future? Initial efforts to answer these questions were sketchy, to say the least. Despite the fact that the intellectual significance of fiction, literature, and the imaginaries has increasingly been recognized by the humanities in general and by science studies in particular, science fiction itself has seemed—until recently—to remain on the disciplinary sidelines.[1] However, in the past few years, this has begun to change. Panels on science fiction (SF) have begun to appear at conferences organized by societies devoted to different aspects of the history and cultures of science; symposia and workshops that have as their focus the relationship between SF and science studies have been held; and the role that science fiction plays in both lay and professional understanding of, and engagement with, scientific knowledge is being seriously interrogated by scholars. This volume of *Osiris*, then, seeks to bring together scholars involved in these recent developments to consider how the history of science should position itself in relation to SF.

The first question that might be asked is, "Why?" Why should historians worry about stories—fantastical, fictional accounts—of the future? There are a number of reasons, but the most important is that the future itself has a history, and that history is deeply entangled in the relationship between science and society.

WANTED—PROFESSORS OF FORESIGHT!

Efforts to predict and assess the future are almost certainly as old as humanity—but modern attempts to do so in an organized and systematic fashion probably date back

---

* Department of Sociology, University of York, Heslington, YO10 5DD, UK; amanda.rees@york.ac.uk.

§ Department of History & Welsh History, 3.06 International Politics Building, Aberystwyth University, Aberystwyth, SY23 3FE, Wales; irm@aber.ac.uk.

The authors would like to thank *Osiris* editors W. Patrick McCray and Suman Seth for their intellectual and emotional contribution to this volume. Their critical commentary was enormously useful and very productive. We are also tremendously grateful to the anonymous reviewers for the staggering amount of work they put in to providing thoughtful, helpful, and stimulating feedback on each of the essays contained herein. Finally, we acknowledge with deep gratitude the financial support of the Arts and Humanities Research Council of the United Kingdom.

[1] Martin Willis, *Literature and Science: Readers' Guides to Essential Criticism* (London, 2015); Charlotte Sleigh, *Literature and Science* (London, 2010); Kanta Dihal, "On Science Fiction as a Separate Field," *Journal of Literature and Science* 10 (2017): 32–6.

to H. G. Wells's call in 1932 for "Professors of Foresight." Their job would be to prepare the country for the changes that would be wrought by technological and scientific development. In one of his many BBC radio broadcasts, Wells used the history of the motor car to illustrate both the desirable and unintended consequences of improved communication and transport technology. Alongside the facilitation of commerce, industry, and the opportunities for contact between kith and kin came, he claimed, the destruction of the railways, the congestion of towns, and the emergence of the "motor bandit." It was now possible, as he pointed out, to do murder in Devon at midnight and take breakfast in Birmingham the next morning.[2] Nothing, Wells felt, was being done to anticipate, or prepare for, the staggering social, economic, or political dislocations that accompanied technological and scientific change. Of course, this broadcast was neither the first nor the last of Wells's efforts to awaken his fellows to the need to think about the future—his nonfiction *Anticipations of the Reaction of Mechanical and Scientific Progress Upon Human Life and Thought: An Experiment in Prophecy* had been serialized in the *Fortnightly Review* from late spring 1901, and his fictional *The Shape of Things to Come* was to appear in 1933—but this was one of the first times that someone had called for institutionalization of the systematic, scientific study of the future.[3] Wells wanted, understandably, to make imagining the future respectable.

Wells was a prolific writer and commentator in many genres, but he is, of course, best remembered today for his science fiction work. Books such as *The War of the Worlds* (1898) and *The Island of Doctor Moreau* (1896) retain their purchase on the public imagination in the West, even if—like Mary Shelley's *Frankenstein* (1818)— they operate now more as cultural resources than as novels. Together with Shelley and Aldous Huxley, Wells is often treated as one of the progenitors of the genre of SF and is credited with establishing classic themes that later writers would return to and reimagine. But what is notable about all three figures is not whether they predicted the emergence of biotechnology, interplanetary travel, rapid communication, or artificial reproduction. Science fiction is not prophecy, and any lucky hits can only be recognized retrospectively and serendipitously. What is important is that all three writers reflected a particular understanding of science that emerged within a given social context. The key point is that they did not explore their scientific imaginings within an abstracted realm, but situated them, together with their consequences, within a living social world. While arguments about the coproduction of science and society might be a philosophical step too far for Wells—although it might be interesting to revisit his Martians from an actor-network perspective—the core of his approach, which was his stress on the necessity of situating scientific, medical, and technological changes in their social, political, cultural, and economic contexts, clearly shares the philosophical and empirical approach of large areas in the history of science.

This makes it all the stranger, then, that historians of science have, historically, paid relatively little attention to science fiction. Partly, we would suggest, this has to do

---

[2] H. G. Wells, "Wanted - Professors of Foresight!" first broadcast on the *National Programme*, BBC, 19 November 1932.

[3] In the United States, *Anticipations* appeared in the *North American Review* from summer 1901. The series appeared in book form in winter of that year as H. G. Wells, *Anticipations of the Reaction of Mechanical and Scientific Progress Upon Human Life and Thought: An Experiment in Prophecy* (London, 1901), and was reissued by Chapman and Hall (London, 1914) just prior to the outbreak of World War 1. Wells, *The Shape of Things to Come* (London, 1933).

with this very question of "respectability," since science fiction has traditionally struggled to be acknowledged as a legitimate genre of creative expression. As Kingsley Amis famously quipped: "'Sf's no good,' they bellow till we're deaf. 'But this looks good.'—'Well then, its (*sic*) not sf.'"[4] Science fiction in the popular imagination can often mean tentacled monsters, knights battling with lasers rather than swords, and brains in jars. Certainly, any examination of the (early) iconography of SF might reinforce such assumptions.[5] Engagement with the content, however, will tell a different story, as this volume will show.

But it is important to note that this relative neglect is also related to the more general problems faced by scholars who want to study events that either have not yet happened, or will never happen outside the imagination. The ongoing methodological and philosophical problems and challenges faced by practitioners of counterfactual histories are a case in point.[6] Efforts to write "what if" history can, like Wells's call for Professors of Foresight, be dated to the early twentieth century with the appearance of *If It Had Happened Otherwise*, edited by J. C. Squires and including essays by Hilaire Belloc, Andre Maurois, and Winston Churchill, among others.[7] Little serious attention was paid to the field, however, until Geoffrey Hawthorn published *Plausible Worlds* in 1991, outlining and exploring three different counterfactual scenarios and prompting an ongoing historiographical debate concerning the methodological significance of this kind of "virtual" history.[8] Of course, counterfactual histories have their counterparts in alternate history, a major subgenre of science fiction, and have also played an interesting role in some recent histories of science.[9] In fact, recent years have seen a series of significant and substantive shifts in the attitude of scholars toward these exercises of the imagination, both in relation to thinking about and planning for the future, and in relation to the way in which fiction, literature, and the "not real" are treated by scholars more broadly.

FICTION AND THE FUTURE

The study of fiction is clearly now a well-established part of the remit of the humanities in any number of ways. In the first instance, it has been a key source of data. Anthropologists, geographers, historians, and sociologists alike have mined the novels of particular times and places for information about habits, practices, manners,

---

[4] Kingsley Amis and Robert Conquest, eds., *Spectrum II: A Second Science Fiction Anthology* (1962; repr., London, 1971), on copyright page.

[5] Frank M. Robinson, *Science Fiction of the 20th Century: An Illustrated History* (Portland, Ore., 1999).

[6] Catherine Gallagher, *Telling it Like it Wasn't: The Counterfactual Imagination in History and Fiction* (Chicago, Ill., 2018).

[7] J. C. Squires, ed., *If It Had Happened Otherwise* (London, 1931).

[8] Geoffrey Hawthorn, *Plausible Worlds: Possibility and Understanding in History and the Social Sciences* (Cambridge, UK, 1991).

[9] Peter Bowler, *Darwin Deleted: Imagining a World without Darwin* (Chicago, Ill., 2013). See also the work of Annie Jamieson and Greg Radick on genetics pedagogies, and the possibilities of thinking counterfactually about Mendel: "Genetic Determinism in the Genetics Curriculum: An Exploratory Study of the Effects of Mendelian and Weldonian Emphases," *Sci. and Educ.* 10 (2017): 1261–90. Those interested in fictional counterfactuals might usefully look at Kim Stanley Robinson's *The Years of Rice and Salt* (New York, N.Y., 2002), which imagines the history of science in a world where the Black Death eliminated ninety-nine percent of Europe's population; or Ted Chiang's story, "Seventy-Two Letters," which describes a Victorian industrial revolution driven by golems, rather than steam engines, in *Stories of Your Life and Others* (2000; repr., London, 2015).

and mind-sets. They have used fiction to obtain ephemeral information about fashion, and for a sense of the variety of uses to which railway timetables were put.[10] Historians of science have also made use of literature and fiction as data, and not least as a means of accessing the ways in which contemporaries understood and acted upon science and technology in both theory and practice—but as Charlotte Sleigh points out and as we have already noted, science fiction has not historically been included within the category of "literature."[11] However, fiction has not just been of interest as a source of data. The structures of fiction—the use of narrative, the deployment of rhetorical devices—has also drawn sustained scholarly attention, as have the wider practices of fiction, and in particular, their political and conceptual deployment.[12] Since the late 1970s, narrative in particular has been a key focus of critique from anthropologists and historians alike, as they have identified a slippage from telling "a" story to telling "the" story. For both disciplines, unproblematized use of narrative structures tends to wend worryingly close to claiming the status of objective neutrality. Hayden White's deservedly influential analysis of the role of narrative in historical writing charted the way in which this style enabled the author to make reality meaningful.[13] Just as White treated history as literature, James Clifford and George Marcus's *Writing Culture* (1986) took the same approach to ethnography, showing the situated and subjective nature of its production across time and space.[14] Historians and sociologists of science have taken inspiration from this approach, showing how narrative in science writing can imbue the account with an air of omniscient objectivity. As Ron Curtis puts it: "We [begin] with unanswered questions. We end with unquestioned answers."[15]

But perhaps of most interest for this volume is the gathering body of interest in fiction as practice, and in particular, the capacity of social collectives to invent both past and future. Interest in this aspect of the imaginary stretches from at least Benedict Anderson's *Imagined Communities* or Eric Hobsbawm and Terence Ranger's *The Invention of Tradition* (both appearing in 1983), to Sheila Jasanoff and Sang-Huyn Kim's more recent *Dreamscapes of Modernity* (2015).[16] Both Anderson, and Hobs-

---

[10] For geography, see Rob Kitchen and James Kneale, eds., *Lost in Space: Geographies of Science Fiction* (London, 2002); Stephen Daniels and Simon Rycroft, "Mapping the Modern City: Alan Sillitoe's Nottingham Novels," *Trans. Inst. Brit. Geogr.* 18 (1993): 460–80; and H. C. Darby, "The Regional Geography of Hardy's Wessex," *Geogr. Rev.* 38 (1948): 426–43. For sociology, see Ruth Penfold-Mounce, David Beer, and Roger Burrows, "*The Wire* as Social Science Fiction," *Sociology* 45 (2011): 152–67; and Ronald P. Corbett, "Novel Perspectives on Probation: Fiction as Sociology," *Sociological Forum* 9 (1993): 307–14.

[11] Sleigh, *Literature* (cit. n. 1).

[12] Gillian Beer, *Darwin's Plots: Evolutionary Narratives in Darwin, George Eliot and Nineteenth-Century Fiction* (London, 1983); N. Katherine Hayles, *Chaos and Order: Complex Dynamics in Literature and Science* (Chicago, Ill., 1991).

[13] Hayden White, "The Historical Text as Literary Artefact," *Clio* 3 (1974): 277–303; White, "The Value of Narrativity in the Representation of Reality," in *On Narrative*, ed. W. J. Mitchell (Chicago, Ill., 1981), 1–24.

[14] James Clifford and George Marcus, *Writing Culture: The Poetics and Politics of Ethnography* (Berkeley, Calif., 1986).

[15] Ron Curtis, "Narrative Form and Normative Force: Baconian Story-Telling in Popular Science," *Social Studies of Science* 24 (1994): 419–61, on 431.

[16] Benedict Anderson, *Imagined Communities: Reflections on the Origin and Spread of Nationalism* (London, 1983); Eric Hobsbawm and Terence Ranger, eds., *The Invention of Tradition* (Cambridge, UK, 1983); Sheila Jasanoff and Sang-Hyun Kim, *Dreamscapes of Modernity: Sociotechnical Imaginaries and the Fabrication of Power* (Chicago, Ill., 2015).

bawm and Ranger, dealt with the capacity of communities—usually nations—to moralize (to draw on Hayden White's term) their histories, and to create a narrative of the past where veridical reality was relatively irrelevant. What mattered for the present was the collective commitment to the possibilities of the past. Jasanoff and Kim's development of the concept of the "sociotechnical imaginary" applies, in contrast, to visions of technological and social futures. In their definition, these imaginaries constitute "collectively held, institutionally stabilised and publicly performed visions of desirable futures, animated by shared understandings of forms of social life and social order attainable through, and supportive of, advances in science and technology."[17] Not limited to nation-state formation, they could be supported and sustained by corporations or social movements—but crucially, they needed to be collectively held, encapsulating both the hope of progress and the fear of harm. What unites all three of these contributions, among many others, is their willingness to treat the imagination not as an exercise in fantasy or daydreaming, but as something that is done. The imaginary here is treated as organized work and performed social practice; it is meant by practitioners to have a clear, if sometimes unacknowledged, purpose.

In contrast, future studies and futurology, together with related fields such as scenario planning, are more concretely aimed at developing a clear vision of potential futures that are usually focused at global, corporate, or local levels. Although Wells's call for Professors of Foresight had initially fallen on stony ground, by the late 1960s and early 1970s, future studies had become an established part of the intellectual landscape. Crucially, however, its background and influence stretched far outside the academy. Just as Wells's original *Anticipations* had earlier caught the public imagination, becoming his first nonfiction bestseller, so Alvin Toffler's 1970 bestseller, *Future Shock*, demonstrated that futurology retained its traction on the public imagination.[18] Additionally, the period between the 1930s and 1970s was marked by increasing governmental interest in the concept of future planning, at least on the part of European nations engaged in postwar reconstruction and development. Futures studies thus had an immediate political and policy-oriented audience, together with specific methodological strategies for developing a detailed, if not necessarily comprehensive, framework within which to construct future policies. For the general public, for governments, for NGOs, and for corporations, this kind of pragmatic planning for the future was both practically essential and of abiding interest.

Futurologists were not, however, the only members of the academy actively working on mapping out—and sometimes destabilizing—visions of what was to come. As early as 1971, Margaret Mead had published her "Note on Contributions of Anthropology to the Science of the Future," encouraging anthropologists to focus on the future as well as the past and present of the societies they studied.[19] Anthropologists had also argued for their discipline's potential to contribute to prospective alien contact and space colonization.[20] Later attention focused on sustainability alongside extraterrestrials, but as Arjun Appadurai pointed out, since the core focus of anthropology

---

[17] Jasanoff and Kim, *Dreamscapes* (cit. n. 16), 4.
[18] Alvin Toffler, *Future Shock* (New York, N.Y., 1970).
[19] Margaret Mead, "The Contribution of Anthropology to the Science of the Future," in *Cultures of the Future*, ed. Magoroh Maruyama and Arthur Harkins (The Hague, Neth., 1978), 3–6.
[20] Magoroh Maruyama and Arthur Hawkins, eds., *Cultures Beyond the Earth: The Role of Anthropology in Outer Space* (New York, N.Y., 1975); Maruyama and Hawkins, *Cultures* (cit. n. 19).

tends to be continuity rather than change, such efforts were unlikely to gain much purchase.[21] In contrast, geographers and sociologists have been rather more successful in establishing and retaining an (inter)disciplinary focus on futures.[22] However, for historians—with a few prominent exceptions—the future and its past tended not to appear on the discipline's agenda.[23] Historians of science have been equally restrained.

From these two different perspectives—the study of fiction and the imaginary, and the efforts to develop concrete, practical plans or possibilities for different futures—one can see why science fiction should be the increasing focus of scholarly attention. The field of science fiction studies has done admirable work in showing, to those who were prepared to see, that bug-eyed monsters were not necessarily characteristic examples of the genre's intellectual and methodological complexity. Journals such as *Science Fiction Studies*, *Foundation*, and *Extrapolation* have analyzed, examined, and interpreted the ways in which SF has deployed metaphors of machine, infection, and laboratory; the contributions that it has (sometimes not) made to the public understanding of science; and the relationship between social structure and the technoscientific imaginary. Refusing to be tied to the Anglo-American centers of intellectual activities, theses journals have specifically examined the different national and ethnic traditions of SF, and have reached out beyond the book to consider how science, technology, and medicine are deployed as source, resource, and prize in games, films, comics, and apps. The fact that, until very recently, there have been only limited efforts by other areas of the academy to engage with these endeavors is surprising—the fact that historians and sociologists of science have largely ignored the field is to be deeply regretted. This volume of *Osiris* hopes to remedy this situation.

SCIENCE, HISTORY OF SCIENCE, AND SCIENCE FICTION

At the most basic level, science fiction should be of interest to historians of science because the two ways of knowing the world have so much in common. Like history, SF is committed to the (re)construction and exploration of an unfamiliar and often alien world; it focuses on the impact that developments in science, medicine, and technology have on political, economic, and cultural relationships; and it is intensely and critically aware of the different ways in which the social (class, gender, race, sex, species) has inflected the experience of the scientific in past, present, and future. Science fiction is, in short, a key source of social critique in contemporary technoscientific society and, as a result, concepts, themes, and strategies drawn from SF have come to be far more widely used in mainstream culture. Donna Haraway, Jean Baudrillard, and other postmodern, feminist theorists have used SF tropes to develop their

---

[21] Arjun Appadurai, *The Future as Cultural Fact* (London, 2013). See also Juan Francisco Salazar et al., eds., *Anthropologies and Futures: Researching Emerging and Uncertain Worlds* (London, 2017).

[22] Wendell Bell and James A. Mau, eds., *The Sociology of the Future: Theory, Cases and Annotated Bibliography* (New Brunswick, N.J., 1971); Barbara Adams and Chris Groves, *Future Matters: Action, Knowledge, Ethics* (Leiden, Neth., 2007); Ben Anderson, "Preemption, Precaution, Preparedness: Anticipatory Action and Future Geographies," *Progress in Human Geography* 34 (2010): 777–98.

[23] Those exceptions include but are not necessarily limited to, Daniel Rosenberg and Susan Harding, eds., *Histories of the Future* (Durham, N.C., 2005); and Reinhart Koselleck, *Futures Past: On the Semantics of Historic Time*, trans. Keith Tribe (New York, N.Y., 2004).

analysis. Much of the mass media consistently relies on SF referents ("Franken-foods," "cyberpunk," "warp drive") to communicate new scientific developments to the public.[24] The genre is also part of the framework through which scientists them-selves can experience their research, as Lisa Messeri's fascinating ethnography of planetary science documents. Not only was the SF writer Kim Stanley Robinson fre-quently referenced in the Mars simulation experiences, but the crew of the Arctic sim-ulation took a photograph of themselves restaging one of space-artist Pat Rawlings's Mars paintings, thus themselves creating "a present depiction of a future of the past."[25] At the very least, therefore, historians of science should be able to use SF as a source of data on how different groups accessed and apprehended different aspects of scientific, technological, and medical material and concepts, as well as how these relationships changed over time. But as the articles included in this volume will show, the potential synergies between SF and the history of science go much more deeply than this, and have far greater epistemological, political, and pedagogical potential.

Before examining this assertion in more detail, we would like to draw attention to some key absences in this volume. In the first place, we have largely concentrated on the fictions and sciences of the twentieth and twenty-first centuries. Although the ar-ticles by Morus, Willis, and Raphals touch on earlier periods in the United States, Britain, and China, for the most part this collection focuses on work published in the last hundred years or so. This does not mean that we treat SF either as synony-mous with modernity, or as a modernist product.[26] Studies of medieval SF, as well as studies of premodern, medieval, and early modern futures have clearly shown the genre's capacity to stretch time on more than one level.[27] More problematically, es-pecially given the significance we attach to SF's capacity to act as cultural critique, we have also chosen not to explicitly foreground race, gender, or sexuality as key themes in this collection. Individual essays (Milam, Radin, Garforth) touch on fem-inist critiques of SF, while non-Western SF is discussed in a number of contributions (Isaacson, Krementsov, Mukharji, Raphals)—but we have not included, for example, an analysis of Afrofuturism and the history of science. This is not because we con-sider these issues insignificant—for one thing, the connection between white mascu-

<hr>

[24] Jon Turney, *Frankenstein's Footsteps: Science, Genetics and Popular Culture* (New Haven, Conn., 2000). The influence, naturally, goes both ways; see Mark Decker, "They Want Unfreedom and One-Dimensional Thought? I'll Give Them Unfreedom and One-Dimensional Thought: George Lucas, THX-1138, and the Persistence of Marcusian Social Critique in *American Graffiti* and the *Star Wars* films," *Extrapolation* 50 (2010): 417–41.

[25] Lisa Messeri, *Placing Outer Space: An Earthly Ethnography of Other Worlds* (Durham N.C., 2016), 65.

[26] See, for example, Andrew Milner, *Locating Science Fiction* (Liverpool, UK, 2012), for a compel-ling challenge to conceptions of SF; as well as Anindita Banerjee's *We Modern People: Science Fic-tion and the Making of Russian Modernity* (Middletown, Conn., 2012). For a consideration of the re-lationship between SF and modernism more generally, see Paul March Russell, *Modernism and Science Fiction* (London, 2015).

[27] Carl Kears and James Paz, eds., *Medieval Science Fiction* (London, 2016), examines the relation-ship between key SF themes and the Middle Ages in past and present. J. A. Burrows and Ian Wei, eds., *Medieval Futures: Attitudes to the Future in the Middle Ages* (Woodbridge, UK, 2000); and Andrea Brady and Emily Butterworth, eds., *The Uses of the Future in Early Modern Europe* (London, 2009) consider the wide range of ways and reasons for imagining the future in the past. See also Adam Roberts, *The History of Science Fiction* (London, 2016). For a sustained and serious effort to imagine the impact that alien contact might have had on the medieval mind, see Michael Flynn's *Eifelheim* (New York, N.Y., 2006), which was twice nominated for a Hugo Award. For an exploration of how space flight might work within Aristotelian physics, see Richard Garfinkle, *Celestial Matters* (New York, N.Y., 1996).

line dominance of STEM fields and the presumption that SF (historically, at least) was both for and about white men, seems very clear.

But are these presumptions accurate? Or do they reflect the problems in defining science fiction as a genre? These are problems that stem precisely from SF's relationship with "real" science, and the centrality of the role played by science as (an, or perhaps the?) ultimate arbiter of knowledge in Western society. This plays out in different ways in relation to both race and gender, with white women and fans assigned certain roles and places within the history of science fiction, while black people are sometimes written out of the story altogether. So, for example, one of the editors of this volume (Rees) finds herself in the same position as SF critic Justine Larbalestier; reading science fiction growing up, Larbalestier writes, "I had no idea that science fiction was generally considered to be a 'boy's own' genre."[28] Histories of the genre have tended to locate the institutionalization, if not the origin, of Western SF in the appearance of the American SF magazines associated with Hugo Gernsback in the 1920s,[29] with women entering the field in large numbers only with the rise of second-wave feminism in the sixties and seventies. But as Larbalestier shows, female readers were engaging with SF from the very first appearance of Gernsback's magazine *Amazing Stories* in 1926.[30] By the sixth issue, a clearly surprised Gernsback could boast that "a great many women are already reading the new magazines," and he quoted readers who described their wives "anxiously waiting" for their turn to read the latest issue.[31] In fact, although Larbalestier herself does not make this argument, it is rather tempting to see the masculinization of SF as produced through the textual performances of particular fans. For example, one Isaac Asimov consistently portrayed himself in the magazine's letters as besieged and victimized by females trying to impose their version of SF on him; he actively campaigned for their exclusion.[32] Clearly, women have read and written SF from the earliest days—but distinctions such as those drawn between "hard" and "soft" SF (mirroring the so-called hard and soft sciences) meant that their work, and their influence, could be treated as peripheral to the "real" work being done in the genre. Many women in the history of SF rightly perceived themselves to be marginalized, and obviously, at least some men felt threatened by their presence—even as those who identified with either or neither gender were finding ways of exploring alternative conceptualizations of sex, sexuality, and identity within the genre.[33]

---

[28] Justine Larbalestier, *The Battle of the Sexes in Science Fiction* (Middletown, Conn., 2002), xi.
[29] Mark Bould and Sherryl Vint, *The Routledge Concise History of Science Fiction* (Abingdon, UK, 2011); John Rieder, *Science Fiction and the Mass Cultural Genre System* (Middleton, Conn., 2017); Edward James, *Science Fiction in the 20th Century* (Oxford, UK, 1994); Roger Luckhurst, *Science Fiction: A Literary History* (London, 2017); Charles Platt, *Dream Makers: The Uncommon People Who Write Science Fiction* (New York, N.Y., 1980); Platt, *Dream Makers Volume II: The Uncommon Men and Women Who Write Science Fiction* (New York, N.Y., 1983).
[30] Larbalestier, *The Battle* (cit. n. 28).
[31] Quoted in ibid., 23.
[32] Ibid., 104–43.
[33] The literature on gender, sex, and SF is too broad to explore effectively here. Useful sources for the early debates about representation can be found in Connie Willis, "The Women SF Doesn't See," *Asimov's Science Fiction Magazine* 16, October 1992, 4–8; Joanna Russ, "'*Amor Vincit Foeminam*': The Battle of the Sexes in Science Fiction," *Sci. Fict. Stud.* 7 (1980): 2–15; and Charles Platt, "The Rape of Science Fiction," *Science Fiction Eye* 1, July 1989, 45–9. For those interested in exploring some of the ways in which writers have used SF to challenge the biology, psychology, and history of gender, the three *James Tiptree Award Anthology* volumes edited by Karen Joy Fowler, Pat Murphy, Debbie Notkin, and Jeffrey D Smith (San Francisco, Calif., 2005, 2006, and 2007) represent excellent introductions to the range and impact of this strategy.

The role of race and ethnicity is harder to deal with, not least because both editors of this volume are white. If issues of gender were pushed to the margin, then questions of race were often treated as invisible, even when—to twenty-first century whites' eyes at least—they seem glaringly obvious. Even when humans of the future seemed all white, the way that those writers described and treated robots or aliens were clearly racially coded, as numerous critics have shown.[34] The emergence of a category of individuals that were—by virtue of their physical appearance—perpetually relegated to servant/serf/slave status, or an encounter with a group that represented an existential threat to the rest of humanity and had to be eradicated root and branch, both appear to reflect European and Euro-American attitudes to other human societies in both past and present.[35] In fact, as John Rieder has suggested, it is just as valid to see SF as the product of colonialism and empire, as to view it as the literature of modernity.[36] What is particularly interesting here, however, is the extent to which explorations of race and the future by nonwhite writers continue to be treated as either liminal to, or distinct from, SF. So, for example, the science fiction stories of W. E. B. Du Bois, despite appearing in his autobiographical writings, were not habitually referenced within the SF tradition for the best part of a century.[37] A similar process can be seen with the recognition of black technoculture as self-conscious Afrofuturism in the 1990s.[38] Here the fact that Afrofuturism is expressed through music, art, and gaming, as much as through textual or digital narratives, could be used to justify marginalizing these forms of future-visions within SF, again with a stress on the real SF directly to engage with the hard sciences.[39] Importantly, as we will suggest below, it is precisely because of this that Afrofuturism might represent an even more fruitful source of methodological innovation for those historians of science who recognize its relevance.

In this way we can see that, far from being marginal to the understanding or development of the genre, race and gender have generated fundamental debates about what "counts" as SF, particularly in relation to the relationship it has with the hard sciences, and the extent to which those are considered to be the domain of white, middle-class men. Issues of definition have bedeviled the genre since it was recognized to exist, as readers and writers contested the texts that would become canonical in par-

---

[34] See, for example, Elisabeth Leonard's pioneering *Into Darkness Peering: Race and Color in the Fantastic* (Westport, Conn., 1997); De Witt Douglas Kilgore's *Astrofuturism: Science, Race and Visions of Utopia in Space* (Philadelphia, Penn., 2003); Isiah Lavender III, ed., *Black and Brown Planets: the Politics of Race in Science Fiction* (Jackson, Miss., 2014). For an examination of how even an ostensively inclusive depiction of the future could become progressively white-washed, see Daniel Leonard Bernardi's *Star Trek and History: Race-ing Toward a White Future* (New Brunswick, N.J., 1999).

[35] The classic example here is obviously Isaac Asimov's Robots (who are often referred to as "boy"). See Asimov, *I, Robot* (New York, N.Y., 1950).

[36] John Rieder, *Colonialism and the Emergence of Science Fiction* (Middletown, Conn., 2008).

[37] The short story "The Comet" was published in W. E. B. Du Bois, *Darkwater: Voices from Within the Veil* (New York, N.Y., 1920), 253–73, in 1920, but received little attention until it appeared in Sheree R. Thomas's *Dark Matter: A Century of Speculative Fiction from the African Diaspora* (New York, N.Y., 2000), 5–18. Another writer whose contribution to science fiction has also recently received more recognition is the journalist and social commentator, George Schulyer; see, for example, *Black No More* (New York, N.Y., 1931).

[38] Mark Dery, ed., *Flame Wars: The Discourse of Cyberculture* (Durham, N.C., 1994); Alondra Nelson, ed., "Afrofuturism," special issue, *Social Text* 20 (2002).

[39] Mark Bould, ed., "Africa SF," special issue, *Paradoxa* 25 (2013).

ticular versions of SF's history.[40] They were vigorously debated by fans in the letters pages of SF magazines, carefully discussed in talks and editorials by writers in the genre, and even questioned at UNESCO meetings, as the articles in this volume by Sleigh and White, Rees, Radin, and Bowler demonstrate.[41] If one looks outside the West—as with, for example, the materials examined here by Isaacson, Mukharji, and Raphals—then the boundaries of the genre become even more porous and negotiable.[42] One could treat this as simply a matter for the writers, readers, and critics of SF to analyze and assess, were it not for the fact that deciding whether something "is" or "is not" science fiction has crucial implications for epistemological debates surrounding the relationship between Western science, knowledge, and truth.

WHAT IS SCIENCE FICTION?

We do not want to define SF in this introduction. Indeed, one reason for using the abbreviation "SF" is to broaden the range of material covered (speculative fiction, slipstream fiction, science fantasy?). But we do want to draw attention to the relationship between SF and the real world, which we believe is a key element in understanding the genre's shape and its cultural influence. Science fiction is fiction. But it is, for these editors at least, not fantastic fiction. This distinction may seem pedantic, but it is absolutely central to the genre's intellectual potential and its capacity to contribute to the history of science. There have been many different efforts to define SF and to draw distinctions between it and other genres, but very early on, Darko Suvin pointed to the fact that, characteristically, SF tended to contain a *novum*, or "new thing."[43] This could be a wormhole, a space station, possibly even a "Big Dumb Object,"[44] a neuroactive chemical that either enhances or inhibits intellectual or emotional response, a self-aware mainframe, or a swarm of nanobots. It could even be a talking dragon. But if the story is to be SF, rather than fantasy, horror, or a fairy story, then there needed to be a coherent explanation that either accorded with current physical laws, or could be plausibly extrapolated from likely future developments. That is to say, Anne McCaffrey's *Dragonflight* (published in 1968 and involving talking dragons) is, according to this interpretation, undoubtedly fantasy, while her *Dragonsdawn* (published in 1988 and involving intelligent dragon-like creatures being bioengineered from small, flying lizards as part of a human colony's biological adaptation to a new planet)

---

[40] As John Rieder shows through an examination of contemporary reviews [Rieder, *Science Fiction* (cit. n. 29), 65–81], the status and the early interpretation of texts like *Frankenstein* were deeply unstable, with the creature's ability to speak being treated as far more significant than its initial creation.

[41] Charlotte Sleigh and Alice White, "War and Peace in British Science Fiction Fandom, 1936–1945"; Amanda Rees, "From Technician's Extravaganza to Logical Fantasy: Science and Society in John Wyndham's Postwar Fiction, 1951–1960"; Joanna Radin, "The Speculative Present: How Michael Crichton Colonized the Future of Science and Technology"; Peter Bowler, "Parallel Prophecies: Science Fiction and Futurology in the Twentieth Century," all in this volume.

[42] Nathaniel Isaacson, "Locating Kexu Xiangsheng (Science Crosstalk) in Relation to the Selective Tradition of Chinese Science Fiction"; Projit Bihari Mukharji, "Hylozoic Anticolonialism: Archaic Modernity, Internationalism, and Electromagnetism in British Bengal, 1909–1940"; Lisa Raphals, "Chinese Science Fiction: Imported and Indigenous," all in this volume.

[43] Darko Suvin, *Metamorphoses of Science Fiction: On the Poetics and History of a Literary Genre* (New Haven, Conn., 1979).

[44] Christopher Palmer, "Big Dumb Objects in Science Fiction: Sublimity, Banality and Modernity," *Extrapolation* 47 (2006): 95–111.

might, at a stretch, be defined as SF.[45] However, it is important to remember that by distinguishing between fantasy and SF in this way, we are implicitly accepting a hierarchical valuation that ranks fantasy as less intellectually worthy than SF. But we do believe that—for the purposes of the history of science, at least—it is a distinction that we need to make.

For our purposes, while science fiction does not need to represent or reflect empirical reality, where it conforms to an empirical methodology that is based on the author's understanding of present scientific and technological practice, then it becomes both relevant to, and inspiring for, the history of science. The extent to which this needs to be an accurate understanding is, however, a matter for debate. Not all writers or readers would accept the dictate of Fred Hoyle, astronomer and SF writer, that he laid down in an interview with the British newspaper *Sunday Dispatch* in March 1958:

> Novels based on science . . . must cross the bounds of probability at only one small point. Otherwise they will not be consistent with reality. Only a scientist nowadays can gauge where he [*sic*] can safely make that break. A scientist can usually judge where a lay writer has gone wrong—and these days, so can a lot of the public. If a chap doesn't want to make a fool of himself on the bookstands, he'd better take a few years off and pick up a degree.[46]

Writing at roughly the same time, another British astronomer, Patrick Moore, would have agreed—the British novelist, John Wyndham would not.[47] Fifty years later, the Canadian writer Margaret Atwood tried to position her books (*The Handmaid's Tale*, *The Year of the Flood*) in relation to genre. They are not realist novels, she argues, but nor are they science fiction. Instead, they are "speculative fiction." Why the distinction? Because, she wrote, science fiction deals with "tentacled Martians shot to Earth in metal canisters—things that could not possibly happen."[48]

All four authors are struggling with the same issue, which is central to the relationship between history of science and science fiction. First, despite being fiction, SF can't just be made up. It must at some level reflect a scientific, realistic understanding of the world, as it tells a story that could, plausibly, eventually come to pass. Second, these understandings will change over time, whether or not the predictions of individual writers prove to be accurate or come to pass. In 1865, it was plausible that a space gun might shoot people to the moon. In 1895, it was possible that life might exist on Mars. The fact that Verne and Wells included in their novels information and experimental practices that have since been shown to be incorrect or inappropriate does not mean that they are not valuable sources for the historian interested in considering how people—especially people who were not professional scientists—thought about and used scientific concepts, methodologies, and institutions. As such, and as we argued

[45] Anne McCaffrey, *Dragonflight* (1968; repr., New York, N.Y., 2005); McCaffrey, *Dragonsdawn* (New York, N.Y., 1988).

[46] Fred Hoyle, "Off to Outer Space Go the Bug-Eyed Monsters," interview by Elizabeth Hickson, *Sunday Dispatch*, 2 March 1958.

[47] Patrick Moore, *Science and Fiction* (London, 1958). See Bowler's discussion of Moore, and Rees's consideration of Wyndham in this volume (cit. n. 41).

[48] Margaret Atwood, *The Handmaid's Tale* (Toronto, Can., 1985); Atwood, *The Year of the Flood* (Toronto, Can., 2009); Atwood, "The Road to Ustopia," *Guardian*, 14 October 2011; see also Atwood "*The Handmaid's Tale* and *Oryx and Crake* in Context," *PMLA* 119 (2004): 513–17, https://www.theguardian.com/books/2011/oct/14/margaret-atwood-road-to-ustopia.

at the outset of this introduction, at the most basic level, SF represents an important source of data for those interested in how different publics understand different kinds of science at different times. This is, for example, how the authors in this volume's section on "Mediating Science" are deploying the genre. Morus uses the imaginary "telectroscope" to examine attitudes to electricity and the future among different communities at the turn of the previous century. Krementsov shows how a scientist in Bolshevik Russia used science fiction to transform specialized biomedical knowledge into a broader cultural resource that other scientists, journalists, and writers could in turn use to create both hope and fear for the future. Kirby examines the way that SF films employed evolutionary and Darwinian concepts to show how different groups sought to control the meaning and morality of evolution in US culture. Raphals asks a rather different question that returns us to the broader political question raised by these discussions of SF: what would modern Chinese SF look like, if it had taken—or took—indigenous Chinese natural philosophy seriously?[49]

This is, of course, the unspoken assumption at the heart of the "realism" of Western SF. To what extent is Western science to be treated as the ultimate arbiter of knowledge? The empirical definition given earlier implicitly downgrades "fantasy" as inferior in comparison to SF because of its nonrealist nature. The relationship between fantasy novels, science fiction, and magical realist literature is often contested precisely because of this question. So how should imaginative explorations that reflect indigenous (that is, non-Western) traditions of interacting with and attempting to manipulate the nonhuman world be treated, especially if they don't operate according to a human/nonhuman binary? The Afrofuturist novelist Nalo Hopkinson, for example, defined speculative fiction as the literature that examines the impact toolmaking has had, and will have, on human societies.[50] What might historians of science learn about non-Western attitudes to nature by examining indigenous fictions of the near and far future?

THE HISTORY OF THE FUTURE

But the question of accuracy and plausibility is still central to the relationship between history of science and science fiction, as the articles in the section "Delineating Science" show. Both Bowler and Isaacson focus on the relationship between popular science and science fiction, and the concern with understanding the role that science could play in creating a better future by educating young and old. Isaacson examines the relationship between Western science, the notion of social progress, and indigenous forms of fiction in China in the 1950s. Bowler considers much the same period in Britain, looking at both the extent to which writers sought to warn the public about the dangers of new technology, and the relationship between scientific accuracy and entertainment in the inspiration of the next generation of scientists.[51] Mukharji investigates the ways in which Bengali writers used Western-inspired SF to challenge Eu-

---

[49] Iwan Rhys Morus, "Looking into the Future: The Telectroscope That Wasn't There"; Nikolai Krementsov, "Thought Transfer and Mind Control between Science and Fiction: Fedor Il'in's *The Valley of New Life* (1928)"; David Kirby, "Darwin on the Cutting Room Floor: Evolution, Religion, and Film Censorship," all in this volume; Raphals, "Chinese Science" (cit. n. 42).

[50] Nalo Hopkinson, "Making the Impossible Possible: An Interview with Nalo Hopkinson," interview by Alondra Nelson, in Nelson, "Afrotuturism" (cit. n. 38), 91–113.

[51] Isaacson, "Locating" (cit. n. 42); Bowler, "Parallel Prophecies" (cit. n. 41).

ropean intellectual authority, thereby creating a post-Western universalism in which modern scientific knowledge was shown to have been prefigured in indigenous knowledge of nature.[52] Slocombe presents intellectual historians and historians of technology with a challenge by showing how certain computer games both implicitly and explicitly provide their players with a profoundly deterministic approach to both technology and the history of civilization. Examining the lessons that can be learned about how different technologies develop, and how they relate to the notion of civilization in the game series *Civilization*, he asks whether these should be borne in mind by those presenting "professional" history of technology to the public.[53] One might also like to consider this in the broader context of the engagement of science and the public, and the kind of work that has been done on the category of "popular science" as a genre in and of itself. Will we—historians, that is—need to do the same work in apprehending the plausibility and accuracy of popular history? Does it matter if the material is inaccurate, as long as it inspires the next generation of historians?

Although every article in this volume deals, in one way or another, with dreams about the future, the articles in the section called "Inspiring Science" take this as their particular focus. In different ways, which include focusing on aspects of the activities of professional scientists, politicians, and the lay public, each article examines strategies for thinking about and actively creating different futures. White and Sleigh examine the culture of SF fandom in Britain in the period immediately before and during World War II, showing how fans debated the definitions of science and science fiction, and how they used SF as a means of formulating and formalizing their ideas of the future—something that was especially important during the years of conflict.[54] Milburn and Milam both deal—in different ways—with the impact that SF had on both the human future and the nature of scientific development. Milburn shows how the physicist Gerald Feinberg was inspired to think about tachyons by reading SF, and demonstrates the extent to which SF represents a set of conceptual and practical resources for scientists when it comes to constructing thought experiments. Milam looks at the relationship between fiction, evolution, and feminism in her examination of the ways in which Elaine Morgan's popularization of the Aquatic Ape theory—a dramatic reworking of the understanding of humanity's evolutionary past—enabled writers and scientists to develop new models and narratives of human futures. Both authors demonstrate, with absorbing clarity, the difficulty of charting the line between fiction and science, real and not-real, in these episodes.[55] Garforth's study is, unfortunately, all too real, as she shows the centrality of science fiction to the understanding of ecological crisis. Moving between science, sociology, popular science, and fiction, she models the different ways in which thinking science-fictionally about "catastrophe" can impact the understanding—both popular and political—of "crisis," both in relation to the history of ecological sciences and to the present day.[56] We may indeed live long—but will our children prosper?

---

[52] Mukharji, "Hylozoic Anticolonialism" (cit. n. 42).

[53] Will Slocombe, "Playing Games with Technology: Fictions of Science in the *Civilization* Series," in this volume.

[54] Sleigh and White, "War and Peace" (cit. n. 41).

[55] Colin Milburn, "Ahead of Time: Gerald Feinberg and the Governance of Futurity"; Erika Milam, "Old Woman and the Sea: Evolution and the Feminine Aquatic," both in this volume.

[56] Lisa Garforth, "Environmental Futures, Now and Then: Crisis, Systems Modeling, and Speculative Fiction," in this volume.

Here, we arrive at what we believe is perhaps the most important reason why historians should consider serious engagement with SF—and that is the affective, emotive capacity of the genre. Science fiction does not just extrapolate from present technological or scientific trends, but places these developments within a moral universe, unsettling its audiences' sense of the everyday while enabling them to recognize the fundamentally familiar within the apparently strange. It is a genre based on the habitual deployment of the sense of cognitive estrangement, usually delivered via technoscientific means (the novum), such as a clock that strikes thirteen, a rather simian-looking Abe Lincoln, or a computer with a split personality.[57] The crucial point here is that these sometimes dramatic, technical, or physical estrangements have—and are immediately shown to have had—important, even if apparently infinitesimal, influences over other aspects of human life and interaction. Changes to the physical landscape are produced, or imply, or demand changes in the social contract. Different social structures influence the kind and type of scientific and technological change within their frameworks, as writers as diverse as Ursula Le Guin, C. J. Cherryh, Adam Roberts, and Stephen Baxter have demonstrated.[58] Science fiction authors long predated Bruno Latour and Michal Callon in assuming equivalences between human and non-human actors, as well as in their exploration of different kinds of agency in action—collective, variously located, sometimes chronologically dislocated—and importantly, without ever losing sight of the fact that power tends to be differentially distributed both within and between groups. In many ways, we would want to argue that SF is a form of science and technology studies (STS) in action, and one that has been far more effective in engaging the public imagination than has the history of science.

The articles in the section "Applied (History of) Science" deal with different aspects of this argument. Rees and Radin examine the work, respectively, of SF authors John Wyndham and Michael Crichton to show how both focus on understanding "science in the making" and the complex interdependence and coproduction of science and society.[59] Both articles wonder what impact these writers would have had if they had undertaken academic, rather than literary, careers, and both examine the extent to which SF writers bear, and are aware of, a wider social responsibility to foster democratic debate in societies dependent on the operation of expertise and technocracy. Servitje analyzes the operation of narratives of apocalypse in relation to the rise of antimicrobial resistance (AMR).[60] He shows how a gaming app developed to teach the history of AMR both recapitulates that history (encouraging players to see this as a battle between bacteria and humanity) and challenges it (the only way to not lose is to learn to live with the bugs). In this way, he demonstrates strategies through which ludological practice can both mirror and subvert historiographical understanding. Willis also focuses on methodologies, using literary analysis alongside ethnography to understand how different communities have understood sleep.[61] Asking whether we can, in fact, accomplish STS science-fictionally, he combines the practices of dif-

---

[57] See Adam Roberts, *Science Fiction: The New Critical Idiom* (Oxford, UK, 2006) for further valuable discussion here.

[58] See Le Guin's *The Dispossessed* (1972); C. J. Cherryh's *Downbelow Station* (1981); Adam Roberts's *Salt* (2000); and Stephen Baxter's *Flood* (2008).

[59] Rees, "From Technician's Extravaganza"; Radin, "Speculative Present" (both cit. n. 41).

[60] Lorenzo Servitje, "Gaming the Apocalypse in the Time of Antibiotic Resistance," in this volume.

[61] Martin Willis, "Sleeping Science-Fictionally: Nineteenth-Century Utopian Fictions and Contemporary Sleep Research," in this volume.

ferent temporalities in order to show how both past and future can be put to work to broaden and deepen the understanding of the present.

THE FUTURE OF THE HISTORY OF SCIENCE?

In many ways, that last point sums up the hope and desire of the editors and contributors to this volume of *Osiris*. Historians of science should engage with SF because these imaginary explorations of how changes in knowledge and developments in technique influence changes in social organization, which in turn provide the context within which scientists, engineers, and technicians work. That is to say, while we argue that SF represents a crucial resource for understanding the historical relationship between the sciences and their different audiences, we also want to suggest that, as a model for historical understanding, it warrants serious historiographical attention. In parallel with recent efforts by scientists to engage with the public, SF has sought to recognize and to deploy scientific knowledge and practice through a multiplicity of genres and media. Historians of science, with their focus on texts and the material cultures, would do well to consider how they might also widen their intellectual focus in a digital age, both in terms of producing and communicating their work. This is particularly important when the potential pedagogic role of SF for the history of science is taken into consideration, and when issues of power and influence are considered in light of the different approaches to defining SF, especially in relation to race, gender, and sexuality.

There are four principal trajectories along which the authors of these articles have aligned their investigations of the relationship between SF and the history of science. To begin with, they have considered how SF has been used to investigate the cultural status and authority afforded to the category of "science" at different times and by different human communities—that is, they have looked at how SF has *delineated*, or defined, science and the scientist. Second, they have examined the role of SF in exploring specific scientific disciplines, topics, or cultures; and in moving scientific concepts, methodologies, and practices between wider cultural arenas. In other words, they have considered SF, in all its different formats, as a means of *mediating* science. Third, they have explored what SF can tell us about the histories of how different communities have envisaged their futures, and thus how it conveys the socioscientific concerns of past presents; that is, how it has *inspired* different futures. Finally, they have investigated how SF has and could function as a resource through which historians and laypeople alike can conceptualize the context and consequences of scientific change— or, how SF might be considered as *applied* history or sociology of science, and then used to develop a greater variety of models and practices for understanding science in society. These imaginary explorations of the futures contained within different times and places are themselves experiments that the editors hope will have consequences, both for the present day and for the time to come.

MEDIATING SCIENCE

Looking into the Future:

The Telectroscope That Wasn't There

by Iwan Rhys Morus*

ABSTRACT

In an 1898 short story titled "From the 'London Times' of 1904," Mark Twain introduced an electrical instrument called the telectroscope. Machines for transmitting vision at a distance, telectroscopes had been speculated about since the invention of the telephone in 1876. Over the next quarter of a century, numerous inventors were credited with its imminent, but never realized, production. No such instrument was ever actually built, and it now usually appears only in footnotes to television's prehistory. Nevertheless, the telectroscope offers useful insights into the way the Victorian future was constructed out of assemblages of fact and fiction. In this chapter I chart the ways in which the instrument moved back and forth across the boundaries of the real. Precisely because it never existed, I suggest, the telectroscope offers an excellent example of the ways the Victorian future was made out of its own material culture.

In 1898, Mark Twain wrote a short story for the *Century Magazine* titled "From the 'London Times' of 1904." It was written as a journalist's dispatch from the near future and revolved around the invention of an amazing new machine called the telectroscope. This was an instrument that made it possible to see at a distance—a sort of ocular telephone, or wireless telegraph, that transmitted images instead of sounds. Twain's imagined future correspondent described how the device was "connected with the telephonic systems of the whole world," noting that it made "the daily doings of the globe . . . visible to everybody, and audibly discussable, too, by witnesses separated by any number of leagues."[1] This was fiction, of course, and that would have been clear to its original readers in any number of ways, not least through its authorship. But the status of the telectroscope itself would have been less clear to those readers. Many of them would have come across references to the device in the popular and scientific press over the previous two decades. It was a machine that seemed to be always on the brink of invention. The machine's inventor in Twain's story—Jan

* Department of History & Welsh History, 3.06 International Politics Building, Aberystwyth University, Aberystwyth, SY23 3FE, Wales; irm@aber.ac.uk.

Earlier versions of this work were presented in seminars at Brighton and Oxford. I am grateful to participants in those seminars for useful discussions. I would like to thank my coeditors, Amanda Rees, W. Patrick McCray, and Suman Seth, as well as two anonymous referees, for helpful suggestions. The research for this chapter was funded by the Arts and Humanities Research Council (AHRC) in connection with the Unsettling Scientific Stories Project, and I thank them for their support.

[1] Mark Twain, "From the 'London Times' of 1904," *Century Magazine* 57, 1898, 100–4, on 101.

Szczepanik—was real enough as well, and did indeed claim to have invented and exhibited a device for seeing at a distance.

The telectroscope in Twain's short story occupied an ambiguous place between fact and fiction. While there was no question about its fictionality in the *Century Magazine*, its status elsewhere was, and is, more difficult to establish. As a matter of fact, no device like the telectroscope was ever built during the final quarter of the nineteenth century. But as we shall see, its invention, or its imminent invention, was repeatedly announced. One of the things I want to do in this article is to look at how instruments such as the telectroscope were put together and how they moved around—how they straddled the worlds of fact and fiction. Even though the telectroscope never existed, it was made out of familiar components. It was the family resemblance between it and other items in the late Victorian panoply of technologies of display and communication that gave verisimilitude to accounts of the telectroscope. To understand how the telectroscope worked for its audiences, we need to understand the extent to which it seemed familiar. Twain could tease his readers with the possibilities of seeing at a distance because the instrument he described seemed to belong to a genre of devices that were already in the process of becoming commonplace.

Late nineteenth-century readers of scientific and speculative literature understood that theirs was a culture in which the boundaries of the real were particularly porous.[2] They were surrounded by the evidence of spectacular discovery and technological innovation. As Twain wrote his story, newspapers were carrying news of Guglielmo Marconi's experiments in wireless telegraphy. In North America, Nikola Tesla was attempting to woo potential investors with spectacular scientific showmanship. Scientific spectacles, such as those at which Tesla excelled, were often aimed at challenging their audiences' capacities to distinguish fact from fiction. Offering ambitious promissory notes about the potential of their inventions was clearly at one level a tactic to entice attention, but was also an indication of the appeal of technological progress. Newspaper and magazine editors hungry for readers were often willing collaborators in this game of pushing at the limits of plausibility. There was an appetite for speculation about just how far away the limits of invention lay.[3] The telectroscope offers a good example of how this culture of speculation and innovation worked as it moved between fact and fiction. It therefore offers a promising site for investigating some of the ways in which late Victorian technologies—and the Victorians' own understanding of those technologies—were assembled.

Key to this was the way in which technologies like the telectroscope embodied futurism. Invoking the future was a vital strategy in selling inventions throughout the nineteenth century. It was often a vital strategy in selling science more generally to its publics too. From the invention of the electromagnetic telegraph onward, inventors and popularizers talked about the ways in which technologies—and electrical technologies in particular—would transform culture and make the future. Electricity and the future were entirely intertwined in these kinds of commentaries and visions. In some contexts at least, talking about the future and talking about electricity were simply the same thing. The future would be wireless electrical communication, electrical lights, and electrical transport. Very often—and again the telegraph offers a key example here—

[2] Iwan Rhys Morus, "In the Ether: Electricity and the Victorian Future," in *Utopianism and the Sciences*, ed. Mary Kemperink and Leonieke Vermeer (Leuven, Belgium, 2010), 17–32.
[3] Peter Bowler, *A History of the Future: Prophets of Progress from H. G. Wells to Isaac Asimov* (Cambridge, UK, 2017).

the imagined future was one in which the relationship between individuals' bodies, culture, and environment would be reconfigured.[4] The future that electricity promised its Victorian audiences was one in which their senses would be extended outside their bodies. It would be a future in which they would be able to act, listen, and see at a distance.[5] This was why the telectroscope was such an appealing instrument.

One way of approaching the telectroscope is to interrogate the instrument as a piece of science fiction itself (or, more accurately maybe, as a piece of scientific romance). Rather than being simply sociotechnical imaginary, though, the telectroscope was both imagined and material.[6] It worked as an imagined fact, because it was firmly grounded in material culture. This was how scientific romances worked as well. Such fictions appealed because they were composed of familiar components. They exploited the porosity of the boundaries of the real during the second half of the nineteenth century to construct convincing and materially grounded narratives that challenged their readers to decide on what side of the boundaries they lay. The telectroscope and its boosters in newspapers and magazines, and lectures and exhibitions, were telling tales in the same sort of way. The writers whose speculations about the technological future—or the technologically imminent present—filled the pages of middle-brow magazines like the *Century Magazine*, *Pearson's Magazine*, or the *Strand Magazine* (home of Sherlock Holmes) were riding on a wave of technological innovation that gave their stories verisimilitude. The telectroscope rode that wave as well. Speculation about the possibilities of seeing at a distance rested on a shared understanding of what looking into the future could plausibly see.

I will therefore begin this article by surveying the cultures of Victorian electrical futurism to which the telectroscope belonged. From Cooke and Wheatstone's patenting of the electromagnetic telegraph in 1837, and through the rest of the century, generating visions of electrical futures was an integral part of electricity's narratives and performances. Through books, magazines, and stage performances—through factual and fictional representations—electricity was built into visions of the future. I will then dissect the telectroscope itself as an exemplary element in this world of electrical futures. Even though such a device was never made, its components bear investigating as examples of the ingredients that went into the making of future electricity. Finally, I will look more closely at the ways in which the telectroscope was negotiated between fact and fiction. It was a subject for stories that spanned the spectrum between reporting and fantasizing. By looking at how the telectroscope was moved around, this paper will illuminate the ways in which Victorian futures, fictional and factional, were put together—and how the narratives of Victorian futurity were constructed through telling tales about technology.

ELECTRIC FUTURES

The future was in the process of being reconfigured throughout the nineteenth century. Increasingly, the future was coming to be imagined as different from the present,

    [4] Iwan Rhys Morus, "No Mere Dream: Material Culture and Electrical Imagination in Late Victorian Britain," *Centaurus* 57 (2015): 173–91.
    [5] Carolyn Marvin, *When Old Technologies Were New: Thinking about Electrical Communication in the Late Nineteenth Century* (Oxford, UK, 1988).
    [6] Sheila Jasanoff and Sang-Hyun Kim, eds., *Dreamscapes of Modernity: Sociotechnical Imaginaries and the Fabrication of Power* (Chicago, Ill., 2015).

just as the present was being differentiated from the past in a variety of new ways. Bestselling popular science books like Robert Chambers's *Vestiges of the Natural History of Creation* pictured a universe governed by progressive, unfolding laws that applied to society as well as nature.[7] If society was progressive in the way that Chambers and others portrayed it, then the future had to be reimagined as a different place. William Robert Grove's lecture to the London Institution on the progress of physical science since its establishment offers a nice example of how a kind of futurism was built into public science during the early Victorian period. Here Grove described to his audience how advances in electricity meant that it was now possible "to fuse the most intractable metals, to propel the vessel or the carriage, to imitate without manual labour the most costly fabrics, and in the communication of ideas, almost to annihilate time and space." If anyone had dared suggest any of this at the beginning of the century, he suggested, "the prophet, Cassandra-like, would have been laughed to scorn."[8] It was understood that the trajectory would continue into the future. Satirists caricatured the faith in a technologized future, drawing speculative images of flying steam engines or fantastical pneumatic railways that could cross continents. The satire is revealing in that it exposes the extent to which these futures were made out of the ingredients of the present. This was a feature of discussions of electricity in particular. It was a pervasive feature of this unprecedented new set of technologies that their projection into imagined futures was central to their production.

Selling the electromagnetic telegraph was an exercise in selling the future. William Fothergill Cooke, one of the new technology's first promoters, echoed Dionysius Lardner's fantasy of rail's capacity to reconfigure space and time: The "concentration of mind and exertion, which a great metropolis always exhibits, will be extended in a considerable degree to the whole realm . . . Towns at present removed some stages from the metropolis will become its suburbs; others, now a day's journey away, will be removed to its immediate vicinity."[9] The telegraph, like the railway, would make tomorrow's world a smaller and more controllable place. That kind of telegraphic shrinkage of the future's space and time was a common argument. Latimer Clark argued to the Society of Telegraph Engineers: "Distance and time have been so changed to our imaginations, that the globe has been practically reduced in magnitude, and there can be no doubt that our conception of its dimensions is entirely different to that held by our fathers."[10] One important consequence of the telegraph for its promoters was that future time would become universal and mechanized: "We may soon expect to see every series of telegraph-wires forming part of a gigantic system of clockwork, by means of which, time-pieces, separated from each other by hundreds of miles, may

---

[7] James Secord, ed. and introduction, *Vestiges of the Natural History of Creation and Other Evolutionary Writings*, by Robert Chambers (Chicago, Ill., 1994), originally published 1844; Secord, *Victorian Sensation: The Extraordinary Publication, Reception, and Secret Authorship of Vestiges of the Natural History of Creation* (Chicago, Ill., 2000). More generally, see Secord, *Visions of Science: Books and Readers at the Dawn of the Victorian Age* (Oxford, UK, 2014).

[8] William Robert Grove, *On the Progress of Physical Science since the Opening of the London Institution* (London, 1842), 24. For Grove and progress, see Iwan Rhys Morus, *William Robert Grove: Victorian Gentleman of Science* (Cardiff, Wales, 2017).

[9] William Fothergill Cooke, *The Electric Telegraph: Was It Invented by Professor Wheatstone?* 2 vols. (London, 1857), 2:259.

[10] Latimer Clarke, "Inaugural Address," *Journal of the Society of Telegraph Engineers* 4 (1875): 1–23, on 2.

be made to keep exactly equal time, and the clocks of a whole continent move, beat for beat, together."[11]

The possibilities of electric locomotion and power also offered plenty of opportunities for futuristic speculation. After listing all the things that could already be done with electricity, a commentator extolled the future: "The prospect of obtaining power which shall supercede steam, exceeds in value all these applications. For to cross the seas, to traverse the roads, and to work machinery by galvanism, or rather electromagnetism, will certainly, if executed, be the most noble achievement ever performed by man."[12] Following an exhibition of Thomas Davenport's patented rotary electromagnetic engine in London in 1837, the *Morning Herald*, quoted in the *Mechanics' Magazine*, declared it could not "discover any good reason why the power may not be obtained and employed in sufficient abundance for any machinery—why it should not supercede steam, to which it is infinitely preferable on the score of expence, and safety, and simplicity." In this future, "half a barrel of blue vitriol, and a hogshead or two of water, would send a ship from New York to Liverpool; and no accident could possibly happen, beyond the breaking of the machinery, which is so simple that any damage could be repaired in half a day."[13] Less than a decade after the Rainhill trials, when Stephenson's *Rocket* triumphed over the competition in a race along the Liverpool to Manchester railway, the steam locomotive was already being banished from the future.

Showmanship was a critical strategy in making these sorts of futures, and exhibitions were important spaces for making and displaying the future. Edward Davy, one of Charles Wheatstone and William Fothergill Cooke's telegraphic competitors, joked to his father, "you did not expect to see a son turned showman."[14] Davy had placed his telegraph apparatus on show at London's Exeter Hall. Another telegraphic competitor, William Alexander, displayed his apparatus first at the Royal Society of Arts and then at the Adelaide Gallery. Places like the Adelaide Gallery and its competitor, the Royal Polytechnic Institution in London, as well as other less well-known metropolitan sites for exhibition, might be understood as spaces for futurity. In places like these and their equivalents elsewhere in Europe and North America, the future was put on show. The Adelaide Gallery advertised itself as promoting "the adoption of whatever may be found to be comparatively superior, or relatively perfect in the arts, sciences or manufactures."[15] The Polytechnic promised "the demonstration, by the most simple and interesting methods of illustration, of those new principles upon which every science is based, and the processes employed in the most useful arts and manufactures effected."[16] In practice, they offered spectacular displays of scientific showmanship that embodied the future for their audiences. They were the spaces that made the future visible.[17]

[11] George Wilson, *Electricity and the Electric Telegraph* (London, 1855), 59–60. See Iwan Rhys Morus, "The Nervous System of Britain: Space, Time and the Electric Telegraph," *Brit. J. Hist. Sci.* 33 (2000): 455–75.

[12] Alfred Smee, *Elements of Electrometallurgy* (London, 1844), 348.

[13] "Davenport's Electro-magnetic Engine," *Mechanics' Magazine* 27, 16 September 1837, 404–5.

[14] Quoted in J. J. Fahie, *A History of the Electric Telegraph to the Year 1837* (London, 1884), 418.

[15] Quoted in Richard Altick, *Shows of London* (Cambridge, Mass., 1978), 377.

[16] "Our Weekly Gossip," *Athenaeum* 11, 4 August 1838, 554–5, on 555.

[17] The classic discussion is Altick (cit. n. 15). See also the more recent Bernard Lightman, *Victorian Popularizers of Science: Designing Nature for New Audiences* (Chicago, Ill., 2007). On the Polytechnic in particular, see Brenda Weeden, *The Education of the Eye: History of the Polytechnic Institution, 1838–1881* (Cambridge, UK, 2008); and Jeremy Booker, *The Temple of Minerva: Magic and the Magic Lantern at the Royal Polytechnic Institution, London 1837–1901* (London, 2013).

Spectacles like the "monster coil" that had been installed at the Polytechnic in 1869 by its proprietor, John Henry Pepper, provided the raw materials out of which futures might be constructed. Watching the instrument in action, audiences were invited to consider what the demonstrations implied for the boundaries of the real and to marvel at the technological ambition it embodied. It was a lesson in the practical deployment of the imagination to investigate technological possibilities. The coil, built by the instrument-maker Alfred Apps, generated a "spark, or rather a flash of lightning, 29 inches in length and apparently three-fourths of an inch in width, striking the disk terminal with a stunning shock." The discharge between the coil's electrodes was like "a gush of waving flame." Newspaper accounts emphasized the instrument's impressive dimensions. It was almost ten feet long and two feet wide; there were one hundred and fifty miles of wire in the secondary coil; and the primary coil weighed forty-five pounds. Making the instrument's mechanism explicit in this way was a strategy for emphasizing its capacity for breaching convention. It was deliberately magnified in its dimensions just so as to demonstrate the possibilities implicit in its technologies: "a source of endless delight and wonder" that would "display effects, beautiful or terrible, such as have never been seen before."[18]

Spaces like the Crystal Palace where the Great Exhibition was held in 1851 offered a spectacle of the future on an even more gargantuan scale. The exhibition itself was organized precisely in order to make a spectacle of the Victorian cult of progress and to demonstrate the superiority of Victorian Britain's own imagined future over those of its industrial competitors.[19] Visitors to the Crystal Palace would have been faced with the sight of the huge electric clock, designed by Charles Shepherd. The clock face was outside the building while the clock mechanism was inside in the south transept. The "whole of the works of this great clock were kept in motion by a series of powerful electro-magnets; and, by means of an immense coil of copper wire, other clocks in the Exhibition were kept going."[20] It was a visible reminder of the Victorian domination of time. Between the steam engines, visitors could inspect electromagnetic engines, which demonstrated that "the attainment of this mysterious motive force will soon be followed by . . . making it available for practical purposes."[21] Frederick Collier Bakewell exhibited his patent electric copying telegraph "for transmitting facsimiles of the handwriting of correspondents, so that their signatures may be identified."[22]

The success of the Crystal Palace inaugurated a new tradition of industrial exhibition that lasted the rest of the century and beyond.[23] At the Great Exhibition's immediate successor, the International Exhibition of 1862, even the entrance was electrical. The "magnetic telltale of Professor Wheatstone" connected to the turnstiles meant that "each visitor, on passing through it, unconsciously and telegraphically announced his

---

[18] Quoted in Iwan Rhys Morus, "More the Aspect of Magic than Anything Natural: The Philosophy of Demonstration," in *Science in the Marketplace: Nineteenth-Century Sites and Experiences*, ed. Bernard Lightman and Aileen Fyfe (Chicago, Ill., 2007), 336–70, on 356.

[19] Jeffrey Auerbach, *The Great Exhibition of 1851: A Nation on Display* (New Haven, Conn., 1999).

[20] John Tallis, *History and Description of the Crystal Palace* (London, 1854), 3:28.

[21] Frederick Collier Bakewell, *Electric Science; Its History, Phenomena and Applications* (London, 1853), 191.

[22] *Official Descriptive and Illustrated Catalogue of the Great Exhibition* (London, 1851), 3:447.

[23] Electrical exhibitionism is described in K. G. Beauchamp, *Exhibiting Electricity* (London, 1997).

or her arrival to the financial officers in whose rooms were fixed the instruments for receiving and recording the liberated current."[24] This was one of the first exhibitions where electric light constituted a significant spectacle as it "reflected so as to be visible all down the nave to the western end."[25] Electrical illumination continued to be a significant part of international exhibitions' spectacular culture in subsequent festivals. The Paris Exhibition of 1889 featured not just the Eiffel Tower but the giant electric arc light at its pinnacle. To one visitor, the "bundle of rays" it generated represented "the most powerful electric force of light so far and darkness is defeated by their power."[26] The battle between Edison and Westinghouse to provide the electrical illumination for Chicago's Colombian Exposition in 1893 offers a vivid demonstration of how much could rest on such showmanship. Westinghouse's victory was as symbolic as it was practical—it helped make his AC system of electrical power transmission the dominant one, but the way it planted his technologies in the imagined future was just as important.[27]

The driving force behind Westinghouse's enlightening victory over Edison had been Nikola Tesla, of course.[28] Tesla was himself a consummate generator of electrical futures. In newspaper interviews he invited his readers to imagine future cities: "Perchance it is one of my cities. Then you see that all the streets and halls are lighted by my beautiful phosphorescent tubes, that all the elevated railroads are propelled by my motors, that all the traction companies' trolleys are supplied with my oscillators, or else that my friends of the Cataract Construction Company are transmitting all the power by my system from a far-off Niagara." He was particularly adept at seizing on the moment: "Signaling to Mars? I have apparatus which can accomplish it beyond any question."[29] His public performances presented a vision of the future in which human bodies were incorporated in technological systems. He demonstrated to his audiences how he could bring his body "into contact with a wire conveying alternating currents of high potential, and the tube in my hand is brilliantly lighted." He was showing how his own and his audiences' bodies could act as conduits for the electricity that would light up the "beautiful phosphorescent tubes" of his imagined future city.[30]

It was at the 1876 Centennial Exhibition in Philadelphia that Alexander Graham Bell announced the invention of the telephone. Like Tesla almost two decades later, he understood very well the role exhibitions could play in displaying possible futures that new inventions might inhabit. Putting the telephone through its paces in Philadelphia not only offered Bell a valuable opportunity for publicity, but also gave the people attending the exhibition an understanding of where the invention belonged.[31] Exhibitions offered a context for their exhibits that made them part of a collective future. The

---

[24] Quoted in ibid., 98.

[25] "The International Exhibition," *Times*, 17 June 1862, 11.

[26] Quoted in Rikke Schmidt Kjoergaard, "Electric Adventures and Natural Wonders: Exhibitions, Museums and Scientific Gardens in Nineteenth-century Denmark," in *Popularizing Science and Technology in the European Periphery, 1800–2000*, ed. Faidra Papanelopoulou, Augusti Nieto-Galan, and Enrique Perdiguero (London, 2009), 135–56, on 138–9.

[27] Thomas Hughes, *Networks of Power: Electrification in Western Society, 1880–1930* (Baltimore, Md., 1993).

[28] W. Bernard Carlson, *Tesla: Inventor of the Electrical Age* (Princeton, N.J., 2013).

[29] Both quotations in ibid., 245–6 and 265.

[30] "Mr. Tesla's Lectures on Alternate Currents of High Frequency and Potential," *Nature* 45 (1892): 345–7, on 346.

[31] Beauchamp, *Exhibiting* (cit. n. 23).

telephone's first appearance was carefully choreographed, with Bell intoning Shake-speare into the transmitter for the judges' edification. William Thomson, acting as one of the Philadelphia exhibition judges, was impressed by the "electrical transmission of speech" and called the telephone "perhaps the greatest marvel hitherto achieved by the electric telegraph" as he speculated about the future instrument's capacity for "making voice and spoken words audible through the electric wire to an ear hundreds of miles away."[32] Repeated exhibition was a key strategy in constructing futures for new instru-ments. William Henry Preece brought an example of Bell's telephone to the next meet-ing of the British Association for the Advancement of Science (BAAS). Bell and his telephone were a major feature of the Exposition Universelle in Paris in 1878.[33]

The telephone, like Tesla's power distribution systems, offered ways of bringing sensation into technology. They were both devices (or potential devices) that incorpo-rated the senses in their operation. Tesla's experiments and exhibitions showed how the human body could itself be incorporated into technologies of the future. Bell's telephone, even more than the telegraph, seemed to offer a future in which the range of bodily sensation could be indefinitely extended. Drawing attention to the homology between the telegraph and the nervous system was already commonplace. Nerves were wires and wires were nerves that extended outside the body and therefore extended the range of sensation. Promising this kind of affinity between future technologies and the body was an increasingly common trope at exhibitions during the second half of the century. At the 1881 Paris Electrical Exhibition, audiences could listen to performances at the opera house transmitted through telephone lines to "a suite of four rooms reserved for the purpose in one of the galleries of the Palais de l'Industrie."[34] Future technologies could transmit sensations as well as being sensations themselves. This was the culture that generated the telectroscope. It was a future technology whose plausibility and de-sirability were underpinned by exhibitionary sensation.

That culture generated fictions more explicit in their fictionality than the telec-troscope. From their origins in the gothic fiction of the beginning of the century, sci-entific romances had by the 1890s become an increasingly important element in the literary outpourings of the new generation of popular magazines aimed at a middle-class market. These were the magazines in which Arthur Conan Doyle, George Grif-fiths, and H. G. Wells first published their futuristic stories. Much of the attraction of these romances lay in the ways their authors grounded their fictions in what their read-ers already knew about the technological present.[35] Their characters started their jour-neys in familiar territories and moved through familiar technological landscapes. The technologies embedded in these romances (electric illumination and weaponry, mys-terious new gases, airships) were easily extrapolated from the existing world of exhi-bition and sensation. There were also technologies not far distant from the telec-troscope in those fictions. These future fictions depended on exhibitionism and the

---

[32] Quoted in Silvanus P. Thompson, *Life of Sir William Thompson, Baron Kelvin of Largs*, 2 vols. (London, 1910), 2:671.

[33] Beauchamp, *Exhibiting* (cit. n. 23).

[34] "The Telephone at the Paris Opera," *Sci. Amer.* 45 (31 December 1881): 422–3, on 422.

[35] Paul K. Alkon, *Science Fiction Before 1900: Imagination Discovers Technology* (New York, N.Y., 1994); Will Tattersdill, *Science, Fiction, and the Fin-de-Siècle Periodical Press* (Cambridge, UK, 2016).

technology of display to populate their futures with a material culture that functioned. In just the same way, accounts of the telectroscope were trustworthy because their origins in a familiar array of future technologies were so explicit.

SEEING AT A DISTANCE

On 29 March 1877, the *New York Sun* published a pseudonymous letter announcing the imminent announcement of an amazing new invention. According to the letter, an "eminent scientist of this city, whose name is withheld for the present, is said to be on the point of publishing a series of important discoveries, and exhibiting an instrument invented by him, by means of which objects or persons standing or moving in any part of the world may be instantaneously seen anywhere and by anybody." The electroscope, as the letter's author called it, would "supersede in a very short time the ordinary methods of telegraphic and telephonic communication." Merchants would be able to demonstrate their wares to potential customers in cities all over the world; criminals could be captured; "mothers, husbands and lovers" would be able to see their absent loved ones; scholars would be "enabled to consult in their own rooms any rare and valuable work or manuscript in the British Museum, Louvre, or Vatican, by simply requesting the librarians to place the book, opened at the desired page, into this marvellous apparatus." Combined with the telephone, the instrument would allow its users "not only to actually converse with each other . . . but also . . . to look into each others' eyes, and watch their every mien, expression, gesture, and motion whilst in the electroscope."[36]

The apparatus was based "on the principle of transmitting the waves of light given out by objects in a manner similar to the transmission of sound waves by the telephone." It consisted of a bundle of wires "of a peculiar make and consistency." Each of these wires transmitted light of a particular color "from some object or person opposite it." At the receiving end, the wires were reconfigured into their appropriate positions and the receiver was filled with a "newly discovered gas, a sort of magnetic electric ether, in which the currents of light or color become resplendent again, and by means of which the objects or persons present at the time in the transmitter are reflected as accurately as in a mirror."[37] It was a well worked out account and, in the context of recent excitement about Bell's telephone and its possible applications, presumably seemed plausible enough to the *Sun*'s readers. Nevertheless, the identity of the "eminent scientist" who had invented the new device was never made public, and the newspaper carried no further announcements about the astonishing invention. The device for seeing at a distance was quickly submerged in the deluge of daily news generated by a newspaper hungry for readers.

Two years later, however, the device resurfaced. The London *Times* for 27 January 1879 carried a notice that "M. Senlecq, of Ardres, has recently submitted to the examination of MM. du Moncel and Hallez d'Arros, a plan of an apparatus intended to reproduce telegraphically at a distance the image obtained in the camera obscura." The article was a translation of a notice published by the Abbé Moigno in *Les Mondes* a

---

[36] Electrician, "The Electroscope," *New York Sun*, 29 March 1877.
[37] Ibid.

few days earlier. The notice went into considerable detail about how Senlecq's device was supposed to work. It was "based on the property possessed by selenium of offering a variable and very sensitive electrical resistance according to the different graduations of light." The apparatus would "consist of an ordinary camera obscura containing at the focus an unpolished glass and any system of autographic telegraphic transmission; the tracing point of the transmitter intended to traverse the surface of the unpolished glass will be formed of a small piece of selenium held by two springs acting as pincers, insulated and connected, one with a pile, the other with the line."[38] Variations on the *Times* notice appeared in a number of other British and North American publications, including *Nature*, the *English Mechanic and World of Science*, and *Scientific American*. In most of these accounts the new instrument was called a telectroscope.

A year later, the *Electrician* published a more detailed account, including diagrams, of the new instrument and its recent improvements. Senlecq's telectroscope had "everywhere occupied the attention of prominent electricians who have striven to improve on it." The journal also provided a list of those efforts, but acknowledged that the "result has been on the one hand confusion of conductors beyond a certain distance, with the absolute impossibility of obtaining a perfect insulation" if bundles of wires were used for transmission, and "an utter want of synchronism" if just one wire was used. The "unequal and slow sensitiveness of the selenium" also caused problems for the proper operation of the apparatus. It seemed clear that "without a relative simplicity in the arrangement of the conducting wires" and "without a perfect and rapid synchronism acting concurrently with the luminous impressions, so as to insure the simultaneous action of transmission and receiver . . . the idea of the telectroscope could not be realised." In the improved version, platinum wire was used in the receiver so that "by the incandescence of these wires, according to the different degrees of electricity we can obtain a picture, of a fugitive kind, it is true, but yet so vivid that the impression on the retina does not fade during the relatively very brief space of time the slide occupies in travelling over all the contacts."[39]

A few months later, *Scientific American* published George R. Carey's account of his own improvements, according to which the "art of transmitting images by means of electric currents is now in about the same state of advancement that the art of transmitting speech by telephone had attained in 1876."[40] Carey offered his own versions of the telectroscope, or "instruments for transmitting and recording at long distances, permanently or otherwise, by means of electricity, the picture of any object that may be projected by the lens of [the] camera upon its disk." The disk contained an array of selenium pieces, each part of an electric circuit, so that when an individual piece of selenium was exposed to light a current would flow. If the disk at the receiving end was configured in the same way as the transmitter and the chemically treated paper placed on it, the disk would reproduce the original image viewed by the camera. He also offered an alternative arrangement where a similar effect was achieved using a selenium point moved by clockwork over a glass plate inside the camera; the idea was that the varying intensity of light shining on different portions of the plate would cause variations in the electric current that would be reproduced at the receiving end.

---

[38] "The Telectroscope," *Times*, 27 January 1879.
[39] "The Telectroscope," *The Electrician* 6, 1881, 141.
[40] George R. Carey, "Seeing by Electricity," *Sci. Amer.* 42 (5 June 1880): 355.

The recently discovered light sensitivity of selenium was the common factor in most of the different proposals for telectroscopes, seeing telegraphs, and electric telescopes that proliferated in the five years or so following Bell's exhibition of his telephone at the Philadelphia exhibition. Selenium's sensitivity to light had been noted in 1873 by the telegraph engineer Willoughby Smith. While experimenting with bars of the metal to see if its high resistance would make it useful for underwater telegraphy, he had noticed that the selenium bars' resistance varied with the intensity of light to which they were exposed. The higher the light intensity, the lower the resistance.[41] In their communication to *Nature* about the prospects of seeing at a distance with electricity, the telegraph engineers John Perry and William Ayrton suggested that the "complete means for seeing by telegraphy have been known for some time by scientific men."[42] Their own proposal consisted of a transmitter composed of a large surface made up of small squares of selenium on which the image to be sent was projected by a lens. Each square of selenium was connected to a wire leading to the receiver, where the image was reconstituted using different methods of electrically opening and shutting apertures through which light was shining.

Clearly, the immediate technological context for this flurry of interest in electrical seeing was provided by telegraphy and telephony. More or less literally, the new devices that were being projected during these years immediately following Bell's telephone were constructed out of the bits and pieces of telegraphic and telephonic material culture. That was what gave the telectroscope its relevance and its air of plausibility. It was projected as part of a future that was already imminent and familiar. It is easy to forget when observing the meticulous detail of at least some of these projections that the devices described were never made. Perry and Ayrton, despite the detail they offered of their instrument's operation, were clear that their plan had "not been carried out because of its elaborate nature, and on account of its expensive character." The two were certain that they would not "recommend its being carried out in this form."[43] Even though it was Willoughby Smith's observation of the illuminating properties of selenium that gave all these projections their plausibility, it was also selenium's properties that made it difficult to put them into practice. While the metal's resistance increased instantaneously when it was exposed to light, it decreased far more slowly when the light was removed. The projected devices could not be sensitive enough for the changes in illumination at the transmitting end.

Despite these difficulties, selenium in some form also featured in the telectroscope proposals put forward by Jan Szczepanik in the late 1890s. On 24 February 1897, Szczepanik and his financial backer Ludwig Kleinberg applied for a British patent for a "method and apparatus for reproducing pictures and the like at a distance by means of electricity." The patent was awarded a year later. The patent specification included the following description:

> The picture that is to be rendered visible at a distance is broken up into a number of points and the several rays corresponding to these points are combined together again to form a picture by means of two pairs of mirrors placed at the transmitting and receiving stations

---

[41] Willoughby Smith, "Effect of Light on Selenium During the Passage of an Electric Current," *Nature* 7 (1873), 303.

[42] John Perry and William Ayrton, "Seeing by Electricity," *Nature* 21 (1880), 589.

[43] Ibid.

and oscillating synchronously; for the purpose of transmission, the differences, of the rays of light passing from the picture-points obtained by breaking up being first converted into differences of current, and the latter being then transmitted through the line conductor to the receiving station and there re-converted into differences of light [*sic*].

The process depended on sensitivity: if "the conversion of the several points of the picture that is to be rendered visible take place in sufficiently rapid succession, the eye of the observer will receive the impression of the entire picture, and if the picture be reproduced repeatedly in sufficiently rapid succession, the observer will receive the impression of a permanent picture, the subject of which may appear to be at rest or in motion according to the nature of the successive pictures thus 'telectroscoped.'"[44]

Telegraphy and telephony remained the key technological reference points for the new invention—though it is tempting to suspect that Marconi's experiments on wireless telegraphy during the second half of the 1890s provided additional context.[45] The patent specification made the newly invented cinematograph part of the technological makeup of the telectroscope as well, however:

> There may be arranged a photographic plate upon which the object situated in front of the transmitting apparatus will appear after a certain time, or the eye itself may be employed as a photographic camera in which case the several light-rays are rendered visible upon the retina. The impressions of images upon the same are, however, of very short duration and disappear after 0.1 to 0.5 of a second. Consequently if the whole of the points of the original picture be telectroscoped, the observer will, from this succession of photographs produced in the eye, have the consciousness of seeing the entire picture just as if the whole of the points of the picture were transmitted simultaneously.

In addition, if

> before the complete disappearance of the picture, a second, third, and so on repetition of the picture be transmitted in the same way, the eye will receive the impression of a permanent picture. If the successively transmitted pictures correspond to different phases of motion of the object, the impression of a moving picture can be made on the retina of the eye, exactly as in the case of the stroboscope, kinematograph, and the like.[46]

Szczepanik was clearly an extremely adept self-promoter and worker of the media. For the next few years, there was a steady stream of articles in newspapers and magazines describing the fabulous new instrument in detail and giving accounts of Szczepanik's own humble and unpromising background. He was a schoolmaster from rural Poland who had staked everything on the success of his invention. It was an account of a self-made inventive genius that tallied well with similar contemporary narratives of what such genius looked like. Most of these accounts were adulatory, though the *Electrical Engineer* expressed deep skepticism about Szczepanik's description of his instrument and its possibilities. There was no detailed account, they noted, of how he proposed to overcome the problem of selenium's slow response to reduced exposure to light, for example, and what that meant for the telectroscope's sensitivity. The

[44] Jan Szczepanik and Ludwig Kleinberg, Method and Apparatus for Reproducing Pictures and the Like at a Distance by Means of Electricity, British Patent 5031, issued 24 February 1898.

[45] Marc Raboy, *Marconi: The Man who Networked the World* (Oxford, UK, 2016).

[46] Sczcepanik and Kleinberg, Method and Apparatus (cit. n. 44).

announcement that the telectroscope would be prominently displayed at the 1900 Paris Exhibition was a conspicuous feature of many of these accounts. It underlined the device's plausibility and its significance. Being part of the exhibition could be understood as making the telectroscope an already established feature of the future.

CIRCULATING THE FUTURE

In the first substantive section of this paper I delineated some of the key features of the technological world in which the telectroscope was produced. In the second section I focused on the details of its construction. In this final section, I want to look at how the telectroscope and instruments like it moved between fact and fiction. The telectroscope is interesting in this respect as a kind of bridging device. None (or at least very few) of the texts and images that discussed and depicted efforts to build a telectroscope during the final quarter of the nineteenth century described a real instrument. No one actually built or exhibited any device that did the things ascribed to its various projected versions. In that sense, the telectroscope was a fictional instrument. It seems clear, at the least, that in the telectroscope's case there was a spectrum, rather than a strict division, between fact and fiction. Some of the accounts already discussed here—such as Twain's short story—were unambiguously fictional and clearly meant to be understood as such by their readers. The place of other accounts on the spectrum can be more difficult to pin down.

The *New York Sun*, in which the first account of a machine for seeing electrically at a distance was published in 1877, had a long history of deception that stretched back more than half a century. In 1835, the newspaper had published a series of articles describing the discovery of lunar life by John Herschel using a powerful new telescope.[47] In 1844, the newspaper published an account of a successful transatlantic balloon crossing that had in fact been written by Edgar Allen Poe.[48] Both these earlier hoaxes bear some similarities to the 1877 account, and it is certainly possible that it should be regarded as a piece of fiction too. Those earlier hoaxes acquired their verisimilitude through strategic use of context. The moon hoax extrapolated on public interest in John Herschel's astronomical expedition to the Cape of Good Hope, for example, and the balloon hoax played with public interest in aeronautics (including an account of an exhibition at the Adelaide Gallery) in a similar way to the framing of the new electrical seeing device in the context of enthusiasm for telephony in the aftermath of Bell's exhibition. Deliberate hoax or not, it cannot have been entirely clear to the original readership how the account they were reading was to be understood.

When Perry and Ayrton published their telectroscope proposal in 1880 they claimed to have received their initial inspiration from a cartoon published in the *Punch* almanack for 1879. The cartoon, "Edison's Telephonoscope (Transmits Light as well as Sound)," had been drawn by the regular *Punch* cartoonist, George du Maurier.[49] The cartoon depicted a Victorian middle-class mother and father sitting comfortably at home while

---

[47] Matthew Goodman, *The Sun and the Moon: The Remarkable True Account of Hoaxers, Showmen, Duelling Journalists, and Lunar Man-Bats in Nineteenth-century New York* (New York, N.Y., 2010), 185–98.

[48] Harold Beaver, ed., *The Science Fiction of Edgar Allen Poe* (Harmondsworth, UK, 1976), 110–23.

[49] Daphne du Maurier, *The Du Mauriers*, new ed. (1937; repr., London, 2004).

gossiping and watching their children play tennis in the antipodes.[50] That cartoon had itself been inspired by rumors of a new invention—the telephonoscope—by Edison. This telephonoscope was not, in fact, a device for seeing at a distance with electricity, as portrayed by du Maurier, but a device for magnifying sound in the same way that the telescope magnified vision.[51] Du Maurier was satirizing the technological fantasizing that accompanied announcements of Edison's activities and the broader culture of electrical futurism that gave it context.[52] Again, the episode offers a nice example of how the telectroscope moved about. Du Maurier was poking fun at the fact that Edison's telephonoscope was not really as he represented it. Perry and Ayrton knew that the cartoon was satire. It could still provide the foundation for a more or less serious technological projection—though it is worth asking the question of just how seriously Perry and Ayrton themselves regarded their telectroscopic proposal in view of its origins. Was their letter to *Nature* satire too?

Dissolving—or at least blurring—the boundary between fact and fiction was clearly an important strategy in futuristic and technological projection. Inventor entrepreneurs like Edison and Tesla routinely made claims about their inventions and their possibilities that played with that boundary. The example of Tesla's claims about communicating with Martians is salutary in this respect. He was putting his projected technologies of the future in contexts that drew on both (supposedly) factual and fictional representations. Mars was topical in the 1890s as a result of Percival Lowell's speculations about its canals and the Martian civilization they implied. *The War of the Worlds*, by H. G. Wells, made Martian communication a more plausible piece of the future too. Tesla's positioning of his experiments on electromagnetic wireless communication at his Wardenclyffe tower in that context was a way of making their place in the future concrete and recognizable to his audiences.[53] The credibility of the fiction produced by Wells and others depended just as much on the performances, speculations, and technologies generated by people like Lowell and Tesla. All of these narratives were encountered in the same sorts of places. Those were the sorts of places where the telectroscope was constructed as well, and it shared the characteristic ambiguity.

Szczepanik's telectroscope became a sudden sensation at the beginning of February 1898 when the London *Daily Chronicle*'s Vienna correspondent announced the new invention to the Anglophone world. Over the next few weeks a number of newspapers on both sides of the Atlantic picked up the story and repeated the essential details in equally enthusiastic terms.[54] Jan Szczepanik was an obscure Polish schoolmaster laboring away in a rural village school who had somehow arrived at the solution to a problem that had confounded the greatest scientists and inventors of the age. He had built a working telectroscope. Anxious to gain fame and fortune with his invention, he had abandoned his post and traveled to Vienna, desperate to find a financial backer who could help him. The newspapers reported that Szczepanik would "introduce his

---

[50] "Edison's Telephonoscope (Transmits Light as well as Sound)," *Punch's Almanack for 1879*, 9 December 1878, 6.

[51] Paul Israel, *Edison: A Life of Invention* (New York, N.Y., 2000), 152.

[52] Ivy Roberts, "Edison's Telephonoscope: The Visual Telephone and the Satire of Electric Light Mania," *Early Popular Visual Culture* 15 (2017): 1–25.

[53] Carlson, *Tesla* (cit. n. 28).

[54] Many of these are listed in R. W. Burns, *Television: An International History of the Formative Years* (London, 1998); and George Shiers, *Early Television: A Bibliographic Guide to 1940* (London, 1997).

discovery in the course of the next few days to a select circle of scientific men and jour-nalists."[55] Not all the press was impressed. The *Electrical Engineer* published a series of notes and editorial commentaries poking fun both at Szczepanik's claims and the cred-ulousness of the popular press, suggesting that "so far Herr Szczepanik has carried out no successful experiment, but has confined his attention to romancing to untechnical reporters."[56] Rumors were soon flying that the organizers of the 1900 Paris Exhibition had offered the inventor huge amounts of money not to make the device public until after the exhibition.

For the next few years the telectroscope was both everywhere and nowhere, hover-ing between fact and fantasy. Throughout the rest of 1898, reports about Szczepanik and his remarkable invention persisted as the inventor and his backers continued to offer tantalizing details about the telectroscope and its operation. The *New York Times* announced Szczepanik's claim that "the telectroscope is, if practicable, as it is claimed to be, something besides a mere toy, an instrument of practical importance, and its ef-ficiency seems to be more considerable than could be judged from particulars hitherto published." The device could "not only transmit pictures to a great distance, and re-produce them there, but . . . it is possible that it will make the present telegraph instru-ments superfluous over short distances." It could operate "to the same distance as a telephone . . ."[57] The *Review of Reviews* marvelled: "It has been left to Jan Szczepanik to startle the world with the apparatus by which objects in the natural colors can be seen hundreds of miles away. Thus while we can now hear the voices of our friends at a distance, we shall in the near future be able to see them as well."[58] Despite the fe-verish speculation, however, the telectroscope itself remained stubbornly invisible.

When the *Century Magazine*'s readers of the November 1898 issue came across Mark Twain's short story, they would almost certainly have already known about the telectroscope and its inventor. This was not even the first time that Twain had written about them. Only a few months earlier the same magazine had published his piece, "The Austrian Edison Keeping School Again." The article described in whimsical terms how "the youthful inventor of the 'telectroscope' [for seeing at great distances] and some other scientific marvels" had found himself a village schoolmaster again.[59] It seemed, according to Twain, that Szczepanik had forgotten to apply for exemption from compulsory military service, but that a benevolent state had found a way of keeping him out of the army in return for his agreeing to spend one morning every two months teach-ing at his old school for the rest of his life. The piece was a further exercise in promotion, with glowing accounts of Szczepanik's triumphant return to his native village, the "body of schoolmasters from the neighboring districts, marching in column" that came to greet him, and the "holiday dance through the enchanted lands of science and inven-tion" through which he led the children.[60] It was a romantic tale of invention and inven-

[55] "Vienna," 19 March 1898, *New York Times*.

[56] "The Fernseher Again," *Electrical Engineer*, 15 April 1898.

[57] "That New Telectroscope," *New York Times*, 3 April 1898.

[58] "The Telectroscope and Its Inventor," *American Monthly Review of Reviews* 18 (1898): 93–4, on 94.

[59] Mark Twain, "The Austrian Edison Keeping School Again," *Century Magazine* 57, October 1898, 630–1, on 631 (brackets in original).

[60] Ibid. Twain's relationship and fascination with Szczepanik is discussed in Tom Burnam, "Mark Twain and the Austrian Edison," *Amer. Quart.* 6 (1954): 364–72.

tors that readers would have found difficult to distinguish from the fiction by the same author they would read a few months later.

In 1899, *Pearson's Magazine* published Cleveland Moffett's "Seeing by Wire," which offered another biography of Szczepanik and his invention. Szczepanik, "as far back as he could remember," had wanted to invent a machine for seeing at a distance. The "stories of Jules Verne, with their dazzle of scientific possibilities" had been his inspiration. He had emerged from "the obscurity of a Galician schoolhouse to a fine eminence with scientists envying him, news-gatherers pursuing him, and capitalists tendering him millions—all at the age of twenty-seven." The backers behind his forthcoming demonstration of the telectroscope at the Paris Exhibition had "bound themselves under a 2,000,000-franc forfeit . . . to make all necessary arrangements for exhibiting the telectroscope at the exhibition, including the construction of a building capable of seating from eight to ten thousand people."[61] Moffett's readers would have known that as well as being a journalist, he was a prolific author of detective stories, historical romances, and speculative fiction. Like Twain's accounts of Szczepanik and his telectroscope, Moffett's contribution hovered somewhere between genres. When the Paris Exhibition opened its doors in August 1900, there was no Szczepanik and no telectroscope. There were no further accounts of the great invention for seeing at a distance.

CONCLUSIONS

Just as the magic lantern has characteristically been understood by its historians as an element in the prehistory of cinema, the telectroscope has usually been understood as an element in the prehistory of television.[62] Neither characterization is very helpful. If we want to understand the magic lantern, we need to place it in the context of optical technologies and performances in which it was used rather than as a step toward a predetermined future technocultural assemblage. Similarly, the best way of making sense of the telectroscope is to look at how it worked as a piece of technological projection that made sense to protagonists who inhabited and moved through a very specific technocultural landscape. The telectroscope is particularly interesting in this respect because its nonexistence actually brings into sharper focus the ways in which new technologies moved back and forth between fact and fiction throughout the second half of the nineteenth century. In this respect, the telectroscope was typical. It was assembled through a patchwork of fictional and factual representations that deployed existing technologies of the future to make new ones. The same could be said of the telephone, the phonograph, or wireless telegraphy. They became real through their place in fictional as much as factual narratives and performances.

The telectroscope, again like many of the nineteenth century's transformative communication technologies, promised a future in which senses could be extended beyond the body. Marshall McLuhan famously suggested that one of the characteristics of the Victorian age's new electrical technologies was that while "all previous technology (save speech itself) had, in effect, extended some part of our bodies, electricity may be said to have outered the central nervous system itself, including the brain."[63] He

---

[61] Cleveland Moffett, "Seeing by Wire," *Pearson's Magazine* 1, October 1899, 490–6, on 490, 495.
[62] Laurent Mannoni, *The Great Art of Light and Shadow: Archaeology of the Cinema* (Exeter, UK, 2000); Burns, *Television* (cit. n. 54).
[63] Marshall McLuhan, *Understanding Media: The Extensions of Man* (New York, N.Y., 1964), 247.

was writing specifically about the telegraph, but the point is a more general one. Electrical technologies gained their hold on Victorian and Edwardian culture through their capacity to make the future sensational. Tomorrow's bodies would be able to experience the world in extended ways, and the future they inhabited would be composed of the spectacular technologies through which the late Victorian and Edwardian middle classes increasingly experienced their own world. The slippage between the factual and fictional that the telectroscope instantiated was typical of this culture of spectacle. While the material culture of exhibitions was made up of perfectly real technologies, the ways in which their deployment at exhibitions pointed them toward the future also made them fictional.

Like the Jules Verne stories that, according to Szczepanik, were one of the inspirations for the telectroscope, the device itself was made out of the bits and pieces of the present. As a piece of fiction its verisimilitude was underpinned by its familiarity as well as its exoticism in just the same way as Verne's stories or H. G. Wells's scientific romances. The telectroscope not only shared the same familiar exhibitionary world as the telegraph and telephone, it was made out of the same things. This was the material culture that was also extended into the future through fiction. Readers inspecting diagrams of telectroscopic transmitters and receivers in popular magazines would recognize their components even if they were arranged in new configurations. These were familiar parts of the exhibitionary technology of display and of electrical communication. It was that familiarity that made this ultimately fictional piece of the future real. The telectroscope made sense as part of the future—in the same way that voyages to the moon or invasions from Mars made sense as part of the future—because it was an extrapolation of the present, rather than a discontinuity. Looked at another way, instruments like the telectroscope, like scientific romances and speculative pieces in popular magazines, made the future real by showing how to get there.

The telectroscope was (and is) a fascinating and illuminating piece of imagined technology in its own right, but it is also fascinating for the light it casts on the practical work of future making—fictional and factual—at the end of the nineteenth century. The telectroscope moved back and forth along a spectrum between fact and fiction. In some of its manifestations—the patent acquired by Szczepanik and his financial backer in 1898, for example—the technology teetered on the brink of realization; in others, such as Twain's short story in *Century Magazine*, it was manifestly fictional. But these are best understood as points on a continuous spectrum rather than as opposites. The telectroscope's nonexistence makes this explicit in its case, but this was the way that other late Victorian and Edwardian technologies of the future moved around their culture as well. At the end of the nineteenth century, wireless telegraphy was as much an imagined future technology as it was a working commercial one. A few years later the same could be said of radium, for example. The one depended on the other. Telling stories about the future of novel technologies was integral to the process of making them real. Their reality underpinned their verisimilitude as components of the technology of the future too.

Thought Transfer and Mind Control between Science and Fiction:

Fedor Il'in's *The Valley of New Life* (1928)

*by Nikolai Krementsov**

ABSTRACT

This essay makes a detailed analysis of the contents and contexts of a science fiction novel published in Moscow in 1928, and written by gynecologist Fedor Il'in (1873–1959) under the title *The Valley of New Life*. The analysis illuminates the process of the transformation of the specialized, and often quite arcane, scientific knowledge generated by biomedical research into an influential cultural resource that embodied acute societal anxieties (both hopes and fears) about the powers unleashed by the rapid development of the biomedical sciences. It explores the future scientific advances—bio- and psychotechnologies—portrayed in Il'in's novel in light of contemporary research, and especially focuses on studies of telepathy. The essay depicts the "translation" of available scientific descriptions and explanations of telepathy into a highly metaphorical language of science fiction, and the resulting formation of a particular cultural resource embedded in such popular notions as "mental energy," "thought transfer," "radio-brain," "nervous waves," "psychic rays," and "mind control." It examines how and for what purposes this cultural resource was utilized by scientists, their patrons, and literati (journalists and writers) in Bolshevik Russia, Britain, and the United States.

> An ancient sage, inadvertently, of course, uttered a wise adage: "Love and hunger rule the world." Ergo: to rule the world man must master its rulers.
> —Evgenii Zamiatin, 1921[1]

In 1928, one of the few remaining private publishers in Moscow released a novel with the intriguing title, *The Valley of New Life*.[2] The book's cover bore a bold announcement:

* Institute for the History and Philosophy of Science and Technology, 91 Charles Street West, Victoria College, University of Toronto, Toronto, Ontario, M5S 1K7, Canada; n.krementsov@utoronto.ca.

Research underpinning this article was supported by an Insight Grant from the Social Sciences and Humanities Research Council of Canada. An earlier version was presented at the "Histories of the Future" workshop organized by Erika Milam and Joanna Radin at Princeton University in February 2015, and I greatly benefited from that input. I am grateful to Amanda Rees, Iwan Morus, and the anonymous reviewers of this volume for their helpful comments and suggestions. I also learned a lot from lively discussions at the "Imagining the History of the Future" conference held on 27–29 March 2018 in York, England.

[1] Evgenii Zamiatin, *My* (Moscow, 1989), 21. Though first published in Russia in 1989, the novel was written in 1921. All translations are mine.
[2] Teo Eli [F. N. Il'in], *Dolina novoi zhizni* (Moscow, 1928).

> Citizens! Do you know that all of you are in great danger? Do you know that a band of determined people, hiding in an isolated valley in the Himalayas—the so-called valley of "New Life"—has long been plotting a horrific coup?! Using powerful technical inventions unknown to us, they have created a millions-strong army of artificial people and equipped it with terrible weapons. All the preparations for the coup are finished. The band's leaders only await the completion of a tunnel that would connect the valley with the outside world. Soon the tunnel will be ready and the iron battalions of half-people-half-machines will attack stunned humankind. The danger is near! Everybody should know about it. Everyone must buy and read this book that describes this outrageous plot.

Despite its clear advertising tenor, the blurb quite accurately describes the book's contents. Narrated by one of the valley's inhabitants, French electrical engineer Rene Ger'e, the novel indeed portrays a new civilization created by a clan of American billionaires, the Queensleys. With the help of leading scientists and engineers, whom they "import" by co-option or coercion to the valley from around the world, the Queensleys populate this new Atlantis with a race of superadvanced human beings. Brought to life by in-vitro fertilization, interspecies hybridization, and ectogenesis (gestation of human embryos in artificial wombs), these people are "educated" by means of telepathic mass hypnosis. According to the Queensleys' vision, this new race—armed with superior science and technology, including the use of atomic energy, synthetic foods, and mind control—is destined to conquer and remake the world. Ornamented with a canonical "love affair" between its main characters, the novel ends with the miraculous escape of its narrator from the valley just a few days before its army of superhumans is to invade the world. The novel's content indicates that the published text was just the first installment of a larger story. However, readers excited by the novel waited in vain for its sequel. In its full form, the two-part novel would only forty years later.[3]

The Valley of New Life is a fine example of a new literary genre—*nauchnaia fantastika*, scientific fantasy (SF) as it was called in Russian, and science fiction in English[4]—that took newborn Soviet Russia by storm. In his lengthy essay of 1922 about H. G. Wells, an acknowledged founder of the genre, one of its Soviet pioneers, Evgenii Zamiatin, foresaw that "Russia, which during the last few years has become the most fantastic country of modern Europe, will undoubtedly reflect this period in her fantastic literature."[5] Zamiatin's prediction quickly came true. As elsewhere in the world, "fantastic literature" flourished in 1920s Russia, filling the pages of daily newspapers, illustrated weeklies, popular-science and adventure magazines, and books.[6] But Alexander

---

[3] F. N. Il'in [Teo Eli], *Dolina novoi zhizni* (Baku, 1967). Published as a single volume, this edition included both the complete text of the first 1928 edition and the previously unpublished "sequel."

[4] The genre's origins and permutations remain the subject of debates that show no sign of abating; compare, for instance, Edward James and Farah Mendlesohn, eds., *The Cambridge Companion to Science Fiction* (New York, N.Y., 2003); Mark Bould and Sherryl Vint, *The Routledge Concise History of Science Fiction* (London, 2011); and Adam Roberts, *The History of Science Fiction* (Basingstoke, UK, 2016).

[5] E. Zamaitin, *Gerbert Uells* (Petrograd, 1922), 47.

[6] The literature on early Soviet SF is quite extensive. For recent analyses, see V. I. Okulov, *O zhuranl'noi fantastike pervoi poloviny XX veka* [On magazine publications of scientific fantasy in the first half of the twentieth century] (Lipetsk, 2008); Muireann Maguire, *Stalin's Ghosts: Gothic Themes in Early Soviet Literature* (Pieterlen, 2012); and Anindita Banerjee, *We Modern People: Science Fiction and the Making of Russian Modernity* (Middletown, Conn., 2013). See also Sibelan Forrester and Yvonne Howell, "Introduction: From Nauchnaia Fantastika to Post-Soviet Dystopia," in "Reading the History of the Future: Early Soviet and Post-Soviet Russian Science Fiction," ed. Forrester and Howell, special issue, *Slavic Review* 72 (2013): 219–23; and especially the essay by Matthias Schwartz, "How Nauchnaia Fantastika Was Made: The Debates about the Genre of Science

Beliaev, one of the most talented and prolific SF writers of the time, explained the following in a 1928 essay expressively titled "Fantasy and Science":

> There are two different kinds of fantastic literature. One is a free play of imagination, which doesn't take into account either reality, or causality. . . . The second type is so-called scientific fantasy [*nauchnaia fantastika*], which is based on scientific data, paints pictures of the future, [and] builds "perspective plans" for a more or less distant future of technical and social progress.[7]

Undoubtedly, *The Valley of New Life* belongs to the second kind. It echoes certain elements of Zamiatin's famous dystopia, *We* (1921), and presages many of the themes of Aldous Huxley's even more famous *Brave New World* (1932). Yet, in one particular respect the novel differs from the multitude of SF writings flooding Soviet periodicals and bookstores at the time; hiding under the pen name Theo Elie, its author was a scientist. The novel was written by Fedor Il'in (1873–1959), who at the time of its publication was the chairman of the gynecology department at Baku University's Medical School.

Il'in's previous career gave little indication that he might aspire to become the author of an SF novel. Born to a family of the Russian nobility, Fedor graduated in 1898 from the country's premier medical school—the Imperial Military–Medical Academy in St. Petersburg (renamed Petrograd in 1914, Leningrad in 1924, and regaining its original name in 1991). He spent nearly a decade as a doctor with various army and navy regimens. During the 1904–05 Russo-Japanese War, he served on a military cruiser and was twice decorated for "bravery and exemplary service under enemy fire." But, after the end of the war, in 1906, Il'in resigned from the military. He chose to continue his career in a field that could not have been more remote from his previous duties; he joined the staff of the Clinical Institute of Obstetrics and Gynecology in St. Petersburg. The retired military doctor then became engaged in experimental research on exciting new subjects, including the use of radium and X-ray irradiation in treatment of cancerous tumors, and the use of artificial insemination as a means of fighting infertility.[8] He also began lecturing on his new specialty at the Petrograd Women's Medical School.

The October 1917 Bolshevik Revolution in Petrograd, and the bloody civil war that followed, disrupted Il'in's research and his steadily progressing career. As did many other inhabitants of the former imperial capital, he fled to the country's southern regions—a stronghold of the White forces fighting against the Bolshevik Red Army. But unlike many others, Il'in did not emigrate when, in late 1920, the Bolsheviks won the civil war. Instead, at the beginning of 1921, he returned to his home city, hoping to resume his research and teaching. Alas, it was not a propitious time for scientific work in Petrograd, and in 1922, Il'in accepted an offer to organize a gynecology department at the medical school of a new university just established in Baku, the center of the oil industry on the Caspian Sea. He would spend the rest of his life in

---

Fiction from NEP to High Stalinism," 224–46. Alas, as these titles make clear, most of this literature focuses on the "fiction" part of science fiction.

[7] A. Beliaev, "Fantastika i nauka," in *Bor'ba v efire* [Struggle in space] (Leningrad, 1928), 309–23, on 309.

[8] F. N. Il'in, *Dal'neishie uspekhi v terapii luchistoi energiei* [Further advances in radiation therapy] (St. Petersburg, 1914); Il'in, *Iskusstvennoe oplodotvorenie v bor'be s besplodiem zhenshchiny* [Artificial insemination in the stuggle against infertility in women] (St. Petersburg, 1917).

Baku, heading the department he had created, teaching successive cohorts of medical students, and attending to numerous patients at the university's clinics.[9]

Il'in's novel raises a number of intriguing questions. Why would this talented researcher and devoted physician, who earned the title of "Women's God" from his grateful patients, pick up his pen to write an SF novel? What did his novel actually say about science and the scientist? How did Il'in's specialized scientific and medical knowledge inform and shape his fiction? How did his vision of humanity's future differ from that pictured by other scientists and by professional writers? What does the very fact that Il'in wrote fiction tell us about the interrelations between science and literature? And how did such interrelations in Bolshevik Russia compare to those in other countries? In the following pages I will look for answers to some of these questions. Il'in's novel, I will argue, illustrates a particular cultural process unfolding in Bolshevik Russia during this period—namely, the transformation of specialized and often quite arcane knowledge generated by biomedical research into an influential cultural resource exploited by a variety of actors to their own ends.

First, I will briefly describe three different revolutions that defined the contexts of Il'in's novel, including the Bolshevik Revolution, the experimentalist revolution in the life sciences, and the revolution in the public visibility and cultural authority of biomedical research. Next, I will survey the future scientific advances—bio- and psychotechnologies—portrayed in Il'in's novel in light of contemporary research, focusing especially on studies of telepathy. Then, I will explore the "translation" of available scientific descriptions and explanations of telepathy into a highly metaphorical language of "scientific fantasy," and the resulting formation of a particular cultural resource embedded in such popular notions as "mental energy," "thought transfer," "radio-brain," "nervous waves," "psychic rays," and "mind control." Finally, I will examine how and for what purposes this cultural resource was utilized by scientists, their Bolshevik patrons, and literati (journalists and SF writers). I will conclude with some general thoughts on science as a cultural resource suggested by the analysis of Il'in's *Valley of New Life*.

SCIENCE IN REVOLUTIONS

In the 1920s, newborn Bolshevik Russia was imbued by what US historian Richard Stites has perceptively termed "revolutionary dreams" of building a "new world."[10] Inspired by these dreams, the country experimented on an unprecedented scale with new ideas and new practices in every facet of life, from arts to sciences, state administration to education, and industry to health care. Nowhere did these dreams find a more pronounced expression than in numerous works of the new literary genre of scientific fantasy. As its very name indicates, this genre embodied the intersection of two major cultural domains—science and literature—with practitioners in both fields actively contributing to its enormous popularity. A large portion of 1920s SF literature engaged with one particular subset of science—today these are called the life sciences, but at the time they were represented by the two closely entwined and rapidly growing fields of experimental biology and experimental medicine.

---

[9] For a short biography of Il'in, see I. P. Tret'iakov, *Zhenskii bog* [Women's God] (Baku, 2005).

[10] Richard Stites, *Revolutionary Dreams: Utopian Vision and Experimental Life in the Russian Revolution* (Oxford, UK, 1989).

In the first decades of the twentieth century, just as Russia was going through its brutal political revolutions, these fields were undergoing their own revolution. The "experimental" modifiers to biology and medicine allude to the essence of this "mini" revolution, which was the introduction of experimentalist spirit and experimental methods, largely borrowed from physics and chemistry, into the study of life—its past, present, and future. Armed with these new methods, scientists investigated mechanisms of basic life processes and their pathological changes, including metabolism, reproduction, heredity, nervous and endocrine regulation, immunity, growth, evolution, and aging. The advances of experimental research quickly found applications in various branches of medicine, leading to new therapeutic treatments, surgical operations, preventive measures, and diagnostic techniques. In the first decades of the twentieth century, successes in serodiagnostics; tissue and organ transplantation; hormone-, vitamin-, X-ray-, and serotherapies; blood transfusion; immunization; and in deciphering the basic mechanisms of heredity and embryonic development generated a euphoric vision that science could control life and create new life forms.[11] Captivated by this "visionary biology," as US historian of science Mark B. Adams has aptly named it, many scientists the world over came to believe that new experimental techniques could provide them with the means not only to radically improve the present well-being of humanity, but to direct humankind's future development and evolution.[12] Russian scientists were no exception. Numerous investigators, including Il'in, became engaged in research on a variety of new, exciting biomedical subjects. After he had resigned from the military, Il'in conducted much of his research on the cutting edge of the growing field of experimental biology and medicine.

The 1917 Bolshevik Revolution gave experimental biology and medicine in Russia a tremendous boost. "Visionary biology" resonated strongly with the "revolutionary dreams" entertained by the Bolshevik government that became the only patron of science in the country. Not only did the Bolsheviks generously fund a massive institutional growth of these fields, they also made knowledge generated by biomedical research an essential part of their huge science-popularization campaign aimed at legitimizing their regime, thus leading to the rapidly increasing public visibility and cultural authority of such research in the country.[13]

The intersections of visionary biology and revolutionary dreams found expression in particular research agendas pursued by Russian scientists after the Revolution.[14] One dominant agenda became the exploration of the possible mechanisms and plausible directions of human physical, mental, and social evolution. Several researchers created novel techniques of in-vitro fertilization, artificial insemination, hormone production, and interspecies hybridization, which seemed to offer simple means of manipulating and directing human reproduction to any desirable end. Others launched extensive experimental research on telepathy, which promised to uncover "secrets of the

---

[11] Philip J. Pauly, *Controlling Life: Jacques Loeb and the Engineering Ideal in Biology* (New York, N.Y., 1987).

[12] Mark B. Adams, "Last Judgment: The Visionary Biology of J. B. S. Haldane," *J. Hist. Biol.* 33 (2000): 457–91.

[13] For a general analysis of science popularization in Soviet Russia, see James T. Andrews, *Science for the Masses: The Bolshevik State, Public Science, and the Popular Imagination in Soviet Russia, 1917–1934* (College Station, Tex., 2003).

[14] For details, see Nikolai Krementsov, *Revolutionary Experiments: The Quest for Immortality in Bolshevik Science and Fiction* (New York, N.Y., 2014).

mind" and to create effective tools of "mind control." Widely covered by the mass media (including daily newspapers, popular magazines, radio broadcasts, and newsreels), these studies generated acute societal anxieties (both hopes and fears) about the powers unleashed by the rapid advances of biomedical sciences. It was these anxieties that postrevolutionary SF literature attempted to explore, and Il'in's *Valley of New Life* undoubtedly was one such attempt.

BIO- AND PSYCHOTECHNOLOGIES OF THE FUTURE

The "valley of new life" described in Il'in's novel exemplifies humanity's triumph over nature through science and technology. The valley's inhabitants enjoy technologies still unheard of in the "old" world. They use atomic energy, ride superfast "magnetic" trains, transmit electricity wirelessly, and fly on "electrical wings." But it is incredible advances in various biotechnologies that make the valley a true place of the future.[15] Although the narrator apologizes tongue in cheek for "possible inaccuracies" in his descriptions of these biotechnologies, for he "lacks fundamental biological training," he asserts that he "cannot avoid biology because it occupies such a prominent place in everything I had lived through."[16] Indeed, the most astonishing feature of the valley—a new race of human beings, which comprises its "native" population—is created by the application of new biotechnologies developed by scientists and engineers the Queensleys had brought to the valley from all over the world, including Russia. In fact, the creator of the new race is a "Russian zoologist, renowned for his work in artificial insemination and animal hybridization," named Petrovskii (37).[17]

The valley's founders master both "love and hunger" with the help of biotechnologies, thereby echoing Zamiatin's assertion (cited in the epigraph): "Love and hunger rule the world. Ergo, to rule the world man must master its rulers." They conquer hunger through the virtually unlimited production of synthetic foods, employing a variety of chemical and biological processes. And they abolish love by first separating sex and reproduction and then abolishing sex altogether, thus "eliminating all the sex characteristics and passions so harmfully affecting modern humans around the globe" (63). They reduce human reproduction to the in-vitro fertilization of sex cells obtained from embryonic tissue cultures and the subsequent cultivation of human embryos in "incubators." They further control the sex drive in adults by the careful application of sex hormones produced by isolated glands of internal secretion grown in

---

[15] Il'in does not use the term *biotechnology* in his novel, and its appearance in this context might look anachronistic. But, in fact, the term was coined and used in 1920s Russia. In 1922, Mikhail Zavadovskii, one of the country's leading researchers in the field of experimental biology, wrote of "the time, when advances in the study of living nature will create conditions for the flowering of biotechnology [*biotekhnii*] alongside the technology of dead materials, [and when] the biologist's tasks in making new life forms, which now seem akin to [H. G.] Wells's fantasy [*The Island of Doctor Moreau* (1896)], will be as mundane as those of a construction engineer"; see M. M. Zavadovskii, *Pol i razvitie ego priznakov* [Sex and the development of its characteristics] (Moscow, 1922), 235.

[16] Eli, *Dolina novoi zhizni* (cit. n. 2), 4. Further page citations appear parenthetically in the text.

[17] Undoubtedly, the "prototype" for the Petrvoskii character was Il'ia Ivanov, a pioneer in artificial insemination and its uses for interspecies hybridization. On Ivanov and his research, see Kirill Rossiianov, "Beyond Species: Il'ya Ivanov and His Experiments on Cross-Breeding Humans with Anthropoid Apes," *Science in Context* 15 (2002): 277–316; and Nikolai Krementsov, "From a 'prominent biologist' to a 'Red Frankenstein': Il'ia Ivanov in Soviet and Post-Soviet Biographies," in *What Is Soviet Now? Identities, Legacies, Memories*, ed. Thomas Lahusen and Peter H. Solomon Jr. (Berlin, 2008), 120–32.

artificial media. After a few generations, they drastically reduce "production" of females, since females "proved to be unsuitable for the Valley of New Life, because
they are physically weaker and mentally less stable than males" (63). As a result, the
family is no longer the main social grouping of the valley population. As in Huxley's
Brave New World, the new human beings in Il'in's novel have no parents, siblings,
spouses, or children, and thus experience no emotions or affections associated with such
biological and social roles.

Furthermore, Il'in goes beyond Zamiatin's vision to answer the questions of who
would rule the world and how, when humanity has mastered love and hunger. As one
might have expected, the valley's governing council includes its leading scientists
and is presided over by their employers/patrons, the Queensleys. And the main instrument of their governance is "mind control." The valley's rulers possess intricate ap-
paratuses that allow them not only to "read" their subjects' thoughts, but also to control their emotions, communicate orders, and inculcate ideas. Indeed, the entire
system of education and upbringing of the valley's "native" population is built upon
these psychotechnologies. A group of imported scientists, however, creates special
devices that shield their own minds from being "read" and controlled by the valley's
rulers, and this becomes a crucial means for the narrator's successful escape.

Undoubtedly, its readers had no trouble recognizing Il'in's novel as "scientific fantasy." Yet, as the narrator tells us, "there was nothing fantastic, in the exact sense of
the word," in all the bio- and psychotechnologies employed in the valley. "There were
only the advances of science, the beginnings of which had been known in Europe" (33).
Indeed, in his novel Il'in compiled a virtual anthology of new research in experimental
biology and medicine conducted during the 1920s by scientists all over the world. He
directly referred to the studies on tissue cultures and organ transplants pioneered by
Franco-American surgeon Alexis Carrel, as well as the investigations into mechanisms
of fertilization and embryonic development initiated by German zoologist Oscar Her-
twig. Il'in used the results of extensive studies on isolated organs, hormones, artificial
insemination, vitamins, interspecies hybridization, cancer, and many other "hot" new
subjects, which at the time captivated the attention of experimental biologists and physicians. Even the seemingly fantastic notions of "thought transfer" and "mind control"
were firmly grounded in extensive contemporary efforts to uncover the possible mechanisms of hypnosis and telepathy.

"HUMAN THOUGHT—ELECTRICITY"

Surprising as it may seem, in the early 1920s, Bolshevik Russia became one of the
world's leading centers in experimental research on "thought transfer."[18] In August

---

[18] Extensive English-language literature on the history of "psychical research" deals predominantly
with Western Europe and the United States and is almost completely silent about similar developments in Russia. See, for instance, Roger Luckhurst, *The Invention of Telepathy, 1870–1901* (Oxford,
2002); Heather Wolffram, *The Stepchildren of Science: Psychical Research and Parapsychology in
Germany, c. 1870–1939* (New York, N.Y., 2009); and Sophie Lachapelle, *Investigating the Supernatural: From Spiritism and Occultism to Psychical Research and Metapsychics in France, 1853–1931*
(Baltimore, Md., 2011). See also a recent special issue of *Stud. Hist. Phil. Sci.* devoted to the subject;
the editor, Andreas Sommer, wrote the introduction, which has the same title as the overall issue:
"Psychical Research in the History and Philosophy of Science. An Introduction and Review," *Stud.
Hist. Phil. Biol. Biom. Sci.* 48 (2014): 38–45. For an overview of the literary connotations of such
research in Russia, see Mikhail Agursky, "An Occult Source of Socialist Realism," in *The Occult*

1918, less than a year after the Bolsheviks had come to power, Vladimir Durov, a renowned animal trainer and founder of the country's most famous animal circus, managed to secure patronage of the People's Commissariat of Enlightenment (*Narkompros*), a top government agency in charge of education, arts, and sciences headed by eminent Bolshevik Anatoly Lunacharsky.[19] Durov's first concern was the survival of his animals and his circus amid the turmoil of the civil war, epidemics, and famine that came on the tails of the Bolshevik Revolution. But he also had other ambitions. Durov believed that his remarkable achievements in training various animals—not just the usual assortment of circus animals such as dogs, pigs, and monkeys, but also seals, lemurs, badgers, and elephants—were based on his ability to command animals to perform particular tricks telepathically. He wanted not merely to continue performances of his animal circus, but to establish a scientific laboratory to investigate telepathic communications between humans and animals. Lunacharsky was clearly taken by this idea. He personally issued a mandate to protect, fund, and expand Durov's enterprise. Narkompros even provided Durov with a large mansion (requisitioned from a Moscow merchant) to house his circus and laboratory.[20]

Furthermore, Lunacharsky apparently asked Vladimir Bekhterev, Russia's foremost psychiatrist and neurologist, to take part in Durov's investigations and check out his claims. At that very time, Bekhterev was actively lobbying Narkompros and Lunacharsky personally to accept his proposal of establishing an Institute of Brain Research in Petrograd for his own studies. He certainly was more than willing to accommodate the prospective patron, especially since Bekhterev regularly used hypnosis in his clinical practice (particularly in the treatment of alcoholism) and was interested in investigating its possible neural mechanisms. Whatever Bekhterev's motivations were, he quickly amended his proposal to include a special "department of zoopsychology" to be directed by Durov.[21] But Durov declined the offer to join Bekhterev's Institute. In the end, Narkompros established the Institute of Brain Research in Petrograd for Bekhterev, and the Practical Laboratory for Zoopsychology in Moscow for Durov. The two institutions and their respective heads soon launched collaborative studies.

---

in *Russian and Soviet Culture*, ed. Bernice Glatzer Rosenthal (Ithaca, N.Y., 1997), 247–72. For a recent, very brief, and highly superficial English-language depiction of Russian research, see Wladimir Velminski, *Homo Sovieticus: Brain Waves, Mind Control, and Telepathic Destiny* (Cambridge, Mass., 2017), 33–52. There is also limited Russian-language literature on the subject, but it deals mainly with the imperial period; see, for instance, V. S. Razd'iakonov, "'Velikoe delo' uchenykh spiritov: Istoriia Russkogo obshchestva eksperimental'noi psikhologii i spiriticheskii kruzhok N. P. Vagnera" [The "great affair" of scientific spiritualists: the history of the Russian society for experimental psychology and N. P. Vagner's spiritualist circle], *Vestnik RGGU*, "*Filosofskie nauki*" (2013): 141–53; and Maikl Gordin [Michael D. Gordin], "Okhota na prizrakov. Dmitrii Ivanovich Mendeleev i spiritizm" [The hunt for ghosts: Dmitrii Ivanovich Mendeleev and spiritism], *Istoricheskie zapiski* 10 (2007): 251–89.

[19] On the establishment of the agency and its head, see Sheila Fitzpatrick, *The Commissariat of Enlightenment: Soviet Organization of Education and the Arts under Lunacharsky, October 1917–1921* (Oxford, 1970).

[20] A large collection of documents related to the establishment of Durov's laboratory is preserved among the Narkompros materials housed in the State Archive of the Russian Federation (**hereafter** cited as GARF) in Moscow, fond 2307. There is also a large, unprocessed cache of original documents at the Museum of the Durov Animal Theater in Moscow.

[21] Materials pertaining to the establishment of Bekhterev's institute can also be found in GARF, fond 2307 and 482.

These first studies were quite simple and consisted of Bekhterev's (and/or his co-workers') observations, conducted in both Petrograd and Moscow, of Durov's favorite dogs named Mars and Pikki performing certain tricks.[22] Typically, Durov wrote down on a piece of paper a sequence of actions for a dog to perform and gave it to Bekhterev. He then looked the dog in the eyes for several minutes and the dog did exactly what Durov had written it should do. Bekhterev was impressed. In 1920, he devoted a considerable portion of his recently established journal, *Issues in the Studying and Upbringing of Personality*, to these investigations.[23] The experiments seemingly confirmed Durov's ability to communicate with his animals without using voice, gestures, facial expressions, or any other observable means. But they did not answer the questions of exactly how he was able to convey the commands and what possible mechanisms could underlie these supposedly telepathic communications.

A plausible hypothesis regarding the mechanisms of "thought transfer" was not long in coming. On 6 January 1922, a large audience gathered in the Red Army House of Enlightenment in Moscow to listen to a "popular-scientific lecture" with the intriguing title "Human Thought—Electricity." The lecturer, young engineer Bernard Kazhinskii, told his audience that human thought was merely a particular form of electromagnetic energy generated by nervous cells, which could probably be registered by a specially designed apparatus. He suggested that a study of this "mental energy" be conducted "under clinical-laboratory conditions" to lay a foundation for "psychotechnology" (*psikhotekhnika*)—a new field of science soon to be advanced by "engineer-psychologists."[24]

Kazhinskii's lecture was but a theoretical speculation inspired by recent advances in radio and wireless telegraphy.[25] It seems unlikely that the thirty-year-old engineer (who had just a few weeks prior come to Moscow from the provinces in search of employment) actually hoped to turn his personal "revolutionary dream" of becoming an "engineer-psychologist" into reality.[26] But his dream did come true. As it happened, Alexander Leontovich, chairman of the physiology department at Moscow Agricultural Academy and staff member of Durov's laboratory, attended Kazhinskii's lecture. Leontovich's personal scientific interests revolved around the anatomy, his-

---

[22] See Bekhterev's reports on these experiments preserved among his personal papers housed in the Central State Historical Archive (TsGIA) in St. Petersburg, fond 2265, op. 1, dd. 385–86.

[23] See, for instance, V. M. Bekhterev, "Ob opytakh nad 'myslennym' vozdeistviem na povedenie zhivotnykh," *Voprosy izucheniia i vospitaniia lichnosti* 2 (1920): 230–65. An abridged English translation of this article appeared in the United States; see W. Bechterev, "'Direct Influence' of a Person upon the Behavior of Animals," *Journal of Parapsychology* 13 (1949): 166–76.

[24] Materials pertaining to Kazhinskii's lecture, including its manuscript, are preserved in the Archive of the Moscow Polytechnic Museum in the collection of documents of the Association of Self-Taught Naturalists, fond 86; the collection is still in the processing stage.

[25] Since Heinrich R. Hertz's proof of the existence of electromagnetic waves, the hypothesis of "brain waves" became a mainstay of "psychical research." Indeed, the analogy between telepathy and radio became especially popular in the 1920s and 1930s. See, for instance, J. Malcolm Bird, "Telepathy and Radio," *Sci. Amer.* 130 (1924): 382; Ferdinando Cazzamalli, "Fenomeni telepsichici e radio onde cerebrali," *Neurologica* 42 (1925): 193–218; Upton Sinclair, *Mental Radio* (New York, N.Y., 1930); Giuseppe Calligaris, *Telepatia. Radio onde cerebrali* (Milan, Italy, 1934); and many others.

[26] In the late 1950s to early 1960s, Leonid Vasil'ev and Bernard Kazhinskii, two surviving members of the original team of researchers involved with Durov's and Bekhterev's investigations, published popular accounts of their 1920s research; see L. L. Vasil'ev, *Tainstvennye iavleniia chelovecheskoi psikhiki* [Mysterious phenomena of the human psyche] (Moscow, 1959); and B. B. Kazhinskii, *Biologicheskaia radiosviaz'* [Biological radiocommunication] (Kiev, 1962).

tology, and physiology of the nervous system. He had no trouble in visualizing Kazhinskii's speculations about condensers, resisters, inductors, and electric circuits in the brain as corresponding to particular structures in different types of nervous cells and their assemblages that he had observed under the microscope in the course of his own research. In Kazhinskii's imaginary apparatus that could register "mental energy," Leontovich saw a perfect instrument for investigating possible neural mechanisms of Durov's "telepathic" communications with animals. Durov apparently shared Leontovich's enthusiasm; just a few months later, Kazhinskii became a staff member of the Practical Laboratory of Zoopsychology. He soon published an updated version of his 1922 lecture as a brochure, enticingly titled *Thought Transfer*. The brochure's cumbersome subtitle revealed that Kazhinskii was searching for "factors that make it possible for the nervous system to emit electromagnetic waves."[27]

The same year Kazhinskii's search received an unexpected boost. Renowned physicist Petr Lazarev, member of the Russian Academy of Sciences and director of the Institute of Biophysics established in 1920 by the People's Commissariat of Health Protection (*Narkomzdrav*), a new top agency in charge of medicine and public health, published his "ionic theory of excitation."[28] Supported by Lazarev's decade-long experiments, this theory explained various electrical phenomena observed in nervous cells and muscle tissues (such as the propagation of electrical impulses along the neuron, for instance) as the result of changing concentrations of ions within and without the living cell. Lazarev's theory thus provided a plausible physicochemical mechanism that could explain the generation of electromagnetic waves by the brain, and hence, telepathy and hypnosis. Lazarev himself explained:

> Every sensory or motor act originating in the brain must be emitted into the outside environment in the form of an electromagnetic wave. . . . [This] allows us to understand the mechanism of hypnosis: . . . an electromagnetic wave generated by the nervous centers of one individual induces in the [nervous] centers of another individual an [electromagnetic] impulse that starts an oscillating reaction in the [nervous] centers and generates excitation.[29]

In 1924, Durov published a 500-page volume titled *Animal Training*, which detailed his concept of telepathic communications between humans and animals and was supplemented by the protocols of his experiments conducted from 1919 through 1923. The volume incorporated Lazarev's ideas about the possible mechanisms of telepathy, as well as Kazhinskii's schemes regarding the practical ways of studying it.[30] It was but a short step from Kazhinskii's proposal to construct an apparatus for receiving and recording "brain waves," to the idea of an apparatus that could amplify and transmit such waves. In an epilogue to his book, poetically titled "My 'scientific fantasy,'" Durov boldly took this step, envisioning a future in which such ap-

[27] B. B. Kazhinskii, *Peredacha myslei: Faktory, sozdaiushchie vozmozhnost' vozniknoveniia v nervnoi sisteme elektromagnitnykh kolebanii, izluchaiushchikhsia naruzhu* [Thought transfer: Factors that make it possible for the nervous system to generate and emit electromagnetic waves] (Moscow, 1923).

[28] P. P. Lazarev, *Ionnaia teoriia vozbuzhdeniia* [Ionic theory of excitation] (Moscow, 1923).

[29] Ibid., 129.

[30] V. V. Durov, *Dressirovka zhivotnykh. Psikhologicheskie nabliudeniia nad zhivotnymi dressirovannymi po moemu metodu (40-letnii opyt). Novoe v zoopsikhologii* [Animal training. Psychological observations upon animals trained by my method (40-years experience). Advances in zoopsychology] (Moscow, 1924).

paratuses would become "a powerful means to create better thoughts and ideas for humanity and leave no place for harmful thoughts and ideas, which run contrary to the interests of the masses, large collectives, and ultimately the laborers of the entire world."[31] Despite certain very critical assessments of their work (both its theoretical underpinnings and actual investigations) by other scientists and engineers,[32] in the next decade, Durov and his collaborators continued to pursue this "scientific fantasy" at the Practical Laboratory for Zoopsychology with Lunacharskii's personal blessing.[33]

FROM LAB TO FICTION . . . VIA THE MASS MEDIA

Il'in's novel captures a particular preoccupation with technological/engineering solutions to the issues of human nature and human destiny that permeated 1920s Russia. Bio- and psychotechnologies together constitute the very foundations of his "valley of new life." Given Il'in's professional interests, he was undoubtedly familiar with a large body of literature on experimental research related to reproduction, internal secretions, tissue cultures, sex, and embryonic development. After all, specialized periodicals he must have read regularly (from the *Journal of Obstetrics and Gynecology* to the *Journal of Experimental Biology and Medicine*) were flooded with publications on these subjects. But it seems unlikely that he read Bekhterev's journal, or Durov's monograph, not to mention Kazhinskii's and Lazarev's treatises. How then could he possibly learn of their studies on telepathy to render their results and promises in his novel?

The answer is simple. The newest advances in various subfields of experimental biology and experimental medicine were extensively covered by the media—daily newspapers, weekly and monthly magazines, newsreels, radio broadcasts, documentary and feature films, theater productions, and so on. This coverage was part of the enormous science-popularization campaign launched by the Bolsheviks to legitimize their own rule and undermine their "sworn enemy"—religion. As one would have expected, research on telepathy became an integral part of this campaign and was extensively covered by the press. In the early 1920s, *Pravda* and *Izvestiia*, the Bolsheviks' major newspapers with countrywide circulation, regularly carried articles on the subject under such enticing titles as "Electricity and Nervous Waves," "Reading Thoughts at a Distance," and "Thought Transfer."[34] *Pravda*, for instance, greeted the

---

[31] Ibid., 306.

[32] For examples of such criticism, see N. Kol'tsov, "V. L. Durov. Dressirovka zhivotnykh" [V. L. Durov. Animal training], *Pechat' i revoliutsiia* 6 (1924): 203–4; V. Arkad'ev, "Ob elektromagnitnoi gipoteze peredachi myslennogo vnusheniia" [On electro-magnetic hypothesis of the transmission of thought suggestion], *Zhurnal prikladnoi fiziki* 1(1–4) (1924): 215–22; S. Ia. Lifshits, "Eksperimental'nye issledovaniia nad myslennym vnusheniem" [Experimental investigations of thought suggestion], *Trudy psikhiatricheskoi kliniki im. S. S. Korsakova* 1 (1925): 173–83; A. A. Petrovskii, "Telepsikhicheskie iavleniia i mozgovye radiatsii" [Tele-psychic phenomena and brain radiation], *Telefoniia i telegrafiia bez provodov* 34 (1926): 61–6; and V. M. Borovskii, "Retsenziia" [Review of A. Leontovich, ed. Trudy Prakticheskoi Laboratorii], *Psikhologiia*, ser. A, 2(1) (1929): 155–7.

[33] A. Leontovich, ed., *Trudy Prakticheskoi laboratorii po zoopsikhologii vedeniia Glavnauki Narkomprosa* [The proceedings of the Narkompros Practical Laboratory for Zoopsychology] (Moscow, 1928); Leontovich, "Neiron kak apparat peremennogo toka" [The neuron as an apparatus of alternating current], *Biologicheskii zhurnal* 11(2–3) (1933): 252–91.

[34] See, for instance, F. Davydov, "Elektrichestvo i nervnye volny" [Electricity and nervous waves], *Izvestiia*, 13 February 1923, 6; "Gipnologicheskoe obshchestvo v Moskve" [Hypnological society in

publication of Kazhinskii's brochure on "thought transfer" with a complimentary review.[35] Another popular daily, *Evening Moscow*, "previewed" the appearance of Durov's 1924 book by publishing interviews with its author and his collaborators, Bekhterev and Leontovich, under the telling title, "Circus in Assistance of Science."[36] Durov's article, expressively titled "'Saints' and Reflexology," provides an illuminating example of antireligious uses of these studies.[37]

Scientists involved in Durov's research took an active part in the campaign by popularizing their works for lay audiences in numerous popular magazines and interviews with the press. For instance, Leonid Vasil'ev, a staff member of Bekhterev's institute and active participant in studies on telepathy, published a lengthy article titled "Thought Transfer Over Distance" in *Herald of Knowledge*, a popular-science magazine.[38] Alexander Chizhevskii, a staff member of Durov's laboratory, published a similar article in another popular weekly.[39] As a rule, scientists were careful to describe their research in tentative terms and support it with results obtained by other investigators; they also often refused to speculate on broader issues raised by their work.[40] But the facts they discovered and the concepts they developed quickly became "public property" and thus ripe for appropriation.

The media coverage of their studies was much less circumspect, and speculated freely on their possible intellectual, social, and cultural consequences, while completely ignoring the doubts and critiques advanced by other scientists regarding theoretical underpinnings, methodology, and actual results of such research. Journalists and newspaper reporters shouted at the top of their lungs on behalf of those scientists who shied away from speculation.[41] As one reporter candidly admitted in his interview with Lazarev regarding the possibility of "reading minds": "the scientist's modesty *cannot stop us* from envisioning those—truly fantastic—perspectives that the brain science is opening up [for humanity]."[42]

Il'in's novel demonstrates that these "truly fantastic perspectives" fired up the imagination of not only journalists, but also authors of the new literary genre scientific fantasy. Indeed, Il'in was far from the only or even the first Russian writer to incorporate contemporary research on telepathy into his fiction. Probably one of the first attempts to use these ideas in fiction was a novel titled *The Machine of Horror*

Moscow], *Izvestiia*, 11 May 1923, 6; "Chtenie myslei na rasstoianii" [Reading thoughts at a distance], *Izvestiia*, 18 February 1926, 4; "Gipnoz i vnushenie" [Hypnosis and suggestion], *Pravda*, 19 August 1924, 8; "V institute biologicheskoi fiziki" [In the Institute of Biological Physics], *Pravda*, 16 March 1927, 8.

[35] See "B. B. Kazhinskii, Peredacha mysli. Faktory, sozdaiushchie vozmozhnost' vozniknoveniia v nervnoi sisteme elektromagnitnykh kolebanii" [B. B. Kazhinskii, Thought transfer: Factors that make it possible for the nervous system to generate and emit electromagnetic waves], *Pravda*, 4 January 1924, 7.

[36] "Tsirk na pomoshch' nauke," *Vecherniaia Moskva*, 25 March 1924, 2.

[37] V. V. Durov, "'Sviatye' i refleksologiia," in Leontovich, *Trudy Prakticheskoi* (cit. n. 33), 34–38.

[38] L. Vasil'ev, "Peredacha myslei na rasstoianii," *Vestnik znaniia* 7, 1926, 457–68.

[39] A. L. Chizhov, "Peredacha mysli na rasstoianii" [Thought transfer at a distance], *Ekho* 26, 1926, 7.

[40] See also L. Vasil'ev, "Luchi, organizm i peredacha myslei" [Rays, organism, and thought transfer], *Vecherniaia Moskva*, 22 May 1926, 2.

[41] In addition to the references given above (cit. nn. 34–36 and 38–40), see Dr. A. N. Dail'ti, "O neposredstvennoi peredache mysli na rasstoianii" [On direct transfer of thought over distance], *Vecherniaia Moskva*, 2 April 1927, 2.

[42] A. Kun, "Ob ugadyvanii chuzhikh myslei" [On guessing others' thought], *Vecherniaia Moskva*, 9 January 1928, 2 (emphasis mine).

published in 1925 by Vladimir Grushvitskii, a chemist by training and occupation, under the pen name of Vladimir Orlovskii.[43] In the novel, an American millionaire and scientist invents an apparatus for recording "brain waves." He builds a machine for projecting waves of specific length to evoke certain emotions (fear, rage, religious fervor) in anyone who happens to be in an affected area. The American attempts to use this "machine of horror" to become the dictator of his homeland. But his attempt is defeated by a Russian scientist who works on the same subject and synthesizes a special material that protects people from exposure to the machine. The next year, Alexander Beliaev used basically the same plot in a novel titled *The Ruler of the World*. Serialized over a period of two months on the pages of *Whistle*, a popular daily with a countrywide circulation,[44] the novel told the story of a young German physiologist named Ludvig Shtirner who builds an apparatus that transmits his thoughts to animals and humans over distance. Shtirner uses his apparatus to "win" the woman he loves, and to become a financial magnate and dictator. But his efforts to extend his "mind control" over the entire world are thwarted by the Russian scientists Kachinskii (a clear reference to Bernard Kazhinskii) and Dulov (a similarly transparent reference to Vladimir Durov), who develop a device for protecting the population from Shtirner's "thought commands."[45] Beliaev had actually consulted with Kazhinskii on some technical details of the latter's research and even reproduced Kazhinskii's diagrams in his text to illustrate the "transmitters" and "receivers" of electromagnetic waves in the brain. Sergei Beliaev (no relation to Alexander Beliaev), another SF writer and, like Il'in, a physician by training, used similar ideas in his novel *Radio-Brain*, published in 1928.[46] Tellingly, a complimentary afterword to the novel was written by none other than Kazhinskii.[47]

There are numerous revealing similarities in the treatment of scientific studies on telepathy and their projected societal implications in *The Valley of New Life*, *Machine of Horror*, *Ruler of the World*, and *Radio-Brain*. In all four novels, a scientist (invariably male) conducts research that leads to the creation of a particular technology embodied in an intricate apparatus that allows its "owner" to control emotions, thoughts, and actions of other people and to use such control for his own (or his patrons') nefarious purposes. In all four, another (also invariably male) scientist involved in similar research creates a device that protects people from the influence of the "mind control" apparatus and allows its use for various humane purposes, such as further research, healing, reeducation of criminals, protection of the Bolshevik proletarian state from a capitalist attack, or taming wild animals.

The novels effectively molded public expectations and imaginations—fanning both hopes and fears—regarding the dangers and advantages of so-called mind control technologies, even though the actual studies by Durov, Bekhterev, Kazhinskii, Lazarev, and others focused exclusively on documenting "telepathic" phenomena and uncovering their possible physicochemical mechanisms. Yet, in accord with the uncritical praise and wild speculations these studies generated in the media, the authors of the four nov-

---

[43] See V. Orlovskii, *Mashina uzhasa* (Leningrad, 1925).

[44] A. Beliaev, "Vlastelin mira," *Gudok*, October 19–24, 26–31; and November 2–6, 10–14, 16–18, 1926.

[45] In 1929, a revised and expanded version of the novel came out in book format as a supplement to the popular adventure magazine *Around the World*; see A. Beliaev, *Vlastelin mira* (Moscow, 1929). It had a print run of 70,000 copies. In the book version Dulov was renamed Dugov.

[46] Sergei Beliaev, *Radio-Mozg* (Leningrad, 1928).

[47] B. Kazhinskii, "Posleslovie" [Afterword], in ibid., 172–80.

els vividly portrayed the tremendous powers and "fantastic" possibilities unleashed by contemporary biomedical research. Together with journalists, SF writers brought this research out from the obscurity of science laboratories into the limelight of public attention, and in the process dramatically raised their public visibility and cultural authority.

The media reports and SF novels effectively "translated" the available scientific descriptions and explanations of telepathy research into a highly metaphorical language, creating numerous allusions to popular images and clichés of literary fiction. To give but one illuminating example, in various popular and literary renderings, "electromagnetic waves" supposedly emitted and received by the brain became transformed into "rays." This seemingly innocuous rechristening generated allusions not only to the actual X-rays discovered by Wilhelm Röntgen in 1895 or "the heat-ray" of H. G. Wells's 1897 *War of the Worlds*. It also linked telepathy research to the popular image of the terrifying "rays of death" that exploded or killed everything in their path.[48] This particular image was propagated by the 1923–24 sensationalist media reports about British engineer Harry Grindell-Matthews, who claimed to have invented a "ray-gun" for exploding enemy airplanes and warships over great distances.[49]

Though Grindell-Matthews's alleged invention never actually materialized, it almost immediately became a staple of contemporary cultural productions, especially in fiction and films worldwide, including those in Russia.[50] In 1924, just as Durov's book came out, several novels, all titled *The Rays of Death*, appeared in book format and were serialized in various periodicals.[51] The next year, a full-length motion picture with the same title was scripted by Vsevolod Pudovkin, a rising film director (who also played one of the lead characters); the film, directed by Lev Kuleshov, a doyen of Soviet cinematography, hit movie screens around the country.[52] The same year, Aleksei Tolstoi, one of the most successful writers of the day, exploited the same theme in his extraordinarily popular "fantastic" novel *Hyperboloid of Engineer Garin* (1925–26).[53] The representation of electromagnetic phenomena observed in physiological research and purportedly undergirding hypnosis and telepathy as "rays"

---

[48] Journalists and fiction writers also alluded to another popular image of the "rays of life" that became highly influential in 1920s Russia as well. For details, see Krementsov, *Revolutionary Experiments* (cit. n. 14).

[49] For Grindell-Matthews's biography, see Jonathan Foster, *The Death Ray: The Secret Life of Harry Grindell Matthews* (self-pub., 2008).

[50] For a detailed examination of the uses of this image in contemporary, mostly English-language, cultural productions, see William J. Fanning Jr., *Death Rays and the Popular Media, 1876–1939: A Study of Directed Energy in Fact, Fiction, and Film* (Jefferson, N.C., 2015). For reports on Grindell-Matthews's works in the Russian press, see, for instance, "Neuzheli ne illiuziia" [Isn't it really not an illusion], *Krasnaia panorama* 11, 1924, 17; Ia. G., "D'iavol'skie luchi" [Devil's rays], *Krasnaia panorama* 18, 1924, 14–15; Envish, "Diavol'skie luchi" [Devil's rays], *Krasnyi zhurnal* 8, 1924, 605–9; and many others.

[51] See, for instance, N. Karpov, "Luchi smerti," in *Biblioteka revolutsionnykh prikliiuchenii* (Moscow, 1924), vols. 1–3. A year later, the same novel was issued in book format: Karpov, *Luchi smerti* (Moscow, 1925). Also in 1924, there appeared a Russian translation of a lengthy novel entitled "The Rays of Death," by the German writer Hans Dominik; see Gans Dominik, *Luchi smerti* (Kharkov, 1924). The German original was published in 1922 under the title *Die Macht der Drei*.

[52] A large segment of the film has survived and is available online at various sites, including YouTube; see, for instance, https://www.youtube.com/watch?v=lmsBNkx586A (accessed 25 December 2014).

[53] A. Tolstoi, "Giperboloid inzhenera Garina," *Krasnaia nov'* 7–9, 1925; 4–9, 1926. In 1927, the novel came out in book format: A. Tolstoi, *Giperboloid inzhenera Garina* (Moscow, 1927). It also appeared in a German translation: Alexej Tolstoi, *Das Geheimnis der infraroten Strahlen* (Berlin,

successfully hybridized the popular notions of thought transfer and death rays, fueling further public expectations of, and anxieties about, contemporary biomedical research. Indeed, in Grushvitskii's novel, the same machine emits both the rays that control human emotions and the rays that detonate various explosives, and the only difference between the two is their wavelength.

SCIENCE AS A CULTURAL RESOURCE

The literary renderings of research on telepathy in concurrent Russian fiction illustrate a particular cultural process through which highly specialized scientific studies that were often quite obscure for the uninitiated were transformed into an influential cultural resource readily available for exploitation by any interested party. This transformation involved two interconnected activities. One was the translation of particular scientific lexicons describing such studies into a language understandable to the lay public. Another was the appropriation of knowledge generated by such studies (facts, claims, hypotheses, controversies, methods, theories, promises, and so on) for uses and purposes far removed from the actual goals and results of the research.

Three major groups of actors participated in this process—scientists, their Bolshevik patrons, and literati (journalists and SF writers). Scientists themselves actively popularized their work through articles in popular magazines, interviews with the press, radio broadcasts, documentary films, and any other available media. Excited, inspired, or troubled by scientists' aspirations and actual discoveries, journalists and writers eagerly joined in the process, freely speculating on the possible intellectual, social, and cultural consequences of scientific studies, and completely ignoring the existing doubts and criticisms these studies generated within the scientific community. By virtue of their position as the one and only patron of science, and through their tight control over the mass media, the Bolsheviks defined what kinds of knowledge would become culturally dominant. They did this by widely disseminating the views and interpretations that confirmed and conformed to their own revolutionary dreams and actual policies, while effectively silencing all dissenting and subversive voices that attempted to question and critique science's promises and projections.

The transformation of science into a cultural resource was undoubtedly an important part of the huge postrevolutionary campaign (conducted jointly by all the three groups) of educating the population and disseminating scientific knowledge. But each of these groups also pursued its own goals. Scientists effectively exploited it as an indispensable tool in negotiations with their patrons regarding research agendas, institutional growth, and the cultural and social importance of their own studies, including their own expert status within the social structures of the newborn Soviet Russia. In their turn, the Bolsheviks employed this resource as a powerful instrument in supporting and propagating their own materialistic ideology and "class" morality. For instance, in an article published in *Izvestiia* under a telling title, "Miracles from a Scientific Viewpoint," no less a figure than the head of Narkomzdrav (People's Commissariat of Health Protection), Nikolai Semashko, used research on telepathy to undermine and ridicule religious notions of the human "soul."[54] Along with exploiting

---

1927). In the next decade it was reprinted multiple times and was translated into English; see Alexei Tolstoi, *The Death Box* (London, 1936).

[54] N. Semashko, "Chudesa s nauchnoi tochki zreniia," *Izvestiia*, 3 February 1922, 2.

its entertainment (and hence sales) potential as an inexhaustible source of wonder, horror, estrangement, amazement, and awe, some literati utilized this resource to probe the role of science as a foundation of social progress, technological development, and moral values; this role was actively promoted by the country's new rulers. As a result of the combined, but not necessarily coordinated, efforts of these three groups during the first years of the Bolshevik regime, the public visibility and cultural authority of science and scientists rose dramatically, particularly in comparison to the exalted cultural authority that literary fiction, philosophy, and religion enjoyed among the educated Russian public in the prerevolutionary era.

The actual studies of "telepathic" phenomena appearing in specialized scientific, medical, and technical publications differed considerably from their representation in the popular media and SF novels. Scientists' professional publications used dedicated terminology and often included complicated mathematical formulas and equations, tables with statistical data, and drawings of electrical circuits and brain structures. They contained references to the work of other investigators on the same and related subjects and discussed the results obtained in universal, impersonal, hypothetical, and tentative terms. Their conclusions relied on the conventional *scientific* armamentarium, including logic; verifiability; statistics; controlled experiment; technical apparatus; and negotiated and agreed upon disciplinary nomenclatures, methods, and standards (of measurements, classifications, presentations, and so on). Presenting new knowledge generated by their research to a broader audience in their popular writings, certain scientists did eschew some of the conventions of professional publications (often to the chagrin and opprobrium of their colleagues). Yet, as we saw in Vasil'ev's and Chizhevskii's popular accounts of their studies on telepathy, they still maintained the general tone of disinterested academic inquiry—"the search for truth"—and habitually refused to speculate on broader issues raised by their work.[55]

Journalists and SF writers, however, were not bound by the conventions of professional science writing. By bringing into play various literary conventions, allusions, symbols, metaphors, images, clichés, and traditions, they effectively "translated" what was often impenetrable for a lay person, such as descriptions of proposed concepts, actual experiments, their results, and theoretical considerations offered by scientists in both professional and popular writing. Under their pens, the arcane knowledge produced by scientific research became greatly simplified, fragmented, sensationalized, dramatized, indisputable, and highly personal, while illustrations accompanying the press reports were largely limited to the photographs of a scientist hailed as the discoverer of yet another "scientific miracle." By placing scientists in the dramatic literary

---

[55] As is the case with science fiction in the history of literature, the genre of popular-science writing has generated lively debates on its origins, purposes, and permutations among historians of science. For illuminating discussions of "popular science" in Western contexts, see, for instance, Stephen Hilgartner, "The Dominant View of Popularization: Conceptual Problems, Political Uses," *Soc. Stud. Sci.* 20 (1990): 519–39; Roger Cooter and Stephen Pumfrey, "Separate Spheres and Public Places: Reflections on the History of Science Popularization and Science in Popular Culture," *Hist. Sci.* 32 (1994): 237–67; and five contributions to the special "Focus" section on "Historicizing 'Popular Science,'" in *Isis* 100 (2009): 310–68. Alas, none of these, as well as numerous other Western publications, even mentions Soviet popular-science writings. Andrews's *Science for the Masses* (cit. n. 13), the only book-length examination of the early Soviet science-popularization campaign in English, focuses largely on its uses by the Bolsheviks, while available Russian-language literature stays narrowly within the traditional views on science popularization; see E. A. Lazarevich, *Populiarizatsiia nauki v Rossii* [Popularization of science in Russia] (Moscow, 1981).

universe of exciting or terrifying adventures, perpetual struggles between good and evil, and unending pursuits of love, money, and power, literati created a very vivid, but significantly distorted image of contemporary science—its practitioners, goals, settings, motivations, practices, and actual results. They focused mostly on science's societal implications, emphasizing its economic utility, political expediency, ideological conformity, and cultural affinity.

As one would have expected, some scientists were dismayed and offended by such a cartoonish portrayal of their profession by journalists and SF writers. Vasil'ev, for instance, forcefully asserted that "artists are unable to assimilate advances of scientific thought," and thus "the so-called 'artistic truth' quite often has very little resemblance to the scientific truth," while artistic "creations of even top-rank masters" appear "naïve, primitive, if not bizarre" to a scientist.[56] One could suggest that some scientists (Il'in among them) took up their pens and wrote SF stories in order to counter such "naïve, primitive, and bizarre" representations of their chosen vocation's actual workings and achievements. But, at the same time, the medium of SF stories allowed them to break away from the customary conventions of the deliberate impersonality, ideological and political neutrality, national and cultural universality, and proclaimed impracticality of "pure" science, which its practitioners had fiercely defended during its professionalization in the nineteenth century.[57] The new medium provided these scientists an opportunity to leapfrog numerous gaps in their experiments and their actual knowledge of the phenomena under study, and to question not only the contents, but also the ethical foundations of their own and their colleagues' research. They could also examine the relations of science to its concrete social environment, be it Bolshevik Russia, imperial Britain, or the democratic United States.

Indeed, Soviet scientists were certainly not alone in their use of scientific fantasy, a.k.a. science fiction, to expose and explore the social underpinnings of their profession and address societal anxieties raised by the rapid development of experimental biology and medicine, including research on telepathy. To give but one pertinent example, Julian S. Huxley, a leading British biologist and an older brother of Aldous Huxley, was one of the most prolific science popularizers of the time, and also tried his hand at writing fiction. In April 1926, he published simultaneously on both sides of the Atlantic a "biological fantasy" evocatively titled "The Tissue-Culture King," which in certain "scientific" details remarkably resembles Il'in's *Valley of New Life*.[58]

Huxley's story is narrated by a British scientist captured by "natives" in a remote area of Africa. Upon arriving at the tribe's "capital," the narrator meets another British captive, a certain Dr. Hascombe, a talented researcher, who is unappreciated in his homeland. In exchange for his life, Hascombe had promised to make the king of his captors an immortal religious ruler of the tribe. For that purpose, the scientist had created a laboratory for experiments with tissue cultures and embryonic development, which "produced" giant bodyguards for the king and an assortment of various "monsters" (two-headed frogs and three-headed snakes, for example) that serve as sacred

---

[56] L. Vasil'ev, "Piatnadtsat' tezisov normal'noi ideologii" [Fifteen theses of normal ideology], *Bessmertie* 1 (1922): 3–4.

[57] For details on the origin and development of these accepted conventions in Russian scientific publications, see Krementsov, *Revolutionary Experiments* (cit. n. 14), 187–93.

[58] Julian Huxley, "The Tissue-Culture King," *Yale Review* 15 (1926): 487–504; and "The Tissue-Culture King: A Biological Fantasy," *Cornhill Magazine* 60 (1926): 422–57.

symbols of the king's so-called divine powers. He had also trained a group of natives to combine their "mental energies" into a "super-conscience" to telepathically transmit the king's "thought commands" over great distances. Hascombe protects himself from such commands by a metal hat, which clearly implied the electromagnetic nature of this "telepathic" communication and which, as in Il'in's *Valley of New Life*, becomes a key means of the narrator's successful escape from captivity.[59]

A year later, "The Tissue-Culture King" was reprinted in *Amazing Stories*, a popular magazine edited by the godfather of American SF, Hugo Gernsback. A brief preface to the story, presumably written by the editor, indicates that telepathy was the story's selling point. It stated optimistically: "While so far science does not countenance telepathy, no real thought transference has ever been proved, that does not mean to say that it never will be achieved." The preface elaborated: "*Amazing Stories*' sister magazine, *Science and Invention*, has a standing prize of $1,000 for an actual proof of telepathy, which so far has never been claimed." But it ended on an enthusiastic note: "Just the same, maybe 10,000 years hence the human race may have progressed far enough to make telepathy a reality."[60]

For the story's author, however, telepathy was just one among numerous possibilities offered to humanity by the new experimental biology and medicine. Indeed, Huxley concluded his story with the following entreaty to his readers:

> The question I want to raise is this: Dr. Hascombe attained to an unsurpassed power in a number of applications of science—but *to what end did all this power serve*? It is the merest cant and twaddle to go on asserting, as most of our press and people continue to do, that increase of scientific knowledge and power must in itself be good. I commend to the great public the obvious moral of my story and ask them to think what they propose to do with the power which is gradually being accumulated for them by the labors of those who labor because they like power, or because they want to find the truth about how things work.[61]

As Il'in's *Valley of New Life* readily attests, SF stories by certain Soviet scientists commended to their readers—though perhaps not in such a direct manner—analogous "obvious morals," and asked them to think of similar questions.

The materials presented above indicate that the transformation of scientific, especially biomedical, knowledge into a cultural resource is not limited to the specific time (the 1920s), particular setting (Bolshevik Russia), or disciplinary content (research on telepathy) suggested by Il'in's novel. Rather, it represents a characteristic feature of the interactions of science and of scientists with the social hierarchies, power structures, and cultural contexts of their contemporary societies. The same transformation process could clearly be seen in historical and contemporary public debates over evolution, race, eugenics, infectious diseases, genetic engineering, human origins, organ transplantation, transgenic foods, cloning, and many other "hot" biomedical subjects.[62] The migration of scientific facts, ideas, hypotheses, techniques, and concepts into the

---

[59] This was probably the first appearance of this simple device that in the form of a "tin foil hat" would later became a staple of numerous SF stories about aliens "reading the minds" of the Earthlings.

[60] Preface to "The Tissue-Culture King," by Julian Huxley, *Amazing Stories* 2, 1927, 451–9, on 451.

[61] Huxley, "The Tissue-Culture," *Yale Review* (cit. n. 58), 504 (emphasis in the original).

[62] Numerous instances of this process in US contexts, though framed in a different language, can be found in a recent volume by Peter Snyder, Linda Mayes, and Dennis Spencer, eds., *Science and the Media: Delgado's Brave Bulls and the Ethics of Scientific Disclosure* (Boston, 2009). For a fine-

public sphere entails much more than translating specialized scientific lexicons into a language comprehensible to the laity, or disseminating scientific findings beyond a narrow professional audience. In the process of becoming a cultural resource, particular, complex, and often highly technical knowledge about concrete natural phenomena is greatly simplified, fragmented, and reduced to a few, often disjointed, generalized statements, like "radio-brain" or "human thought—electricity." In this process, the complexity of knowledge and the need for its continuous refinement and verification are erased and replaced by a "vivid picture" (to borrow Ludwig Fleck's astute characterization[63]), as exemplified by numerous "human interest" stories about science and its "breakthroughs" propagated by the press and SF literature.[64] Scientific knowledge thus loses much of its conventionality, its embeddedness in a particular mode of production, and its hypothetical, uncertain, relative, and provisional character. Instead, it acquires dogmatic formulations and doctrinaire imagery embodied in such popular cultural memes as "mental radio," "Frankenstein," "the brain in a vat," "Dolly the sheep," "mind control," or the "mushroom cloud," which move with ease from one publication to another and from one cultural production to the next. This process also obscures the historicity of scientific knowledge and masks the multitudes of scientists involved in its production, instead emphasizing one heavily fictionalized personality hailed as the "great discoverer," be it Isaac Newton, Charles Darwin, Ivan Pavlov, or James Watson. Refracted through the medium of popular press and SF writings, scientific knowledge thus transmutes into a resource readily appropriated by any interest group, such as politicians, public health officials, theologians, writers, TV and movie producers, philosophers, journalists, business executives, jurists, artists, or state bureaucrats. Each of these groups exploits it to accomplish their own goals—often completely divorced from the aims and aspirations of the scientists who actually produce such knowledge.

Of course, for a variety of reasons, a number of scientists themselves did (and still do) write both popular accounts of their professional work and science fiction stories, thus willingly or unwittingly contributing to the process of transforming scientific knowledge into a cultural resource. Both Il'in and Huxley, along with a few other scientists in Russia, Britain, and elsewhere, turned to fiction as a way to present their own "vivid picture" of where science belonged in their own societies, delivering both a promise and a warning of the possible futures contemporary science might bring about, while challenging and/or subverting conventional views. Still, whatever their authors sought to accomplish, their stories could not help but foment societal anxieties (both hopes and fears) about the powers unleashed by the rapid advances of biomedical sciences—anxieties fanned by the news media and broadcast through numerous SF writings, cartoons, and films. Yet the stories also helped recruit new generations of scientists and shaped the agendas of future research by firing up people's imaginations with fantastic tales of scientific pursuits, be it a search for a cancer cure, a quest for immortality, a conquest of time and space, or a dream of telepathic communications.

---

grained depiction of the role of the US media in publicizing "medical breakthroughs" in the late nineteenth and early twentieth century, see John H. Warner, "Medicine, Media, and the Dramaturgy of Biomedical Research: Historical Perspectives," in *Science and the Media*, ed. Snyder, Mayes, and Spencer, 13–25.

[63] Ludwig Fleck, *Genesis and Development of a Scientific Fact* (Chicago, Ill., 1979), 122–3.

[64] For a pioneering analysis of the popular imagery of medical science produced by the mass media, see Bert Hansen, *Picturing Medical Progress from Pasteur to Polio* (Newark, N.J., 2009).

Darwin on the Cutting-Room Floor:
Evolution, Religion, and Film Censorship

by David A. Kirby*

ABSTRACT

In the mid-twentieth century, film studios sent their screenplays to the Hays Office, Hollywood's official censorship body, and to the Catholic Church's Legion of Decency for approval and recommendations for revision. This essay examines how filmmakers crafted stories involving evolutionary biology and how religiously motivated movie censorship groups modified these cinematic narratives in order to depict what they considered to be more appropriate visions of humanity's origins. I find that censorship groups were concerned about the perceived impact of science fiction cinema on the public's belief systems and on the wider cultural meanings of evolution. By controlling the stories told about evolution in science fiction cinema, censorship organizations believed that they could regulate the broader cultural meanings of evolution itself. But this is not a straightforward story of "science" versus "religion." There were significant differences among these groups as to how to censor evolution, as well as changes in their attitudes toward evolutionary content over time. As a result, I show how censorship groups adopted diverse perspectives, depending on their perception of what constituted a morally appropriate science fiction story about evolution.

In June 1942, screenwriters William Bruckner and Robert Metzler completed their script for the first science fiction (SF) film produced by Twentieth Century Fox—*Dr. Renault's Secret*. The film was based on Gaston Leroux's 1912 novel *Balaoo*, whose plot involves a scientist teaching an ape to speak, read, and write, in order to prove that humans are glorified apes.[1] In the novel, the "ape-man" is physically indistinguishable from an educated human.[2] But when the ape-man's bestial nature surfaces and he murders several people, the scientist realizes that the "human conscience" is

* Centre for the History of Science, Technology, and Medicine, University of Manchester, Manchester, M13 9PL, UK; david.kirby@manchester.ac.uk.

I want to thank all the editors and two anonymous reviewers for this volume whose invaluable suggestions have improved the work substantially. A debt of gratitude is particularly owed to Laura Gaither, who read and commented on many versions of this manuscript. This work was supported by the Wellcome Trust (100618) through an Investigator Award entitled "Playing God: Exploring the Intersections between Science, Religion, and Entertainment Media."

[1] *Dr. Renault's Secret*, directed by Harry Lachman (1942; Century City, Calif.: Twentieth Century Fox Home Entertainment, 2008), DVD.

[2] Gaston Leroux, *Balaoo* (Paris, 1912). Leroux is best known for his 1911 novel *The Phantom of the Opera*. *Balaoo* was also adapted in 1913 as *Balaoo the Demon Baboon* and in 1927 as *The Wizard*. The full text of the 1912 edition of *Balaoo*, from a translation by Alexander Teixeira de Mattos, can be found at Project Gutenberg, March 2008, gutenberg.net.au/ebooks08/0800281h.html. The scientist's experiment is discussed in chapter 21.

what separates humans from the apes. Bruckner and Metzler's initial script modified the story to emphasize its SF and horror aspects. To that end they depicted the ape-man, Noel, as more of a mentally deficient brute rather than the articulate gentleman portrayed in the novel. They also changed Dr. Renault's motivations for transforming an ape into a human. His new experimental goal is to "prove the descent of man from an ape,"[3] making him not just a "mad scientist," but a "mad evolutionist."[4] The scriptwriters believed that if Renault's objective was to invalidate the biblical account of human origins, audiences would have an even greater desire to see him punished.

The movie that was made, however, didn't include anything about evolution, Darwin, or the descent of humans from the apes. Why did the filmmakers choose not to include these elements of the narrative story line? Did they believe that it took up too much screen time? Did Dr. Renault's motivations clash too much with other plot elements? Did the story perform poorly with test audiences? In actuality, Hollywood's self-censorship organization at the time, the Motion Picture Producers and Distributors of America—popularly known as the "Hays Office"—effectively took the choice of using that narrative out of the filmmakers' hands.

The Hays Office was established by Hollywood in 1922 as a response to protests by the Catholic Church and other Christian organizations over what they considered immoral content in movies.[5] By 1930, studio heads agreed to abide by a code of standards called the Motion Picture Production Code, which was written by two prominent Catholics.[6] The standard procedure in 1942 was to send screenplays to the Hays Office for review and also for recommendations on how to alter the script in case it did not meet the standards of the Production Code. In the case of *Dr. Renault's Secret*, the censors at the Hays Office judged that such frank discussions of Darwin and human evolution made the whole story line in "violation of the provisions of the Production Code."[7] They believed that such open discussions of human evolution would "undoubtedly offend the sensibilities of large groups of religious people of different faiths," and hence were inappropriate for movies meant for entertainment. The censors approved of the script's portrayal of evolution as immoral, but feared that even a critical account of the theory would legitimate the topic. From the Hays Office's perspective in 1942, the only appropriate story about evolution was no story.

---

[3] Harry Brand, "Synopsis of 'Dr. Renault's Secret,'" undated, *Dr. Renault's Secret* file, Production Code Administration Collection, Margaret Herrick Library, Beverly Hills, California (**hereafter** cited as PCAC).

[4] For a history of the mad scientist stereotype, see Roslynn Haynes, *From Faust to Strangelove: Representations of the Scientist in Western Literature* (Baltimore, Md., 1994); and Christopher Frayling, *Mad, Bad and Dangerous* (London, 2005). On the concept of the mad evolutionist, see David A. Kirby, "The Devil in Our DNA: A Brief History of Eugenic Themes in Science Fiction Films," *Lit. Med.* 26 (2007): 83–108.

[5] On the formation of the Hays Office in 1922, see Leonard J. Leff and Jerold L. Simmons, *The Dame in the Kimono: Hollywood Censorship and the Production Code* (Lexington, Ky., 2013); Lee Grieveson, *Policing Cinema: Movies and Censorship in Early-Twentieth-Century America* (Berkeley, Calif., 2004); and Ruth Vasey, *The World According to Hollywood, 1918–1939* (Madison, Wisc., 1997). The Catholic Legion of Decency changed its name to National Legion of Decency in 1935, but I will refer to it as Legion of Decency throughout. For its history, see Frank Walsh, *Sin and Censorship: The Catholic Church and the Motion Picture Industry* (New Haven, Conn., 1996); and Gregory D. Black, *Hollywood Censored: Morality Codes, Catholics, and the Movies* (Cambridge, UK, 1994).

[6] I will use the term "Production Code" to refer to the Motion Picture Production Code of 1930. For its history, see Leff and Simmons, *The Dame* (cit. n. 5).

[7] This and the subsequent quote are in a letter from Joseph Breen to Jason Joy, 30 June 1942, *Dr. Renault's Secret* file, PCAC.

This judgment left the studio with limited choices. It could use the existing script and release the film without the Hay Office's "seal of approval." But that would likely be a financial kiss of death, since the film would be barred from most movie theatres.[8] Alternatively, it could produce an entirely new script. But the studio had already invested resources into the project and also felt that the story had strong box office potential. Finally, it could negotiate with the Hays Office to develop a story acceptable to the censors. Thus, one of the film's producers and the studio's director of public relations met with the Hays Office.[9] The Hays Office explained that the script violated the category of "religion" in the Production Code by presenting evolution as a provable alternative to the account of creation in Genesis.

While the studio would have preferred the added horror of a mad evolutionist, the general concept of a scientifically created killer ape-man was more important than the scientist's justification of his experiments. Therefore, to the gratification of the Hays Office, the studio "agreed to eliminate any reference to Darwin or his theory, and to establish the ape as a throwback." They used pseudoscientific words in the script to render the scientist's motivations nebulous. For example, he said the following: "My experiment in transmutation begins tomorrow. The throwback is an excellent candidate for experimental humanization."[10] Even an SF story line as ludicrous as the one in the draft script for *Dr. Renault's Secret* was off limits if the censors believed it would legitimate an inappropriate message about the origin of the human species.

In this article, I explore the ways filmmakers tried to craft SF stories with evolutionary themes and how religiously motivated movie censorship groups modified these narratives before, during, and after production in order to tell what they considered to be more acceptable stories about humanity's origins. Moving pictures had a level of realism never previously exhibited by mass media. Censorship groups worried that this realism would normalize and legitimate messages in movies. Especially worrying was the fact that cinema's visuality meant that audiences did not have to be literate. Censors were concerned this meant they would lack the mental capacity to critically evaluate a film.[11] Realism, for the censors, made film an extremely powerful and potentially dangerous medium for mass communication.

Concerns about film realism made SF movies particularly problematical for censors because of the role that science plays in the genre. The cultural authority of science provided an added level of realism, which censors feared might mislead audiences into accepting what they could see as legitimate representations of the world; they also worried that SF would exaggerate the power of science without acknowledging its limitations. Fundamentally, the censors did not trust the public to interpret the complexities or the implications of science in cinema.[12]

For the Christian organizations that motivated the censors, evolution was an exceptionally dangerous scientific topic since it conflicted with literal interpretations of scripture. Evolution challenged many of Christianity's fundamental beliefs about the

[8] The potential economic impact of not having approval from the Hays Office is discussed in Black, *Hollywood Censored*; and Leff and Simmons, *The Dame* (both cit. n. 5).

[9] Unless noted otherwise, the information in this paragraph and the next comes from T. A. Lynch, "Memorandum for the Files," 1 July 1942, *Dr. Renault's Secret* file, PCAC.

[10] William Bruckner and Robert F. Metzler, script, 2 July 1942, Mike Mazurki Papers, 4-f.59, Margaret Herrick Library, Beverly Hills, California.

[11] Walsh, *Sin and Censorship* (cit. n. 5).

[12] David A. Kirby, "Harnessing the Persuasive Power of Narrative: Science, Storytelling, and Movie Censorship," *Science in Context* 31 (2018): 85–106.

elevated place of humanity in creation, the existence of free will, moral responsibility for one's actions, and a rational and immortal human soul.[13] Recent historical work on representations of Darwin and evolution in late nineteenth-century popular culture reveals a Victorian fixation on the implications of the human connection to other primates.[14] Gorillas and other ape-like human progenitors such as cavemen and "missing links" became symbolic images of the proposed evolutionary relationship between humans and apes. Popular cultural products used these images for satiric purposes by making fun of evolutionary advocates but also for more ominous purposes to show the dangers of such beliefs.

Before 1925, these Victorian motifs and thematic content could be used relatively easily in Hollywood movies to poke fun at those who took Darwin's claims seriously.[15] The same evolutionary plot that the Hays Office forced Twentieth Century Fox to remove in 1942—a scientist attempting to prove Darwin's theory by transforming an ape into a human—was a standard story line for SF comedies before 1925. The highly publicized Scopes "Monkey Trial," which galvanized the antievolution movement, marked the decline of the comedic human/primate category of SF films.[16] Evolution—especially for fundamentalist Protestants and conservative Catholics—wasn't funny anymore. After the Scopes Trial, antievolutionists' perception of evolution and its supporters turned from comedic to horrifying. With the adoption of the Production Code in 1930 and the subsequent restructuring of the Hays Office in 1934, as well as the creation of the Catholic Legion of Decency in 1933, it became more difficult for SF filmmakers to include evolutionary ideas.

The goal of this article is to understand the struggle between studios and censorship groups over the production of meaning in SF films featuring human evolution. The "censors" did not aim to stop studios from releasing films, but rather to ensure that the movies were not blasphemous or indecent, and that they did not legitimate dangerous ideas. The Hays Office did not seek to ban films, but instead operated "at the level of the text" and "at the level of representation."[17] Subsequent to the development of the Production Code, censorship organizations took the approach of closely analyzing, commenting upon, and recommending changes to every story treatment and script in order to control "the production of meaning" in films.[18] By altering a film script before production, censors could control a film's dialogue, visuals, individual scenes, character motivations, plot points, and overall narrative. Thus, they could help create movies that they believed would have a positive influence on audiences.

---

[13] A discussion of the potential conflicts between evolution and Christianity is found in Ron Numbers, *The Creationists: From Scientific Creationism to Intelligent Design* (Cambridge, Mass., 2006).

[14] For example, see Janet Browne, "Darwin in Caricature: A Study in the Popularisation and Dissemination of Evolution," *P. Am. Philos. Soc.* 145 (2001): 496–509; Jane R. Goodall, *Performance and Evolution in the Age of Darwin: Out of the Natural Order* (New York, N.Y., 2002); Bernard Lightman, *Victorian Popularizers of Science: Designing Nature for New Audiences* (Chicago, Ill., 2007); Constance Areson Clark, "'You Are Here': Missing Links, Chains of Being, and the Language of Cartoons," *Isis* 100 (2009): 571–89; and Lightman and Bennett Zon, eds., *Evolution and Victorian Culture* (Cambridge, UK, 2014).

[15] Oliver Gaycken, "Early Cinema and Evolution," in Lightman and Zon, *Evolution* (cit. n. 14), 94–120.

[16] Edward Larson, *Summer for the Gods: The Scopes Trial and America's Continuing Debate over Science and Religion* (New York, N.Y., 1997); Michael Lienesch, *In the Beginning: Fundamentalism, the Scopes Trial, and the Making of the Antievolution Movement* (Chapel Hill, N.C., 2007).

[17] Lea Jacobs, "Industry Self-regulation and the Problem of Textual Determination," *Velvet Light Trap* 23 (1989): 4–15, on 8.

[18] Ibid., 4.

Censorship groups saw movies as a battleground over evolution's impact on morality. By controlling the stories told about evolution in SF cinema, they could regulate the meaning of evolution itself. This study will thus focus on the negotiations found in correspondence held in the archives of the Hays Office, the Legion of Decency, and the studios.[19] The battle over evolution in SF cinema was a battle over the perceived impact of film on the public's belief systems and the wider cultural meanings of evolution. But this is not a straightforward story of "science" versus "religion."[20] The censors did not reject every SF story involving evolution. Instead, I show how different censorship groups adopted diverse perspectives, depending on their perception of a morally appropriate SF story about evolution.

DEBATING THE BLASPHEMY OF CINEMATIC EVOLUTION IN THE PRE-CODE ERA

In 1930, the Studio Relations Committee (SRC) was the branch of the Hays Office tasked with enforcing the Production Code. Colonel Jason Joy and his successor James Wingate ran the SRC from the adoption of the Production Code in 1930 until its reformation as the Production Code Administration (PCA) in 1934.[21] The SRC advised studios on how to alter their scripts so that they met the standards of the Production Code. However, the SRC could not force studios to accept their suggestions. This meant that despite their agreement to abide by the Production Code, studios frequently ignored the SRC's recommendations in the early 1930s.[22] In addition, Joy and Wingate took a particularly lenient approach to the Production Code, which they viewed as a flexible set of guidelines rather than a rigid set of rules.[23] Confusingly, the time period between when the SRC adopted but laxly enforced the Production Code in 1930 and the creation of the PCA in 1934 is referred to as the "pre-Code" period.

For Joy and Wingate, censorship was primarily about protecting audiences from indecent depictions of sex and violence or from overtly immoral stories. They did not think the SRC should prevent studios from producing films with controversial topics as long as they were handled appropriately. According to film scholar Gregory Black, "Films that were considered immoral by religious clergy and other guardians of morality were often seen by Joy and Wingate as good entertainment, satire, comedy, or legitimate commentary on contemporary social, moral, or political issues."[24] In essence, they believed that adult audiences could handle thought-provoking stories as long as they did not blatantly incorporate immoral, offensive, or blasphemous messages. Joy and Wingate also took into account the cinematic context of potentially controver-

---

[19] This article is part of a larger book project exploring film censorship of science titled *Indecent Science: Religion, Science, and Movie Censorship.*

[20] My analysis of science, religion, and movie censorship is in line with John Brooke's "complexity thesis." See John Brooke, *Science and Religion: Some Historical Perspectives* (Cambridge, UK, 2014).

[21] For a history of the SRC and the PCA, see Black, *Hollywood Censored* (cit. n. 5); and Thomas Patrick Doherty, *Pre-Code Hollywood: Sex, Immorality, and Insurrection in American Cinema 1930–1934* (New York, N.Y., 1999). Both the SRC and the PCA were referred to as the Hays Office. However, I will use the designations for the SRC and PCA to differentiate between the pre-Code Hays Office and the Hays Office after the creation of the PCA in July 1934.

[22] The SRC's ineffectiveness is discussed in Doherty, *Pre-code Hollywood* (cit. n. 21); and Marvin N. Olasky, "The Failure of Movie Industry Public Relations, 1921–1934," *J. Pop. Film TV* 12 (1985): 163–70.

[23] Joy and Wingate's stance on the Production Code is discussed in Black, *Hollywood Censored* (cit. n. 5).

[24] Ibid., 53.

sial topics; they believed they were less problematic if found in nonrealist films genres such as SF, fantasy, horror, or comedy, where audiences would not take them seriously.

Joy and Wingate's permissive stance meant that, in the pre-Code era between 1930 and 1934, studios were able to tell stories in which characters contradicted the creation story in Genesis. In contrast, the various regional and international censor boards pushed for stronger measures at that time, and SRC-approved films still had to successfully pass through these censor boards before they could be shown in that area's or country's movie theaters.[25] Between 1930 and 1934, a number of major studio SF films incorporating evolution concepts played a central role in the struggle between the SRC, moral reformers, and the regional censor boards over appropriate movie content.

THE IRREVERENCE OF SCIENTIFIC CREATION IN *FRANKENSTEIN*

Universal Studios's 1931 adaptation of *Frankenstein* was the first SF film involving thematically problematic implications of evolution that the SRC had to grapple with after their adoption of the Production Code.[26] Although the film did not overtly incorporate evolution as a process, its theme of scientific creation conflicted with the biblical creation story. Joy's response to the script, as well as to the resulting controversy, shows how the SRC attempted to mitigate religious concerns about cinema during the pre-Code era while also allowing studios to produce movies with challenging narratives.

The SRC requested a single change in the script, replacing the line "In the name of God! Now I know what it feels like to be God!" with the less specific "Now I know how it feels to be a god."[27] Joy thought this modification would distinguish Frankenstein's scientific creation from God's divine creation. The studio, however, ignored this request and filmed the scene as written. Outside of that dialogue, the SRC was certain that the film would be "reasonably free from censorship action" by the regional censor boards.[28]

They were completely wrong. The film ran into significant opposition from these boards. Censors banned the film in numerous locations because they considered a film about a scientist usurping God's role in life's creation to be blasphemous.[29] The Kansas censor board's initial rejection of the film makes this objection clear:

> The reason [for this rejection] is given because of BLASPHEMY which [the dictionary] says is . . . "the act of claiming the attributes or prerogatives of the deity. Besides being an ecclesiastical offense, blasphemy is a crime at the common law, as well as generally by statute, as tending to a breach of the peace and being a public nuisance or destructive of the foundations of civil society.[30]

If a scientist could create life merely using organic materials, then God's act was not sacred or unique. In fact, the Kansas censor board considered the nondivine origin

---

[25] The SRC referred to the various regional censor boards, including city and state boards in the United States as well as international censors, as the "political censor boards." However, for clarity I will refer to them collectively as the "regional censor boards."

[26] *Frankenstein*, directed by James Whale (1931; Universal City, Calif.: Universal Pictures Home Entertainment, 2002), DVD.

[27] Letter from Jason Joy to Carl Laemmle Jr., 18 August 1931, *Frankenstein* file, PCAC.

[28] Letter from Fred W. Beeston to Carl Laemmle Jr., 2 November 1931, *Frankenstein* file, PCAC.

[29] The regional boards were also concerned about the horrific nature of the monster and the film's visuals.

[30] "Censorship Decision on *Frankenstein*," 17 December 1931, Box 35-06-02-08, Kansas State Board of Review Collection, Kansas State Historical Society, Kansas State Archives, Topeka, Kansas.

of humanity such a threat to civil society that they warned against disseminating these beliefs through a movie.

Censors in the Canadian province of Quebec also condemned *Frankenstein* because the film "foster[ed] creating by Man and it being a dogma of the Catholic Church that only God can create it is not advisable to be shown on screen."[31] Generally, the film-makers accommodated regional censor boards by removing dialogue or by eliminating problematic scenes before the film's release. But the blasphemous notion of a scientist creating life is central to *Frankenstein*. Even the dialogue about "feeling like God" was just an overt expression of the film's underlying theme. There was no way for the studio to overcome the objection without fundamentally changing the film's narrative, which was an unacceptable solution for both the studio and the SRC's Jason Joy.

Eventually, the studio satisfied censors' objections by simply adding a short prologue to remind audiences that the story they were about to see was a fantasy.[32] This might seem to be a perplexing solution given that the prologue does nothing to change the blasphemous nature of the story. The film still went on to show a scientist creating human life. But the censors' problems were not just with the story, but how it was presented. This particular SF story had existed in book form for over a century without being a threat. Their anxiety stemmed from their conviction that the persuasive power of the audiovisual medium of cinema would lead audiences—especially less educated working-class audiences—to believe that a scientist could create life in a laboratory.

The studio suggested the resulting prologue be presented by tuxedo-clad actor Edward Van Sloan (figure 1), who plays Frankenstein's teacher Doctor Waldman: "The story of 'Frankenstein' is pure fiction. It delves into the physically impossible and the fantastic. For almost a hundred years the story has furnished diversion and the picture like the story is for entertainment only. Although no moral is intended it shows what might happen to Man if he challenges the unknowable."[33] For Universal Pictures, a prologue was an ideal solution because it required minimal resources and kept their films mostly intact. For Joy, prologues were the perfect result of censorship because they allowed studios to maintain the integrity of their films while reminding people not to take these cinematic stories seriously. The success of *Frankenstein*'s prologue quickly became a strategy whereby studios tried to fend off major cuts to films in the pre-Code era, as was the case with another film that contradicted the biblical origins of humanity—Universal Pictures's *Murders in the Rue Morgue* (1932).

"OBJECTIONABLE AND ANTI-RELIGIOUS": MAD EVOLUTION AND *MURDERS IN THE RUE MORGUE*

In *Murders in the Rue Morgue*, Bela Lugosi plays Dr. Mirakle, who attempts to prove his evolutionary theories by mixing human and ape blood.[34] In one scene, Mirakle stands in front of a chart detailing the evolutionary "tree of life" (figure 2), alongside Erik the gorilla, whose cage is labeled "Le Gorille au Cerveau Humain" (The beast

---

[31] Telegram from D. Leduc to Ted Fithian, 12 January 1932, *Frankenstein* file, PCAC.

[32] Letter from D. Leduc to Jason Joy, 12 April 1932, *Frankenstein* file, PCAC. The studio also suggested "a fade out on the windmill burning, leaving the impression that Frankenstein was dead, for this would help the moral purport of the story."

[33] "Suggested Preface for Frankenstein," undated, *Frankenstein* file, PCAC. The version used in the film uses a variation on this preface.

[34] *Murders in the Rue Morgue*, directed by Robert Florey (1932; Universal City, Calif.: Universal Pictures Home Entertainment, 2012), DVD.

Figure 1. *In this film still, actor Edward Van Sloan reminds the audience of* Frankenstein *(cit. n. 26) that the story of scientific creation they are about to witness is only a fiction. (Courtesy of Universal Pictures.)*

with a human soul). Mirakle explains to an unbelieving carnival crowd his desire to prove that humans evolved from apes:

> MIRAKLE: The shadow of Erik the ape hangs over us all. In the darkness before the Dawn of Man. . . . In the slime of chaos there was a seed that rose and grew into the tree of life. Life was motion. Fins changing into limbs, limbs grew into ears. Crawling reptiles grew legs. Aeons of ages passed. There came a time when a four legged thing walked upright. Behold the first man.
> (Mirakle points to Erik; audience erupts in disbelief)
> AUDIENCE MEMBER: Heresy!
> MIRAKLE: Heresy? Do they still burn men for heresy? Then burn me monsieur, light the fire! Do you think your little candle will outshine the flame of truth? My life is consecrated to a great experiment. I tell you I will prove your kinship with the ape. Erik's blood shall be mixed with the blood of Man![35]

The image of an unruly crowd threatening a scientist over his scientific beliefs, coupled with the film's allusion to burning at the stake was clearly meant to evoke images of the Catholic Church stifling scientific inquiry.

---

[35] Ibid.

Figure 2. Dr. Mirakle, played by Bela Lugosi, explains his theories about human evolution in front of a "tree of life" in this film still from Murders in the Rue Morgue *(cit. n. 34). (Courtesy of Universal Pictures.)*

Given *Frankenstein*'s censorship difficulties, Universal Pictures and the SRC might have been concerned about censor boards' responses to an SF movie that even more overtly challenged the notion of humanity's divine creation. However, after examining the initial outline for *Murders in the Rue Morgue* in June 1931, the SRC told the film's producer Carl Laemmele Jr.: "we find nothing concerning which we feel we need to caution you at the time."[36] Jason Joy came to a similar conclusion after watching the completed film in January 1932, telling Laemmele that outside of some scantily clad dancers and some screaming sounds, "it is our belief that it is satisfactory from the standpoint of the Code."[37]

Despite the SRC's confidence, the studio recognized the story might offend some people. Laemmele sent Joy a letter wondering if the film might offend the French because it takes place in France. Joy's response illustrates his approach to controversial topics in certain genres: "Because of the source of the story, its imaginative and fantastic qualities and because it is laid in a period so far away, we doubt if anything in the picture will offend French sensibilities."[38] Reviewer J. V. Wilson mirrored Joy's rea-

---

[36] Letter from John V. Wilson to Carl Laemmele Jr., 8 June 1931, *Murders in the Rue Morgue* file, PCAC.
[37] Letter from Jason Joy to Carl Laemmle Jr., 8 January 1932, *Murders in the Rue Morgue* file, PCAC.
[38] Ibid.

soning in his "Synopsis and Code Review" by asserting that the story's "fantastic" elements would protect the film from action by the censor boards.[39]

It was clear that under Joy's administration, controversy was offensive in direct proportion to its realism. From his perspective *Murders in the Rue Morgue* was not a serious film; it was a ludicrous SF horror film set in a fantastic past where an extremely unsympathetic character championed ideas about human evolution. In his opinion nobody would take Mirakle's claims seriously. Despite the experience with the similarly fantastical *Frankenstein*, Joy felt confident that *Murders in the Rue Morgue* would pass through the censor boards without a problem.

Again, Joy and Universal Pictures were blindsided by the hostile response from some regional censor boards to the film. They had not foreseen the implied bestiality in Mirakle's desire to "mix" Erik's blood with that of human women. As Barbara Creed argues, cinematic bestiality embodies evolutionary themes, invoking "de-evolutionary fears" that expose the blurred border between humans and animals, and indicating general moral collapse.[40]

It was clear that Joy had fundamentally misjudged regional censor boards' tolerance for evolutionary ideas, even in movies that were fantastical in nature. British Columbia's board, for example, initially banned the film because they believed the evolutionary theme would offend religious people. Universal's Canadian representative informed the movie studio by telegraph: "Rue Morgue condemned by British Columbia censor Board. The reason 'theme of the story Man's descent from ape objectionable and anti-religious, action macabre in the extreme.' Censor action here, in my estimation, ridiculous."[41] The SRC recommended the same solution they had implemented with *Frankenstein*: "We suggested that we would be willing to put a foreword on the picture and order the deletion of a number of the scenes in the hope that the Appeal Board would approve it."[42] But by 1932 censor boards considered prologues an unsatisfactory solution. They had decided movie magic was so powerful that audiences would be certain of the story's veracity, even if they were told it was fiction. Ultimately, Joy and his staff convinced British Columbia's board to accept that the fantastic nature of the SF story and the punishment of the heretic neutered the impact of any evolutionary ideas. Afterward Joy's assistant on foreign markets wrote to them: "It seemed to me so imaginative and fantastic, and I felt that no one could be seriously affected by the incidental idea of the doctor's experiments."[43]

THE BLASPHEMY OF EVOLUTION IN *ISLAND OF LOST SOULS*

Paramount Pictures's *Island of Lost Souls* (1932) was based on H. G. Wells's 1895 novel *The Island of Dr. Moreau*.[44] In the novel, a scientist tries to scientifically evolve "perfect" humans but instead creates half-human/half-animal monsters called the

---

[39] John V. Wilson, "Synopsis and Code review," 11 January 1932, *Murders in the Rue Morgue* file, PCAC.

[40] Barbara Creed, *Darwin's Screens: Evolutionary Aesthetics, Time and Sexual Display in the Cinema* (Melbourne, Aust., 2009), 10–12, 32.

[41] Telegram from R. A. Scott to Ted Fithian, undated, *Murders in the Rue Morgue* file, PCAC.

[42] Memorandum from John V. Wilson, 27 February 1932, *Murders in the Rue Morgue* file, PCAC.

[43] Letter from John V. Wilson to Anne Begley, 11 April 1932, *Murders in the Rue Morgue* file, PCAC.

[44] H. G. Wells, *The Island of Doctor Moreau*, 15th ed. (1896; repr., New York, N.Y., 1996); *Island of Lost Souls*, directed by Erle C. Kenton (1933; New York, N.Y.: The Criterion Collection, 2011), DVD.

A
B

C
D

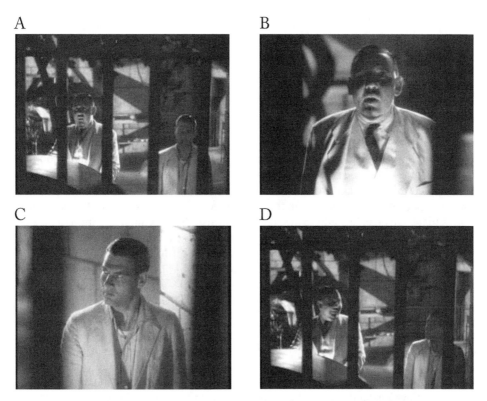

Figure 3. *Film stills from the four shots in this film sequence from* The Island of Lost Souls *(cit. n. 44) show how filmmakers anticipated censure. They could easily remove shots* b *and* c, *in which Moreau is saying his potentially offensive line, then splice together shots* a *and* d *with no loss of continuity. (Courtesy of Paramount Pictures.)*

"Beast-People." Given the recent negative reactions to *Frankenstein* and *Murders in the Rue Morgue* the SRC might have been hesitant to endorse another script with such obvious evolutionary themes. Yet, as with those two films, Joy only challenged a single line of dialogue. The line occurs when Moreau explains his evolutionary experiments and asks his companion: "Mr. Parker, do you know what it means to feel like God?" Knowing the difficulty *Frankenstein* had with a comparable line, Joy warned the studio that the film would have trouble with state censor boards.[45]

Although the studio recognized the blasphemy might offend regional censor boards, they were unwilling to remove the line.[46] Instead they anticipated potential censure by filming the scene in such a way that the shot could easily be removed without harming the story. The stage direction in the final dialogue script shows how they arranged filming so that the line could be edited out seamlessly if necessary (see figure 3):

56. Medium shot Moreau and Parker
57. Closeup Moreau
Moreau: Mr. Parker, do you know what it means to feel like God?

[45] Letter from Jason Joy to Harold Hurley, 26 September 1932, *Island of Lost Souls* file, PCAC.
[46] Letter from Tom Bailey to Jason Joy, 3 October 1932, *Island of Lost Souls* file, PCAC.

58. Closeup Parker
59. Medium shot Moreau and Parker.[47]

It would be a fairly easy editing operation to cut out closeup shots 57 and 58, and then splice together medium shots 56 and 59 without any loss of continuity. In this way, the studio could tell the story the way they wanted, but they were protected should censor boards object to this dialogue.

By the time the finished film was sent to the SRC for approval, James Wingate had taken over from Joy as director. Aside from the one line, Wingate's approval letter stated: "we see nothing in the picture to which any objection could be made. . . . In our opinion it is satisfactory from the standpoint of the Code."[48] Wingate concluded by writing that he "enjoyed this picture thoroughly." He reiterated this sentiment in his monthly censorship report to Will Hays, saying that the film struck them "as one of the best horror stories that have been brought to the screen." He also drew explicit attention to the evolutionary theme by saying, "Charles Laughton provides another fine performance as the doctor whose dream it is to hasten evolution by turning animals into human beings."[49] With that level of endorsement Paramount Studios felt safe sending the film to regional censor boards.

Unfortunately for Paramount, censor boards did not share Wingate's enthusiasm. Many boards insisted on cutting Moreau's line that compared himself to God.[50] But this was only one of the issues. More worryingly for Paramount, censor boards considered the plot's overt reliance on evolutionary theory and the insinuation of bestiality to be unacceptable; removing a few lines of dialogue would not be enough. The film was "rejected *in toto* by fourteen state censor boards" and was banned in Germany, Italy, Hungary, New Zealand, the Netherlands, Latvia, India, South Africa, Tasmania, and Singapore.[51] The British Board of Film Censors (BBFC) banned the film until 1958 because of religious concerns over its evolutionary story being "repulsive" and "unnatural"; it also objected to its vivisection scenes.[52] While Joy and Wingate felt that audiences could handle such "unnatural" themes, most censor boards believed the public was best served by not being exposed to these ideas.

Island of Lost Souls was the third evolution-themed SF film in two years that had been endorsed by the SRC and subsequently ran into strong resistance from regional censor boards. Joy and Wingate's approach to evolution themes was in line with their

[47] "Release of Dialogue Script," Cutting Department Office, 17 December 1932, *Island of Lost Souls* file, PCAC.

[48] This and subsequent quote from letter from James Wingate to Harold Hurley, 8 December 1932, *Island of Lost Souls* file, PCAC.

[49] Letter from James Wingate to Will Hays, 9 December 1932, *Island of Lost Souls* file, PCAC.

[50] Massachusetts, Maryland, British Columbia, Pennsylvania, Quebec, and Australia were among the boards making this request. See the various censor board reports in the *Island of Lost Souls* file, PCAC.

[51] Letter from Joseph Breen to John Hammell, 18 September 1935, *Island of Lost Souls* file, PCA. See also the various national censor board reports in the *Island of Lost Souls* file PCAC.

[52] Britain's censorship of *Island of Lost Souls* is discussed in David Skal, *The Monster Show: A Cultural History of Horror* (New York, N.Y., 1993); and Karen Myers, "Case Study: *Island of Lost Souls* (1932)," in *Behind the Scenes at the BBFC*, ed. Edward Lamberti (Basingstoke, UK, 2012), 27–8. Although the original judgment letter from 1932 is lost, the existing evidence suggests that a major part of their decision was the board's belief that the concept of evolving humans from animals was too "repulsive." When the studio resubmitted the film for approval in 1951 and 1957, the BBFC reaffirmed their unease with the evolutionary aspects of the story. For example, see letter from John Nicholls to Sheila White, 29 October 1957, *Island of Lost Souls* file, British Board of Film Censors Archive, British Board of Film Classification, London.

response to other provocative ideas that were not in direct violation of the Production Code. Scholars refer to 1930 to 1934 as the "pre-Code era" because the SRC then only loosely enforced the Production Code. Joy and Wingate believed that it was better to allow the public to grapple with controversial themes like evolution as long as films did not present biased versions or did not treat the ideas seriously.

Regional censor boards' responses to SF films significantly involving evolution showed how Joy and Wingate's views were out of sync with the majority of moral reformers. Religious reformers believed that the danger of motion pictures demanded an approach that prevented disturbing narratives from even reaching audiences. These moral crusaders, along with regional censor boards, considered evolution an inappropriate topic for motion pictures no matter how delicately the film handled the supposedly sinister idea, how ridiculous the film's plot, or whether the film punished any characters advocating human evolution. From their perspective, Joy and Wingate's failure to censure evolutionary elements was further evidence that Hollywood's self-regulation of content was not working and that movies were just as morally problematic as they were before the adoption of the Production Code.

THE PRODUCTION CODE ADMINISTRATION AND THE BLASPHEMY OF EVOLUTION IN SF CINEMA

Religious reformers were frustrated with the SRC's refusal to rigidly enforce the Production Code because they believed that cinema's reality effect was normalizing and legitimating scientific ideas like evolution, as well as other messages. Sociological studies of movies in the early 1930s, such as the Payne Fund studies, seemingly provided scientific evidence for religious reformers' assumptions that individual movies could directly affect a viewer's attitudes and behaviors.[53] This research made the practice of closely analyzing and commenting upon minute details in every script seem a rational approach for censors. If individual films could have such a powerful effect on audiences, then it made sense to censorship organizations for them to only permit what they considered "acceptable" narratives to reach the screen.

This moral outrage resulted in the Catholic Church forming its own censorship organization in 1933—the Catholic Legion of Decency. Will Hays also dissolved the SRC in 1934 in order to curtail calls by religious groups for governmental movie censorship. In its place, Hays created the PCA with the tough-minded Catholic Joseph Breen as director. Although the Catholic Church was not the only religious voice in these organizations, it certainly had an immense influence.[54] Studios' agreement not to release any films without the PCA's seal of approval gave Breen the power he needed to force studios to alter their scripts if they did not meet the standards of the Production Code.[55] Breen's more inflexible approach meant that—after 1934—evolution was no longer a permissible theme.

[53] For a history of the Payne Fund Studies, see Gareth S Jowett, Ian C. Jarvie, and Kathryn H. Fuller, *Children and the Movies: Media Influence and the Payne Fund Controversy* (New York, N.Y., 1996).

[54] See Walsh, *Sin and Censorship* (cit. n. 5). Breen's Catholic influences are also discussed in Thomas Doherty, *Hollywood's Censor: Joseph Breen and the Production Code Administration* (New York, N.Y., 2007).

[55] Leff and Simmons, *The Dame* (cit. n. 5). It should be noted that films approved by the PCA could still face significant censorship trouble from the various regional censor boards, including city, state, and international censors. Unlike the SRC, the PCA used this threat to leverage changes to scripts before production.

In Breen's 1938 "Annual Report of the President of the Association," he warned that the film industry "must continue to resist the lure of propaganda."[56] For Breen, motion pictures were for entertainment, and "the distinction between motion pictures with a message and self-serving propaganda is one determinable only through the processes of common sense." Francis Harmon, who was in charge of the Eastern Division of the PCA, was uncomfortable with Breen's conception of what made something propaganda.[57] Harmon felt that the line between a message and propaganda was too indistinct and subjective to be determined by "common sense." To demonstrate this, he asked colleagues in the PCA to decide if certain books were propaganda.

Almost ninety percent of respondents to Harmon's survey felt strongly that Darwin's *Origin of the Species* was propaganda—the same percentage that categorized Hitler's *Mein Kampf* as such. More PCA employees believed that Darwin's scientific book was propaganda than those who considered the *Communist Manifesto*, Karl Marx's *Das Capital*, or Jean Jacques Rousseau's "Social Contract" to be propaganda. Harmon defined propaganda as "a deliberate distortion or manipulation of facts or realities." From the PCA's perspective, Darwin's text was propaganda because they considered it dishonest. For them, the book was not presenting an alternative hypothesis in a debate over human origins; it was proliferating a dangerous lie and claiming it was truth. Their belief that Darwin, and by extension evolutionary biology, was propaganda thereby justified, to them, their caution regarding evolution themes in SF movies.

SCIENTIFIC CREATION IN *BRIDE OF FRANKENSTEIN* AND A CHANGE IN POLICY FOR THE PCA

Unlike Joy's lenient examination of *Frankenstein*'s script, Breen found numerous violations of the Production Code in *The Bride of Frankenstein*'s script.[58] Most prominent were the story's comparisons of Frankenstein's creation of the monster to God's creation of man. Breen noted that "all such references should be deleted or changed in a manner which will avoid all possible objection."[59] He reminded the studio that the first *Frankenstein* film had been censored because of an "alleged irreverent attitude on the part of some of the characters, particularly wherever they even suggested that their actions were paralleling those of the Creator."[60] Breen requested two modifications to appease regional censor boards. First, the film should include punishment of the scientists to convey a moral lesson. Second, the studio must remove dialogue suggesting any equivalence with God's creation, as well as dialogue mocking the biblical creation story.

The PCA convened a meeting with director James Whale to discuss Breen's requests.[61] Whale proposed minor alterations to dialogue that he felt would diminish the

---

[56] Joseph Breen, Annual Report of the President of the Association, "Entertainment vs. Propaganda," 28 March 1938. Cited in Francis Harmon, "Memorandum Commenting Upon Document Entitled 'Code, Extra-Code, and Industry Regulation in Motion Pictures,'" 22 June 1938, Motion Picture Producers and Distributors of America (MPPDA) Digital Archives, Flinders University, Adelaide, South Australia, Australia.

[57] Information in this and subsequent paragraph from Francis Harmon, "Memorandum" (cit. n. 56).

[58] *Bride of Frankenstein*, directed by James Whale (1935; Universal City, Calif.: Universal Pictures Home Entertainment, 2000), DVD.

[59] Letter from Joseph Breen to Harry H. Zehner, 24 July 1934, *Bride of Frankenstein* file, PCAC.

[60] Quotes and information in this paragraph in letter from Joseph Breen to Harry H. Zehner, 5 December 1934, *Bride of Frankenstein* file, PCAC.

[61] Letter from Joseph Breen to Harry H. Zehner, 7 December 1934, *Bride of Frankenstein* file, PCAC.

film's blasphemous premise.[62] He also offered to insert a prologue and epilogue where an actor portraying Mary Shelley would clearly state that her "purpose was to write a moral lesson [and present] the punishment that befell a mortal man who dared to emulate God," and that "no one is meant to know these things—it is blasphemous and wicked." Breen told Whale his proposed dialogue changes ameliorated the blasphemy sufficiently to avoid "the censorship difficulties from which your first Frankenstein picture unfortunately had to contend with."[63] He still believed, however, that it was "more than likely that this picture will meet with considerable difficulty at the hands of [regional] censor boards both in this country and abroad."[64] As predicted, the regional censor boards severely edited the film. Gruesomeness was the primary issue, but a number of boards insisted on removing apparently blasphemous dialogue. Pennsylvania and Quebec, for example, eliminated the line that the studio had already adjusted slightly: "After twenty years of secret scientific research, and countless failures, I have also created life as we say—in God's own image."[65]

As they had done previously with Joy/Wingate and the SRC, the studio asked Breen to intervene with the censor boards to minimize cuts.[66] In response, Breen decided that it was not the PCA's role to convince censors to retain scenes and dialogue that the PCA had warned the studio would be a problem for the regional censor boards. *Bride of Frankenstein* thus represented a major turning point for Breen and the Hays Office. The studio's failure to heed his advice led Breen to enact a new policy. He would guarantee that the PCA would fight for studios against the regional censor boards, but *only* if the studios agreed to first make changes to their scripts when requested by the PCA. Breen's new policy gave him the leverage he needed to force studios to remove what he considered problematic themes and contexts, thus enabling him to become one of the most powerful men in Hollywood.

REMOVING EVOLUTION AND REAFFIRMING CREATION NARRATIVES IN SF FILMS

After 1935, the PCA began to routinely ask studios to modify SF scripts that could be interpreted as containing ideas about evolution. In the original script of Warner Bros. Studio's *The Walking Dead* (1936), for example, a scientist resurrects Boris Karloff's character Dopey by transforming him into a "half-man/half-animal."[67] The PCA rejected the script because the evolution connotations were too obvious. They instructed the studio that there should be "no suggestion in the character of Dopey—after his death—that he is a half-man/half-animal. In our judgment this would be most offensive."[68] The studio

---

[62] For example, Whale suggested changing the line "After twenty years of secret scientific research, and countless failures, I have also created life as *they* say—in God's own image" to: "After twenty years of secret scientific research, and countless failures, I have also created life as *we* say—in God's own image" (emphasis mine); and the line "If you are fond of your fairy tales" to: "If you are fond of your scriptures." Letter from James Whale to Joseph Breen, 7 December 1934, *Bride of Frankenstein* file, PCAC.

[63] Letter from Joseph Breen to James Whale, 11 December 1934, *Bride of Frankenstein* file, PCAC.

[64] Letter from Joseph Breen to Harry H. Zehner, 15 April 1935, *Bride of Frankenstein* file, PCAC.

[65] Pennsylvania and Quebec censor board reports, *Bride of Frankenstein* file, PCAC.

[66] The remainder of the information in this paragraph is in the letter from Joseph Breen to Will Hays, 8 May 1935, *Bride of Frankenstein* file, PCAC.

[67] *The Walking Dead*, directed by Michael Curtiz (1936; Burbank, Calif.: Warner Home Video, 2009), DVD.

[68] Letter from Joseph Breen to Jack L. Warner, 26 September 1935, *Walking Dead* file, PCAC.

Figure 4. *Instead of Boris Karloff's character becoming a half-man/half-animal, Dopey's resurrection was indicated by a white streak of hair as shown in this film still from* The Walking Dead *(cit. n. 67). (Courtesy of Universal Pictures.)*

removed the half-man/half-animal element.[69] In its place they indicated Dopey's resurrection from the dead by putting a white streak in Karloff's hair (figure 4).

The PCA was also concerned about more subtle representations or dialogue that could be read as supporting an evolutionary origin for humanity. For example, in the original script for *The Monster and the Girl* (1941) the scientist delivers this line: "Look—the human brain! Even more than the human heart it is *nature's* greatest handiwork."[70] Although these words might seem neutral, the use of the word "nature"

[69] Letter from Joseph Breen to Jack L. Warner, 5 November 1935, *Walking Dead* file, PCAC.
[70] Stuart Anthony, script of *D.O.A.*, 5 July 1940, B-12, p. 72, Paramount Script Files, 650-f.M-823, Margaret Herrick Library, Beverly Hills, California (emphasis mine). *D.O.A.* was the original title of the film before it was changed to *The Monster and the Girl*. *The Monster and the Girl*, directed by Stuart Heisler (1941; Universal City, CA: Universal Pictures Home Entertainment, 2014), DVD.

implies that humans were the result of an organic evolutionary process. After a phone call with the PCA, the studio agreed to make several major alterations to the script, including its evolutionary elements.[71] One minor change involved substituting a single word so that the previous dialogue now read: "Look—the human brain! Even more than the human heart it is *God's* greatest handiwork." This alteration made explicit who was responsible for the origin of the human brain.[72]

REFRAMING MOREAU'S EVOLUTION EXPERIMENTS IN *ISLAND OF LOST SOULS*

The PCA even rejected SF films the SRC approved before 1934 that studios later put forward for re-release. When Paramount tried to re-release *Island of Lost Souls* in 1941, a much more stringent PCA rejected the film because of the evolutionary basis of Moreau's experiments.[73] For Breen, *Island of Lost Souls* violated the spirit, if not the precise wording, of the Production Code by including an evolutionary concept of human origins that he believed would be deeply offensive to Christians. Breen explained to the studio: "The general unacceptability of this picture is suggested by the blasphemous suggestion of the character, played by Charles Laughton, wherein he presumes to create human beings out of animals."[74]

Paramount responded by saying they would eliminate dialogue suggesting that Moreau was "creating" humans by accelerating the evolution of animals.[75] The studio was particularly attentive to dialogue indicating that Lota the panther woman was Moreau's creation. Unlike the other Beast-People, Lota, played by Kathleen Burke, appears to be human (figure 5). Any suggestion that she was Moreau's creation legitimated the idea that humans could be created using the "lower" animals. The studio removed dialogue such as, "Lota is my most nearly perfect creation" to make it appear that she was the scientist's daughter instead.

The studio assured the PCA that "these cuts eliminate from the picture the suggestions that Moreau considers himself on par with God as a creator, and reduces him to the status of a scientist conducting bio-anthropological experiments." In the edited film, there is no longer any indication that Moreau made the creatures on the island; he is now merely an anthropologist studying their behaviors. In this way, the Beast-People simply were another of God's creations. That was a "scientific" notion that the PCA deemed appropriate for audiences.

FEARS OF GORILLA LOVE: THE EVOLUTIONARY
THREAT OF BESTIALITY IN SF CINEMA

It is important to note that the PCA did not remove every reference to evolution from SF films. Although Breen cast a long shadow over the organization, it was not a monolithic entity but a collection of individuals, and the PCA's response to evolutionary

---

[71] Letter from Joseph Breen to Luigi Luraschi, 24 July 1940, *Monster and the Girl* file, PCAC.

[72] Stuart Anthony, script of *Monster and the Girl*, 29 January 1940, Reel 2B, pages 9–10, Paramount Script Files, 650-f.M-826, Margaret Herrick Library, Beverly Hills, CA (emphasis mine).

[73] Letter from Joseph Breen to Luigi Luraschi, 4 March 1941, *Island of Lost Souls* file, PCAC. The PCA had also rejected the film for re-release in 1935. See letter from Joseph Breen to John Hammell, 18 September 1935, *Island of Lost Souls* file, PCAC.

[74] Letter from Joseph Breen to Luigi Luraschi, 4 March 1941, *Island of Lost Souls* file, PCAC.

[75] All quotes and information in this and subsequent paragraph are in letter from Luigi Luraschi to Joseph Breen, 15 March 1941, *Island of Lost Souls* file, PCAC.

Figure 5. *Lota the panther woman, played by Kathleen Burke, shows no signs of her nonhuman origins in this film still from* Island of Lost Souls *(cit. n. 44). (Courtesy of Paramount Pictures.)*

aspects sometimes depended on which censor read the script.[76] There was one element related to evolution, however, that every PCA reviewer flagged as problematic. They automatically rejected scripts that even subtly suggested bestiality.

Julia Voss observes that there has been a long association between gorillas, evolution, and bestiality: "the gorilla embodied the essence of bestiality, and thus represented an affront to humanity."[77] Gorillas were a staple of SF and horror films from the 1930s through the 1950s. The animal featured in a wide variety of "mad science" plots, such as scientists turning humans into gorillas or gorillas into humans, or putting human brains into gorillas or vice versa. The original scripts for these films frequently involved the human to gorilla, or gorilla to human, character developing ro-

[76] Although censors approached each script differently, there is little historic information on the individuals. One exception is Jack Vizzard's book on his time at the PCA, but there are some questions about the reliability of his memoir; Jack Vizzard, *See No Evil* (New York, N.Y., 1970).
[77] Julia Voss, "Monkeys, Apes and Evolutionary Theory: From Human Descent to King Kong," in *Endless Forms: Charles Darwin, Natural Science and the Visual Arts*, ed. Diana Donald and Jane Munro (New Haven, Conn., 2009), 214–34, on 216.

mantic feelings for a human character. From the PCA's perspective the implications of such a relationship, even if unrequited, made this type of plotline unsuitable. It rejected a large number of scripts with this narrative element including *White Pongo* (1945), *Bride of the Gorilla* (1951), and *The Bride and the Beast* (1958).[78] The love triangle was a basic plot element in this time period, but it became problematic for the PCA when one point of the triangle was a gorilla, as was the case in *Jungle Woman*'s (1944) original script.[79]

EVOLUTION IN SF SHIFTS FROM OBJECTIONABLE
TO ACCEPTABLE AFTER *INHERIT THE WIND*

The PCA contended with evolutionary themes within SF movies well into the 1950s. But their response to scripts for the non-SF film *Inherit the Wind* (1960) reveals how the organization altered their approach to evolution in general by the end of the 1950s.[80] *Inherit the Wind* is the most well-known fictional film that dealt overtly with human evolution. The film is based on a stage play that debuted in 1955 and was written as a response to US Senator Joseph McCarthy's high profile anti-Communist activities. Its story consists of a highly fictionalized retelling of the 1925 Scopes Trial.[81]

Given that the story openly supported the teaching of evolution and mocked antievolutionists, it is not surprising that the PCA initially rejected Twentieth Century Fox's script submitted for approval just after the play's opening on Broadway in March 1955: "A story such as this violates the portion of the code which states that 'No film . . . may throw ridicule on any religious faith.' This material contains an attack on Christian doctrines and, in general, presents religious thinking people in an extremely unfavorable light."[82] The PCA judged the story was making fun of Christian doctrines about creation and implying that all Christians were biblical literalists. Burt Lancaster's production company, Hecht-Hill-Lancaster, subsequently bought the screen rights from Twentieth Century Fox and resubmitted the script in June 1955. While the PCA did not immediately reject the script as they had in March, they requested extensive changes that the studio could not accommodate.[83]

So, why after two initial rejections did the PCA approve a script in 1959 that was still clearly about the realities of evolution? Four years after rejecting those scripts, the PCA's power was in serious decline. The post-World War II period saw a general loosening of moral strictures, and the American public began questioning the group's control over movie content.[84] Protestants also began complaining that the Production Code was bi-

[78] Letter from Joseph Breen to Sigmund Neufeld, 14 March 1945, *White Pongo* file; letter from Breen to Jack Broder, 27 June 1951, *Bride of the Gorilla* file; letter from Geoffery Shurlock to Louis Weiss, 17 October 1956, *The Bride and the Beast* file; all of foregoing in PCAC.

[79] Letter from Joseph Breen to Maurice Pivar, 27 January 1944, *Jungle Woman* file, PCAC; *Jungle Woman*, directed by Reginald LeBorg (1944; Universal City, Calif.: Universal Pictures Home Entertainment, 2015), DVD.

[80] *Inherit the Wind*, directed by Stanley Kramer (1960; Beverly Hills, Calif.: MGM Home Entertainment, 2007), DVD.

[81] For a history of the play, see Albert Wertheim, "The McCarthy Era and the American Theatre," *Theatre Journal* 34 (1982): 211–22.

[82] Letter from Geoffery Shurlock to Frank McCarthy, 24 March 1955, *Inherit the Wind* file, PCAC.

[83] Albert Van Schmus, "Memo for the Files, Re: Inherit the Wind," 10 June 1955, *Inherit the Wind* file, PCAC.

[84] Leff and Simmons, *The Dame* (cit. n. 5), 173.

ased against non-Catholic themes.[85] In addition, the Supreme Court's decision to give movies First Amendment protection in 1952 put significant restrictions on film censorship.[86] Breen retired in 1954, and the PCA replaced him with Geoffrey Shurlock, a more lenient regulator who questioned the contemporary relevance of the Production Code.[87] These changes led to a revision of the PCA's philosophies that allowed some previously banned topics, like evolution, in SF films, as long as the treatments were "tasteful."

This is not to say that *Inherit the Wind* was awarded a seal of approval without comment. The story still violated the Production Code. But the modifications director Stanley Kramer negotiated with the PCA did not alter the tenor of the film in any way. The PCA asked that the film depict Reverend Brown as "earnest and sincere but misguided" and that a "sympathetic character" promote the idea that "religion, as practiced in this one community, was not representative of the true Christian faith."[88] They also required the removal of irreverent phrases including "hot for Genesis" and "belching beatitudes." These were minor changes overall, especially compared to the script's hostile rejection just four years earlier. The PCA had become less worried about the theological implications of scientific topics, focusing instead on retaining some influence over increasingly explicit depictions of sex and violence. Ultimately, the PCA dissolved in 1968 and was replaced by the Code and Ratings Administration, which provided guidance to audiences but did not censor content.

EVOLUTIONARY NARRATIVES, SF FILMS, AND THE CATHOLIC LEGION OF DECENCY

The PCA had been Hollywood's official self-censorship organization, but there was another important group that also began operating in the United States in the early 1930s. As previously mentioned, the Catholic Church formed its own censorship organization in 1933 called the Catholic Legion of Decency.[89] Although these two organizations shared a similar philosophy about cinema's supposed moral dangers, the two were completely distinct censorship groups with very different operating procedures, relationships to the film industry, and parent organizations (the Hays Office for the PCA, and the Catholic Church for the Legion of Decency).

The Legion of Decency's primary means for compelling studios to modify their scripts was their film classification system, which advised Catholics as to a film's moral acceptability.[90] They had three levels of classification: "A" for morally acceptable, "B" for morally objectionable in part, and "C" for condemned.[91] Studios believed that a B or C classification would seriously affect their box office by driving significant numbers of Catholics away from the film, and they were willing to negotiate to avoid

[85] Andrew Quicke, "The Era of Censorship (1930–1967)," in *The Routledge Companion to Religion and Film*, ed. John Lyden (New York, N.Y., 2009), 32–51.

[86] Ellen Draper, "'Controversy Has Probably Destroyed Forever the Context': *The Miracle* and Movie Censorship in America in the Fifties," *The Velvet Light Trap* 25 (1990): 69–79.

[87] See Leff and Simmons, *The Dame* (cit. n. 5).

[88] All quotes and information in this paragraph are in a letter from Geoffrey Shurlock to Stanley Kramer, 15 October 1959, *Inherit the Wind* file, PCAC.

[89] Walsh, *Sin and Censorship* (cit. n. 5).

[90] After receiving the PCA's seal of approval, the studio would submit a film to the legion for classification prior to its theatrical release. The organization changed their name to the National Catholic Office of Motion Pictures (NCOMP) in 1966. This title change did not alter its mode of operation. On the change to NCOMP, see Walsh, *Sin and Censorship* (cit. n. 5), 318.

[91] Ibid., 135.

this.[92] Their approach could include sending scripts to the legion for approval before filming or submitting recommendations for scenes to cut from finished films.

Before 1950, movie depictions that even implied an evolutionary connection between humans and nonhuman primates led to B classifications, even if the films had already been passed by the PCA. Twentieth Century Fox's removal of dialogue concerning the scientist's evolutionary goals in *Dr. Renault's Secret* may have satisfied the PCA, but the Legion of Decency gave the film a B classification because "the subject material and treatment reflects some acceptance of the possibility of changing an ape into a human being."[93] The Legion of Decency was even more concerned about films that visualized a connection between humans and apes through graphic transformation scenes, as in *Jungle Woman* (1944).[94] From their standpoint, these SF films were theologically unpalatable because they legitimated a biological connection between apes and humans and indicated a natural progression from one to the other.

LOOKING FOR A FEW CATHOLIC APES: EMBRACING EVOLUTIONARY NARRATIVES IN SF CINEMA

Catholic reaction to evolutionary stories changed significantly after Pope Pius XII's papal encyclical *Humani Generis* in August 1950 affirmed that there was no inherent conflict between biological evolution and Catholic doctrines.[95] Catholic views on evolution were further influenced by Jesuit Pierre Teilhard de Chardin's 1955 book *Phenomenon of Man*.[96] After the 1950 papal encyclical the Legion of Decency no longer considered stories involving human evolution to be inherently problematical. Instead, the legion frequently used these films as an opportunity to congratulate themselves for being open to evolutionary ideas and to express bemusement at their fellow Christians who did not accept this science.

The legion's reviewers, a combination of clergy and lay Catholics, were extremely complimentary toward *Inherit the Wind*, and several wanted to give the film a special award. One reviewer's comments sum up the legion's general opinion: "The theory of evolution as shown in this film is allied with our acceptance. I doubt that anyone holds any longer the literal seven day interpretation of Genesis. At least among Catholics. The Fundamentalists, as well shown in this film, make a pain of religion when it is really a joy."[97] Another reviewer relished the film's portrayal of Protestant fundamentalists: "The Bible gets a whalloping in 'Inherit the Wind' that is wholly justified and convincing."[98] Catholics were sensitive to Protestant characterizations of their religion as antibiblical. One of the main differences between the two is that Catholics reject the

---

[92] Movie studios took the threat of a Catholic boycott seriously. In the 1930s one in five Americans was Catholic. Concentrated in eastern urban areas like Boston and New York, Catholics were essential for a successful box office. See Black, *Hollywood Censored* (cit. n. 5), 162–70.

[93] Classification of *Dr. Renault's Secret*, in *Motion Pictures Classified by the Legion of Decency, February 1936–October 1959* (New York, N.Y., 1959), 60. Available online: https://archive.org/details /motionpicturescl00nati.

[94] Classification of *Jungle Woman*, in ibid., 118.

[95] Pius XII, Encyclical Letter, *Humani Generis*, Rome, St. Peter's, 12 August 1950.

[96] Pierre Teilhard de Chardin, *The Phenomenon of Man*, trans. Bernard Wall (Glasgow, UK, 1959).

[97] Sal Miraliotta, "Reviewer's Comments," 27 June 1960, *Inherit the Wind* file, National Legion of Decency files, Records of the Office of Film and Broadcasting of the United States Conference of Catholic Bishops Communications Department, The American Catholic History Research Center and University Archives, Catholic University, Washington, DC (**hereafter** cited as LDA).

[98] Jim Scovotti, "Reviewer's Comments," undated, *Inherit the Wind* file, LDA.

doctrine of *sola scriptura*, the reliance on the Bible alone as sufficient for spiritual guidance.[99] It was gratifying for these Catholic reviewers to see a film questioning this element of Protestant doctrine. *Inherit the Wind*'s treatment of Protestant Fundamentalists seemed like a vindication for Catholic beliefs, as well as for Darwinian evolution.

By the late 1960s, the Legion of Decency's perception that "nobody can seriously doubt evolution now" emerges in their response, and subsequently that of the National Catholic Office of Motion Pictures (NCOMP), to several SF films. The central conflict in *Planet of the Apes* (1968) concerns a science/religion debate about evolution.[100] The stranded human astronaut's ability to talk leads to a parody of the Scopes Trial, in which literalist orangutan religious leaders oppose freethinking chimpanzee scientists; the latter believe the human's speech supports their contention that apes evolved from humans. Even NCOMP reviewers who found the film silly considered it a "provocative and ironic statement about mankind and evolution."[101] Several reviewers thought the film was behind the times regarding perceptions of the religion/evolution issue. One asked the following: "Is it not a fact that the evolution 'problem' is dated? Surely there is not an intelligent person alive today who does not at least suspect the possibility of evolution?"[102] In fact, NCOMP's only problem with the evolutionary theme in *Planet of the Apes* was that the movie legitimated the perception that all religious groups had a problem with evolution. The organization would have preferred that the film include a few Catholic apes who sided with the scientists in this debate.

Reviewers from NCOMP also embraced *2001: A Space Odyssey* (1968) because of its inclusion of supreme beings, whether metaphorical, alien, or divine.[103] Despite the fact that the film never refers to religion, most of NCOMP's reviewers felt that the film conveyed a spiritual message about God's role in human evolution. The vague nature of the black monolith in the film allowed for multiple interpretations of its meaning (figure 6). Director Stanley Kubrick and scriptwriter Arthur C. Clarke may have considered the monolith advanced alien technology, but many NCOMP reviewers believed it "symbolizes God, who is unchangeable, always present and powerful."[104]

For NCOMP, the monolith was an acknowledgement that human evolution was too complex to occur without external intervention. The monolith appears at crucial stages of human evolution; these include the point when humanity's ancestors first learned to use tools, when humanity was on the cusp of space travel, and when *Homo sapiens* evolved into *Homo superior*, as represented by the "star child" at the film's end. The monolith's influence on these events supported the concept of directed evolution, or so NCOMP argued. The moment hominids learned how to use tools is the moment in which the monolith/God infused humanity with an immortal soul. Because of this interpretation, NCOMP eventually awarded the film its "Best Film of Educational Value"

[99] For a history of the disputes between Protestants and Catholics in the United States, see Jay P. Dolan, *The American Catholic Experience: A History from Colonial Times to the Present* (New York, N.Y., 1985).
[100] See Amy C. Chambers, "The Evolution of *Planet of the Apes*: Science, Religion, and 1960s Cinema," *Journal of Religion and Popular Culture* 28 (2016): 107–22; *Planet of the Apes*, directed by Franklin J. Schaffner (1968; Century City, Calif.: Twentieth Century Fox Home Entertainment, 2014), DVD.
[101] Lois Trella, "Reviewer's Comments," undated, *Planet of the Apes* file, LDA.
[102] Benjamin Colimore, "Reviewer's Comments," undated, *Planet of the Apes* file, LDA.
[103] *2001: A Space Odyssey*, directed by Stanley Kubrick (1968; Burbank, Calif.: Warner Home Video, 2001), DVD.
[104] John C. Connelly, "Reviewer's Comments," 2 April 1968, *2001: A Space Odyssey* file, LDA.

Figure 6. *Many reviewers for National Catholic Office of Motion Pictures (NCOMP) considered the black monolith, shown in this film still from* 2001: A Space Odyssey *(cit. n. 103), to be symbolic of God's influence on human evolution. (Courtesy of Metro-Goldwyn-Mayer.)*

in 1968.[105] For NCOMP, the SF film not only confirmed that humanity's future will be awesome, it established that God had made this future possible.

REJECTING "INAPPROPRIATE" CINEMATIC NARRATIVES ABOUT EVOLUTION IN SF CINEMA

I do not want to give the impression that the Legion of Decency simply embraced every evolutionary story in SF films after 1950. The Legion of Decency's response to evolution post-1950 actually reveals how the Catholic Church remained conflicted about the moral implications of evolutionary narratives. The 1957 movie *I Was a Teenage Werewolf* illustrates this.[106] The Legion of Decency gave the movie a B rating because it "tends to give credence to certain philosophical theories whose acceptance can lead to serious moral harm."[107] In the movie, a scientist explains that his goal is to re-evolve humanity by regressing a teenager back to his primitive state:

> DR. BRANDON: Through hypnosis, I'm going to regress this boy back . . . back into the primitive past that lurks within him. I'll transform him and release the savage instincts of life hidden within.
> Dr. WAGNER: And then?
> DR. BRANDON: Then I'll be judged a benefactor. Mankind is on the verge of destroying itself. The only hope for the human race is to hurl it back to its primitive dawn . . . to start all over again.[108]

---

[105] "The National Catholic Office for Motion Pictures Presents its 1968 Award for Best Film of Educational Value to *2001: A Space Odyssey*," undated, File: Awards 1968–1969, Box 175, LDA.

[106] *I Was a Teenage Werewolf*, directed by Gene Fowler Jr. (1957; Los Angeles, Calif.: RCA/Columbia Pictures Home Video, 1991), VHS.

[107] Classification of *I Was a Teenage Werewolf*, in *Motion Pictures Classified* (cit. n. 93), on 108.

[108] *I Was a Teenage Werewolf* (cit. n. 106).

Several other SF films of this period involved scientists releasing humanity's inherited but submerged animal instincts. The scientist in *Blood of Dracula* (1957), for example, argues that our bestial ancestry left us with "a power strong enough to destroy the world buried within each of us."[109] Her experiments unleash this power in her teenage assistant who then kills several classmates. The Legion of Decency gave the film a B classification because the "film tends to give credence to an erroneous philosophy of the origins of human life."[110]

It might seem surprising for the Legion of Decency to respond so negatively to such absurd low-budget SF horror films, especially since they had recently commended films with far more realistic views of evolution. But the Catholic Church's acceptance of human evolution was conditional; suggesting that evolution could explain sinful behaviors such as murder or adultery was unacceptable. The scientists' use of hypnosis in both films points to a link between psychology and Darwinism that the Catholic Church found a particularly alarming misapplication of evolutionary thought.[111] Darwin's writings were a major influence on the field of psychology and on Sigmund Freud's theories in particular. According to Lucille Ritvo, Freud made it clear that Darwinian theory was "essential to psychoanalysis" and that it "has always been present in Freud's writings, albeit never explicitly."[112] Frank J. Sulloway similarly argues that "Freud, ethology, and sociobiology share a common evolutionary heritage."[113] Freud derived concepts such as the libido, id, and psychosexual stages from the Darwinian notion that behaviors are the result of a few basic animal instincts produced by natural selection to facilitate survival.

The Catholic Church took issue with scientific explanations for processes, including human evolution, that removed personal responsibility from an individual's actions. For Catholics the root of sin is God's gift of free will.[114] Sinful actions represent choices by individuals, and are not the product of "primitive instincts" inherited from our animal ancestors as scientists in these films claimed. From the Legion of Decency's perspective, these fictional scientists appeared to be offering absolution through scientific explanations based on Darwinian-informed psychiatry. The fact that these were frivolous SF films actually made the underlying ideas more worrisome. The Catholic Church viewed the films as exploitation aimed squarely at teenagers who might not be able to think through the moral implications of this philosophical position. The Legion of Decency's film reviewers thus used the lens of Catholic theology to think through all the potential moral implications of evolutionary narratives in a film, even in grade Z movies like *I Was A Teenage Werewolf* or *Blood of Dracula*.

---

[109] *Blood of Dracula*, directed by Herbert L. Strock (1957; Santa Monica, Calif.: Lions Gate Films Home Entertainment, 2006), DVD.

[110] Classification of *Blood of Dracula*, in *Motion Pictures Classified* (cit. n. 93), 26.

[111] See Robert Kugelmann, *Psychology and Catholicism: Contested Boundaries* (Cambridge, UK, 2011).

[112] Lucille B. Ritvo, *Darwin's Influence on Freud: A Tale of Two Sciences* (New Haven, Conn., 1990), 2.

[113] Frank J. Sulloway, *Freud, Biologist of the Mind: Beyond the Psychoanalytic Legend*, 2nd ed. (Cambridge, Mass., 1992), xii.

[114] The *Catechism of the Catholic Church* covers the relationship between Catholicism and free will; *Catechism of the Catholic Church*, 2nd ed. (Vatican, 2000). The specific section on free will can be found online: "Article 3, Man's Freedom," Catechism of the Catholic Church, Catholic Church, www.vatican.va/archive/ccc_css/archive/catechism/p3s1c1a3.htm (accessed 3 September 2018).

CONCLUSIONS

The administrators of the SRC, PCA, and Legion of Decency were not trying to destroy the movie industry. They strongly believed that they were acting in a parental role by protecting audiences from immoral and indecent films. From their perspective, films told linear stories using a heightened visual realism that conveyed easily understandable narratives to a monolithic audience. From this simplistic viewpoint cinema seemed to be a powerful medium of persuasion. By controlling the content of scripts, these religiously oriented, elitist groups believed they could ensure that movies disseminated only morally appropriate messages, including those about human origins. I have shown in this essay that some censors found even the idea of evolution to be a moral danger, so they modified cinematic narratives in SF films in order to tell what they considered more appropriate stories about evolution as a social, cultural, and moral force.

Censorship organizations favored SF narratives that recognized God's role as creator, and they prohibited stories implying a materialistic explanation for humanity's origins. Science fiction stories about evolution in films were not just about humanity's past, they were also about our visions for the future. Therefore, censors rejected many scripts that included evolutionary connections between humans and other animals. They rejected stories that included attempts by fictional scientists to "prove" Darwin's theories, to find the "missing link," or that encompassed irreverence toward the notion of a divine creator. They also deemed evolutionary ideas indecent when they included implications of bestiality in many human/gorilla films. The threat of censorship forced filmmakers to make decisions about including evolutionary themes based on reasons that had nothing to do with artistic merit. This means that we need to look beyond the finished film, or we miss the story of how and why organizations outside Hollywood studios shaped and influenced cinematic depictions of evolution in SF cinema.

Significant differences among the groups persisted regarding how to censor evolution, in addition to their changes in attitudes over time toward evolutionary content. For the SRC, the context was crucial as to whether or not they removed a script's evolutionary aspects. The organization felt that evolution was only dangerous if the film itself took the idea seriously. But their handling of cinematic evolution was not in line with moral reformers who demanded more stringent censorship. Moral crusaders considered evolutionary themes to be inappropriate no matter how delicately a film handled them, how ridiculous the film's plot was, or if characters were punished for advocating human evolution.

Science fiction movies, in particular, led to disagreements between the SRC and regional censor boards. Science fiction film scholar Christine Correa argues: "science fiction is a genre that is demonstrably located between fantasy and reality."[115] The different approaches taken by the SRC and regional censor boards to evolution in SF cinema could be viewed as a debate over which aspects were stronger—the fantasy or the realism. The SRC felt that SF cinema's fantasy elements rendered its messages harmless, while regional censor boards believed that SF's realism made the genre a potential moral danger if it contained inappropriate ideas. The SRC's handling of evolutionary themes in SF films was one factor that led to its replacement by the PCA in 1934. The head of the PCA, Joseph Breen, rejected any distinction between fantasy and realism in

---

[115] Christine Cornea, *Science Fiction Cinema: Between Fantasy and Reality* (New Brunswick, N.J., 2007), 4.

SF cinema. Breen believed that cinema was a problematic medium no matter what the genre or how fantastic the stories were. From his perspective, the potential danger of motion pictures demanded an approach that prevented what he considered unacceptable stories from even reaching audiences. Ultimately, any theological implications of evolution in SF films became far less significant than the growing depiction of graphic sex and violence as the PCA lost its power to compel studios to change film content in the early 1960s.

Until 1950, the Legion of Decency concurred with the PCA's approach to excluding evolution from SF movies. But institutional and cultural changes in the 1950s and 1960s led to them changing their opinion of what constituted an inappropriate story involving evolution. The Legion of Decency's censorship decisions for post-1950 SF films illustrate the Catholic Church's continuing conflict with the moral implications of evolutionary narratives. Although the Legion of Decency's reviewers often congratulated themselves for their progressive views on evolution, they also reacted strongly to evolutionary narratives that they believed removed moral responsibility for a person's actions; an example was the notion of inherited animal instincts or the idea that evil is a legacy of evolution. Science fiction narratives including evolution may have become acceptable to the Catholic Church, but they still wanted to control how stories about the implications of evolutionary thought were told in movies.

Chinese Science Fiction:
Imported and Indigenous

by Lisa Raphals*

ABSTRACT

The relation of science to science fiction in the history of Chinese science fiction has been closely linked to both the influence of Western science and to ideals of progress, nationalism, and empire. But when we turn to China's long history of philosophical speculation, a rather different story needs to be told. This article examines the ways in which the indigenous Chinese sciences have fed into fiction, and considers the consequences for our understandings of the genre of science fiction itself and its broader social and historical contexts, as well as relationships between modernity, progress, and science in a non-Western, but globally crucial, context.

In the debates surrounding Chinese science fiction, scholars have often concerned themselves with the question of "origins." At what point can "Chinese science fiction" be said to have emerged?[1] Crucially, was it a modern phenomenon that arose toward the end of the Qing dynasty (1644–1911) in the late nineteenth/early twentieth century, or was it rooted in earlier literary genres? If we adopt the "modernist" view, further questions arise. What did Qing writers and readers understand as "Western" SF? Which writers were available in translation? How did "Western" SF evolve, and what did the Chinese authors of works we (retrospectively) recognize as SF consider themselves to be doing? How did SF become indigenous to China? Answers to these questions, some of which are explored elsewhere in this volume, are complicated by genre definitions and the status of SF in China—until recently relegated to children's pedagogic literature, and considered of little literary or even popular interest. In China today, the genre of "science fiction" (*kehuan* 科幻) is considered distinct from "fantasy" (*qihuan* 奇幻), which includes both *xuanhua*n 玄幻, fantastic fiction with Chinese supernatural elements, and *mohuan* 魔幻, magical fiction with Western elements.[2] Both are made even more complicated when the history of indigenous Chinese science is taken into consideration.

* Department of Comparative Literature and Languages, University of California Riverside, Riverside, CA 92521, USA; lisa.raphals@ucr.edu.

[1] I use the terms science fiction and SF interchangeably, to include both clearly "science"-oriented and speculative fiction.

[2] Pinyin transliteration of Chinese words is used throughout, except for personal names, where I follow authors' own usages. For consistency, all Chinese references are in traditional characters (*fanti*), including contemporary pieces originally published in simplified (*jianti*) characters. Chinese names are cited surname first in accordance with Chinese name conventions. Chinese characters are included because Chinese SF Anglophone literature often omits them, making names and titles harder to find in Chinese sources.

Chinese scientific and philosophical literature offers a long and rich parallel history of speculation on topics that are now staples of science fiction; these span several genres of Chinese writing since the fourth century BCE. This article explores the way in which these topics, distinct threads that appear in Chinese philosophical and historicoliterary texts from the fifth or fourth centuries BCE to roughly the sixth century CE, were expressed. In particular, it considers accounts of travels in time and space (above or beyond the earth), and accounts of immortality or extreme longevity. It also considers "transformative accounts," including contact with sentient nonhuman entities, descriptions of changing species, and the indigenous Chinese genre of "tales of the strange" (*zhiguai* 志怪). While these texts can retrospectively be recognized as science-fictional, from the viewpoints of their creators, these genres address practices and theories defined by indigenous Chinese sciences. Their relationship with indigenous science fiction, however, is considerably more difficult to define. Nonetheless, Chinese indigenous literary, religious, and scientific traditions informed Chinese SF in important ways. This article thus surveys what might be called a "parallel SF context" for relationships between science and fiction in a Chinese context, one that demonstrates that speculations on science in fiction need not necessarily lead to the Western modes of "science fiction" that have dominated the literature.

A MODERNIST HISTORY OF CHINESE SF

Modernist accounts of Chinese SF describe three distinct phases: the utopian, science fictional writings of the late Qing dynasty; pedagogical, didactic stories produced in the Maoist period; and the rise of speculative, often dystopian, science fiction since 1989. The first part of this article will briefly review these developments before moving on to consider the role of the indigenous Chinese sciences.

A range of contemporary scholars are actively exploring the role of SF in the cultural life of the late Qing dynasty.[3] During this period, Chinese SF and utopian texts explored notions of "Chineseness," modernity, and human nature.[4] Several key intellectual figures of the period concerned themselves with SF. Both the great Qing statesman Liang Qichao 梁啟超 (1873–1929) and the great writer Lu Xun 魯迅 (1881–1936) thought that SF, particularly that by Jules Verne, would help spread modern Western knowledge into China.[5] By 1919, at least fifty SF titles had been translated into Chinese in both books

---

[3] See, in particular, David D. Wang, *Fin-de-Siècle Splendor: Repressed Modernities of Late Qing Fiction, 1849–1911* (Redwood City, Calif., 1997); as well as Nathaniel Isaacson, *Celestial Empire: The Emergence of Chinese Science Fiction* (Middleton, Conn., 2017); Song Mingwei, ed., "Chinese Science Fiction: Late Qing and the Contemporary," special issue, *Renditions* 77–78 (2012); Wang Dun, "The Late Qing's Other Utopias: China's Science-Fictional Imagination, 1900–1910," *Concentric: Literary and Cultural Studies* 34 (2008): 37–61; Wu Xianya 吳献雅, "Kexue huanxiang yu kexue qimeng: Wan Qing 'kexue xiaoshuo' yanjiu" 科學幻想与科学启蒙: 晚清'科学小说'研究 [Science fiction and the scientific enlightenment: a study of "science fiction" in the Late Qing dynasty], in *Jia Baoyu zuo qianshuiting: Zhongguo zaoqi kehuan yanjiu jingxuan* 賈寶玉坐潛水艇: 中國早期科幻研究精選 [Jia Baoyu by submarine: Selected research on China's early science fiction], ed. Wu Yan 吳岩 (Fuzhou: Fujian shaonian ertong chubanshe, 2006), 37–91; and Zhang Zhi 張治, "Wanqing kexue xiaoshuo chuyi: dui wenxue zuopin ji qi sixiang beijing yu zhishiye de kaocha" 晚清科學小説芻議:對文學作品及其思想背景與知識野的考察 [On science fiction in the Late Qing dynasty: A study of literary works and their ideological background and field of knowledge], *Kexue wenhua pinglun* 科學文化評論 [Scientific and cultural commentary] 6 (2009): 69–96.

[4] Jiang Jing, "From the Technique for Creating Humans to the Art of Reprogramming Hearts: Scientists, Writers, and the Genesis of China's Modern Literary Vision," *Cult. Critique* 80 (2012): 131–49, on 131.

[5] Han Song, "Chinese Science Fiction: A Response to Modernization," in "Chinese Science Fiction," special issue, *Sci. Fict. Studies* 40 (2013): 15–21. Liang translated Jules Verne's *Deux ans de vacances*

and magazines; they appeared under the rubric of "science fiction"—*kexue xiaoshuo* 科學小說—a term that was not yet in general use in the West.[6] These translations focused on particular authors and themes, and especially on technological fantasies. As the SF author and editor Xu Nianci 徐念慈 (1875–1908) remarked, their plots all originated in a scientific ideal of transcending nature and promoting evolution. Readers of this literature rejected indigenous literature grounded in the traditional sciences as "unscientific."[7] Another important element was utopianism, prompted by early translations of Edward Bellamy's *Looking Backward* (1888).[8] Utopianism was central to Liang Qichao's vision of a revitalized Confucian China, and his unfinished novel *Future of New China* has been considered the origin of Chinese SF.[9] Liang's novel drew on theories of evolution and confidence in national rejuvenation, which began to dominate modern Chinese intellectual culture at the beginning of the twentieth century. It influenced several other early twentieth-century utopian works.[10] Common to Liang and these other works was the view that fiction could both civilize and imagine a future for a China defeated by the Opium War and partitioned by Western powers. As Li Boyuan 李伯元 (1867–1906) put it in his founding manifesto for the magazine *Xiuxiang xiaoshuo* 繡像小說 (Illustrated fiction), Western countries used fiction to "civilize their people" through writers who analyze the past, predict the future, and use their insights to awaken the populace.[11]

Over the ensuing years, utopian and science fiction drifted apart. By the 1950s, SF was "science fantasy fiction" (*kexue huanxiang xiaoshuo* 科學幻想小說), itself a sub-

---

[Two years' vacation, 1888] into Chinese, and Lu introduced Verne's *De la terre à la lune* [From the earth to the moon] (1865) to Chinese audiences. Other translations of Verne included *Journey to the Center of the Earth* (1864), *Twenty Thousand Leagues Under the* Sea (1871), *The Mysterious Island* (1875), *and Five Weeks in a Balloon* (1865). By contrast, H. G. Wells's *Outline of History* (1920) was translated into Chinese, but not his science fiction.

[6] This term first appeared in the table of contents of Liang Qichao's literary magazine *Xin Xiaoshuo* 新小說 [New fiction], first published in Japan in 1902. See Issacson, *Celestial Empire* (cit. n. 3), 7–8.

[7] See Wang, *Fin-de-Siècle Splendor* (cit. n. 3), 256. Translation of SF into Chinese declined, but then revived after 1949, with yet more translations of Jules Verne and a substantial number of Russian works. It was only in the early 1980s that a wider range of SF was translated into Chinese, including Ray Bradbury, Arthur C. Clarke, and Isaac Asimov. See Jiang Qian, "Translation and the Development of Science Fiction in Twentieth-Century China," *Sci. Fict. Studies* 40 (2013): 120–1.

[8] Jiang, "Translation" (cit. n. 7), 116–7. Further details appear in Jiang's 2006 PhD dissertation from Fudan University, which is cited in "Translation," but unavailable to this author.

[9] See Song Mingwei, "After 1989: The New Wave of Chinese Science Fiction," *China Perspectives* 1 (2015): 7–13; Liang Qichao, *Xin Zhongguo weilai* 新中國未來 [Future of new China] (unpublished manuscript, 1902; Taipei, Taiwan: Guangya chuban youxian gongsi, 1984, posthumously published).

[10] Wu Jianren 吳趼人, *Xin shitou ji* 新石頭記 [The new story of the stone] (1908; repr., Guangzhou: Huacheng chubanshe, 1987). Also first published in 1908 was Yang Ziyuan 楊致遠, *Xin jiyuan* 新纪元 [New era] (1908; repr., Nanning: Guangxi shifan daxue chubanshe, 2008); this was written under the pseudonym Bi heguan zhuren 碧荷馆主人 [Master of the Sapphire Lotus House]. Wu Jianren lived from 1866 to 1910, and Yang Ziyuan from 1871 to 1919. Lu Shi 陸士諤 (1878–1944) wrote *Xin Zhongguo* 新中國 [New China] (Beijing: Zhongguo youyi chuban gongsi, 2009), which was originally published in 1910. For further details, see Lorenzo Andolfatto, "Paper Worlds: The Chinese Utopian Novel at the Beginning of the Twentieth Century, 1902–1910" (PhD diss., Università Ca' Foscari Venezia [Ca' Foscari University, Venice], 2015); Douwe Fokkema, *Perfect Worlds: Utopian Fiction in China and the West* (Amsterdam, Neth., 2011); Mikael Huss, "Hesitant Journey to the West: SF's Changing Fortunes in Mainland China," *Sci. Fict. Studies* 27 (2000): 92–104; Jiang Jing, "Creating Humans" (cit. n. 4); Jiang, "Translation" (cit. n. 7); Song, "Chinese Science Fiction" (cit. n. 3); Wang, *Fin-de-siècle Splendor* (cit. n. 3); and Wang Dun, "The Late Qing's Other Utopias" (cit. n. 3).

[11] Li Boyuan 李伯元, *Xiuxiang xiaoshuo* 繡像小説 [Illustrated fiction] 1, no. 1, 1903, 1; quoted in Wang Xiaoming, "From Petitions to Fiction: Visions of the Future Propagated in Early Modern China," in *Translation and Creation: Readings of Western Literature in Early Modern China, 1840–1918*, ed. David E. Pollard (Philadelphia, Penn., 1998), 50.

genre of "science belles-lettres" (*kexue wenyi* 科學文藝). Both were distinct from the category of "utopian fiction" (*lixiang xiaoshuo* 理想小說).[12] After the founding of the People's Republic of China in 1949, the agenda of Chinese science fiction was set first by Marxism, and then Maoism. Marxist priorities drew on Soviet theories, according to which science fiction should concentrate on describing two things: (1) the scientific imagination as the source of technoscientific development, and (2) the imagined future of communist society.[13] Government campaigns for "Marching toward Science" (*xiang kexue jinjun* 向科学进军) in the mid-1950s promoted both science fiction and popular science, although between 1949 and 1966 Chinese science fiction focused on short stories aimed at young readers, with few works for adults being published.[14] Since the late 1950s, science fiction has been designated in Chinese by the term *kexue huanxiang xiaoshuo* 科學幻想小說 or science fantasy fiction. It falls under the broader category of *kexue wenyi*, "science belles-lettres," which included all "artistic" science propaganda. *Kexue wenyi* in turn was a subcategory of "science popularization" (*kexue puji* 科學普及). All were linked to the Chinese Association for the Popularization of Science (*Zhongguo Kexue Puji Xiehui* 中國科學普及協會), founded in Shanghai in 1978. This science popularization was aimed at an audience of children and young people. The narratives were linear and action oriented, conspicuously included children, and first appeared in specialized children's magazines and publishing houses. However, the "fantasy" content of this genre was strictly limited to the scientifically plausible, a point that is significant for comparison with the use of indigenous Chinese "tales of the strange" (discussed below).[15] Given its primary interest in science education, this period produced little in the way of either indigenous SF or translations of foreign work. As Wu Dingbo has observed, these productions share several common traits: providing science education through a cast of scientist characters, using patriotism and optimism to resolve conflicts, and settings in a near and—by implication—possible, future.[16]

During the Cultural Revolution (1966–76), SF disappeared from China, reappearing with a vengeance in the 1980s. This was the Chinese "new wave," which rejected both propaganda and utopianism.[17] Song Mingwei identifies the key year as

---

[12] Rudolf G. Wagner, "Lobby Literature: The Archaeology and Present Functions of Science Fiction in the People's Republic of China," in *After Mao: Chinese Literature and Society 1978–1981*, ed. Jeffrey C. Kinkley (Cambridge, Mass., 1985).

[13] See Liuluepu Luofu 利亞普諾夫 [B. Liupulov], *Jishu de zui xin chengjiu yu Sulian kexue huan xiang du wu* 技术的最新成就與蘇聯科學幻想讀物 [The latest technological achievements and Soviet science fiction], trans. Yu Shixiong 余士雄 (Beijing: Kexue jishu chubanshe, 1959), as summarized in Wu Yan, "'Great Wall Planet': Introducing Chinese Science Fiction," trans. Wang Pengfei and Ryan Nichols, *Sci. Fict. Studies* 40 (2013): 1–14.

[14] For accounts of this situation, see Wu Yan, "'Great Wall Planet'" (cit. n. 13); and Rui Kunze [Wang Rui 王瑞], "Displaced Fantasy: Pulp Science Fiction in the Early Reform Era of the People's Republic Of China," *East Asian History* 41 (2017): 25–40.

[15] Wagner, "Lobby Literature" (cit. n. 12), 19–23.

[16] Wu Dingbo and Patrick D. Murphy, *Science Fiction from China* (New York, N.Y., 1989), xxxvi.

[17] This period also saw the publication of Rao Zhonghua's 饒忠華 *Zhongguo Kexue Xiaoshuo Daquan* 中國科幻小說大全 [Compendium of Chinese science fiction] (Beijing: Haiyang chubanshe, 1982), which subsequently became a standard sourcebook for the subject. Examples of this new wave include Tong Enzheng 童恩正, "Shanhu Dao Shang de Siguang" 珊瑚岛上的死光 [Death ray on a coral island], *Renmin wenxue* 人民文学 (August 1978): 41–58; Zheng Wenguang 鄭文光, "Feixiang Renmazuo" 飞向人马座 [Flying to Sagittarius], Shijie kehuan bolan 世界科幻博览 [World scifi expo], May 2005, originally published 1979; Wang Xiaoda 王晓达, "Bing xia de meng" 冰下的梦 [Dream under the ice], in *Wang Xiaoda juan: Bing xia de meng* 王晓达卷: 冰下的梦 [Collected short stories of Wang Xiaoda: Dream under the ice], eds. Dong Renwei 董仁威 and Yao Haijun 姚海军 (Beijing: Shijie huaren kehuan xie huizu bian 世界华人科幻协会组编 [World Chinese Science Fiction Associ-

1989—the year of the Tiananmen massacre and the collapse of the democracy movement. Several of these writers pursue "hard science" themes deployed in socially complex and nuanced settings. The three most prominent of these writers are Liu Cixin 劉慈欣, Wang Jinkang 王晋康, and Han Song 韩松. Liu Cixin's *Three-Body Problem* (Santi 三体) trilogy imagines a disastrous scenario of the consequences of reckless alien contact, beginning during the Cultural Revolution and ending in a distant future of inter-civilizational and interdimensional warfare.[18] Wang Jinkang also focuses on science, but in the context of ethics.[19] Han Song addresses problems of society and culture, and the implications of science for society.[20] China and Chinese nationalism remain prominent themes, with China literally saving the world in several plot lines.[21] But younger new wave Chinese science fiction also offers an indirect critique of government policies (in sometimes dystopian visions), the human implications of technology, and issues of censorship and government control.[22]

Other authors are more experimental, and draw on history and legend, sometimes combined with time travel. For example, Zhao Haihong 趙海虹, one of the few women authors of Chinese SF, also has a longstanding interest in Chinese short story genres and martial arts fiction.[23] Fei Dao 飛氘 imagines Confucius returning to Mount Tai

---

ation], 2012); Wei Yahua 魏雅华, "Wenrou zhixiang de meng" 温柔之乡的梦 [Conjugal happiness in the arms of Morpheus], originally published in Chinese in 1982, translated into English in *Science Fiction from China*, ed. Patrick D. Murphy and Wu Dingbo (New York, N.Y., 1989), 9–52; and Jiang Yunsheng 姜云生, "Wubian de juanlian" 无边的眷恋 [Boundless love], originally published in Chinese in 1987, translated into English in Murphy and Wu, *Science Fiction from China*, 157–164. For details on these writers' publications, please see the appendix, "A Brief Bibliography of Contemporary Chinese Science Fiction." For surveys of "new wave" SF, see Han, "Chinese Science Fiction" (cit. n. 5); Isaacson, *Celestial Empire* (cit. n. 3); Leung Laifong, *Contemporary Chinese Fiction Writers: Biography, Bibliography, and Critical Assessment* (London, 2017); and Song, "After 1989" (cit. n. 9). The Francophone website "SinoSF Carnet de recherche sur la littérature de science-fiction chinoise" (OpenEdition), 16 January 2019, sinosf.hypotheses.org, provides a useful inventory of translations of Chinese SF into English, French, German, and Italian, to which the present discussion is indebted.

[18] See Liu Cixin 劉慈欣, *Santi* 三体 (Chongqing, 2007); translated by Ken Liu, *The Three-Body Problem* (New York, N.Y., 2016). The next two volumes of this trilogy are *Heian sanlin: Santi di er bu* 黑暗森林: 三體第二部 (Chongqing, 2008); and *Santi. Di III bu, Si shen yong sheng* 三體第三部, 死神永生 (Taipei, 2011); Ken Liu also translated these two books of the series as *The Dark Forest* (New York, N.Y., 2015) and *Death's End* (New York, N.Y., 2016). For an excellent review of the trilogy, see Nick Richardson, "Even What Doesn't Happen Is Epic," *London Review of Books* 40, 8 February 2018, 34–6.

[19] This point is indebted to Ken Liu, "China Dreams: Contemporary Chinese Science Fiction," *Clarkesworld* 88, December 2014, http://clarkesworldmagazine.com/liu_12_14/.

[20] Han Song 韩松, *Yuzhou mubei* 宇宙墓碑 [Cosmic tombstones] (Beijing, 1991); Han, *Renzao ren* 人造人 [Artificial people] (Beijing, 1997); Han, *Wo de zuguo bu zuomeng* 我的祖国不做梦 [My homeland never dreams] (unpublished manuscript, 2007). While the last never appeared in print in China, it was published online in *Impressions d'Extrême-orient*, June 2016, "Censures et littératures d'Asie" [Censorship and the literatures of Asia], https://journals.openedition.org/ideo/471?file=1.

[21] Wang Jinkang 王晋康, *Yu Wu Tong Zai* 與吾同在 [Being with me] (Chongqing, 2011); He Xi 何夕, "Yiyu" 異域 [Foreign land], *Kehuan Shijie* 科幻世界 (Science fiction world), August 1999; *Liudao Zhongsheng* 六道眾生 [Six lines from Samasara] (2002; repr., Changsha, 2012).

[22] Ma Boyong 马伯庸, "The City of Silence" [Jijing zhi cheng 寂静之城], in *Invisible Planets: 13 Visions of the Future from China*, trans. Ken Liu (London, 2016), 153–96; Chen Qiufan 陈楸帆 [Stanley Chan], "Shu nian" 鼠年 [The year of the rat] (location unknown, 2009), translated by Ken Liu in *Magazine of Fantasy & Science Fiction*, July–August 2013, 68–92, and reprinted in Ken Liu, *Invisible Planets*, 21–49; Chen, "The Flower of Shazui" [Shazui zhi hua 沙嘴之花], *Interzone*, trans. Ken Liu, November–December 2012, 20–9, also published in Ken Liu, *Invisible Planets*, 69–87; and Chen, *The Waste Tide* (Huang chao 荒潮), trans. Ken Liu (London, 2019).

[23] Zhao Haihong 赵海虹, *Shijian de bifang* 时间的彼方 [The other side of time] (1998; repr., Wuhan, 2006); and Zhao, "Yi'ekasida" 伊俄卡斯达 [Jocasta], *Kehuan Shijie* 科幻世界 [Science fiction world], March 1999.

to understand the history of Chinese civilization.[24] Xia Jia 夏笳, another female SF writer, draws on themes from legend and religion in novels such as *The Demon-Enslaving Flask* (2004), *Carmen* (2005), and *Dream of Eternal Summer* (2008); the last is a love story between an immortal and a time traveler.[25] Zhao Haihong, Fei Dao, and Xia Jia have academic backgrounds, which link them both to the writing of SF and to its reception through both teaching and critical studies.[26] While these authors do not explicitly draw on the traditional Chinese sciences, they come closer to it. At the same time, in a parallel SF universe in Hong Kong, Ni Cong 倪聰 (1935–), writing under the name Ni Kuang 倪匡, published the Wisely (or Wesley, *Wei si li* 衛斯理) series of some one hundred and fifty novels between 1963 and 2004. They included both SF and martial arts elements, and Ni was a close friend of the martial arts writer Louis Cha (discussed below).[27]

In summary, early twentieth-century SF in China, both indigenous and in Western translations, was focused on themes of evolution and technology, with specific interest in helping China gain scientific and technological expertise in the wake of its defeats in the Opium War. While it included some of the recognized "classics" of Western SF, others were conspicuously absent, including Mary Shelley's *Frankenstein* (*Kexue guairen* 科學怪人, literally "science madman"), which would have been entirely inconsistent with the priorities behind early Chinese interest in SF. The Maoist period was equally preoccupied with (Western) science, but with a marked shift from interest in the power of fiction to promote social change, toward a narrower view of SF as a tool of science education. Liang Qichao's call for a science-fictional literature of national renewal had all but disappeared, leaving only the science. Since 1989, SF has flowered in China, free from earlier constraints. In some cases—notably Liu Cixin's *Three-Body Problem*—it has retained an orientation toward science. But in all these periods, "science" is unquestionably understood as modern Western science. What happens if we look again at themes from Chinese philosophy and from the indigenous Chinese sciences?

SF AND THE TRADITIONAL CHINESE SCIENCES

While a defining feature of the late Qing emergence of Chinese SF was its close engagement with modern science, as Fan Fa-ti has pointed out, historical actors—including early twentieth-century Chinese scientists, and readers and writers of early Chinese SF—wrestled with binary concepts such as traditional/modern and Chinese/Western; these binaries and categories informed their practices as producers and as

---

[24] Fei Dao 飛氘, "Yilan zhong shan xiao" 一览众山小 [A list of small mountains], *Kehuan shijie* 科幻世界 [Science fiction world], August 2009.

[25] Xia Jia, "Yongxia zhi meng" 永夏之夢 [Dream of eternal summer], *Kehuan Shijie* [Science fiction world], September 2008.

[26] Zhao Haihong is a professor at the Institution of Foreign Languages of Zhejiang Gongshang University. Jia Liyuan (Fei Dao) holds a PhD in comparative literature from Tsinghua University. Wang Yao (Xia Jia) holds a PhD in comparative literature and world literature from the Department of Chinese, Peking University (2014), and is currently a lecturer of Chinese literature at Xi'an Jiaotong University.

[27] I am grateful to Bill Mak for calling this series to my attention. Several of the books have been adapted for film and television, including the films *The Legend of Wisely* (1987), *The Cat* (1992), and *The Wesley's Mysterious File* (2002); and the television series *The New Adventures of Wisely* (1998) and *The "W" Files* (2004).

readers of both science and SF.[28] Given these orientations, it is no surprise that the authors and audiences of late Qing SF described other veins of fantastic indigenous literature as "mythology" or "superstition." But we are not obliged to base a contemporary history of Chinese SF solely on these assessments.

Debates about the history of science in China, including the question of where "science" stood in indigenous hierarchies of knowledge, began with the pioneering work of Joseph Needham (1900–95).[29] Needham approached the problem of the history of science in China by trying to fit the Chinese scientific tradition into the categories of twentieth-century Western science. Many later historians of science in China, including several of Needham's own close collaborators, later rejected this "universalist" approach as anachronistic and culturally inappropriate.[30] Part of the problem, as Nathan Sivin argues, is that indigenous Chinese accounts focused on specific sciences—quantitative and qualitative—rather than on any unified notion of "science." In his taxonomy of the traditional Chinese sciences, Sivin distinguishes three quantitative and three qualitative sciences.[31] Most important for Chinese SF is the qualitative science of medicine, which included "nurturing life" (*yangsheng* 養生), longevity, and alchemy (discussed below). These three themes link directly to immortality themes in SF, but—crucially—they also link to other Chinese literary genres. Philosophical texts include references to interspecies transformation and space/time travel.

The historical development of the Chinese sciences has been extensively explored by historians of science. Some of these studies do not inform SF but are important to mention in order to demonstrate the variety and development of the Chinese sciences.

[28] Fan Fa-ti, "Redrawing the Map: Science in Twentieth-Century China," *Isis* 98 (2007): 524–38, on 526. For other important studies of Western science in early twentieth-century China and how these problems were understood and addressed, see Benjamin Elman, *On Their Own Terms: Science in China, 1550–1900* (Cambridge Mass., 2005); Fan Fa-ti, "Science, State, and Citizens: Notes from Another Shore," *Osiris* 27 (2012): 227–49; Grace Shen, "Murky Waters: Thoughts on Desire, Utility, and the 'Sea of Modern Science,'" *Isis* 98 (2007): 584–96; Wang Zuoyue, "Saving China through Science: The Science Society of China, Scientific Nationalism, and Civil Society in Republican China," *Osiris* 17 (2002): 291–322; and Wang Zuoyue, "Science and the State in Modern China," *Isis* 98 (2007): 558–70.

[29] Much of Needham's early work focused on questions of why the Scientific Revolution in Europe did not also take place in China. See Joseph Needham and Wang Ling, *Science and Civilization in China*, vol. 1, *Introductory Orientations* (Cambridge, UK, 1954); Needham and Wang, *Science and Civilisation in China*, vol. 2, *History of Scientific Thought* (Cambridge, UK, 1956); Needham and Wang, *Science and Civilisation in China*, vol. 3, *Mathematics and the Sciences of the Heavens and Earth* (Cambridge, UK, 1959); Needham, *The Grand Titration: Science and Society in East and West* (Boston, Mass., 1979). Needham's monumental encyclopedic series *Science and Civilisation in China* (1954–) is an ongoing project of the Needham Research Institute, currently under the direction of the archeometallurgist Mei Jianjun 梅建军.

[30] Nathan Sivin, "Why the Scientific Revolution Did Not Take Place in China—Or Didn't It?" *Chinese Science* 5 (1982): 45–66; Robin Yates, "Science and Technology," in *Encyclopedia of Chinese Philosophy*, ed. A. S. Cua (New York, N.Y., 2003), 657–63.

[31] Sivin classes the three quantitative sciences as mathematics (*suan* 算), mathematical harmonics (*lülü* 律吕), and mathematical astronomy (*lifa* 歷法). The three qualitative sciences were astronomy (*tianwen* 天文), medicine (*yi* 醫), and siting (*fengshui* 風水). *Tianwen* included celestial and meteorological observation and astrology. Medicine also included materia medica (*bencao* 本草), and internal (*neidan* 内丹) and external (*waidan* 外丹) alchemy. For further discussion see Nathan Sivin, "Science and Medicine in Imperial China—The State of the Field," *J. Asian Stud.* 47 (1988): 41–90; Sivin, "Science and Medicine in Chinese History," in *Heritage of China: Contemporary Perspectives on Chinese Civilization*, ed. Paul S. Ropp (Berkeley, Calif., 1990), 164–96; Sivin, "State Cosmos and Body in the Last Three Centuries B.C.E.," *Harvard J. Asia. Stud.* 1 (1995): 5–37; and Lisa Raphals, "Science and Chinese Philosophy," in *The Stanford Encyclopedia of Philosophy Archive*, ed. Edward N. Zalta, spring 2017 (entry first published April 28, 2015), https://plato.stanford.edu/archives/spr2017/entries/chinese-phil-science/.

For example, in mathematics, they include studies of early Chinese notions of demonstration and proof (Karine Chemla's work is an example), as well as the translation and study of complex Chinese mathematical works, including texts excavated from tombs.[32] In astronomy, they include studies of archeoastronomy and the unparalleled record of Chinese astronomical records and observations.[33] And in medicine, they include studies of nurturing life and longevity practices; Chinese botany and materia medica; the history of medicine; and translation of key texts, both classical and contemporary.[34] The rich history of indigenous Chinese sciences puts us in a position to ask a different set of questions, and to imagine an alternative past. What would Chinese SF have looked like if writers had taken the indigenous sciences of their own tradition seriously?

TRAVEL IN SPACE AND TIME

Space and time travel are classic motifs of Western SF. They first appear in China in the great Daoist fourth-century (BCE) classic, the *Zhuangzi*. One passage describes a "spirit person" who appears to practice diet and breath regulation—"He does not eat the five grains but sucks in the wind and drinks dew." These practices are powerful: "He roams beyond the world" (literally, "the four seas"), and when he concentrates his spirit, it protects living things from plagues and makes the grain ripen.[35] Time travel also features, and is expressed as a paradox:

> For there to be a "right" and "wrong" [or alternately, "it's so; it's not so"] before they are formed in the heart-mind would be to go to Yüe today and arrive yesterday.[36]

This time paradox is introduced as an analogy to a category of error, rather than as a description of actual time travel. The same paradox reappears in a list of paradoxes attributed to Zhuangzi's friend (and logical sparring partner) Hui Shi: "I go to Yue

---

[32] Andrea Bréard, "How to Quantify the Value of Domino Combinations: Divination and Shifting Rationalities in Late Imperial China," in *Coping with the Future: Theories and Practices of Divination in East Asia*, ed. M. Lackner and E.-M. Guggenmoos (Leiden, Neth., 2018), 499–529; Karine Chemla, ed., *The History of Mathematical Proof in Ancient Traditions* (Cambridge, UK, 2012); Chemla and Guo Shushun, *Les neuf chapitres: Le classique mathématique de la Chine ancienne et ses commentaires* (Paris, 2004); Christopher Cullen, *Astronomy and Mathematics in Ancient China: The Zhou bi suan jing* (Cambridge, UK, 1996); Joseph Dauben, "Suan shu shu. A book on Numbers and Computations," translated from the Chinese and with commentary by Joseph W. Dauben, *Arch. Hist. Exact Sci.* 62 (2008): 91–178.

[33] See Needham and Wang, *Science and Civilisation*, vol. 3 (cit. n. 29); and David Pankenier, *Astrology and Cosmology in Early China: Conforming Earth to Heaven* (Cambridge, UK, 2013)

[34] For longevity and nurturing life, see Donald Harper, *Early Chinese Medical Literature* (London, 1998); Vivienne Lo and Christopher Cullen, *Medieval Chinese Medicine: The Dunhuang Medical Manuscripts* (London, 2005). For botany and materia medica see Georges Métaillié, *Science and Civilisation in China*, vol. 6, *Biology and Biological Technology*, pt. 4, *Traditional Botany: An Ethnobotanical Approach* (Cambridge, UK, 2015); Joseph Needham, Ho Ping-Yu, and Lu Gwei-djen, *Science and Civilisation in China*, vol. 5, *Chemistry and Chemical Technology*, pt. 3, *Spagyrical Discovery and Invention: Historical Survey, from Cinnabar Elixirs to Synthetic Insulin* (Cambridge, UK, 1976); and Paul U. Unschuld, *Medicine in China: A History of Pharmaceutics* (Berkeley, Calif., 1986).

[35] *Zhuangzi jishi* 莊子集釋 [Collected explanations of the Zhuangzi] (Beijing, 1961), 1:28; compare A. C. Graham, *Chuang tzu: The Inner Chapters* (London, 1981), 46. Translations from the *Zhuangzi* are my own, but are indebted to Graham's translation.

[36] 未成乎心而有是非，是今日適越而昔至也。是以無有為有，*Zhuangzi jishi* 2:56; compare Graham, *Inner Chapters*, 51 (both cit. n. 35).

today yet arrive yesterday."[37] Here the travel to Yue (contemporary Vietnam) is posed as a paradox, and the idea of time travel is clearly articulated. Time travel plots feature in some new wave Chinese SF, although very infrequently in comparison to the West, where time travel is an early and prominent theme.[38] The genre in which time travel comes into its own for China is, in fact, the historical television drama. The first show of this genre, Shen Hua 神話 (*The Myth*, 2010), was based on the 2005 Jackie Chan movie of the same name.[39] Time travel plots became so popular that the State Administration of Radio, Film, and Television eventually banned them on the grounds that they "treated [*sic*] serious history in a frivolous way."[40] But as *New Yorker* columnist Richard Brody pointed out, Chinese time-travel plots offer escape—escape from contemporary China to the China of an earlier and perhaps preferable time.[41]

TRANSFORMATION STORIES

If time travel can be found in modern historical Chinese drama, the idea of one species or sex transforming into another also emerges in an unexpected genre. This motif is found in several different early stories, which range from accounts of human origins, the genre of *zhiguai*, and Buddhist reincarnation stories. Early twentieth-century Chinese interest in theories of evolution also prompted a debate on "spontaneous generation." Hu Shih 胡適 (1891–1962), one of the great scholar-diplomats of twentieth-century China, and a student of John Dewey at Columbia University, identified a narrative from the *Zhuangzi* as an example of a possible early theory of biological evolution. In this elaborate account of how species transform into one another under the influences of different environments, organisms start from "minute beginnings," and when they reach the boundary of water and land they become algae. They transform again when they germinate in elevated places, and when they reach fertilized soil they develop into plants. Here more transformations occur; roots become grubs, and their leaves become butterflies, which in turn transform into insects, which transform into birds. A further series of transformations between plants and insects leads to the green *ning* 寧 plant. It produces panthers, panthers produce horses, horses produce humans, and humans return to minute beginnings.[42] Hu argued that theories of *qi* introduce issues of potentiality and actuality; if all organisms arise from some kind of elemental and generative *qi*, it must contain the potential for all later forms, providing the conceptual groundwork for a theory of evolution. Hu used this *Zhuangzi* passage to argue

[37] 今日適越而昔來, *Zhuangzi jishi* 33:1102; compare Graham, *Inner Chapters*, 283 (both cit. n. 35).

[38] Qian Lifang 钱莉芳, *Tian yi* 天意 [The will of heaven] (Chengdu, 2004); and Qian, *Tian ming* 天命 [The command of heaven] (Changchun, 2011). I am grateful to my PhD student Fan Yilun for calling these works to my attention.

[39] *The Myth* was produced in Hong Kong in 2005, directed by Stanley Tong, and starred Jackie Chan, Tony Leung Ka-fai, Kim Hee-sun, and Mallika Sherawat (Culver City, Calif., Sony Pictures Home Entertainment, 2007), DVD.

[40] See Robert G. Price, *Space to Create in Chinese Science Fiction* (Kaarst, Ger., 2017), 70.

[41] Richard Brody, "China Bans Time Travel," *New Yorker*, 8 April 2011, http://www.newyorker.com/culture/richard-brody/china-bans-time-travel.

[42] *Zhuangzi jishi* 18:624–5; compare Graham, *Inner chapters*, 21–2 (both cit. n. 35); and Hu Shih, "The Development of the Logical Method in Ancient China" (PhD diss., Columbia University, 1917), 135–6, also published in Shanghai, 1922.

that Warring States (475–221 BCE) thinkers recognized organic continuity throughout the gradations of the animate world.[43]

Other types of transformation stories appear in the indigenous genre, *zhiguai* 志怪, "tales of the strange" or "records of anomalies." These accounts became prominent toward the end of the Han, Six Dynasties, and Tang periods (200–900 CE), and are classified by Chinese sources as history, not fiction.[44] They extend the boundaries of the human by portraying both humans and other animals as part of a continuous moral community, depicting transformation between humans, animals, plants, and spirits, in which animals in particular are capable of reward and retribution. They include narratives of reincarnation and human interactions with gods, ghosts, and spirits; but many are focused on crossing boundaries of both species and sex.[45] Some describe partial transformations, such as an animal growing extra or inappropriate body parts, while others describe transformations between species. Still others recount cross-species matings and anomalous births, such as one species giving birth to another, or babies born with multiple heads or feet. Other transformations involve gender.[46] Especially interesting are the highly normative accounts of reward and retribution between humans and animals, where animals behave according to the standards of human morality, sometimes better than humans do.

Again, these stories do not relate easily to Western categories of SF. *Zhiguai* was one of several genres that dealt with the supernatural in late imperial China. The Ming dynasty bibliographer Hu Yinglin 胡應麟 (1551–1602) rethought the traditional genre of "fiction," literally "small talk" (*xiaoshuo* 小說), which he considered too vague and reworked into six distinct genres. The first two deal with the supernatural—anomaly accounts (*zhiguai*) and "prose romances of the marvellous" (*chuanqi* 傳奇), a genre that dates from the Tang dynasty.[47] Several influential figures link martial arts fiction (*wuxia*

---

[43] Hu, "Logical Method" (cit. n. 42), 121–2; compare Hu Shi 胡適, "Xian Qin zhuzi jinhualun" 先秦諸子進化論 [Theories of evolution in the philosophers before the Qin period], *Kao xue* 考學 [Study of antiquity] 3, no. 1 (1917): 19–41. This point is explicitly echoed in Joseph Needham and Donald Leslie, "Ancient and Mediaeval Chinese Thought on Evolution," *Bulletin of the National Institute of Science of India* 7 (1955): 1–18, which quotes Hu. For discussion of debates on spontaneous generation in early twentieth-century China, see Fan Fa-ti, "The Controversy over Spontaneous Generation in Republican China: Science, Culture, and Politics," in *Routes of Culture and Science in Modern China*, ed. Benjamin Elman and Jing Tsu (Leiden, Neth., 2014), 209–44.

[44] For introductions to *zhiguai* and anomaly literature, see Robert Ford Campany, *Strange Writing: Anomaly Accounts in Early Medieval China* (Albany, N.Y., 1996); and Hu Ying, "Records of Anomalies," in *The Columbia History of Chinese Literature*, ed. Victor H. Mair (New York, N.Y., 2001), 542–54. For English translations, see Karl S. Y. Kao, ed., *Classical Chinese Tales of the Supernatural and the Fantastic: Selections from the Third to the Tenth Century* (Bloomington, Ind., 1985). For comparison with science fiction, see Fontaine Lien "The Intrusion Story and Lessons from the Fantastic: A Cross-Cultural Study" (PhD diss., Univ. of California, Riverside, 2014); Liu Mingming, "Theory of the Strange towards the Establishment of Zhiguai as a Genre" (PhD diss., Univ. of California, Riverside, 2015); and Lisa Raphals, "The Limits of 'Humanity' in Comparative Perspective: *Cordwainer Smith* and the Soushenji," in *World Weavers: Globalization, Science Fiction, and the Cybernetic Revolution*, ed. Wong Kin Yuen, Gary Westfahl, and Amy Kit-sze Chan (Hong Kong, 2005), 143–56.

[45] Campany, *Strange Writing* (cit. n. 44), 52–79.

[46] The most important text that includes gender transformation accounts is the fourth- century *Soushen ji* 搜神記 [Records of an inquest into the spirit realm] by Gan Bao 干寶 (335–49 CE), in *Congshu jicheng chubian* 叢書集成初編 [Complete collection of books from various collectanea], vols. 2692–2694, ed. Wang Yunwu 王雲五 (Shanghai: Shangwu yinshuguan, 1935–40). For details of these story types, see Raphals, "The Limits" (cit. n. 44).

[47] The other four genres are anecdotes (*za lü* 雜錄), miscellaneous notes (*cong tan* 叢談), researches (*bian ding* 辯訂), and moral admonitions (*zhen gui* 箴規). See Lu Xun 魯迅 (Zhou Shuren 周樹人), *Zhongguo xiaoshuo shilüe* 中國小說史略 [A brief history of Chinese fiction] (1930; repr., Shanghai 2001), 4; and Lu Xun, *A Brief History of Chinese Fiction*, trans. Gladys Yang and Yang Xianyi (Beijing, 1959), 5.

xiaoshuo 武俠小說) with *chuanqi*; these include Jin Yong 金庸 (Louis Cha), author of many of the most influential contemporary martial arts novels, which form the basis of many martial arts films.[48] What is important for the present discussion is the strong distinction between anomaly accounts and prose romances. This distinction occurs in several ways. First, contemporary martial arts fiction, such as the writings of Jin Yong, privileges realism, and has tended to eclipse nonrealist narratives derived from the traditional genres of *zhiguai* and *chuanqi*. Second, contemporary martial arts fiction also uses vernacular language (*baihua* 白花), which further distances it from the literary language of the older genres. Li Tuo suggests that these two phenomena are related. He argues that nonrealist genres dominated Chinese premodern literary history in *zhiguai*, *chuanqi*, Yuan drama, and Ming novels, and asks whether the Europeanization of the modern Chinese language has inexorably connected modern Chinese narratives and Western models of representation.[49] The point for purposes of this discussion is that modern Chinese narratives conspicuously include science fiction, and Chinese SF inherently includes the nonrealism of *zhiguai* and *chuanqi*. We can nuance the distinction further by arguing that realist martial arts fiction draws on *chuanqi*, and SF draws on *zhiguai*.

In contrast to the theme of space and time travel, both late Qing dynasty and contemporary new wave science fiction draw on *zhiguai*. Human-animal hybrids appear in the cat-headed citizens of the novel by Lao She, *Mao cheng ji* 貓城記 (Cat country).[50] Another important SF element that first appears in *zhiguai* are some of the important seeds of utopian literature in China. Several *zhiguai* stories involve nonexistent utopias or dystopias. For example, one story in *Liaozhai zhiyi* 聊齋誌異 (Strange tales from a Chinese studio), by Pu Songling 蒲松齡 (1640–1715), opposes a dystopian "City of Ogres" to an idealized "City under the Ocean."[51] Another example is the novel by Wu Jianren 吳趼人 (1866–1910), *Xin shitou ji* 新石頭記, 1908 (New story of the stone), which is one of the great novels of late Qing China—a utopian account of the continuing travels of Jia Baoyu, the main character of the original *Shitou ji* 石頭記 (Story of the stone), after the end of the original novel.[52] Baoyu encounters a "Civilized

---

[48] Cao Zhengwen 曹正文, *Xia ke xing: Zong tan Zhongguo wuxia* 俠客行: 縱談中國武俠 [Chivalry: On the Chinese knight-errant] (Taipei, 1998), 45–51; Fei Yong 费勇 et al., *Jin Yong chuanqi* 金庸传奇 [*Chuanqi* in Jin Yong's stories] (Guangzhou, 1995). Jin Yong's martial arts novel, *Xia ke xing* 俠客行 [Ode to gallantry], ed. Zhou Yushun 周育顺, 4 vols. (Tianjin, 2010), includes a group of traditional woodcuts called *Sasan jian ke tu* 卅三劍客圖 [Portraits of thirty-three swordsmen]. The portraits include four heroes from Tang *chuanqi*: no. 2, *Qiuran Ke Zhuan* 虬髯客傳) [Romance of the curly-bearded stranger]; no. 9, *Nie Yinniang* 聶隱娘; no. 11, *Hong Xian* 紅線 [The red-thread maid]; and no. 13, *Qunlun Nü* 崑崙奴 [The Qunlun maidservant].

[49] Li Tuo, "The Language of Jin Yong's Writing: A New Direction in the Development of Modern Chinese," in *The Jin Yong Phenomenon: Chinese Martial Arts Fiction and Modern Chinese Literary History*, ed. Ann Huss and Jianmei Liu (Youngstown, N.Y., 2007), 39–54, esp. 42 and 45; see also Huss and Liu, "Introduction," in *Jin Yong Phenomenon*, 1–22, on 15.

[50] See Lao She 老舍 (Shu Qingchun 舒慶春), *Maocheng ji* 貓城記 [The city of cats] (1932; repr., Hong Kong: Zhongsheng shuju, 1963), translated by William A. Lyell as *Cat Country* (Columbus, Ohio, 1970). For a discussion of this book, see Lisa Raphals, "Alterity and Alien Contact in Lao She's Martian Dystopia, Cat Country," in "Chinese Science Fiction," special issue, *Sci. Fict. Studies* 40 (2013): 73–85. Anomaly accounts also inform martial arts fiction, which has also retained ongoing popularity in both text and film.

[51] Pu Songling, "City of Ogres and the City under the Ocean" [Luosha hai shi 羅刹海市], in *Strange Tales from a Chinese Studio* [*Liaozhai zhiyi* 聊齋誌異], trans. Herbert Allan Giles, 2 vols. (London, 1880), 2:1–16. Giles translates the story title as "The Lo-cha Country and the Sea Market."

[52] Also known as the *Dream of the Red Chamber* [Honglou meng 紅樓夢], the *Story of the Stone*, by Cao Xueqin 曹雪芹 (1791), is considered one of the four classic novels of China.

Realm" (*wenming jingjie* 文明境界) with futuristic technology, including medical lenses that image bone and soft tissue, underground trains, underwater wireless telephones, and submarines that fire "silent electric cannons" (*wusheng dianpao* 無聲電炮). Yet its vision of moral governance is Confucian—its ruler is the benevolent monarch Dongfang Qiang 東方強 (Strength of the East). This realm's districts are named for traditional Confucian virtues, including compassion (*ci* 慈), filiality (*xiao* 孝), loyalty (*zhong* 忠), benevolence (*ren* 仁), and trustworthiness (*xin* 信).[53]

MODES OF IMMORTALITY

Extreme longevity and immortality are staples of Western SF, and unsurprisingly, the *Encyclopedia of Science Fiction* describes immortality as one of the basic motifs of speculative thought.[54] Chinese medical and scientific traditions have been concerned with a spectrum ranging from health to longevity, and including—in some cases— attempts at literal, physical immortality, for some two millennia. Two particular areas are important here. One is nurturing life (*yang sheng* 養生), an interest of common ground for philosophers and practitioners of medical arts, which encompassed a range of practices for preserving one's person, self, or essential nature. The other are accounts of Daoist adepts who were also physicians.

The term *yang sheng* first appears in the *Zhuangzi*, which makes fun of a tradition of exercise for therapy and health known as *daoyin* 導引 (pulling and guiding). The *Zhuangzi* contrasts real sages to *yang sheng* practitioners who "blow out, breathe in, old out, new in." Caring only for longevity, these are the practitioners of "guide-and-pull" (*dao yin*) and "nourish the body" (*yang xing* 養形).[55] In the Han dynasty, "nurturing life" techniques became a major concern of the so-called Recipe Masters (*fang shi* 方士) of the Han court. This group included physicians, compounders of elixirs of immortality, and magicians whose spells and recipes involved curses, love charms, and poisons. *Fang* texts on nurturing life include methods for absorbing and circulating *qi* in the body, for example, and breathing and meditation exercises, diet, drugs, and sexual techniques. Medical texts excavated from Han dynasty tombs also document these practices. A corpus of medical manuscripts excavated from Mawangdui 馬王堆 (Changsha, Hubei, 169 BCE) includes six medical manuscripts specifically concerned with nurturing life.[56] One, "Drawings of guiding and pulling" (*Daoyin tu* 導引圖), is a series of forty-four drawings of human figures performing exercises. Some are described in another excavated text, the "Pulling book" (*Yin shu* 引書), ex-

---

[53] See Wang Dun, "The Late Qing's Other Utopias" (cit. n. 3), 38. *The New Story of the Stone* was initially serialized under the pen name "Lao Shaonian" 老少年 [Old youth] in the newspaper *Nanfang Bao* 南方報 in 1905 (July 21 to November 29) as *shehui xiaoshuo* 社會小說 [Social fiction]. It was published as an illustrated book in 1908 by Shanghai Reform Fiction Press [Shanghai gailiang xiaoshuo she 上海改良小說社] under the pen name Wofo Shanren 我佛山人, and labeled as *lixiang xiaoshuo* 理想小說 [fiction of ideals].

[54] Brian M. Stableford and David Langford, "Immortality," in *The Encyclopedia of Science Fiction*, ed. John Clute et al., 3d ed. available online, Gollancz, 19 January 2019, http://www.sf-encyclopedia .com/entry/immortality. For the immortality theme in SF, see George Edgar Slusser, Eric S. Rabkin, and Gary Westfahl, *Immortal Engines: Life Extension and Immortality in Science Fiction and Fantasy* (Athens, Ga., 1996).

[55] *Zhuangzi jishi* (cit. n. 35), 15:535.

[56] For a translation of these medical texts, see Harper, *Early Chinese* (cit. n. 34).

cavated from a tomb at Zhangjiashan 張家山 (Jiangling, Hubei, tomb no. 247), which describes exercises modeled on the movements of animals. The Mawangdui medical texts also include a "recipe" (*fang* 方) text; this is "Wushier bing fang" 五十二病方 (Recipes for fifty-two ailments). Other recipe texts have been excavated from tombs at Zhangjiashan and Wuwei 武威 (Gansu, first century CE). The *"yang sheng* culture" of these texts emphasized control over physiological and mental processes, both understood as self-cultivation, through the transformation of *qi.*[57]

To summarize, most of these texts can be described as part of a *yang sheng* culture, which offered and emphasized control over physiological processes of the body and mind that were understood as transformations of *qi*. These technical arts form a continuum with philosophy because their transformations were understood as self-cultivation in the coterminous senses of moral excellence, health, and longevity (rather than medical pathology), and physiological transformation through the manipulation of *qi*. Such views informed Warring States accounts of dietary practices, exercise regimens, breath meditation, sexual cultivation techniques, and other technical traditions associated with *fang shi*.

Longevity practices were closely linked with traditional Chinese medicine. Three important early physicians were also Daoist scholars who were concerned with longevity practices. Ge Hong 葛洪 (283–343 or 363 CE) was the first of several explicitly Daoist physicians to write about the practice of alchemy. He initially studied the Confucian classics but became interested in immortality techniques. Ge Hong gave up a military career and political life in order to devote himself to immortality practices, and eventually settled at Mt. Luofu 羅浮 in Guangxi where he studied alchemy until his death.[58] He was the first to systematically describe the history and theory of Daoist immortality techniques such as "preserving unity" (*shou yi* 守一), circulating energy (*xing qi* 行氣), "guiding and pulling" (*dao yin*), and sexual longevity techniques (*fang zhong* 房中).[59] As an alchemist, he experimented with drugs and minerals.[60] Tao Hongjing 陶弘景 (456–536), the effective founder of Shangqing 上清 (Highest Clarity) Daoism, held several court positions under the Liu Song and Qi dy-

---

[57] There is also evidence for *yang sheng* in lists of lost book titles. The *Hanshu* Bibliographic Treatise includes titles of lost medical works on nurturing life, health, and longevity, such as *Shen Nong Huangdi shi jin* 神農黃帝食禁 [Food prohibitions of Shen Nong and Huangdi], *Huangdi sanwang yangyang fang* 黃帝三王養陽方 [Recipes of Huangdi and the three sage-kings for nurturing Yang], and the *Huangdi zazi bu yin* 黃帝雜子步引 [Stepping and pulling book of Huangdi and other masters]. See *Hanshu* 漢書 [Standard history of the Han dynasty] (Beijing, 1962), 30:1778–79. For further information on this material, see Lisa Raphals, "Chinese Philosophy and Chinese Medicine," in Zalta, *Stanford Encyclopedia* (cit. n. 31) (entry first published 28 April 2005), https://plato.stanford.edu /archives/spr2017/entries/chinese-phil-medicine/.

[58] See J. Sailey, *The Master Who Embraces Simplicity: A Study of the Philosopher Ko Hung, A.D. 283–343* (San Francisco, Calif. 1978), 242–72 and 277–8; and J. R. Ware, *Alchemy, Medicine and Religion in the China of A.D. 320: The Nei P'ien of Ko Hung* (Cambridge, Mass., 1996), 6–21.

[59] See Lai Chi-tim, "Ko Hung's Discourse of Hsien-Immortality: A Taoist Configuration of an Alternate Ideal Self-Identity," *Numen* 45 (1998): 183–220, esp. 203–4.

[60] The "Jin dan" 金丹 [Gold elixir] and "Huang bai" 黃白 [Yellow and white] chapters of the *Baopuzi neipian* 抱朴子內篇 [Book of the master who embraces simplicity: Inner chapters] (ca. 320 CE), survey the history of alchemy and describe in detail a method for "alloying cinnabar," quoting from ancient recipes and "cinnabar methods." The "Xian yao" 仙藥 [Immortal herbs] chapter gives information on medical herbs. See Gao Riguang 高日光, "Ge Hong" 葛洪, in *Zhuzi baijia da cidian* 諸子百家大辭典 [Dictionary of philosophers], ed. Feng Kezheng 馮克正 and Fu Qingsheng 傅慶升 (Shenyang, 1996), 87; and Qing Xitai 卿希泰, *Zhongguo daojiao* 中國道教 [Chinese Daoism] (Shanghai, 1994), 1:236–38.

nasties. He was educated in both early Daoist traditions and the works of Ge Hong, and was actively engaged in mostly unsuccessful attempts to produce alchemical elixirs.[61] Sun Simiao 孫思邈 (581–682), the author of two major medical works that are still consulted and a work on Daoist longevity prescriptions, is still worshiped as the "Medicine Buddha" and "King of Medicine" (yao wang 藥王).[62] He also wrote several works on Daoist alchemy, which he is believed to have practiced (he died at the age of 101). Taiqing Danjing Yaojue 太清丹經要訣 (Essential instructions from the Scripture of the Elixirs of Great Clarity), ca. 640, is an anthology of some thirty selected methods.[63] Sun Simiao's Sheyang lun 攝養論 (Essay on preserving and nourishing life) gives monthly advice on food, sleeping habits, and good and ill auspices for various types of action.[64] In summary, Sun Simiao, like Ge Hong and Tao Hongjing, combines explicit interests in Daoist philosophy, medicine, materia medica, and alchemy. All these authors focused, to varying degrees, on longevity and immortality.

But despite its importance in philosophy and medicine, immortality is not a major trope in Chinese SF, with the single exception of Xu Nianci's New Tales of Mr. Braggadocio (Xin faluo xiansheng tan 新法螺先生譚), the story of a man whose body and soul are separated in a typhoon.[65] His body sinks toward the center of the earth, and his soul travels to Mercury and Venus.[66] On Mercury, it watches the transplantation of brains as a method of rejuvenation. (It also discovers on Venus that rudimentary plants and animals appear at the same time, refuting biologists' claims that plants preceded animals in evolution.) Immortality appears when the protagonist's body, hav-

---

[61] Michel Strickman, "On the Alchemy of T'ao Hung-ching," in Facets of Taoism: Essays in Chinese Religion, ed. Holmes Welch and Anna Seidel (New Haven, Conn., 1979), 123–92, on 152.

[62] Sun Simiao's Qian jin fang 千金方 [Prescriptions worth a thousand in gold] (652 CE) was a comprehensive treatise on the practice of medicine, including herbal remedies, the history of medicine, and the first Chinese treatise on medical ethics. The supplement (Qian jin yi fang 千金翼方, 682 CE) records some thirty years of Sun's own experience, with special interest in folk remedies.

[63] For a translation see Nathan Sivin, Chinese Alchemy: Preliminary Studies (Cambridge Mass., 1968), 262–4.

[64] Fabrizio Predagio, ed., The Encyclopedia of Daoism, 2 vols. (London, 2013), 2:928.

[65] Xu Nianci 徐念慈, Xin faluo xiansheng tan 新法螺先生譚 [New tales of Mr. Braggadocio] (Shanghai, 1905); reprinted in Zhongguo jindai wenxue daxi 1840–1929: xiaoshuoji 中國近代文學大系 1840–1929: 小說集 [A treasury of modern Chinese literature 1840–1929: fiction collection], ed. Wu Zuxiang 吳組緗 et al., vol. 6 (Shanghai, 1991), 323–43, on 325; Xu Nianci's New Tales of Mr. Braggadocio was translated by Nathaniel Isaacson in Renditions 77/78 (2012): 15–38; see this publication for bibliographic details on additional publications of New Tales and stories of Mr. Braggadocio preceding it. It was also published under the pseudonym Donghai Juewo 東海覺我 (Awakened One from the Eastern Sea), and included in The New Braggadocio [Xin faluo 新法螺], by Fiction Forest Press [Xiaoshuo lin she 小說林社] in Shanghai in 1905. A founder of the press, Xu was one of the editors of the journal Xiaoshuo lin 小說林 [Forest of novels]. The novel is a sequel to the novel by Iwaya Sazanami 岩谷小波 (1870–1933), Tales of Mr. Braggadocio (Hora Sensei 法螺先生), translated into Chinese by Bao Tianxiao 包天笑 (1876–1973) as Faluo xiansheng tan 法螺先生譚 [Tales of Mr. Braggadocio] and Faluo xiansheng xu tan 法螺先生續譚 [Continued tales of Mr. Braggadocio]. For details see Isaacson, "New Tales."

[66] Chinese has no word that is equivalent to the English "soul." Two terms in use in popular religion and medicine refer to a "cloud soul" [hun 魂] and a "white soul" [po 魄]. For "soul," Xu Nianci uses the unusual term linghun 靈魂, which he explains thus: "I have no word for it, but it would be referred to in the common language of religion as the 'soul.'" Xu uses the terms "soul-body" [linghun zhi shen 靈魂之身] and "corporeal-body" [quqiao zhi shen 余軀殼之身]. He avoids the strong mind-body dualism of some Western traditions and apparent mind-body holism of traditional China by clearly referring to the linghun soul as a kind of body [shen 身]. For the semantic field of Chinese words for mind, soul, and spirit, and for discussion of issues of mind-body dualism in China, see Lisa Raphals, "Body and Mind in Early China and Greece," Journal of Cognitive Historiography 2 (2015): 132–82.

ing arrived at the center of the earth, encounters a quasi-immortal man.[67] The story imitates—and quotes—the *Zhuangzi* (discussed above), but immortality and rejuvenation are passing plot elements and not the central concern of the text. Immortality themes sometimes appear in new wave writing, but in new guises that are still closely linked to science—for example, in Fei Dao's "The Demon's Head," where a scientist preserves the brain of a general assassinated conveniently close to a neurological research institute. But even here, his "immortality" arises from neuroscience, and not from immortality practices.[68]

CONCLUSION

In conclusion, we are now in a position to answer some of the questions with which we began. First, what counts as Chinese SF? Second, what "science" did it draw on? Third, how, if at all, can we connect a Chinese SF born from scientific modernity with a possible alternate Chinese SF that arises from deeper Chinese traditions of natural philosophy?

In all its forms, from its origins in the late nineteenth century, at least until the new wave, Chinese SF—*kehuan xiaoshuo*—has been a modernist phenomenon that emerged out of interaction with the West; and it seems clear that the original Qing dynasty readership understood it this way. Nor was the Maoist version of Chinese SF any less indebted to Western science and technology, although for slightly different reasons. That situation at least partially continues under the new wave, though some authors have begun to explore more widely. As for the second question, the science that Chinese SF drew from was clearly both the "science" (a unified notion of science) and particular sciences of the modern West. So, Chinese SF has been inextricably linked to modern science in ways that largely preclude the indigenous sciences. And there are further demarcations of genre, which contrast the "Chinese SF" (*kehuan*) based on Western science to "fantasy" (*qihuan*). *Qihuan* includes both Chinese supernatural (*xuanhuan*) and Western magical fiction (*mohuan*). *Xuanhuan* might appear to accommodate the traditional Chinese sciences, but it excludes Daoist immortality stories (*xiuzhen* 修真), an important part of the landscape of the traditional Chinese sciences. And both *kehuan* and *qihuan* exclude time travel stories (*chuanyue* 穿越).[69] Some of these issues parallel arguments about the boundaries between SF, fantasy, and horror in Western genres. But there seems to be no easy mapping between Chinese and Western SF and the other genres with which they coexist.

These problems bring us to the third question of how, if at all, a Chinese SF born from scientific modernity might connect to a Chinese SF rooted in Chinese traditions of natural philosophy. One possibility is that at least some new wave authors will reject the premises of early Chinese SF from the late Qing and early twentieth century, and extend their reach beyond its concerns. Here an interesting divide arises between the very different training of contemporary Chinese "hard SF" and new wave writers.

---

[67] Wu Dingbo, "Chinese Science Fiction," in *Handbook of Chinese Popular Culture*, ed. Wu Dingbo and Patrick D. Murphy (Westport Conn., 1994), 257–78, on 257–9, esp. 260.

[68] Fei Dao, "The Demon's Head" [*Mogui de Toulu* 魔鬼的头颅], trans. David Hull, *Renditions* 77/78 (2012): 263–71.

[69] For these distinctions, see Regina Kanyu Wang, "A Brief Introduction to Chinese Science Fiction," *MITHILA Journal of International Science Fiction & Fantasy* 9 (2016), http://mithilareview .com/wang_11_16/.

The most prominent hard SF writers, Liu Cixin and Wang Jinkang, were trained as engineers—hydroelectric and civil, respectively. By contrast, several new wave writers hold advanced degrees and university positions in literature, and their interests open the prospect of themes from philosophical writing, including that of time travel, transformation, and Daoist immortality motifs.

But neither group has had obvious exposure to themes from the indigenous Chinese sciences, so it is perhaps not surprising that explicit themes from these areas do not appear. Nor do debates about what counts as Chinese SF help clarify the absence of the indigenous Chinese sciences. Some scholars try to trace Chinese SF back to *zhiguai*, while others date it to the 1930s, 1950s, or even the post-Mao period. An interesting middle ground is offered by Isaacson, who dates it to the late Qing, arguing that it arose in response to an epistemological crisis due to subjugation to European powers and the translation into Chinese of newly emergent Western science fiction.[70] He also argues that an adequate account of the history of SF in China requires an understanding of its relationship to earlier genres. His approach helps explain the absence of the indigenous sciences, which could have been lost under pressure of the two developments noted above that were fundamentally responses to the West. But that situation may change as the agenda of Chinese SF changes.

My goal here is not to argue for *zhiguai* or *chuanqi* as the origins of Chinese SF, or to claim that the indigenous sciences inform Chinese SF, which they clearly do not. Rather, Chinese indigenous natural philosophy and sciences suggest an alternative literary path that could happen, and perhaps already is happening. In conclusion, relations between Chinese literary genres, their indigenous scientific traditions, the introduction of Western science, and the introduction and development of science fiction all form a complex network that warrants further study and should not be oversimplified.

---

[70] For discussion of translations of Western science fiction into Chinese in the early twentieth century, see Jiang, "Translation" (cit. n. 7); and Isaacson, *Celestial Empire* (cit. n. 3), 2, 26, 146–80.

Appendix: A Brief Bibliography of Contemporary Chinese SF

Bao Shu 宝树 (pen name of Li Jun 李峻, 1980–)

———. 2015. "Preserve Her Memory" [Liuxia ta de jiyi 留下她的記憶]. Translated by Ken Liu. *Clarkesworld* 108, September 2015. http://clarkesworldmagazine.com/bao_09_15/.
———. 2016. "Everybody Loves Charles" [Renren dou ai Chaersi 人人都爱查尔斯]. Translated by Ken Liu. *Clarkesworld* 112, January 2016. http://clarkesworldmagazine.com/bao_01_16/.

Chen Qiufan 陈楸帆 [Stanley Chan] (1981–)

———. 2011. "The Fish of Lijiang" [Lijiang de yu'er men 丽江的鱼儿们]. Translated by Ken Liu. *Clarkesworld* 59, August 2011. http://clarkesworldmagazine.com/chen_08_11/. Reprinted in Liu, *Invisible Planets*, 51–68. London, 2016.
———. 2012. "The Flower of Shazui" [Shazui zhi hua 沙嘴之花]. Translated by Ken Liu. *Interzone* 243, November–December 2012, 20–9. Reprinted in Liu, *Invisible Planets*, 69–87. London, 2016.
———. 2013. "The Year of the Rat" [Shu nian 鼠年]. Translated by Ken Liu. *The Magazine of Fantasy & Science Fiction*, July–August 2013, 68–92. Reprinted in Liu, *Invisible Planets*, 21–49. London, 2016.

———. 2014. "The Mao Ghost" [Mao de guihun 猫的鬼魂]. Translated by Ken Liu. *Lightspeed* 46, March 2014. http://www.lightspeedmagazine.com/fiction/the-mao-ghost/.

———. 2015. "The Smog Society" [Mai 霾]. Translated by Carmen Yiling and Ken Liu. *Lightspeed* 63, August 2015. http://www.lightspeedmagazine.com/fiction/the-smog-society/.

———. 2017. "A Man Out of Fashion" [Guoshi de ren 过时的人]. Translated by Ken Liu. *Clarkesworld* 131, August 2017. http://clarkesworldmagazine.com/chen_08_17/.

———. 2019. *The Waste Tide* [Huang chao荒潮]. Translated by Ken Liu. London: Head of Zeus.

Fei Dao (pen name of Jia Liyuan 賈立元, 1983–)

———. 2009. "Yilan zhong shan xiao" 一览众山小 [A list of small mountains], *Kehuan shijie* 科幻世界 [Science fiction world], August 2009.

———. 2012. "The Demon's Head" [Mogui de Toulu 魔鬼的头颅]. Translated by David Hull. *Renditions* 77/78: 263–71.

Han Song 韩松 (1965–)

———. 1991. *Yuzhou mubei* 宇宙墓碑 [Cosmic tombstones]. Beijing: Xinya chubanshe.

———. 1997. *Renzao ren* 人造人 [Artificial people]. Beijing: Zhongguo renshi chubanshe.

———. 2007. *Wo de zuguo bu zuomeng* 我的祖国不做梦 [My homeland never dreams]. Never printed in China, but available online in Loïc Aloisio. "Ma Patrie ne rêve pas." Impressions d'Extrême-Orient (website), "6 | 2016." 2 December 2016. http://journals.openedition.org/ideo/470.

———. 2012 "The Passengers and the Creator" [Chengke yu chuangzaozhe 乘客与创造者]. Translated by Nathaniel Issacson. *Renditions* 77/78: 144–72.

He Xi 何夕 [He Hongwei 何宏偉] (1971–)

———. 1996. "Pan Gu" 盘古. *Kehuan Shijie* [Science fiction world], June 1996.

———. 1999. "Yiyu" 異域 [Foreign land]. *Kehuan Shijie* [Science fiction world], August 1999.

———. 2002. *Liudao Zhongsheng* 六道眾生 [Six lines from Samasara]. Changsha: Wenyi chubanshe, 2012. There is no further information on the 2002 composition.

La La 拉拉 (1977–)

———. 2012. "The Radio Waves That Never Die" [Yongbu xiaoshi de dianbo 永不消逝的电波]. Translated by Petula Parris-Huang. *Renditions* 77/78: 210–238.

Ling Chen 凌晨 (1972–)

———. 2013. "A Story of Titan" [Taitan de gushi 泰坦的故事]. Translated by Joel Martinsen. In *Pathlight: New Chinese Writing*, May 2013, pagination unknown.

Liu Cixin 劉慈欣 (1963–).

———. 2016. *The Three-Body Problem*. Translated by Ken Liu. New York, N.Y.: Tor. Originally published as *Santi* 三体 (Chongqing: Chongqing chubanshe, 2007).

———. 2015. *The Dark Forest*. Translated by Ken Liu. New York, N.Y.: Tor. Originally published as *Heian sanlin: Santi di er bu* 黑暗森林: 三體第二部 (Chongqing: Chongqing chubanshe, 2008; and Taipei, Taiwan: Maotouying chubanshe, 2008).

———. (1998) 2010. *Wei ji yuan* 微纪元 [The micro-age]. Shenyang: Shenyang chubanshe.

———. 2016. *Death's End*. Translated by Ken Liu. New York, N.Y.: Tor. Originally published as *Santi. Di III bu, Si shen yong sheng* 三體第三部, 死神永生 (Taipei, Taiwan: Maotouying chubanshe, 2011).

Liu, Ken (translation)

———. ed. and trans. 2016. *Invisible Planets: 13 Visions of the Future from China*. London: Head of Zeus.

Mao Boyong 马伯庸 (1980–)

———. 2016. "The City of Silence" [Jijing zhi cheng 寂静之城]. In Liu, *Invisible Planets*, 153–196.

Qian Lifang 钱莉芳 (1978–)

_____. 2004. *Tian yi* 天意 [The will of heaven]. Chengdu: Sichuan Keji chubanshe.
_____. 2011. *Tian ming* 天命 [The command of heaven]. Changchun: Shidai wenyi chubanshe.

Wang Jinkang 王晋康 (1948–)

———. 2011. *Yu Wu Tong Za*i 與吾同在 [Being with me]. Chongqing: Chongqing chubanshe.
———. 2012. "The Reincarnated Giant" [Zhuansheng de juren 转生的巨人]. Translated by Carlos Rojas. *Renditions* 77/78: 173–209.

Xia Jia 夏笳 (pen name of Wang Yao 王瑶, 1984–)

———. 2005. *Kamen* 卡门 [Carmen]. *Kehuan Shijie* [Science fiction world], August 2005.
———. 2008. "Yong xia zhi meng" 永夏之梦 [Dream of eternal summer]. *Kehuan Shijie* [Science fiction world], September 2008.
———. 2012. "The Demon-Enslaving Flask" [Guan yaojing de pingzi 關妖精的瓶子]. Translated by Linda Rui Feng. *Renditions* 77/78: 272–82.
———. 2016. "A Hundred Ghosts Parade Tonight" [Bai gui yexing jie 百鬼夜行街]. In Liu, *Invisible Planets*, 91–109.
———. 2016. "Night Journey of the Dragon-Horse" [Long ma yexing 龙马夜行]. In Liu, *Invisible Planets*, 131–49.

Zhao Haihong 赵海虹 (1977–)

———. (1998) 2006. *Shijian de bifang* 时间的彼方 [The other side of time]. Reprint of the first edition. Wuhan: Hubei shaonian ertong chubanshe.
———. 1999. *Yi'ekasida* 伊俄卡斯达 [Jocasta]. *Kehuan Shijie* [Science fiction world], March 1999.
———. 2012. "1923—a Fantasy" [Yijiuersan nian kehuan gushi 一九二三年科幻故事]. Translated by Nicky Harman and Pang Zhaoxia. *Renditions* 77/78: 239–54.

DELINEATING SCIENCE

Hylozoic Anticolonialism:
Archaic Modernity, Internationalism, and Electromagnetism in British Bengal, 1909–1940

by Projit Bihari Mukharji*

ABSTRACT

Imperial ideology identified "science" and "progress" as the prerogative of the "West," while "religion" and "spirituality" were located in the "East." Yet, in practice, these neat dichotomies were far more difficult to sustain. Science and religion were braided together by spiritually inquisitive scientists as much in the West as in the East. Various strands of hylozoic thought that undermined the dichotomy of matter and spirit were located in the liminal space between these orthogonal categories. One such strand of hylozoism, engendered in electromagnetic ideas, was articulated in the early science fiction of authors like Edward Bulwer-Lytton. Dinendrakumar Ray, a Bengali novelist, inserted this science-fictional hylozoism into his translations of the novels of the Australian author Guy Boothby. By selectively adapting and blending Boothby's international plot lines with Lytton-like electromagnetic hylozoism, Ray was able to craft a "hylozoic anticolonialism" that resonated emphatically with the thought of Sri Aurobindo, a revolutionary nationalist turned neo-Hindu spiritual master.

INTRODUCTION

Science fiction (SF) and colonialism share a complex and intimate history. Though the nature of the connection has been a matter of debate, several scholars have noted both the chronological and tropic intimacies between the two. The majority of scholarly works that have engaged with this connection, however, have done so through an archive of SF written in European languages by European authors.[1] By contrast, the scholarship on non-European SF has overwhelmingly focused on postcolonial narra-

* History and Sociology of Science, University of Pennsylvania, 326 Claudia Cohen Hall, 249 S. 36th St., Philadelphia, PA 19104-6304, USA; mukharji@sas.upenn.edu.

I am thankful to the editors and the referees for their comments and suggestions. A previous draft of this article was presented at the Max Planck Institute for Human Development in Berlin. The feedback I received there was most useful. Conversations with Lorraine Daston were immensely helpful. I am especially grateful to her for suggesting that I locate my argument in relation to the hylozoic tradition. I am also grateful to Manjita Mukharji who read multiple drafts of the article. Above all, I am deeply indebted to Iwan Rhys Morus for his numerous enormously helpful suggestions. Unless otherwise noted, all translations are my own.

[1] John Rieder, *Colonialism and the Emergence of Science Fiction* (Middletown, Conn., 2012); Patricia Kerslake, *Science Fiction and Empire* (Liverpool, UK, 2011).

tives.[2] Happily, this trend is beginning to change, and an increasing number of scholars are now exploring a richer, older seam of non-European SF.[3]

Defining SF has long been a vexed issue. Defining it in a non-European context, such as in Bengal, is even more complicated. John Rieder argues that SF emerged at the moment of "coalescence of a set of generic expectations into a recognizable condition of production and reception that enables both writers and readers to approach individual works as examples of a literary kind that in the 1920s and after came to be named science fiction."[4] Working with this definition of the genre is particularly helpful in the Bengali context, since many of the nineteenth-century authors, such as Edgar Allan Poe, Edward Bulwer-Lytton, Jules Verne, and H. Rider Haggard, who were explicitly invoked as the pioneers of SF in the 1920s, were also explicitly cited as influences by some of the early Bengali authors. Yet, three important caveats are worth keeping in mind; Rieder makes each of these points. First, any new genre emerges by violating or pushing the limits of existing genres.[5] Thus, if the preexisting genres in one language are different from those in another language, the pioneering generic innovations also will look different from each other. Second, a genre acquires its identity, status, and limits within the entire landscape of other available genres. Mary Shelley's *Frankenstein* in 1818 and *Frankenstein* in the late nineteenth century were not the same thing.[6] Thus, once more, a genre that is adopted from one language into another—even if the author wished to remain faithful to the original genre—would morph. Finally, a genre is nothing more than "a web of resemblances."[7] This is as true for works identified as belonging to the same genre in a single language, as it is for cross-cultural identifications.

Thinking of genre as a web of resemblances and recalling the explicit acknowledgment of European authors by the early Bengali authors of SF does more than simply clarify which texts can rightfully be called SF and which cannot. It reminds us of the braided nature of intellectual traditions developed in the colonial "contact zones."[8] All too often, linguistically defined literary traditions and the quest for ahistorical forms of alterity unfortunately lead even the most sophisticated scholars to posit historically inaccurate binaries. Recognizing the webs of resemblances that stretch across literary traditions, linguistic silos, and cultural boundaries serves to relocate colonial difference firmly within networks of specific historical actors.

Exploring these webs of resemblances also forces us to rethink the very idea of science. Scholars writing on Bengali SF have noted the tendency to render "native knowledge" as science in Bengali SF narratives. This has necessarily meant recoding religious myths in the vocabularies of science. Relative insensitivity to the webs of resemblances across the literary, cultural, and colonial divides has meant that many scholars have has-

---

[2] Ericka Hoagland and Reema Sarwal, eds., *Science Fiction, Imperialism and the Third World: Essays on Postcolonial Literature and Film* (Jefferson, N.C., 2010); Jessica Langer, *Postcolonialism and Science Fiction* (New York, N.Y., 2011); Eric D. Smith, *Globalization, Utopia and Postcolonial Science Fiction: New Maps of Hope* (Basingstoke, UK, 2012).

[3] Debjani Sengupta, "Sadhanbabu's Friends: Science Fiction in Bengal from 1882 to 1974," *Sarai Reader* 3 (2003): 76–82; Bodhisattva Chattopadhyay, "Kalpavigyan and Imperial Technoscience: Three Nodes of an Argument," *Journal of the Fantastic in the Arts* 28 (2017): 102–22.

[4] Rieder, *Colonialism* (cit. n. 1), 15.

[5] Ibid., 18.

[6] Ibid., 19.

[7] Ibid., 17.

[8] Mary Louise Pratt, *Imperial Eyes: Travel Writing and Transculturation*, 2nd ed. (London, 2007).

tened to read this tendency as a peculiar feature of Bengali SF that has sought to marry "religious nationalism" with "scientific materialism."[9]

One such scholar, Bodhisattva Chattopadhyay, labels the efforts to recode native knowledge as science as a "peculiar tendency" and says it leads to a particular pitfall. This pitfall, Chattopadhyay continues, is the political expediency, driven by "nationalist imperatives . . . to claim a greater scope for native knowledge than can be factually justified."[10] In his haste to locate an alternate essence for Bengali SF, Chattopadhyay overlooks two important issues. First, "scientific facts" are not transparently obvious apriori and ahistorical givens. They are produced through carefully crafted experiments, disseminated and established through a range of "technologies of witnessing."[11] Second, Western science, or simply science, was neither a single monolith, nor was it devoid of complicated entanglements with religion in the late nineteenth century. After all, around the time that Chattopadhyay finds the Bengali chemist P. C. Ray attempting to recode precolonial Tantric knowledge as "ancient Hindu science" in Calcutta, a host of leading British physicists were sitting at séances in Cambridge, London, and elsewhere, wondering about the existence of the immortal soul.[12]

Victorian and Edwardian science, for many, was above all a universal language. The ghost-hunting physicist Sir William Crookes was as much a universalist as the Tantra-admiring chemist Sir P. C. Ray. Authors of SF partook of this aspiration to imagine the universal as much as "scientists" did. Dinendrakumar Ray, who will be the main focus of my essay, introduced one of the two novels to be discussed in this way: "Since the European novelist whose footsteps I have followed in writing this novel has [managed to] fascinate both his own countrymen as well as foreigners by his subject and variety, I am hopeful that readers in . . . [Bengal] will not ignore this book."[13] In his next novel, Ray was even more explicit: "The imaginative (*kalpana-kushal*), [and] talented (*pratibhaban*) European novelist to whom I am indebted for the plan (*parikalpana*) of . . . [this novel] is well-known and adored beyond expectations in Great Britain, America, Australia and other British colonies by readers who love novels. Therefore I dare to hope that . . . [this novel too] will be admired by Bengali readers throughout British India and Burma. For while castes/nations (*jati*), religions (*dharma*) and social mores might vary from one country to another, the habits of man's heart (*manober hridoybritti*) are the same everywhere and ruled by the same laws."[14]

By invoking Ray's claims to a universal set of literary tastes and his explicit acknowledgement of his debt to an unnamed "talented European novelist," I do not mean to say that his novels, or for that matter any other Bengali SF, were not ethnocentric.

---

[9] Chattopadhyay, "Kalpavigyan" (cit. n. 3), 118.

[10] Ibid., 113–4.

[11] Steven Shapin and Simon Schaffer, *Leviathan and the Air-Pump: Hobbes, Boyle, and the Experimental Life* (Princeton, N.J., 1989). See also Jan Golinski, *Making Natural Knowledge: Constructivism and the History of Science* (Chicago, Ill., 1998); Bruno Latour, *Laboratory Life: The Construction of Scientific Facts* (Princeton, N.J., 1986).

[12] Jason A. Josephson-Storm, *The Myth of Disenchantment: Magic, Modernity, and the Birth of the Human Sciences* (Chicago, Ill., 2017); Janet Oppenheim, *The Other World: Spiritualism and Psychical Research in England, 1850–1914* (Cambridge, UK, 1988); Richard Noakes, "Spiritualism, Science, and the Supernatural in Mid-Victorian Britain," in *The Victorian Supernatural*, ed. Nicola Bown, Carolyn Burdett, and Pamela Thurschwell (Cambridge, UK, 2004), 23–43.

[13] Dinendrakumar Ray, *Jal mohanto* [The false abbot] (Calcutta, 1909), 2.

[14] Dinendrakumar Ray, *Pishach purohit* [The zombie priest] (Calcutta, 1910), ii.

I merely want to emphasize the webs of resemblances that the authors themselves constructed through the idea of a universal literary taste shared across the British Empire from America to Australia, including India. This is similar to the universal claims made on behalf of Yoga by scientists, which have been studied by Joseph Alter. Alter calls these claims a "post-Western universalism," where the West is no longer by default at the center of any imagination of a modern universal. Such a universalism is, however, quite distinct from the postcolonial emphasis on cultural difference.[15]

It is the contours of this post-Western universalism of Bengali SF that I wish to map here. I will do so by exploring in depth two novels of Dinendrakumar Ray. The rest of the essay is organized into six sequential sections. The first of these will introduce the author and the text as well as outline why his novels merit particular attention. The second section will take up the colonial ideology of "progress," something that scholars of SF often insist is an "absolute prerequisite" for the genre,[16] and Ray's response to it. Third, I will argue against the tendency to analyze colonial texts within the narrow confines of the colonizer-colonized binary. I will show that Ray in fact explicitly invoked other colonial and independent nations in his stories and articulated a complex internationalism that easily went beyond merely responding to Britain. The fourth section is in some ways the heart of the essay. It will demonstrate the crucial role played by theories of electromagnetism in structuring Ray's novels. It will also describe, develop, and define one of the central concepts of this essay—that is, hylozoic anticolonialism. The penultimate section will trace how Ray's fiction became recast as an important strand within modern Hindu theology. It will thus relocate the debate about the relationship between religious nationalism and scientific materialism within a concrete intellectual milieu. Finally, I will offer a summary of the arguments in a brief conclusion.

LOCATING DINENDRAKUMAR RAY

Dinendrakumar Ray (1869–1943) was born in the small village of Meherpur in present-day northern Bangladesh. The family was affluent without being wealthy. They had been in clerical service to a succession of local feudal administrations, but crucially did not belong to the upper castes that dominated the clerical professions. The Rays belonged to the relatively low-ranking *Tili* caste. Meherpur's own importance as a minor administrative center had declined since its heyday in the eighteenth century. The Rays had also taken up Western education quite early, and Dinendrakumar's uncles were educated in Western-style schools that exposed them to English. As Dinendrakumar's biographer points out, "it was lucrative to learn English in those times."[17]

Naturally, Dinendrakumar was taught English in keeping with the family's traditions and aspirations. This took him beyond his village to the administratively and culturally important northern Bengali town of Krishnanagar, with even a brief stint in Calcutta, which was then the capital of British India. There, he was further exposed to English language and literature, and even became attracted to the Brahmo movement that had

---

[15] Joseph S Alter, *Yoga in Modern India: The Body Between Science and Philosophy* (Princeton, N.J., 2010), 77.
[16] Rieder, *Colonialism* (cit. n. 1), 29.
[17] Satanjeeb Raha, *Kathachitrakar Dinendrakumar Rai* [An artist of stories: Dinendrakumar Ray] (Calcutta, 1959), 10.

sought to reform Hinduism along Unitarian lines before seceding from the Hindu fold altogether and establishing itself as a distinct religion around the 1860s.[18] Despite the family's hopes that he would become a lawyer, Dinendrakumar repeatedly failed to qualify for the legal profession. Eventually, he had to settle for a clerical position at the Rajshahi District Court in 1893. He resented the dreary monotony of the job and, in 1898, obtained a job that allowed him to quit the clerkship.

The job was an unusual one. He was to be a private tutor and companion for a young man named Aurobindo Ghose. Ghose had been raised in England and could not speak or read any Indian language. He was otherwise highly educated and had attended King's College, Cambridge. He had passed the Indian Civil Service examinations in London but was rejected from service after failing to pass the physical tests. As a result, Ghose had returned to the land of his birth and taken a job in the administration of one of the many nominally independent princely states. He was keen to learn his mother tongue and sought a tutor who would stay with him as a companion and teach him Bengali.[19]

Ghose would later go on to become involved with a secret underground nationalist organization, inspired in part by the Irish Republican Army,[20] before renouncing it and becoming one of the best-known and most influential modern Hindu spiritual leaders. It is as Ghose's tutor, friend, and biographer that Dinendrakumar is usually remembered today.

As his own biographer admits, Dinendrakumar never achieved true literary fame. After his death, what little reputation he had acquired quickly dissipated. Yet, in his day, he was well connected with the contemporary literary world and was one of the most prolific authors.

Ray's literary interests had been cultivated from an early age by his father, an unsuccessful poet who had published one volume of poems. The journalist, novelist, and champion of peasant rights, Harinath Majumdar, also influenced him.[21] During his college days he became very close to two contemporaries who would go on to become significant literary figures. One, with whom Dinendrakumar once even boarded, was Jaladhar Sen, who became editor of the prestigious literary journal *Bharatbarsha*.[22] The other was Jagadananda Roy, Dinendrakumar's closest friend in his college years, who is today credited with having authored *Shukro Bhromon* [Travels in Venus], now identified as the first proper SF story in Bengali. Dinendrakumar himself tasted some early, short-lived, critical success by authoring a set of sketches of rural Bengali life.[23] It was this minor literary fame and his connections to the world of high literature that got him the position with Ghose.

After two years of residence with Ghose in the central Indian state of Baroda, Dinendrakumar returned to Bengal and was able to obtain the editorship of the highly

[18] David Kopf, *The Brahmo Samaj and the Shaping of the Modern Indian Mind* (New Delhi, 1979).

[19] On Ghose, see Peter Heehs, *The Lives of Sri Aurobindo* (New York, N.Y., 2008).

[20] Michael Silvestri, *Ireland and India: Nationalism, Empire and Memory* (Basingstoke, UK, 2009), 18.

[21] Raha, *Kathachitrakar* (cit. n. 17), 18. On Majumdar's life and politics, see Bipasha Raha, "Harinath Majumdar and the Bengal Peasantry," *Indian Historical Review* 40 (2013): 331–53.

[22] On Jaladhar Sen, see *Banglapedia: National Encyclopedia of Bangladesh*, rev. 2nd ed. (Dhaka, 2012), s.v. "Raibahadur Jaladhar Sen," by Shipra Dastider, http://en.banglapedia.org/index.php?title=Sen,_Raibahadur_Jaladhar.

[23] On the first SF story in Bengali, see Sengupta, "Sadhanbabu's Friends" (cit. n. 3). On Ray's sketches, see Raha, *Kathachitraka* (cit. n. 17).

respected literary journal *Bharati*. But the editorship did not produce the kind of literary prestige and fame it could have. Later, a bitter autobiography of his resulted in a very public and long-drawn-out confrontation with his friend, the much more successful Jaladhar Sen, whom he accused of plagiarism during their early careers. Many of Sen's allies, who were all well placed in the literary world, attacked Dinendrakumar in retaliation. This further alienated Dinendrakumar and pushed him to the margins of the literary canon in Bengali.

Part of the reason for his lack of success, however, has to be connected to his choice of genre. While his sketches of rural Bengal have remained popular with readers, and a biography of Aurobindo that he wrote remained an important reference work,[24] his prolific writings in more popular genres have sunk to oblivion. Between roughly 1898 and 1914, he wrote 217 novels.[25] Published simultaneously from Calcutta and Meherpur, these novels spanned the spectrum from detective stories and romance quests to stories of international intrigue. They borrowed heavily from English pulp fiction. For instance, Dinendrakumar's detective, who appeared in several of his novels, was called Robert Blake and was modeled on the popular English pulp fictional character Sexton Blake.

Dinendrakumar's appetite for English detective novels had been nurtured in his student years, when a headmaster stocked the school library with several such novels.[26] Later, when he was working as a court clerk in Rajshahi, his friend, the poet Rajanikanta Sen, encouraged him to translate a French detective novel via an English translation that was available at the local public library. This was Dinendrakumar's first published novel, and it went into two editions.[27]

His two Bengali novels to be discussed here, *Jal mohanto* (The false abbot) of 1909 and *Pishach purohit* (The zombie priest) of 1910, appeared in a series alongside these detective novels.[28] Dinendrakumar explicitly stated in the preface to both that they were modeled on the same original plan and were distinct from his detective novels. He also obliquely referred to the influence of an unnamed "talented European novelist" whose footsteps he had followed.

Notwithstanding his silence about the name of the author whose lead he had taken, within the second of the two novels he provided a further clue to the web of resemblances that he had mobilized. *Pishach purohit* commenced with a framing narrative wherein we read of a Bengali gentleman engrossed in a book at his home. This book we are told is either a "Hull Caine [*sic*], a Mary Corelli or a Rider Haggard."[29] All three were popular English authors at the time.

---

[24] Dinendrakumar Ray, *Aurobindo prasanga* [About Aurobindo] (Chandannagar, 1923).

[25] Subodhchandra Sengupta and Anjali Basu, *Bangali charitabhidhan* [Biographical dictionary of Bengalis] (Calcutta, 1976), s.v. "Dinendrakumar Ray," 202–3.

[26] Dinendrakumar Ray, *Sekaler smriti* [Memories of bygone times] (Calcutta, 1953), 98.

[27] Ibid., 138.

[28] Unfortunately, circulation figures for these two novels, like most Bengali fiction catering to a more popular market, are missing. These novels are also very poorly archived, and hence it is not clear whether some of them went into multiple editions or not. We know that the first edition of *The False Abbot* had a print run of three thousand copies. We can also assume that it did well, because the author wrote the second novel, *Pishach purohit*, on the same acknowledged plan within months of the former. For *Pishach purohit*, unfortunately, no estimates of the print run are to be had. We do know, however, that a separate publisher published an undated second edition of the work, possibly some time in the 1920s. See Dinendrakumar Ray, *Pishach purohit* [The zombie priest], 2nd ed. (Calcutta, n.d.). See also Ray, *Jal mohanto* (cit. n. 13); Ray, *Pishach purohit* (cit. n. 14).

[29] Ray, *Pishach purohit* (cit. n. 14), 10.

Marie Corelli (1855–1924) was well known to Victorian readers for her "scientific romances" such as the two-volume *A Romance of Two Worlds* (1886). H. Rider Haggard (1856–1925) mostly wrote romance and adventure tales, but also made prominent use of the lost race motif that is a classic SF genre. Haggard's novels like *King Solomon's Mines* (1885), *Allan Quatermain* (1887), and *She* (1886) became sensational hits across the anglophone world. Sir Thomas Henry Hall Caine (1853–1931), popularly referred to simply as Hall Caine, was mainly known for his stories of love, marriage, and adultery. However, at the time of Dinendrakumar's writing, he published a two-volume work set in Egypt and sympathetic to the Egyptian cause, titled *The White Prophet* (1909).[30]

It is unclear why Dinendrakumar chose to remain silent about the name of the "talented European novelist" mentioned in his preface. Indeed, one of the reviewers of *Pishach purohit*, Sureshchandra Samajpati, wondered about this curious omission in his lengthy review of the novel. Samajpati congratulated Dinendrakumar for not having tried to pass off a European work as his own, "as so many others did these days," but then, he queried, "why did Dinendrakumar not disclose the name of the original author?"[31] Nevertheless, an old catalogue in the Imperial Library of India gives the full title as *Pishach purohitra* [*sic*]: *An Adaptation of Guy Boothby's Pharos, the Egyptian*.[32] It is unclear if there had indeed been an edition with the extended title, but the entry does clear up the matter of identifying the source texts.

Both the novels I discuss here were in fact adaptations of Guy Boothby's (1867–1905) works. *Jal mohanto* was an adaptation of Boothby's *Dr. Nikola* (1896) and *Pishach purohit* was an adaptation of *Pharos, the Egyptian* (1899).[33] While Boothby was certainly one of the most popular and prolific authors of Victorian genre fiction, he was not exactly European, at least not entirely, as Dinendrakumar suggested. Boothby had been born in Australia and spent more than half his short life in Australasia.

Precisely what drew Dinendrakumar to Boothby is difficult to explain. Boothby had distinctive imperialist tendencies. Scholars who have studied his oeuvre are explicit in pointing out the strong imperial sentiments articulated in his works. Ailise Bulfin, for instance, maps the "imperial paranoia" in Boothby's gothic images of Egypt in his fiction.[34] Likewise, Clare Clarke identifies what she terms the "fears of reverse colonization," where undesirable colonials wreak havoc on imperial London. Moreover, the key work upon which Clarke bases her analysis has a rogue protagonist who hails from Calcutta itself.[35] Such writings naturally did not endear their author to the Bengali nationalist press. The leading nationalist newspaper of the time, the *Amrita Bazar Patrika*, when reviewing one of Boothby's novels set in India, stated, "Mr. Guy

---

[30] Marie Corelli, *A Romance of Two Worlds*, 2 vols. (London, 1886); H. Rider Haggard, *King Solomon's Mines* (New York, N.Y., 1885); Haggard, *Allan Quatermain* (New York, N.Y., 1887); Haggard, *She* (London, 1886); Thomas Henry Hall Caine, *The White Prophet*, 2 vols. (London, 1909).

[31] Sureshchandra Samajpati, "Samalochana," *Pishach purohit*, by Dinendrakumar Ray [critical discussion of The zombie priest], *Sahitya* [Literature] 22 (1911): 412.

[32] *Imperial Library Author Catalogue of Printed Books in Bengali Language*, vol. 1, A–F (Calcutta, 1941), 249.

[33] Guy Boothby, *Dr. Nikola* (New York, N.Y., 1896); Guy Boothby, *Pharos, the Egyptian* (London, 1899).

[34] Ailise Bulfin, "The Fiction of Gothic Egypt and British Imperial Paranoia: The Curse of the Suez Canal," *English Literature in Transition, 1880–1920* 54 (2011): 411–43.

[35] Clare Clarke, "Imperial Rogues: Reverse Colonization Fears in Guy Boothby's *A Prince of Swindlers* and Late-Victorian Detective Fiction," *Vict. Lit. Cult.* 41 (2013): 527–45.

Boothby . . . is, sometimes, a very clever and entertaining writer, but not always is he in
the vein . . . when he reaches India he throws probability to the winds and his story
becomes a tissue of absurdity."[36]

In contrast, Dinendrakumar's fiction always had a strong nationalist appeal. Several trials of militant young nationalists often involved with underground organizations seeking to overthrow British rule revealed their ownership of Dinendrakumar's
books, alongside proscribed political works. For instance, members of the famous
Dhaka Anushilan Samiti, an underground militant organization, were described as
having read books on "Napoleon, Mysteries of Nihilism and other novels."[37] The
second of these was written by Dinendrakumar. He is also known to have authored
a popular Bengali biography of Napoleon.[38] It is therefore very possible that the other
unnamed novels were also by Dinendrakumar. Members of another small and more
obscure underground outfit known as the Howrah Political Gang were also caught in
possession of *Mysteries of Nihilism* along with other political literature.[39] Both these
trials took place in 1911, less than a year after *Pishach purohit* was published.

Notwithstanding this militant nationalist association, Dinendrakumar chose to lav-
ishly praise and adapt the writings of Boothby, a writer with pronounced imperialist
views. One of the things that might have appealed to Dinendrakumar was his straightforward literary style. Samajpati, in his review of *Pishach purohit*, wrote the following: "There is no dissection of human psychology here, neither is there the analysis or
resolution of any deep moral, social or political problem. It is just a novel. It is a varied, strange, mysterious novel [and] it is a pleasurable read . . . It is true that the heart
here is bathed by the strange sport of the imagination, yet [happily] there is nothing
here that is soaked in the blood of the macabre, disgusting or erotic."[40] Samajpati's
review was a glowing testimonial to the success of such a literary venture. He was
deeply appreciative of a novel that avoided analytic complexities and was instead
propelled by action and plot. Dinendrakumar himself had written about his own aspiration to write plot-driven narratives for the common people who did not care for
deep philosophical arguments.[41]

Nevertheless, Dinendrakumar had to radically transform the novels to make them
resonate with his militant nationalist readers. Transforming explicitly imperialist narratives into anticolonial fictions was not easy, particularly as Dinendrakumar sought
to preserve the tight-knit plot of the original novels he so admired. In what follows,
I will map the process by which Dinendrakumar transformed Boothby's imperialist
novels into anticolonial narratives that strongly resonated with one emerging strand
of neo-Hindu theology.

PROGRESS TO THE PAST

The ideology of progress was central to both Victorian SF and imperial rule in South
Asia. As Thomas Metcalf points out, the liberal view of British imperialism in India

---

[36] "Review of 'My Indian Queen,'" *Amrita Bazar Patrika*, 26 April 1901.
[37] "Dacca Conspiracy Case," *Amrita Bazar Patrika*, 20 May 1911.
[38] Raha, *Kathachitrakar* (cit. n. 17), 42.
[39] "Howrah Political Gang Case," *Amrita Bazar Patrika*, 14 February 1911. See also, "Howrah
Gang Case," *Amrita Bazar Patrika*, 9 February 1911.
[40] Samajpati, "Samalochana" (cit. n. 31), 413.
[41] Raha, *Kathachitrakar* (cit. n. 17), 61.

found its fullest expression in James Mill's *History of British India* (1818). Mill vigorously disputed earlier, eighteenth-century claims of an advanced civilization in ancient South Asia, arguing that the British should endeavor to "free India from stagnation and set it on the path to progress," for progress alone was the yardstick of civilization.[42] But it was Sir Henry Maine, through his enormous influence on Indian administrative policy in the second half of the nineteenth century, who really converted the notion of progress from a social theory to a moral and political ideology. A dichotomy was set up between India's alleged stasis and Britain's social progress.[43]

Rieder's analysis of the ideology of progress in Victorian SF is more detailed. He argues that it is constituted of three "ideological fantasies": the "discoverer's fantasy," the "missionary's fantasy," and the "anthropologist's fantasy." The first of these denies the existence of complex cultural and social histories of the colonized and pretends to "discover" the land and its customs anew. The second posits that the colonizer's culture is necessarily preferable and that all colonized will be willing to embrace it if given the right opportunity. Finally, the anthropological fantasy postulates that the colonized and the colonizer exist in different temporal epochs. Despite living side-by-side, it is assumed that the colonized are "backward" and "primitive," living in a time that is identical to the remote past of the colonizer's own being. It is the combination of these three ideological fantasies that constitute the ideology of progress as a whole.[44]

In both *Jal mohanto* and *Pishach purohit*, Dinendrakumar partially subverts the ideology of progress. In both novels, the central plot involves an ancient priesthood that possesses knowledge of physical phenomena far in advance of the science of the Edwardian era. The central characters are at first, very much like their Western counterparts, oblivious of these advances, but eventually come to realize that this is knowledge that had been known to their forefathers. The ideological fantasy of discovery thus becomes reworked into a fantasy of rediscovery of what one has lost rather than one of finding something anew in a new land. By having switched the narrative voice from European to Bengali, Dinendrakumar transformed the discoverer's fantasy. Crucially, Dinendrakumar had already written a number of works in which the chief protagonist was a European, such as in his Robert Blake mysteries, yet he chose in these novels to make the narrative voice Bengali. This was an attempt to "own" or, more accurately, "repossess" the narrative.[45]

Similarly, the missionary fantasy is reworked to portray the prodigal son getting reacquainted with the values of his own culture, rather than natives accepting the superior culture of the colonizer. This is one of the ways in which Dinendrakumar distances his novels from Boothby's. Unlike Boothby's characters, such as Dr. Nikola, who are Europeans seeking to appropriate ancient Oriental knowledges and artifacts, Dinendrakumar's characters are men who reclaim their lost cultural heritage.

Finally, the anthropological fantasy of bifurcated temporal existences is inverted. Rather than the linear chronology of progress, where the more modern is always better than its predecessors, in Dinendrakumar's novels the ancient is superior to the modern. His novels go beyond the usual "Golden Age" myths positing that a certain specific past had attained greatness and been compatible with the present. Instead, in

---

[42] Quotation in Thomas R. Metcalf, *Ideologies of the Raj* (Cambridge, UK, 1997), 30.
[43] Ibid., 67–9.
[44] Rieder, *Colonialism* (cit. n. 1), 29–33.
[45] On "owning the narrative," see ibid., 53.

Jal mohanto and *Pishach purohit*, the past has actually attained knowledge that allows it to trounce the modern, and in order to comprehend it, the modern must play catch-up. Whereas Golden Age myths compare the societies of the past to their temporal contemporaries, Dinendrakumar compares the past to the present and finds the former's knowledge superior.

This particular orientation, where the ancient past is seen to have anticipated and then surpassed contemporary scientific and technical achievements, remains a major theme in South Asian modernities. It is today particularly prominent in right-wing Hindu nationalist politics, where claims of ancient space travel and plastic surgery are regularly made by leading politicians, including the current prime minister of India. Banu Subramaniam has dubbed this "archaic modernity." This archaic modernity is not a simple rejection of modernity and a return to a premodern past, though that is what is claimed for it. Rather, it inserts a modern vision of science and technology and their goals into a newly constructed genealogy: "Religious nationalists thus bring together a modern vision with an archaic vision, that is, an archaic modernity. By strategically employing elements of science and religion, orthodoxy and modernity, the Hindu Right is attempting to create a 'modern Hinduism' for a Hindu India. They suggest that we need to return to Hindu values while incorporating Western and Vedic sciences. Contrary to their claims, religious nationalists are not merely reverting to 'tradition' or decolonizing India or Indian history but appropriating modernity and Western science into a Hindu agenda."[46]

Thus, archaic modernity, despite its appearances, is not a total subversion of the ideology of progress. It is a subtle reworking of that ideology—a reworking that constantly defines the past in terms of a positive excess of whatever present science and technology have to offer. Instead of the colonial ideology of progress that frames the past in terms of a lack vis-à-vis the present, archaic modernity sees the present as forever approaching the past, but falling slightly short of it.

Jal mohanto features an ancient Buddhist monastery tucked away somewhere in Tibet. The monks, and especially the abbot, possess highly advanced knowledge that outpaces contemporary scientific achievements. They can cure paralysis. They can raise the dead. They can even communicate with those long dead whose corpses have ceased to exist. This, we are told, is only a fraction of what they can do. Their true powers include the secret to immortality. Likewise, in *Pishach purohit* an ancient Egyptian priest can return from the dead, assume another body, and then proceed to wreak havoc on the great imperial capital of London itself. He can control people's minds and bodies and get them to do his bidding despite their opposition. He can cure victims of bubonic plague, but he can also unleash the plague at will. His powers, like the Tibetan abbot's, are boundless, and in their presence contemporary powers pale into oblivion.

What is striking about the way these powers are described, however, is their similarity to the actual technologies that were already known to Edwardian readers. The abbot raises the dead using a machine that explicitly resembles an electric generator. When he invokes the long dead, they appear—again explicitly—like characters in silent movies. The way in which the Egyptian high priest cures or spreads plagues overtly resembles vaccinations. His ability to control others also bears much more

---

[46] Banu Subramaniam, "Archaic Modernities: Science, Secularism, and Religion in Modern India," *Social Text* 18 (2000): 67–86, on 74.

than a family resemblance to mesmerism, a science that had enjoyed official support and status in colonial Calcutta around the middle of the nineteenth century.[47] The ancient powers might be superior, but they are always represented as a superior version to what was already known.

Far from being antimodern, archaic modernity is in some ways modernity on steroids—always pushing for further enhancements of available science and technology. But instead of seeing these advances as a linear teleology, it casts them as an endless return to an ever-receding perfect past.

INTERNATIONALISM IN THE AGE OF EMPIRE

The existing scholarship on early Bengali SF frames the issue of archaic modernity almost entirely in terms of a colonial situation and a response to it. Such an impact-response paradigm produces a dichotomy between the colonizer and the colonized that leaves little room for any other actors. Dinendrakumar's novels trouble this dichotomy from the very outset and break with this narrow, binary model of the world. Besides the imperial British and the colonized Bengalis, several other people and places populate these novels. Samajpati, writing of *Pishach purohit*, noted, "the entire world is, in a way, the playfield of these characters."[48]

While much of the larger spatial canvas of these plots derives from Boothby's original novels, Dinendrakumar reworks them in crucial ways. Unlike Boothby, for whom an imperial set of characters traverse the world while the locals, so to speak, stay local, in Dinendrakumar's novels, it is the colonized Bengalis who roam the world. From Cairo to Shanghai, they travel, live, befriend, and even marry non-Bengalis. Nor are the Bengalis the only ones to move. The Japanese and Egyptians move in ways that do not always conform to the routes established for them by European colonialists. The mobile imperialist traversing the world as his petty stage is evicted from his central role, and a wider, more diverse set of characters move in a more complicated set of directions.

In *Jal mohanto*, for instance, the Bengali hero's employer and mentor is a Japanese doctor. The people whose knowledge they seek are Tibetans. In *Pishach purohit*, the villain who seeks throughout the novel to co-opt the Bengali hero to his project, and from whom the hero gains knowledge of the ancient powers, is an Egyptian. In contrast, all the central characters, barring villains, in Boothby's originals were Europeans.

Both of Dinendrakumar's novels draw upon existing Orientalist discourses in the Victorian and Edwardian worlds. *Jal mohanto* clearly resonates with myths of Shangri-La and the Victorian cult of Tibetan mysticism, which found one particularly explicit and intimate articulation in South Asia through Theosophy. Peter Bishop has argued that during the last quarter of the nineteenth century, Victorian curiosity about Tibet was transformed "into a fascination bordering on compulsion, . . . a sober concern for scientific observation became attenuated into an attitude of intense spiritual expecta-

---

[47] Alison Winter, *Mesmerized: Powers of Mind in Victorian Britain* (Chicago, Ill., 2000), 187–213; Waltraud Ernst, "Colonial Psychiatry, Magic and Religion. The Case of Mesmerism in British India," *Hist. Psychiat.* 15 (2004): 57–71.
[48] Samajpati, "Samalochana" (cit. n. 31), 413.

tion."[49] This transformation was rooted in the faltering imperial confidence and spiritual doubt that ensued from both military conflicts, such as the "Indian Mutiny" of 1857, and the publication of such works as Charles Darwin's *Origin of Species* and John Ruskin's *Modern Painters*. In a mid-Victorian society increasingly assailed by doubt, Tibet emerged as a place above "global catastrophe . . . to seed the civilization of the future."[50] This romantic image of Tibet was not opposed to science and technology. In fact, its development depended crucially upon new technologies like photography and scientific explorers such as Joseph Dalton Hooker. Bishop names this an "empirical imagination" of Tibet.[51] This spiritualized, romantic image of Tibet that was nurtured through contemporary science and technologies is precisely what we notice in Dinendrakumar's novel.

Interestingly, the British romance about Tibet had, even before the Victorian era, been frequently connected to images of ancient Egypt. From the East India Company employee and traveler Samuel Turner, in 1783, to Madame Blavatsky, the charismatic founder of Theosophy, several compared ancient Egypt to Tibet.[52] But there was an independent tradition of Egyptology, which also flourished in the late Victorian empire, along with a more popular fascination with ancient Egyptian culture.[53]

I have written in detail elsewhere about the Bengali appropriation of Egyptology and Egyptomania.[54] Several Bengali authors and intellectuals were fascinated by ancient Egypt, and some of them even imagined a direct, biological connection between the two regions. Dinendrakumar was clearly an early exponent of this Bengali Egyptomania. In *Pishach purohit*, we hear the Egyptian high priest Ra-Tai exhort the Bengali protagonist Naren Sen with redolent words: "You may not know this, but the ancient Hindus and the ancient Egyptians are the same people. They are scions of the same family. They are two branches of the same great tree."[55]

Unfortunately, the field of Bengali Tibetology remains underexplored. Some scholars have, however, pointed toward the importance of the Himalayas in general for the constitution of a modern Bengali sense of nation-space. These authors, interestingly, point to Jaladhar Sen's travelogue of the Himalayas as one of the key constitutive elements in the articulation of a new, secularized nation-space.[56] This is interesting because Sen's travelogue was precisely the text over which Dinendrakumar fell out with Sen and that he continued to claim was coauthored.[57]

Egyptology/Egyptomania and Tibetology were both prominent Victorian trends, but in Dinendrakumar's hands they acquired a distinct valence. They exalted the greatness of particular colonized cultures and posited an intimate connection between them

---

[49] Peter Bishop, *The Myth of Shangri-La: Tibet, Travel Writing and the Western Creation of Sacred Landscape* (Berkeley, Calif., 1989), 99.

[50] Ibid., 101.

[51] Ibid., 103–7.

[52] Ibid., 109–10.

[53] Elliott Colla, *Conflicted Antiquities: Egyptology, Egyptomania, Egyptian Modernity* (Durham, N.C., 2007), 179.

[54] Projit Bihari Mukharji, "The Bengali Pharaoh: Upper-Caste Aryanism, Pan-Egyptianism, and the Contested History of Biometric Nationalism in Twentieth-Century Bengal," *Comp. Stud. Soc. Hist.* 59 (2017): 446–76.

[55] Ray, *Pishach purohit* (cit. n. 14), 99.

[56] Sandeep Banerjee and Subho Basu, "Secularizing the Sacred, Imagining the Nation-Space: The Himalaya in Bengali Travelogues, 1856–1901," *Modern Asian Studies* 49 (2015): 609–49.

[57] Raha, *Kathachitrakar* (cit. n. 17), 84–96.

and Bengal. When Ra-Tai attempted to educate Naren about the ancient connections between "the Hindu" and ancient Egyptians, his speech was neither a dispassionate Egyptological discourse nor a fevered Egyptomaniacal one. Unlike Victorian Egyptomaniacs who sought to acquire ancient Egyptian artifacts without recognizing the claims of contemporary Egyptians, Ra-Tai was explicitly exercised by what he saw as the desecration of ancestral graves. His exhortations to Naren were to stop such desecration and recognize the deep affinities of the Egyptians and Indians: "If we can unite the power I possess with the power you have, then the combined force of us, both denizens of eastern lands, will easily be able to stun Europe."[58]

In the Tibetan context, the politics of anticolonialism blended seamlessly into the rising wave of Pan-Asianism. Bengalis had been deeply appreciative of Japan since their 1905 victory in the Russo-Japanese War, and Pan-Asianism had, for a time, found a strong resonance in Bengali intellectual and youth circles. Especially influential in this regard was the Japanese art critic Kakuzo Okakura. Okakura gave several talks in Calcutta around 1902, and a number of young men were deeply influenced by him. Dinendrakumar's friend and student, Aurobindo Ghose, was particularly impressed. According to some uncorroborated statements, he even attributed the founding of the revolutionary secret society he was said to be involved with directly to Okakura.[59] This Pan-Asianist influence is also attested to in the name of "Dr. Okuma" in *Jal mohanto*. One of the most important pre–World War I Japanese politicians, who twice became prime minister of Japan, was named Count Okuma Shigenobu. Count Okuma was also president of the Japan-India Society, and his speeches on the question of Indian independence were reported in the Bengali press at length. Editors of leading newspapers discussed Okuma's call for national self-awakening and social reform prior to demanding political independence.[60]

While the novels' main tropes and trends were frequently drawn from Victorian and Edwardian discourses, Dinendrakumar molded them into a very specific shape. The reworked tropes reflected the complex political location of a particularly important strand of Bengali nationalist opinion within an international context. This international context was not defined by a simple us-versus-them binary that pitted colonial Britain against colonized Bengal. Rather, it was refracted through the politics of Pan-Asianism on the one hand, and the emerging solidarity with Egyptian nationalists on the other. Neither Bengali nationalism nor Bengali SF was defined merely with reference to the colonizer. They were worked out within an emerging welter of international political alliances and sympathies.

Adding a gendered aspect to these alliances, in both novels the heroes fall in love with and marry non-Bengali women: in the first novel, a woman from a European Jewish family, and in the second novel, the daughter of a Japanese family settled in China. The representation of such alliances was complicated by fairly explicit ideologies of patriarchy and race, and drew upon the motif of rescuing the lost princess, which was popularized by European authors such as Rider Haggard.[61] Unlike the European models, the lady being rescued was not a Haggardian "white princess," nor did she belong to

---

[58] Ray, *Pishach purohit* (cit. n. 14), 101.
[59] Peter Heehs, "Foreign Influences on Bengali Revolutionary Terrorism 1902–1908," *Modern Asian Studies* 28 (1994): 533–56, on 542.
[60] Birendra Prasad, *Indian Nationalism and Asia (1900–1947)* (New Delhi, 1952), 45–6.
[61] Rieder, *Colonialism* (cit. n. 1), 41–3.

the same community as the hero. She acted instead as a very human and personal bridge between two clearly distinct cultural groups.

The structure of these gendered alliances was in fact one of the most prominent points where Dinendrakumar broke with Boothby. Where Boothby uses representations of gendered alliances to police racial boundaries, Dinendrakumar clearly uses them to build bridges with other "Oriental" peoples such as the Jews and Japanese. Gender, race, and nationalism all intersect to create a complex web of affinities and oppositions, sympathies, and anxieties that define the plot, politics, and perspective of Dinendrakumar's SF. The worlds engendered by these narratives and the sciences implicated in those imagined worlds were neither narrowly local nor seamlessly universal. Least of all were they hemmed in by the dichotomies created by the "colonial rule of difference."[62]

ANTICOLONIAL ELECTROMAGNETISM

The tremendous powers that the ancient priests controlled in Dinendrakumar's SF novels were frequently likened to electromagnetic forces. In one of the earliest scenes in *Jal mohanto*, Dr. Okuma displays his powers to the protagonist Nalini Karforma, first by having a pet cat seemingly perform successful acts of divination, and then by creating a set of tangible and realistic illusions. Finally, he is able to control Nalini's movements. When asked to explain what he had just witnessed, Nalini suggests that all can be explained by Okuma possessing a stronger willpower (*iccha-shakti*) than his own. Okuma refuses to accept this explanation because it does not clarify how the willpower actually operated. He then instructs Nalini to carefully observe the tips of his fingers, and Nalini sees "the emission of a bluish light, akin to electricity, from his fingers."[63] Likewise, in one of the earliest encounters between Naren Sen and Ra-Tai in *Pishach purohit*, Naren tries unsuccessfully to stop the latter from drowning a man in the River Thames. Perplexed by his own inability to stop the old man, Naren reflects upon Ra-Tai's mysterious ability to subvert his will and render him an automaton. Naren muses, "upon the touch of this man, quite inexplicably, the state of my body became akin to the way one's hand becomes immobilized and insensible upon [accidentally] touching an electric battery."[64]

Remarkably, these references to electromagnetism are almost entirely absent in Boothby's novels. Dinendrakumar is in fact fairly faithful in his translation of the two episodes I have cited in the foregoing paragraph, with the sole exception of inserting the explicit references to electromagnetic energies. The single stray reference to such energies in either of Boothby's novels is at a much later point in *Dr. Nikola*, which Dinendrakumar also reproduces far along in *Jal mohanto*. Hence, the insertion of the extra references earlier in the narrative was clearly a deliberate interpolation. Why, then, did Dinendrakumar choose to thus amplify the references to electromagnetic forces?

These references are to the notion of "animal electricity." Iwan Rhys Morus points out that, "at the beginning of the nineteenth century it seemed indisputable to many

---

[62] Partha Chatterjee, *The Nation and Its Fragments: Colonial and Postcolonial Histories* (Princeton, N.J., 2007).

[63] Ray, *Jal mohanto* (cit. n. 13), 80.

[64] Ray, *Pishach purohit* (cit. n. 14), 30.

natural philosophers that there was a strong connection between electricity and animal and human bodies."[65] Electricity was often referred to as a "galvanic fluid," and it was widely believed that there was a particular type of galvanic fluid that was "peculiar to the animal machine, [independent] of the influence of metals, or of any other foreign cause."[66] Morus argues that electricity, for the Victorians, was a deeply paradoxical entity: "It was invisible, ethereal and no one really knew what it was, but at the same time its effects were unambiguously physical and thrillingly visual."[67]

These paradoxes drew many to dabble in galvanism, from natural philosophers seeking to unravel the mysteries of the universe, to hack showmen out to shock and awe the public with an eye on their wallets. "By the end of the nineteenth century, electricity had been joined by a plethora of other previously unheard-of fluids and forces. These new powers held out the promise of communication without wires, of miracle cures for diseases, and even of ways of talking to the dead."[68] Researchers in the physical sciences, such as Oliver Lodge and William Crookes, attempted to deploy electrical theory to investigate psychical phenomena.[69] Richard Noakes argues that the relationship between electrical engineering, physical sciences, and spiritualism in the latter part of the Victorian era was particularly intimate.[70]

Given these paradoxes and associations, galvanism became a favorite among the early authors of what would eventually come to be SF. There are early references to galvanism in Mary Shelley's *Frankenstein* (1818), though its relationship to the body is only alluded to. A much more prominent SF use of electromagnetic theories can be witnessed in one of the most popular of these early works: Edward Bulwer-Lytton's *The Coming Race* (1871). Bulwer-Lytton imagined an underground utopia built upon the control of a "super-force" called "vril": "I should call it electricity, except that it comprehends in its manifold branches other forces of nature, to which, in our scientific nomenclature, different names are assigned, such as magnetism, galvanism, &c."[71] Bruce Clarke suggests that Bulwer-Lytton "modeled vril, at least in a roundabout way, on the same all-pervasive luminiferous or electromagnetic ether through which Maxwell had recently produced a unified theory of radiant energies."[72] *The Coming Race*, writes Morus, "was a massive hit, and vril entered the popular imagination."[73] More strikingly, in an even earlier novel, *Zanoni* (1842), Bulwer-Lytton had brought together electromagnetic theories with an ancient Oriental priesthood and the secret of immortality.[74]

The significance of these electromagnetic theories in Dinendrakumar's novels can only be fully grasped by locating the novels within the ideologies of British imperial rule in South Asia. "India" was imagined through a series of essentialist contrasts with

[65] Iwan Rhys Morus, *Shocking Bodies: Life, Death & Electricity in Victorian England* (Stroud, UK, 2011), 29.

[66] Ibid.

[67] Ibid., 9.

[68] Iwan Rhys Morus, *When Physics Became King* (Chicago, Ill., 2005), 156.

[69] Ibid., 158–9.

[70] Richard Noakes, "Cromwell Varley FRS, Electrical Discharge and Victorian Spiritualism," *Notes Rec. Roy. Soc. Lond.* 61 (2007): 5–21.

[71] Edward Bulwer-Lytton, *The Coming Race* (London, 1886), 53.

[72] Bruce Clarke, *Energy Forms: Allegory and Science in the Era of Classical Thermodynamics* (Ann Arbor, Mich., 2001), 49.

[73] Morus, *When Physics* (cit. n. 68), 266.

[74] Edward Bulwer-Lytton, *Zanoni* (Leipzig, Ger., 1842).

Britain in particular and Europe more generally. One of the most redolent of these contrasting essences was between "European science" and "Indian religion." As Inden points out, both the image of "European science" as an "exemplar of world-ordering rationality[,] the science of the heavens or natural philosophy," and of "Indian religion"—usually equated with a problematic category called "Hinduism" or "Brahmanism"—as its absolute negation, were equally fictitious.[75] Yet, through the powerful apparatus of Victorian knowledge production, both as part of the imperial-state apparatus and as seemingly dispassionate academic analysis, this contrast emerged as one of the fundamental cornerstones of imperial ideology.

This dichotomy between European science and Indian religion was part of a series of oppositions. Other dichotomies in this series included materialist/spiritualist, mechanical/vitalist, quantifiable/qualitative, and so forth. Together these binaries constructed two essentialized and diametrically opposite geocultural entities.[76] Peter van der Veer points out that though terms such as "spirituality" emerged in part out of Orientalist scholarship, being pitted against terms such as "science" and "materialism" also transformed each of these latter categories and gave them a specifically South Asian resonance.[77] One of the outcomes of this was the emergence of a set of allegedly more spiritualized sciences that did not accept the European Enlightenment as their *fons et origo*.[78] Gyan Prakash argues that one fundamental concern for elite South Asian nationalists was to find a way of undoing the binary posited between the colonizer and the colonized in terms of science/magic or materialism/spiritualism, and so forth.[79] Electromagnetism provided a pliable and convenient middle term that allowed this undoing by inhabiting a liminal space between the allegedly mutually opposed categories.

The two terms that Dinendrakumar uses most frequently in his novels to describe contemporary European sciences and the sciences of ancient Oriental priests were, respectively, *jorobadi* and *oloukik*. In the case of *oloukik*, moreover, Dinendrakumar asserts that it is an inadequate term and that the true nature of the science of the ancients far exceeds it. The two terms are usually translated in contemporary Bengali as "materialist" and "supernatural." The former, however, is not the only commonly used term for materialism. There are a range of others, including words such as *bostubad*, *bastobbad*, *dehatmobad*, *prakritik*, and so forth. Each of these emphasizes slightly different aspects of the assemblage of philosophical and popular understandings that come unevenly together in the common usage of the English term materialism. *Bostubad*, for instance, emphasizes the centrality of "things," or *bostu*. *Bastobbad* emphasizes the centrality of the "real" or "true." *Dehatmobad* emphasizes the centrality of the body and bodily effects. *Prakritik* accents the "natural" or "characteristic." *Jorobad* likewise focuses upon insensateness and inarticulateness.[80]

Jorobadi is, therefore, a person or a doctrine that holds the world to be comprised of insensate and inarticulate entities. Etymologists suggest two possible roots for the

[75] Ronald B. Inden, *Imagining India* (Bloomington, Ind., 2000), 87, 85–9.

[76] J. Lourdusamy, *Science and National Consciousness in Bengal: 1870–1930* (New Delhi, 2004).

[77] Peter van der Veer, *Imperial Encounters: Religion and Modernity in India and Britain* (Princeton, N.J., 2001).

[78] Alter, *Yoga* (cit. n. 15). See also Projit Bihari Mukharji, *Doctoring Traditions: Ayurveda, Small Technologies, and Braided Sciences* (Chicago, Ill., 2016).

[79] Gyan Prakash, *Another Reason: Science and the Imagination of Modern India* (New York, N.Y., 2000).

[80] Gyanendramohan Das, *Bangla bhashar abhidhan* [Dictionary of the Bengali language] (Allahabad, India, 1916), 614.

word *joro*: either it derives from a root that means joining, jointed, knotted, or the like, or it derives from a set of roots that signifies that which covers up.[81] Any science imagined in opposition to this is therefore one that would hold the world to be comprised of entities that are, at least potentially, sensate and articulate. This is what the ineffable science of the ancients, by implication, signifies—namely, a world that is populated by entities that are all sensate and articulate. Electromagnetism acted as a kind of master code that allowed the imagination of such a world.

The idea of enspirited, sensate, and articulate matter was not unique to Dinendra-kumar. There was in fact a much older, Greek tradition, which had been reinvigorated in early modern English natural philosophy, of conceptualizing sensate matter.[82] The Greek term that is used in philosophy to refer to this type of sensate matter is "hylo-zoic," which is derived by joining the Greek words for matter, *hule*, and for life, *zoe*. This hylozoic tradition in Europe was very much a part of the history of galvanism, which in turn informed both the English SF and theosophical traditions, and, finally through them, Dinendrakumar.[83] This is why I use the term hylozoic to describe the type of anticolonial vision that is crafted in Dinendrakumar's SF. I also use the term to highlight what I have elsewhere dubbed the "braided" nature of colonial sciences and images of science, and to eschew any implication that ideas of sensate matter were somehow an "essential" feature of "mystic eastern" ontologies.[84] I want to emphasize that colonial contact brought multiple, internally heterogeneous intellectual and practical traditions into contact, and specific protagonists creatively braided particular strands of those traditions together to craft their own scientific repertoires.

A number of physical scientists of the fin-de-siècle era embraced electromagnetic theories to develop more coherent and full-blown worldviews that challenged the mechanistic viewpoints prevailing earlier in the century.[85] Dinendrakumar's formulations neither had the coherence of those world views, nor were they based on any actual physical science research. Their roots lay in the English SF tradition of Mary Shelley, Bulwer-Lytton, and others, and in the Theosophical Society of Madame Blavatsky. Dinendrakumar braided this hylozoic strand of European thought together with a certain image of Indian religions that especially accented the yogic tradition and "spirituality," as well as the emergent trends in international politics, to produce a distinctive form of hylozoic anticolonialism.

FICTION TO THEOLOGY

Dinendrakumar's hylozoic anticolonialism would have likely remained a mere exotic footnote to history had it not been for its amplification by his student, Aurobindo Ghose. Aurobindo's recent biographer Peter Heehs writes the following: "In the half century since Sri Aurobindo's death, his reputation has continued to grow. Discussed by historians, philosophers, literary scholars, and social scientists, and admired, even

[81] Ibid.

[82] Robert Crocker, *Henry More, 1614–1687: A Biography of the Cambridge Platonist* (Dordrecht, Neth., 2003), 167–82.

[83] I am grateful to Lorraine Daston for drawing my attention to the hylozoic roots of Dinendra-kumar's vision.

[84] Mukharji, *Doctoring Traditions* (cit. n. 78).

[85] Suman Seth, "Quantum Theory and the Electromagnetic World-View," *Hist. Stud. Phys. Biol.* 35 (2004): 67–93.

worshipped, by thousands of spiritual seekers, he regularly is numbered among the most outstanding Indians of the twentieth century."[86]

Describing the philosophy of Aurobindo, or rather Sri Aurobindo, as he came to be called after his spiritual awakening, Sugata Bose writes that "Aurobindo's metaphysics was based on 'a reconciliation between Spirit and Matter, between God and Creation' and a synthesis of yoga—the union of the human and the divine. . . ."[87] This reconciliation of spirit and matter and the union of the human and the divine were precisely the themes Dinendrakumar pursued through his hylozoic anticolonialism. The similarities between Sri Aurobindo's metaphysics and Dinendrakumar's novels are striking and obvious. Given their intimate personal acquaintance the parallels are not surprising.

The importance of electromagnetism in Sri Aurobindo's thought has not yet received attention. Like Dinendrakumar's novels, Sri Aurobindo's metaphysics was constructed upon an electromagnetic code. Arguably, the most complete vision of his metaphysics is presented in his 24,000-line epic poem, *Savitri: A Legend and a Symbol*. Though *Savitri* was finally published in the 1950s, its earliest drafts date from 1916, just six years after the publication of *Pishach purohit*. In *Savitri*, Sri Aurobindo makes several clear references to electromagnetic theories as the code by which to achieve the reconciliation of spirit and matter that he sought. Describing the cosmology of matter, for instance, Sri Aurobindo wrote the following:

> An ocean of electric Energy
> Formlessly formed its strange wave-particles
> Constructing by their dance this solid scheme.[88]

Elsewhere in the same book he described how he "heard strange voices cross the ether's waves."[89] In more Maxwellian tones, Aurobindo wrote of "sight's sound-waves breaking from the soul's great deeps,"[90] or a "cave where coiled World-Energy sleeps."[91] Such language would have been easily recognizable to readers of Bulwer-Lytton's *Zanoni*. Electromagnetic images, concepts, and metaphors abound not only in *Savitri* but also in several other writings of Sri Aurobindo's. Like Dinendrakumar's SF, Sri Aurobindo also articulated a hylozoic anticolonialism. They both did so by positing an enspirited, sensate matter that subverted the central ideologies of colonial rule by undermining essentialist dichotomies.

I do not claim that Dinendrakumar invented this hylozoic anticolonialism and Aurobindo received it from him. Indeed, there were several other Bengali intellectuals of the period who articulated such ideas. Moreover, the causal arrow might have easily pointed the other way. What is worth underlining, however, is the fact that an almost identical version of this hylozoic anticolonialism appeared about the same time in the metaphysical writings of Aurobindo and in the SF novels of Dinendrakumar.

While clearly seeking to undermine colonial ideologies of progress and difference, both authors were also resolutely cosmopolitan and internationalist. Dinendraku-

[86] Heehs, "Foreign Influences" (cit. n. 59), 413.
[87] Sugata Bose, *His Majesty's Opponent* (Cambridge, Mass., 2011), 26. No source is cited for the quotation within Bose's statement.
[88] Sri Aurobindo, *Savitri: A Legend and a Symbol* (Pondicherry, India, 1997), 155.
[89] Ibid., 401.
[90] Ibid., 383.
[91] Ibid., 665.

mar's explicit acknowledgement of his debts to European authors and the complex international vision of his plots, mirror Aurobindo's own movement around this period toward a "cosmopolitan universalism" and away from a "nationalist particularism."[92]

Notwithstanding the remarkable similarities, we must remember the generic differences between the two authors. One was writing in a fictional genre and the other in a philosophical one. This is not to suggest that these genres are somehow quintessentially distinctive or that either has some kind of distinguishing innate value. As Rieder points out, genres acquire their specific connotations within a system that includes the entire set of neighboring genres. Therefore, the emergence of a genre of philosophical writings on what I have called hylozoic anticolonialism, alongside the SF novels articulating the same basic ideas, will in fact change the generic horizons of the latter—the horizon of expectations within which people produce and consume these texts.

Hence, the role of Sri Aurobindo's writings in interrogating Dinendrakumar's SF served to alter the horizon of expectations that consumers would have brought to these novels. Likewise, the popular awareness of these novels would have shaped the horizon of expectations within which some of these readers consumed Aurobindo's philosophy. We do not have explicit testimonies of how, or if, readers interpreted these two authors together, but it is safe to assume that the coexistence of these works and the relationship between their authors would have had some impact upon the expectations brought to either of the writer's works.

CONCLUSION

In this essay I make five interrelated points. I argue first that Dinendrakumar's novels explicitly acknowledged their debts to specific European authors and were, in fact, adaptations of two novels by Australian author Guy Boothby. Dinendrakumar's novels therefore cannot be located within any insular cultural genealogy. Nor indeed can they be read as giving voice to some deeper or specifically Bengali mode of imagining or writing. They must be located historically within the web of resemblances that the author himself acknowledges.

Second, I argue that the novels rework the "ideology of progress" that is central to early British SF. Analyzing the three ideological fantasies of the discoverer, the missionary, and the anthropologist, I demonstrate how Dinendrakumar neither embraces nor rejects all of these ingredients of the ideology of progress. Rather, by strategically refiguring Boothby's narratives, he articulates a form of archaic modernity, where progress means returning to an ever-receding past that needs to be repossessed as a legacy.

Third, I call attention to how, in focusing heavily upon the colonial fault lines, extant scholarship has missed the rampant internationalism of early SF novels. Moreover, this internationalism itself was not based on any simple equality of all nations of the world. It was in fact shaped by the emerging tides of Pan-Asianism, and to a lesser extent, Pan-Egyptianism. Dinendrakumar's novels sought to create a new scientifically oriented future for Bengalis, and by extension, for Indians, but did so by locating them within multiple lines of political force ranging from the anticolonial to the Pan-Asiatic.

---

[92] Andrew Sartori, "The Transfiguration of Duty in Aurobindo's Essays on the Gita," *Modern Intellectual History* 7 (2010): 319–34. See also Sugata Bose, "The Spirit and Form of an Ethical Polity: A Meditation on Aurobindo's Thought," *Mod. Int. Hist.* 4 (2007): 129–44.

My fourth argument is actually the crux of the essay. It posits that electromagnetic theories played a fundamental role in Dinendrakumar's vision. I demonstrate how he inserted a range of new electromagnetic references into the novels that were absent in Boothby's original works. I argue further that he had encountered these electromagnetic ideas either through a tradition within English SF, or the related but very distinct networks of the Theosophical Society. Irrespective of the source, I argue that the utility of these theories for Dinendrakumar was that they subverted the central binaries that structured British imperial ideology in South Asia. By cutting across such oppositions as that between matter and spirit, or science and religion, electromagnetism offered Dinendrakumar a pliable, "fluid" resource, if you permit the pun, by which to undermine these Orientalist dichotomies. This form of anticolonialism that seeks to dismantle imperial ideologies by subverting mutually exclusive dichotomies through a liminal electromagnetic fluid, I call "hylozoic anticolonialism."

Finally, I demonstrate that Dinendrakumar's hylozoic anticolonialism bore a strong resemblance to the metaphysical writings of his student and friend Sri Aurobindo. I argue, however, that we should not take this to necessarily mean that either one of them borrowed from the other. Having said this, I still emphasize that the coexistence of metaphysical and science-fictional articulations of hylozoic anticolonialism, a coexistence created by authors who were known to be personally close, must have shaped the horizon of expectations with which readers read both the metaphysics and the SF. Or, to put it differently, we might say the existence of both the authors within each other's fields of force would necessarily have changed the galvanic potential carried by hylozoic anticolonialism.

Parallel Prophecies:
Science Fiction and Futurology
in the Twentieth Century

by Peter J. Bowler*

ABSTRACT

Science fiction and popular science writing sometimes intersect—authors have written in both genres, and magazines have published their writing side by side. Producing "hard" science fiction depends on getting the known science right and ensuring that predictions seem plausible. There have even been claims that science fiction should be used to make science itself accessible to a wider public. This article uses case studies from mid-twentieth-century Britain to illustrate the range of attitudes adopted by authors and publishers who sought to associate the two genres and to probe the notion of "plausibility" in this context.

Science fiction comes in many forms, but what is known as "hard" science fiction involves setting a story in a world transformed, often by the development of new technologies. Enthusiasts of the subgenre insist that the predicted technologies are at least plausible given the current state of science; speculating too far ahead leads to imaginative scenarios more suited to the invocation of transcendental issues. There is no sharp division, of course, and eminent writers such as Arthur C. Clarke and Isaac Asimov worked across the whole spectrum. Both were also involved in popular science writing and sometimes used a nonfiction format to predict future developments and their implications. It is the potential for interaction between these two genres—science fiction and popular science writing—that forms the topic of this study.

In a recent book I used both science fiction and nonfiction futurological speculations to chart competing visions of future developments through the mid-twentieth century.[1] Most writers specialized in one area or the other, but a few followed Clarke and Asimov and chose to work in both. This article uses lesser-known figures from British literature of the period to suggest that, from certain viewpoints, we can see science fiction and popular science writing as two sides of the same coin. There are obvious differences created by the fiction-writers' need to set a human-interest story within the projected future. But since both genres require an imaginative extrapolation of the tech-

---

* School of History and Anthropology, Queen's University, Belfast, BT7 1NN, UK; P.Bowler@qub.ac.uk.
[1] Peter J. Bowler, *A History of the Future: Prophets of Progress from H. G. Wells to Isaac Asimov* (Cambridge, UK, 2017).

nical developments, they have a shared foundation. Looking at authors who wrote in both genres illuminates the interests and purposes that drove them to speculate about the future. Did they intend to educate, to inspire, or to entertain? And how did these functions relate to each other?

Parallels can also be seen in the formats in which predictions were published. Fiction and nonfiction sometimes appeared in the same magazine or book, and could even intermingle. The pulp science fiction magazines of interwar America emerged from popular science publishing, and the same process is visible to a lesser extent in British magazines. As Iwan Morus shows in the case of the "telectroscope," presentation of some plausible inventions hovered between fact and fiction.[2] There are cases in which fictional accounts set in the future were embedded in otherwise didactic futurological speculations. The boundary between fiction and nonfiction is also blurred in stories presenting an account of future states of the world as a historical document somehow transmitted back to the present, a technique used by both H. G. Wells and Olaf Stapledon.

This study looks at the different ways in which science fiction authors approached their writing, at the various objectives they hoped to achieve, and the formats in which they published. It focuses on the extent to which authors worried about the plausibility of the imagined future—that is, whether its scientific background was accurate by the standards then accepted and whether any new technologies seemed at least within reach of current developments. Isaac Asimov once insisted that "real science fiction" is defined as "those stories that deal with scientific ideas and their impact written by someone knowledgeable in science."[3] Clarke agreed, and, as Colin Milburn shows, his fiction actually went on to inspire Gerald Feinberg's scientific research.[4] This in turn raises the question of expertise: How was the author to gain the technical knowledge needed in order to be plausible, and was it necessary for the reader to be informed of any qualifications the author possessed? For popular science writing beyond the level of mass-market journalism it was usual for the author to have some technical background, even if gained informally. But was this also necessary to write and promote hard science fiction, where the element of plausibility really counted? We shall see that there were some authors who viewed their fiction as an indirect way of instructing readers who might not be willing to read popular science books or magazines. For them, the element of plausibility was crucial.

At the same time, there were authors who had no technical expertise but who still opted to imagine technologically advanced futures. Some were experienced writers of conventional adventure stories who saw the prospect of space travel as a source of potentially exciting backgrounds. They might feel little or no compulsion for accuracy, and instead provided exciting "space operas" that populated the solar system with planets inhabited by aliens long after astronomers had realized that none had conditions suitable for advanced life. Still other authors were literary figures with strong moral or ideological commitments who did not share the popular science writers' enthusiasm for technological progress—but who could still mirror their passion for plausibility, if not accuracy. Aldous Huxley, for example, was fully aware of the latest scientific developments—his brother Julian, a biologist, worked with J. B. S. Haldane, and *Brave*

---

[2] Iwan Rhys Morus, "Looking into the Future: The Telectroscope That Wasn't There," in this volume.
[3] Isaac Asimov, "A Literature of Ideas," in *Towards Tomorrow* (London, 1983), 161–9, on 167.
[4] Colin Milburn, "Ahead of Time: Gerald Feinberg and the Governance of Futurity," in this volume.

New World (1932) grew out of direct knowledge of the latter's *Daedalus*.[5] The writers who predicted catastrophic future wars drew on genuine fears apparent in the popular press about futuristic weapons, some of which were highly speculative; but a few books about conflict in the future were also written by military experts who wanted to drive home their warnings. Amanda Rees and Joanna Radin show in this volume how authors later than Aldous Huxley (John Wyndham and Michael Crichton, respectively) continued this tradition, using science fiction to publicize their concerns about the misuse of scientific expertise.[6]

This study will begin by considering how science fiction (SF), to a certain extent, emerged from popular science writing, and will then reflect on how, and for what purposes, fiction and fact coexisted in the construction of possible futures. It will go on to examine the backgrounds of some of the lesser-known writers, and look at the sometimes uneasy coexistence of entertainment and education in their stories, before focusing closely on the juvenile SF market and its audiences. How did experts manage to inspire and to teach at the same time? Is it possible to combine the creation of an imaginary world with accurate scientific information? How much inaccuracy can be tolerated before plausibility is lost? And does that matter, if the writer continues to inspire the audience to dream of a space-based future?

BLURRING THE BOUNDARIES

To begin, we need to appreciate how thin the boundary was between popular science writing and science fiction. Both routinely appeared side by side in the same publications and were very occasionally combined in the same text. The pioneering Hugo Gernsback initially introduced fiction into his popular science magazines aimed at young men hoping to improve themselves through technical education. When he began publishing science fiction magazines, beginning with *Amazing Stories* in 1928, Gernsback ensured that these too would report on the latest developments in science and have their material authenticated by expert advisory boards.[7] The American pulp magazines were avidly read by British fans, and the "fanzines" included some factual reports on scientific developments.[8] Of the small number of dedicated popular science magazines published in interwar Britain, one—*Armchair Science*—included a few science fiction stories along with numerous more practical efforts at prediction; the latter were mostly in the regular monthly column written by the magazine's science advisor and later editor, "Professor" A. M. Low. A serialized story written by Low was later published as the novel *Mars Breaks Through*.[9]

---

[5] J. B. S. Haldane, *Daedalus; or Science and the Future* (London, 1924); Aldous Huxley, *Brave New World* (London, 1932).

[6] Amanda Rees, "From Technician's Extravaganza to Logical Fantasy: Science and Society in John Wyndham's Postwar Fiction, 1951–1960"; Joanna Radin, "The Speculative Present: How Michael Crichton Colonized the Future of Science and Technology," both in this volume.

[7] See, for instance, John Cheng, *Astounding Wonder: Imagining Science and Science Fiction in Interwar America* (Philadelphia, Penn., 2012), 101. See also Mike Ashley, *The Time Machine: The Story of the Science-Fiction Pulp Magazines from the Beginning to 1950* (Liverpool, 2000).

[8] Charlotte Sleigh, "'Come on You Demented Modernists, Let's Hear from You': Science Fans as Literary Critics in the 1930s," in *Being Modern: The Cultural Impact of Science in the Early Twentieth Century*, ed. Robert Bud et al. (London, 2018), 147–65, on 5.

[9] A. M. Low's story, "The Great Murchison Mystery," was published in 15 parts in *Armchair Science*, beginning in September 1936, 246–8 and 283; and ending in November 1937, 342–3 and

Many juvenile periodicals routinely combined fiction and nonfiction, allowing predictions to appear in both formats. By the 1920s, the long-established *Boy's Own Paper* ran adventure stories alongside accounts of modern inventions in areas such as
radio and aviation. Occasional stories had a science fiction element based on amazing
new aircraft and other technologies. In 1937, a prediction of flights to New York at
one thousand miles per hour described how the passengers would actually experience
the journey. Throughout the 1930s, the weekly *Modern Boy* and its companion *Modern Boy's Annual* also combined the two formats and included many predictions of
future technical developments. They published stories on aviation themes and science fiction stories based on the adventures of "Captain Justice," which impressed
the young Brian Aldiss. In 1935, *Modern Boy's Annual* included a story outlining a
day in the life of a boy living in the year 2500. There was no adventure component—
the fictional format was used to predict the kind of inventions that might transform
everyday life in the future.[10] In the postwar era, *Boy's Own Paper* published accounts
of futuristic aircraft and serialized a space exploration story, "The Lord of Luna," in
1961.[11] Other magazines and comics specialized in this theme, along with a host of
writers of juvenile science fiction novels discussed in detail below.

Some authors worked in both formats, sometimes even including them in the same
text. They occasionally used a hybrid style of writing, presenting serious predictions—
which might easily have been written in a didactic style—as fiction, in order to gain
greater effect. The result is not fiction in the form adopted in space operas or other science fiction works with a dramatic human-interest content. In its most common form,
it appears instead as an account of history or "contemporary" social conditions written
in the future and somehow transmitted back to the present. It is meant to read as a factual account, even though it is obviously fiction. Olaf Stapledon's *Last and First Men*
of 1930 is presented as an historical account "by one of the Last Men" who was relaying his words to the present-day narrator's mind.[12] *The Shape of Things to Come*, by
H. G. Wells (published in 1933 and soon the basis of a spectacular movie), introduced
its account of the future as "The Dream Book of Dr. Philip Raven," with the narrator
asking whether Raven's story is fiction, imagination, or "is it really what he believed it
to be?" The book was a part of a universal history for students of the year 2106. This is

---

378. The novel was later published as *Mars Breaks Through: The Great Murchison Mystery* (London,
1938). See also "A Fairy Tale of the Future: A Medical Fantasy," *Armchair Science*, December 1930,
507–8. On British popular science magazines, see Bowler, *Science for All: The Popularization of Science in Early Twentieth-Century Britain* (Chicago, Ill., 2009), 161–94, and on A. M. Low, see 261–2.
Low's academic credentials were very slender—he was an inventor and popular science writer; Ursula
Bloom, *He Lit the Lamp: A Biography of Professor A. M. Low* (London, 1959).

[10] For a futuristic aviation story, see S. T. James, "White Heather," *Boy's Own Paper*, August and
September 1932, 1–23 and 65–88. For the article "London to New York in A.D. 2000" (by "an aircraft
designer"), see *Boy's Own Annual*, 1936–37, 447–8 (the annual is a compilation of material originally
issued in the monthly magazines). *Modern Boy's Annual* was also filled with images of futuristic technologies; see, for instance, 1931, 7, 34, 137; 1933, 11–13, 117; 1934, 94; 1935, 23, 25, 188. See also
"A Modern Boy in the Year 2,500!" *Modern Boy's Annual*, 1935, 23–5. On the "Captain Justice" stories, see Brian Aldiss and David Wingrove, *Trillion Year Spree: The History of Science Fiction* (London, 1986), 596–7.

[11] H. B. Gregory, "The Lord of Luna," *Boy's Own Paper*, January 1961, 20–3 and 55–6.

[12] See the preface to Olaf Stapledon, *Last and First Men: A Story of the Near and Far Future* (1930;
repr., London, 1999), xvii–xx. On his work, see Leslie Fielder, *Olaf Stapledon: A Man Divided* (Oxford, UK, 1983); and Robert Crossley, *Olaf Stapledon: Speaking for the Future* (Liverpool, UK,
1994).

a technique used to great effect more recently by Erik Conway and Naomi Oreskes to warn of the dangers of climate change.[13]

Several authors embedded future histories within otherwise straightforward efforts to predict the future consequences of technological and social development. The best-known example is J. B. S. Haldane's *Daedalus*, which attempted to shift the focus of prediction to the biological sciences and inspired Huxley to write *Brave New World*. After outlining his vision of how new reproductive technologies will affect society, Haldane sought to give his readers a more realistic image of the consequences by offering them an essay written by "a rather stupid undergraduate" a hundred and fifty years in the future. He used the same technique again in his broader vision of the future, "The Last Judgement." To drive home his prediction that the earth will eventually become uninhabitable, he offers us an account that "will be broadcast to infants on the planet Venus some forty million years hence."[14]

Another example of an embedded future narrative occurs in J. P. Lockhart-Mummery's *After Us* of 1936. Lockhart-Mummery was an eminent surgeon and a friend of H. G. Wells. His book presents an optimistic vision of the benefits that technology might bring in the future. Amidst his predictions he presents two chapters written as the diary of a family from New Zealand visiting London in the year 2456. There is no real drama, but the family's reactions give the reader an effective view of what the future might actually be like for our descendants.[15]

Future histories were not the only means of interleaving fact and fiction when making predictions. During the postwar era of enthusiasm for space exploration, astronomer Patrick Moore and science fiction writer William F. Temple both published popular science texts that included sections of fiction to bring home the potential implications of space travel. Both intended their books to be read by the younger generation, presumably hoping that exciting stories set in an imaginary near-future world would make their educational texts more palatable. Their effectiveness must have been limited, because neither returned to this format despite producing extensive bodies of fiction and—in Moore's case—popular science (both authors are discussed in more detail below).

CELEBRATION OR WARNING?

Much of the popular science literature celebrated the progress being achieved and predicted the benefits of future developments. A magazine such as *Armchair Science*

---

[13] H. G. Wells, *The World Set Free: A Story of Mankind* (London, 1914), 1–29; Wells, *The Shape of Things to Come* (London, 1933), 6–7. There are many studies of Wells; those focusing on his predictive writings include Roslynn D. Haynes, *H. G. Wells: Discoverer of the Future* (London, 1980); John Huntington, *The Logic of Fantasy: H. G. Wells and Science Fiction* (New York, NY, 1982); and Patrick Parrinder, *Shadows of the Future: H. G. Wells, Science Fiction and Prophecy* (Liverpool, UK, 1995). For the modern example, see Erik H. Conway and Naomi Oreskes, *The Collapse of Western Civilization* (New York, N.Y., 2014), which looks back from the year 2393.

[14] Haldane, *Daedalus* (cit. n. 5), 57–68; also Krishna R. Dronamraju, ed., *Haldane's Daedalus Revisited* (Oxford, 1995), 39–43. "The Last Judgement" appears in Haldane's *Possible Worlds and Other Essays* (1927; repr., London, 1932), 287–311; for the Venus broadcast see 292–309. See Mark Adams, "Last Judgment: The Visionary Biology of J. B. S. Haldane," *Journal of the History of Biology* 33 (2000): 457–91; Huxley, *Brave New World* (cit. n. 5).

[15] J. P. Lockhart-Mummery, *After Us: Or the World as It Might Be* (London, 1936), 113–38 and 12. Clarke would use the same "future diary" technique much later on; Arthur C. Clarke, *July 20, 2019: A Day in the Life of the 21st Century* (London, 1986).

was aimed at a readership that was assumed to be young, white, and probably male, although some efforts were made to encourage women to get involved. There was a strong focus on the convenience and comfort to be gained from new and proposed inventions, but also an awareness that readers would find some topics exciting. Articles frequently hailed the prospect of wider travel opportunities that might become available from civil aviation. It was a small step from this theme to imagining the adventures that might ensue if humankind could eventually travel into space.

This leap was made by the editor of *Armchair Science*, "Professor" A. M. Low. In addition to his monthly contributions to that magazine (often speculating about future developments), he was a prolific author of popular science books. These included two that conjectured how future science and technology would affect everyday life: *The Future* (1925) and *Our Wonderful World of Tomorrow* (1934). He also wrote two science fiction novels in the late 1930s—a space adventure under the title *Adrift in the Stratosphere*, and a story in which Martians seek control of earth via telepathy (which Low believed was a real phenomenon that could be artificially enhanced).[16] Introducing *Adrift in the Stratosphere*, Low argued that science fiction could play a significant role in predicting future developments: "The line between possibility and imagination is not easy to define. Wild dreams of flying a few centuries ago have become the commonplace of travel. Electricity, radio, synthetic chemistry, and the study of the atom have shown us new worlds, more striking than any fiction. To deny the existence of other worlds is even more vain than to conceive their eventual conquest. For knowledge should teach us above all that every fact is a matter of opinion."[17] The last sentence might give philosophers of science food for thought, but Low's purpose is clear; he wanted his readers to believe that speculations about space travel were pointing the way toward developments in the real world.

Low was generally optimistic about the future, but he was aware of the potential dangers that might arise from some new technologies—his writing was celebratory but not utopian. In particular, he was aware that new weapons were being developed that might have catastrophic, destructive potential. Others were more inclined to worry about the danger these weapons might pose if used in a future war, which was a fear that seemed all too realistic given the international tensions of the period. In *The Shape of Things to Come*, Wells imagined a global conflict that almost destroys civilization, but offered the hope that this would clear the way for the emergence of a World State. The interwar years saw the publication of many novels predicting future wars leading to the destruction of civilization, and even human extinction. Unsurprisingly, historians who focus on these novels as the only popular predictions about the future tend to depict the period as a time of unrelieved gloom.[18] The theme of new weapons including poison

---

[16] A. M. Low, *Adrift in the Stratosphere* (London, 1937). He was not the only writer to use the term "stratosphere" to mean more or less the same as "outer space." See also his *Mars Breaks Through* (cit. n. 9), which involves telepathy; *The Future* (London, 1925); and *Our Wonderful World of Tomorrow: A Scientific Forecast of the Men, Women and the World of the Future* (London, 1934). On Low's career, see n. 9.

[17] Low, *Adrift* (cit. n 16), foreword (unpaginated).

[18] See especially I. F. Clarke, *The Pattern of Expectation, 1644–2001* (London, 1979), 225–51; and Clarke, *Voices Prophesying War: Future Wars, 1763-3749* (Oxford, UK, 1992). Richard Overy's *The Morbid Age: Britain and the Crisis of Civilization* (London, 2009) uses a much wider range of evidence but also fails to take note of the more positive visions of those who welcomed the development of science.

gas, germs, and even atomic bombs added to the sense of threat—Harold Macmillan later remembered that the foreboding was equivalent to that generated by the threat of a nuclear holocaust during the Cold War.[19]

Some worried about the dehumanizing effects of new inventions. Enthusiasts such as Low and Lockhart-Mummery imagined a future world in which everyday life benefited from a host of conveniences. Others were not so sure the effects would always be positive. Haldane's *Daedalus* was enthusiastic also, but he made no effort to conceal the disruptive social effects of the reproductive technologies he predicted. Haldane himself welcomed the disruption of cloying traditional values, but he can hardly have been surprised when Huxley pointed out that the effect might be to destroy what had traditionally been assumed to be the central purposes of life. *Brave New World* has become a classic example of the way in which science fiction can be used to ward against the assumption that all new technologies will have beneficial consequences.[20] The perversion of civilization might turn out to be even worse than its total destruction by war.

PLAUSIBILITY AND EXPERTISE

A prediction of future inventions made as an extrapolation from current developments by a popular science writer was intended to have some degree of plausibility. This would in part be conferred by the expertise of the writer—someone actually familiar with current research was in a better position than a total outsider to foresee the next steps. It would also be important for readers to be made aware that the source of the prediction was someone with relevant experience or qualifications. This would be a straightforward extension of the common practice by which the editors and publishers of popular science literature printed acknowledgements of their authors' expertise. However, this would not necessarily be the case for science fiction. If written purely for entertainment, the public might prefer to know that the author had a reputation for producing exciting fiction even in areas unrelated to science and technology. There were indeed plenty of novelists who tried their hand at science fiction having already gained a reputation for producing crime stories or even westerns. Such a writer could replace stagecoaches with exotic spaceships and bandits with bug-eyed monsters and still hope to excite and entertain the readers. They could get their science background totally wrong and have little fear that the average reader would notice.

Somewhere between these two extremes we have those who—as Asimov recommended—wrote science fiction from their own knowledge of current science and technology. Here there might be a real desire to retain as much plausibility as possible in the background against which an adventure story was set. There might even be a genuine desire to inform the reader about science and current technical developments, thereby slipping an element of education into what was presented as pure entertainment. This kind of literature could be written from both sides of the "celebration versus warning" dichotomy outlined above. A space opera might be written to inform readers about our knowledge of the solar system and the opportunities that might be opened up by the latest rocket technologies. A future-war novel might convey a warning from

---

[19] Harold Macmillan, *Winds of Change, 1914–1939* (London, 1966), 575.
[20] Haldane, *Daedalus*; Huxley, *Brave New World* (both cit. n. 5).

someone whose knowledge of military affairs gave them real insight into the impact of potentially devastating advances in military technology. In both cases, the publisher might want readers to know that the author was writing with a foundation of genuine experience in the relevant fields.

This expertise could be gained in various ways, ranging from a career as a fully qualified scientist to less formal immersion in the popular science literature of better quality. As a biologist, Haldane would represent the most formal level of qualifications, and his readers would have been aware that his increasing involvement in popular science writing was based on a genuine desire to educate the public. At the opposite end of the spectrum, Aldous Huxley was an established novelist who did not need to provide evidence of technical expertise to write *Brave New World*—although he and Haldane were part of the same social circle.

In the applied sciences, expertise could be derived either from formal qualifications or from experience in industry. Young men also gained a technical training in the military during both of the world wars. Many of the science fiction "fans" studied by Charlotte Sleigh came from this background, as did Patrick Moore and Arthur C. Clarke, although Clarke went on to gain a degree in physics after the war. Both became widely recognized as "experts"—and both published widely in both popular science and science fiction. They also played key roles in the British Interplanetary Society, which served as a forum where experts and enthusiasts could interact.[21] A few popular science writers simply became well read in their chosen field, often gaining extra credibility by assuring their readers that they consulted friends who had more formal qualifications. This would be a route much more easily followed by those beginning a career writing science fiction.

FUTURE WARS AS WARNINGS

The various routes by which popular science writers sought credibility were all open to those who moved sideways into science fiction and wanted to write fiction that had some degree of plausibility. Two areas of SF clearly illustrate these points—the first dealing with those future-war novels actually written by military or aviation experts, and the second with the space operas that became popular in the 1950s, especially those aimed at juvenile readers.

Most anticipations of the horrors in future wars were written by novelists with little technical knowledge, and often invoked futuristic weapons that seem highly implausible today (although it should be remembered that reports of "death rays" were commonplace in the more sensationalist newspapers at the time). But there were also a few experts dealing with military technologies who sought to use fiction as a means of driving home their warnings about the potential impact of new weapons. For them, the use of a fictional account of a future war's impact on real people offered a way to emphasize to the general public the same warnings they offered to the military and political elites in their more technical works.

Hector C. Bywater was a naval expert who wrote on British sea power in the Great War, and commented on the situations that would face both Britain and America as

---

[21] See, for instance, Oliver Dunnett, "Patrick Moore, Arthur C. Clarke and 'British Outer Space' in the Mid-Twentieth Century," *Cultural Geographies* 19 (2012): 505–22.

military technologies developed. His *Sea Power in the Pacific* of 1921 warned of the potential for conflict between the United States and Japan. The Washington naval conference that year defused some tensions, but Bywater still thought conflict might be triggered, and in 1925 he wrote *The Great Pacific War* to explore the consequences. In this fictional account he imagined a surprise attack by the Japanese on Panama and the Philippines, followed by a naval battle in which aviation plays a crucial role—including poison gas delivered from airplanes. Bywater also argued that torpedoes launched from the air would be more effective than bombs. A second edition of *Sea Power in the Pacific* drew attention to these predictions and forecast that the development of aircraft carriers would create a major problem for the Japanese navy.[22]

A similar argument was advanced by Edward Frank Spanner, an enthusiast of heavier-than-air flight, who wrote extensively in opposition to the lobby promoting airships. In his *Armaments and the Non-Combatant* of 1927, he stressed the potential for devastating air attacks on civilian targets, pointing out that the navy was powerless to protect Britain from a continental power with a strong air force. He wanted the Royal Navy to be given control of the country's air force and to train its own pilots. To drive home his point, Spanner wrote a series of novels in which these pilots became the "naviators" who defended the country in a series of imaginary future wars. *The Broken Trident* (1926) had already paved the way for *Armaments and the Non-Combatant*, using Germany as the hypothetical enemy. In the same year, *The Naviators* explored the possibility of a naval war to defend Australia against a Japanese attack, with retaliatory air strikes against the Japanese fleet being launched from secret bases in the Pacific. A third novel, *The Harbour of Death*, imagined a Turkish air attack on the Suez canal.[23]

Another novelist who eventually gained a substantial reputation began his career with a future-war story conveying a serious warning. This was Nevil Shute, the pen name of Nevil Shute Norway, who had worked originally as an airship and aircraft designer. In 1929, he contributed a single chapter to a book written by Sir Charles Burney in which he explored the possibility of future developments in heavier-than-air machines. He was pessimistic about the rate of progress that could be expected, predicting it would take fifty years to produce a machine that could carry a substantial payload at ninety-five miles per hour.[24]

By the late 1930s, as he began his second career as a novelist, Shute realized that developments had advanced much more rapidly, and that bombing from aircraft posed a serious threat if war broke out. His novel *What Happened to the Corbetts* of 1939 recounts the experiences of a family driven from their home in a country virtually paralyzed by German bombing. He imagined the attacks being delivered from cloud cover,

---

[22] Hector C. Bywater, *The Great Pacific War: A History of the American-Japanese Campaign of 1931–33* (London, 1925), esp. 280 and 287. See also Bywater, *Sea Power in the Pacific: A Study of the American-Japanese Naval Problem*, 2nd ed. (London, 1934), xxi–xxii. Bywater was naval correspondent of the London *Daily Telegraph* and author of *Navies and Nations: A Review of Naval Developments since the Great War* (London, 1927); and *A Searchlight on the Navy* (London, 1934).

[23] E. F. Spanner, *Armaments and the Non-Combatant: To the 'Front-line Troops' of the Future* (London, 1927); *The Broken Trident* (London, 1926); *The Naviators* (London, 1926); *The Harbour of Death* (London, 1927). Spanner's campaign against airships included *This Airship Business* (London, 1927); *About Airships* (London, 1929); and *Gentlemen Prefer Aeroplanes!* (London, 1928).

[24] For Nevil Shute's experiences in aviation, see his *Slide Rule: The Autobiography of an Engineer* (1954; repr., London, 2009). See also Nevil Shute Norway, "Heavier-than-air Craft," in *The World, the Air and the Future*, by Sir Charles Dennistoun Burney (London, 1929), 259–90, on page 279; Burney himself added a footnote suggesting that the prediction of very slow progress was too pessimistic.

safe from British defending fighters and antiaircraft fire, due to superior navigational equipment. The publisher distributed a thousand copies free to workers already being recruited to Air Raid Precautions, and Shute's prediction of the devastating effects of conventional bombing proved all too accurate in the war that soon broke out. In the Cold War era, he returned to the same genre with *On the Beach*, imagining the gradual extinction of the human race after a nuclear war that used bombs designed to maximize radioactive fallout. Published in 1957 and filmed a few years later, the story drove home the threat of a nuclear holocaust. In this case, the predictions were never put to the test, but Shute's novel again showed how effective a fictional account of a future war could be in exposing the dangers of progress in military technology.[25]

SPACE OPERAS AS EDUCATION

By the 1950s, rocket technology had become a reality, and the space race began, prompting an explosion of public interest. Fictional accounts of interplanetary explo-ration, often dismissed as space operas by critics, became widely available. Some of these, as we shall see in the next section, were written by authors with no real concern for scientific accuracy. Their stories imagined undiscovered planets, usually inhabited by aliens either humanoid or bug-eyed. A few authors took the scientific and technical background more seriously. Some actually contributed to the popular science litera-ture and saw their fiction as an additional way to inform people about the facts of as-tronomy and the latest technical developments. A considerable proportion of this lit-erature was aimed at juvenile readers (mostly assumed to be boys). The aim was to produce material that would encourage interest in the space program without making exaggerated claims about what might be discovered. Authors who had some degree of technical expertise wrote space adventures that were intended, in theory at least, to ed-ucate the young about scientists' understanding of the solar system and how it might be explored.

The best-known author to combine science fiction with popular science writing is Arthur C. Clarke, whose first two science fiction novels—*Prelude to Space* and *The Sands of Mars*—came out in 1951, alongside his nonfiction survey *The Exploration of Space*. Another early novel, *Islands in the Sky*, was intended for teenage readers and explored his vision of how space travel would work in a period when the planets had been colonized. This was actually written for a US publisher, although also issued in Britain.[26] Here the need to ensure plausibility was especially important if his young readers were to understand the potential of the space program. Clarke's work is al-ready the subject of considerable scholarly attention, so the following survey focuses on two other British writers who shared the ambition of using both genres to promote

[25] Nevil Shute, *What Happened to the Corbetts* (1939; repr., London, 2009); for details of the pub-lication, see the preface to this 2009 edition. See also Shute, *On the Beach* (1957; repr., London, 1990).

[26] For Clarke's views on the link between science and fiction, see his *Profiles of the Future: An En-quiry into the Limits of the Possible* (London, 1962), 10. *Islands in the Sky* (Philadelphia, Penn., 1952) was commissioned for a series issued by the John C. Winston Company. See also Clarke, *The Explora-tion of Space* (London, 1951); and *2001: A Space Odyssey* (London, 1968). The extensive literature on Clarke includes Neil McAleer, *Odyssey: The Authorized Biography of Arthur C. Clarke* (London, 1992); Joseph D. Olander and Martin Harry Greenberg, eds., *Arthur C. Clarke* (Edinburgh, 1979); and Robert Poole, "The Challenge of the Spaceship: Arthur C. Clarke and the History of the Future, 1930–1970," *Hist. Technol.* 28 (2012): 255–80.

the space program to juvenile readers. These were the astronomer Patrick Moore and the aviation writer Captain W. E. Johns. Both became renowned figures in Britain, although not for their science fiction. Moore was the country's best-known writer and broadcaster on astronomical topics and to a lesser extent on space exploration. He moved into science fiction to encourage interest in space and to educate his younger readers. Johns was an aviation expert who became a popular author of stories aimed at teenage boys. Later in his career he acquired an interest in space, and he too made at least limited efforts to base his science fiction on known astronomical facts.

Patrick Moore served as a navigator in the Royal Air Force during World War II, but had no academic training in science; instead, he had the enthusiasm of the dedicated amateur and the ability to communicate it directly to the public. His passion was astronomy, but he also took a strong interest in space flight. Moore wrote both popular science and science fiction for younger readers and was unusually articulate about his belief that the fiction could serve a valuable function in helping to educate teenagers about the universe. In 1954, he published *The Boys' Book of Space*, a well-illustrated survey that explained current and pending developments in the space programs, and also speculated on what might become possible in the more distant future. The book was revised several times to accommodate new developments. Moore took on the job of editing *Spaceflight* in 1956; this was a new magazine aimed at general readers and published by the British Interplanetary Society. There had been some debate in the society about how its more formal journal could be made more accessible, and the creation of a separate publication for the general public was the solution adopted.[27]

By this time Moore had already begun to write science fiction stories for younger readers. As he later admitted, his early efforts, including *The Master of the Moon* and *The Frozen Planet*, were highly speculative and included plenty of contact with alien races on the moon and planets.[28] He soon began to realize, however, that if his fiction was to have any hope of meshing with his efforts to promote interest in science, he would have to avoid these more exotic scenarios. His coauthored *Out into Space* provides another example in which fiction and popular science are combined in the same text. Written with A. L. Helm and published in 1954, this book describes the visit of two teenagers to their uncle, an astronomer who is working with a friend involved with the space program. The two experts give informal talks about their work, allowing Moore to adopt the popular science mode. But the adults also decide to spice up their lectures by telling stories about what they expect will be discovered when we really do venture into space. These chapters are pure science fiction. This is the only book in which Moore used this literary device, so perhaps he found it too artificial. He continued to write prolifically in the two genres—but kept them separate. From this point onward, his stories about the exploration of the solar system made a greater effort to stay within the bounds of what was scientifically plausible.

Moore's new approach was spelled out explicitly in a 1957 book on the history and purpose of science fiction; it was entitled *Science and Fiction*. The book probably had little appeal to the newer generation of fiction writers, since it made clear his dislike of the more speculative approach. He argued that distrust of the literary establishment

---

[27] Patrick Moore, *The Boy's Book of Space*, 3rd ed. (London, 1959). For details of how the magazine *Spaceflight* was founded, see Moore's editorial in volume 1, October 1956, 1.
[28] Patrick Moore, *The Master of the Moon: An Enthralling Science-Fiction Story* (London, 1952); Moore, *The Frozen Planet* (London, 1954).

was slowly being overcome, but insisted that the kind of science fiction that could be identified with horror comics and their bug-eyed monsters was to be avoided. He identified two kinds of science fiction that were appropriate—stories in which the scientific background was completely sound, and those that were as accurate as possible but took some license with what might be discovered in the future. The need for accuracy was important, because the stories should have an educational purpose: "Many people who would refuse to study a popular scientific book are all too ready to read science fiction; if they learn some genuine science in the process, so much the better."[29] In a concluding chapter on the future of science fiction, he complained about the inaccuracy of the science in so much of the genre, and about its tendency to adopt a pessimistic vision of the future. Moore hoped that the planned launch of artificial satellites would encourage a more realistic and optimistic approach, spelling the end of the bug-eyed-monsters. An appendix records an address he made to a 1955 UNESCO meeting in Madrid in which he expounded his vision of science fiction as a means of educating the public, and proposed a "selection board" of experienced writers that would grant official approval for serious works. He admits that his proposal lost by a vote of eight votes to seven.[30]

Moore began to put his vision of hard science fiction into practice with his *Mission to Mars* of 1955. Here he gave a realistic account of how a voyage to Mars might be made and projected an image of the red planet that was more or less within the limits of what was known at the time: it had a thin atmosphere and supported some vegetation near the "canals." The dust jacket insisted that the story was "all vividly told against the accurate and authentic background of the author's scientific knowledge" and included an advertisement for *The Boy's Book of Space*.[31] To generate excitement Moore did, however, go so far as to predict primitive flying animals that were a danger to the human explorers. A series of later books followed the process of colonizing Mars and involved the creation of a breathable atmosphere.[32]

A slightly more exotic vision of the solar system came in *Quest of the Spaceways*, with a follow-up imagining a project to give Venus a breathable atmosphere. *Invader from Space*, originally issued in 1963, warned of dangerous microorganisms being brought back from Venus. It was reissued in 1975 with comments on developments in astronomical knowledge of Venus and a dust jacket insisting on the book's "accurate background detail." None of these books involved intelligent aliens—the excitement came from the threats posed by more natural phenomena or human activity. *The Wheel in Space* and *Wanderer in Space* gave fairly realistic accounts of the building of a space station and the efforts made by the staff of a moon base to head off the threat of an exploding asteroid.[33]

---

[29] Patrick Moore, *Science and Fiction* (London, 1957), 10.

[30] Ibid., 182–3 and 185; for the appendix, see 186–9. Moore insisted that he was not proposing censorship, because the board would have no power to prevent publication.

[31] Patrick Moore, *Mission to Mars* (London, 1955); dust-jacket information is from the Cambridge University Library copy.

[32] Patrick Moore, *The Domes of Mars* (London, 1956); Moore, *The Voices of Mars* (London, 1957); Moore, *Peril on Mars* (London, 1958); Moore, *Raider of Mars* (London 1959).

[33] Patrick Moore, *Quest of the Spaceways* (London, 1955). Moore's book about Venus is *World of Mists* (London, 1956); a later book, *Planet of Fire* (London, 1959), is also set on Venus. See also *Invader from Space*, rev. ed. (London, 1975). Moore's story of the building of the space station invokes human threats; *Wheel in Space: The Amazing Story of how a Satellite was Built in Space in Spite of Treachery and Danger* (London, 1956); and *Wanderer in Space* (London, 1961).

William Earle Johns had flown as a fighter pilot in World War I, and by the 1930s had built a reputation as a writer on aviation-related issues, including fictional accounts of air combat. His nonfiction included a coauthored manual for amateur flyers, and he edited the magazine *Popular Flying*.[34] His war stories began to appear in magazines including *Modern Boy*, which also listed him as its "aviation expert." Many of them focused on the exploits of "Biggles," otherwise the fictional Major James Bigglesworth of the Royal Flying Corps (later the Royal Air Force); many of these stories were soon being collected in book form. The books were increasingly directed at juvenile readers, and Biggles was gradually transformed into an "air police" pilot who grappled with wrongdoers around the world. A few of these stories featured advanced technologies, including mysterious power sources and "death rays" capable of bringing down aircraft.[35] By the 1950s, the Biggles books were enormously popular among teenage boys (the present author included).

Via a serendipitous introduction to postwar rocket programs, Johns moved from Biggles to stories of space exploration. He explained this later, in 1960:

> My introduction to astronautics was accidental. I happened to dine with some pioneers of rocketry and found myself listening to talk of vertical flight (as opposed to horizontal) in terms so casual that it might already have been achieved. It was apparently merely a question of time and money.
>
> From that moment I was "sold" on astronautics. It seemed time the younger generation was told something of what the future had in store, so events could be followed intelligently. The problem was how to do this. Text books with their terrifying mathematics, and astronomical figures to make the brain reel, were unlikely to appeal. . . . I resolved to "sugar the pill" by wrapping fact and surmise in fiction.
>
> Even this was not easy, for the modern boy has an inquiring mind and technical details cannot be glossed over. . . . No. In a book intended to be instructional as well as entertaining this was not easy.[36]

The result was Johns's first science fiction novel, *Kings of Space*, published in 1954. In a foreword, he explained why he chose to circumvent the slow development of rocket technology by imagining the invention of an entirely new technology: "It is unlikely that a rocket will answer the question of space flight, for although it has taught us much its limitations are already apparent."[37] There will either be very slow improvement, he thought, or some unexpected breakthrough. In the story, the problems of rocket propulsion are explained by the inventor Professor Brane, who takes a group of youngsters on a journey around the solar system in his "flying saucer" powered by cosmic rays. They encounter a Martian civilization devastated by disease, and "pre-

---

[34] William Earle Johns, *Fighting Planes and Aces* (London, 1932); Johns, *The Air V.C.'s* (London, 1935); Johns, *Thrilling Flights* (London 1935); Johns, *Some Milestones in Aviation* (London, 1935); H. M. Schofield and W. E. Johns, *The Pictorial Flying Book* (London 1932). On Johns's life, see Peter Berresford Ellis and Piers Williams, *By Jove, Biggles! The Life of Captain W. E. Johns* (London, 1981).

[35] For a collection of the early Biggles stories with a preface explaining their origins, see W. E. Johns, *The Camels are Coming* (London 1932); this was reprinted with additions as *Biggles: Pioneer Air Fighter* (London, 1954). *Biggles Hits the Trail* (London, 1941) featured new power sources and death rays, while *No Rest for Biggles* (London, 1956) has him tracking down a stolen American prototype for an antiaircraft ray.

[36] W. E. Johns, *To Worlds Unknown: A Story of Interstellar Exploration* (London, 1960), 5–6.

[37] W. E. Johns, *Kings of Space: A Story of Interplanetary Exploration* (London, 1954), 11; see also 35.

historic" monsters on Venus. Ten more books in a similar vein appeared over the next decade.

Johns's stories are highly fanciful, but he did try to use correct terminology where appropriate. He seems to have gained some information from his brother Russell, who had some knowledge of science.[38] In *Now to the Stars*, he apologizes for some looseness in his language but provides correct definitions of terms such as star, planet, and satellite. *The Edge of Beyond* includes a foreword correctly explaining the nature of the galaxy and conceding that the idea of a Martian civilization responsible for the planet's so-called canals was now largely discounted. *To Worlds Unknown* had a list of "Definitions You should Know," again all accurate.[39] Johns seems to have felt that the liberties he was taking with scientific fact were justified, because they allowed him to arouse the interest of teenagers in space travel and help them appreciate that real progress was being made. As he wrote in 1962: "However remarkable, or even fantastic, some of these adventures may seem, it is safe to predict that when men of Earth have succeeded in sending space ships into outer space (as they will) the discoveries made by the crews will be even harder to believe." In the following year he welcomed the launch of the first man into space: "so space fiction is no longer far removed from space fact, if, indeed, it has not become the same thing." Soon, he believed, there would be men on the moon—and we shall probably watch them on television.[40]

DEGREES OF PLAUSIBILITY

Moore and Johns were by no means the only British writers aiming their space fiction at juvenile readers. This became a popular genre that attracted authors from a wide variety of backgrounds, many with little or no knowledge of science or technology. Their efforts display a range of plausibility, with some expressing at least the intention of keeping close to the known facts, and others venturing into highly speculative accounts of mysterious planets and aliens; this range was also reflected in the popular comics, most of which by now included at least some science fiction content.

Space fiction had begun to appear in comics and juvenile magazines during the interwar years and became commonplace with the popularity of figures such as Flash Gordon and Buck Rogers. By the 1950s, these magazines began to link science fiction with popular science articles, often dealing with futuristic themes such as the space program. We have already noted the interleaving of fact and fiction in the *Boy's Own Paper*, which expanded its coverage of space in the 1950s. Its efforts were paralleled in several comics; the most successful of these was the weekly *Eagle*, launched on 14 April 1950. The editor was clergyman Marcus Morris, who intended it as an alternative to the American horror comics flooding into the country. It provided wholesome adventure stories in a comic-strip format, but also included a significant amount of nonfiction devoted to hobbies and technology.

Prominent on the front page were the adventures of space-pilot Dan Dare, which combined futuristic technology with occasional nostalgic allusions to British traditions.

[38] Ellis and Williams, *By Jove, Biggles!* (cit. n. 34), 213–4.
[39] W. E. Johns, *Now to the Stars* (London, 1956), 11n; Johns, *The Edge of Beyond* (London, 1958), 11–15; and Johns, *To Worlds Unknown* (cit. n. 36), 11–12.
[40] W. E. Johns, *Worlds of Wonder: More Adventures in Space* (London, 1962), 7; *The Man Who Vanished into Space* (London, 1963), 5–6.

There was no attempt to invoke current developments in the space program—Dan Dare's rockets did not need clumsy booster stages, and the planets of the solar system were inhabited by intelligent aliens (the Treens of Venus were a particular threat under their evil leader, the Mekon). Yet *Eagle* also became famous for its centerfold illustrations of technological innovations featuring "exploded" drawings of the interior mechanisms by Leslie Ashwell Wood. Some of these depicted futuristic aircraft, rockets, and space stations. When Sputnik was launched, *Eagle* hailed it as the fulfillment of its own prophecy of an artificial moon. If Dan Dare was meant to arouse enthusiasm for the space program, the nonfiction contents kept speculation down to a more realistic level.[41]

Eagle was the most successful of a number of comics, several of which featured space adventures on their front pages. One rival flagrantly copied *Eagle*'s format but focused more closely on the theme of space travel. This was *Rocket*, under the editorship of wartime air-ace Douglas Bader, which launched on 21 April 1956, as "The First Space-Age Weekly." The Dan Dare lookalike on the cover was Captain Falcon, along with notices of a number of other science fiction stories in both comic and text formats, including Flash Gordon. Even more than *Eagle*, *Rocket* took the element of science education seriously. In the first issue, Bader declared: "In *Rocket* we hope to give you all the Space-Age factual information together with entertainment and exciting adventure." This issue had an article on plans for an artificial satellite, and the comic ran a regular "Science with a Smile" column with stories about "St. Rockets: The Science College of Tomorrow."[42] A series of articles on "Frontiers of Space" by Commander J. Yunge-Bateman, Royal Navy, provided information on planned developments, and later issues contained articles on nuclear energy. The comic only ran for thirty-two weeks, suggesting that there was a limit to how far the space-exploration theme could be pushed. Even so, it provides an illustration of how closely the themes of science fiction and educational science could become associated when juvenile readers were the target.

Along with the comics came a flood of books aimed at both adult and juvenile readers. British editions of the second series of the long-running American series of "Tom Swift" books for teenagers were printed, with several devoted to space travel.[43] Some

[41] For a selection of the contents of *Eagle*, see Daniel Tatarsky, ed., *Eagle Annual: The 1950s* (London, 2007); futuristic centerfolds can be found on 30–1, 120–1, 128–9, 144–5. The anonymous "Eagle Moon Prophecy Comes True," reproduced on page 148, notes that an artificial satellite had been predicted in the comic in September 1955. For more details, see Sally Morris and Jan Hallwood, *Living with Eagles. Priest to Publisher: The Life and Times of Marcus Morris* (Cambridge, UK, 1998); and James Chapman, *British Comics: A Cultural History* (London, 2011), 60–71. On Dan Dare, see Tony Watkins, "Piloting the Nation: Dan Dare in the 1950s," in *A Necessary Fantasy? The Heroic Figure in Children's Popular Fiction*, ed. Dudley Jones and Tony Watkins (New York, N.Y., 2000), 153–76; and Oliver Dunnett, "Framing Landscape: Dan Dare, the *Eagle* and Post-War Culture in Britain," in *Comic Book Geographies*, ed. J. Dittmer (Stuttgart, 2014), 23–40.

[42] Douglas Bader, "Why I am Editing *Rocket*," *Rocket*, 21 April 1956, 5; Herbert O. Johnson, "Plastic Moon to Circle the Earth," *Rocket*, 21 April 1956, 14; also in the same issue, see "St. Rockets: The Science College of Tomorrow," 11–2; and "Frontiers of Space," 15. "The First Space-Age Weekly" was the masthead on the front page of every issue.

[43] See, for instance, Victor Appleton II, *Tom Swift and his Rocket Ship* (London, 1960); as well as *Tom Swift and his Outpost in Space* (London, 1961). See also Appleton II, *Tom Swift in the Race to the Moon* (London, 1969), originally published in the United States in 1958 and presumably issued in Britain in response to the actual moon landing. The books were actually the work of James Duncan Lawrence.

locally produced equivalents were equally imaginative, originating from writers with an established reputation for conventional adventure stories who ventured into the realm of space opera. They may have acquired a genuine desire to arouse enthusiasm for the topic of space exploration, but their primary concern was entertainment, not scientific plausibility. One author, Hereward Ohlson, imagined his hero "Thunderbolt" exploring interstellar space, conveniently ignoring the question of how travel to the stars would be achieved. Most of the books concentrated on the solar system, even if it was rather imaginatively depicted.[44] Phillip Briggs and Angus Macvicar produced a series of books in which teenage boys somehow became involved in the exploration of mysterious planets inhabited by alien races. Macvicar did write a separate book titled *Satellite 7*, which described the construction of a space station with some degree of plausibility—apart from the fact that the launch site for the rockets was set on a remote Scottish island. At this point in time, the British still hoped to have their own space program, and stories in which rockets were built through private enterprise or by eccentric inventors abounded. Some of Macvicar's later books had appendixes entitled "Diagrams and Technical Data," but these were merely drawings of imaginary rocket ships, space stations, and so forth.[45]

Implausibility, even downright inaccuracy, were no barriers to achieving publishing success, as illustrated by the work of E. C. Elliot, one of several pen names used by Reginald Alec Martin. Previously, he had written a wide range of adventure stories for juveniles, including (under the name Reg Dixon) the "Pocomoto" series set in the American West. When he turned to science fiction in 1954, he published, under the name Rafe Bernard, an imaginary account of the building of the first space station. This was reasonably plausible apart from the suggestion that the "wheel in the sky" had to be rotated in order to maintain its orbital velocity.[46] The same misunderstanding reappeared in the series he wrote under the E. C. Eliott pen name, in which he described the adventures of Kemlo, a teenager born to parents living on a space station. Fifteen of these stories were published over a period of a decade (and the author of the present study confesses to having read early contributions to the series with enthusiasm). All of the children born aboard the stations had somehow been adapted so they could breathe in outer space—the author was presumably thinking in terms of the old idea of the "ether." The rotation mistake also appears in a later contribution to the series, while Kemlo's adventures around the solar system include a journey to a "minor galaxy" consisting of a cloud of debris. The Milky Way is defined as "a group of insignificant satellites formed around a minor star."[47]

Other authors specializing in juvenile literature usually strove to avoid such flagrant inaccuracies, even when they had no scientific training. This allowed their work to be

---

[44] Hereward Ohlson, *Thunderbolt of the Spaceways: The Story of a Daring Pioneer of the Twenty-Second Century* (London, 1954); *Thunderbolt and the Rebel Planet: The Captain of the Spaceways Leads an Expedition to the Strange World of Pluvius* (London, 1956).

[45] Angus Macvicar, *Satellite 7* (London, 1958); *Secret of the Lost Planet* (London, 1955); *Red Fire on the Lost Planet* (London, 1959). For the "technical" appendixes, see Macvicar, *Super Nova and the Rogue Satellite* (Leicester, 1969); and *Super Nova and the Frozen Man* (Leicester 1970). Phillip Biggs wrote *Escape from Gravity: The Gripping Story of a Journey to a New Planet and the Adventures that Followed* (London, 1955); and *The Silent Planet: The Story of Two Boys and a Daring Rescue from Outer Space* (London, 1957).

[46] Rafe Bernard, *The Wheel in the Sky* (London, 1954), 149–50.

[47] E. C. Eliott, *Kemlo and the Star Men* (London, 1955), 30. On rotating the stations, see Eliott, *Kemlo and the Satellite Builders* (London, 1950), 11.

presented by the publishers as a genuine effort to instruct the young about the realities of astronomy and space exploration, even when substantial liberties were taken regarding the possibility of life elsewhere in the solar system. Mary Elwyn Patchett was already an established author when she turned to the theme of space exploration in 1953. Her *Kidnappers of Space* included a map of the moon, and its dust jacket proclaimed the following: "This thrilling story is not just a fantasy. The Rocket lore, the interplanetary flavor of the book, and the astronomical information are as accurate as present knowledge allows." A sequel's dust jacket again claimed that "all incidental astronomical information is correct" and carried an advertisement for the British edition of an authoritative American account of the space program by Lynn Poole. A later book imagined a Russian probe to Venus bringing back alien life that became dangerous when exposed to our own radioactive fallout.[48]

The tension between plausibility and excitement was even more evident in the writings of William F. Temple, who produced a serious book on the space program as well as a highly imaginative series of space operas. Temple was a one-time roommate of Arthur C. Clarke and a fellow member of the British Interplanetary Society. He was a prolific writer of science fiction, best known for his novel *The Four-Sided Triangle* about a device that duplicated human beings and was tested by two men in love with the same woman. His "Martin Magnus" space-opera series, aimed at teenage readers, had the solar system inhabited by intelligent aliens, including telepathic Venusians. But Temple also produced *The True Book about Space Travel*, paralleling Clarke's popular descriptions of the space program and its prospects, and giving details of the solar system, rocket technologies, and the prospect of atomic propulsion. It was well illustrated and obviously appropriate for a juvenile readership. The book concluded with a fictional chapter written in the form of a letter from a teenager to his friend in 1983 describing a journey to the moon, where there is already a well-established base. Here again, then, we see the elision of the distinction between fiction and popular science. In this case, the purpose was to inspire the young enthusiast with what might become possible if the developments described in the factual section of the text come to pass.[49]

CONCLUSIONS

Patrick Moore's ambition to provide hard science fiction that would be exciting enough to arouse the interest of younger readers and provide them with some reasonably accurate information at the same time illustrates both the potentials and the limitations of the link between popular science and fiction. His intention to produce fiction with a degree of scientific plausibility was paralleled in the work of Johns, Patchett, and Temple—although Johns's technical expertise was confined to aviation, and both Patchett and Temple seem to have been largely self-taught. Their efforts were products of a trend among authors and magazine editors who sought to popularize science and

---

[48] M. E. Patchett, *Kidnappers of Space: The Story of Two Boys in a Spaceship Abducted by the Golden Men of Mars* (London, 1953); *Lost on Venus: The Thrilling Story of Two Boys who Land on the Planet and Explore a Fantastic World* (London, 1954). Dust-jacket information is from the Cambridge University Library first editions; see Lynn Poole, *Your Trip into Space* (London, 1954). See also Patchett, *The Venus Project* (Leicester, 1963).

[49] William F. Temple, *The True Book about Space Travel* (London, 1954), 136–44. See also Temple, *The Four-Sided Triangle* (London, 1949); Temple, *Martin Magnus, Planet Rover* (London, 1954); Temple, *Martin Magnus on Venus* (London, 1955); and Temple, *Martin Magnus on Mars* (London, 1956).

also inform the public, and who saw fiction as a way of sugarcoating their educational efforts. Their work provided a significant alternative to that of the increasing number of science fiction writers who were critical of applied science and its impact on society. Yet the work of Bywater, Spanner, and Shute shows that even at this opposite end of the ideological spectrum, expertise could still have a role to play when fiction was used to emphasize warnings about developments in military technology.

Serious efforts to predict future developments in a popular science format and a hard science fiction story both require the creation of an imaginary world shaped by plausible scientific and technical progress. Whether the two activities could be made fully compatible was a question only a few, most notably Moore, bothered to explore in detail. Writers who strove to ensure a degree of accuracy—as well as their publishers—were keen to let their readers know that their imaginary futures were conceived from a foundation of real experience with current science and technology. The authors who openly promoted the space program claimed to have both entertainment and education in mind when they wrote, but the efforts of Johns (and the earliest fiction by Moore) show how easy it was for even a writer with some expertise to slip into a level of exaggeration that ignored current scientific knowledge. There was certainly some common ground between the popular science predictions and science fiction; both needed to extrapolate from what was known to what might happen. But the demands of fiction writing made it hard to keep within the boundaries of scientific plausibility. There could be useful parallels between the two activities, but there was also an inevitable tension between their requirements.

Meanwhile, authors whose main concern was to provide entertainment had no interest in retaining any serious level of plausibility and sometimes simply ignored the correct use of technical information and terminology. But perhaps this did not really matter. The books by E. C. Elliot were as popular as those of more conscientious authors, and played a similar role in leading young readers toward an interest in space travel. Most of them probably did not look back to see how unreliable their guide had been. But ultimately, did that make a difference if it took them to where they wanted to go?

Locating *Kexue Xiangsheng* (Science Crosstalk) in Relation to the Selective Tradition of Chinese Science Fiction

by Nathaniel Isaacson*

ABSTRACT

Kexue xiangsheng (science crosstalk) features comic dialogues aimed at popularizing knowledge in the physical and social sciences. This genre emerged in the late 1950s in the People's Republic of China (PRC) as part of a massive effort in the state-supervised culture industry to promote science. The genre shared many of the hallmarks of PRC instrumentalist science fiction, as both were based on a Soviet model. Authors and literary theorists like Guo Moruo, Ye Yonglie, and Gu Junzheng reiterated developmental narratives of socialism and of the power of science as a tool for mastery of nature developed by authors like Maxim Gorky and Mikhail Il'in. These works of socialist realism narrated transformations in the consciousness of their characters as they came to understand guiding principles of the world around them, including basic science, evolution, and dialectical materialism. Dramatic forms like *kexue xiangsheng* worked in concert with other socialist-realist representative modes, including popular performance, reportage, fiction, film, song, and reappropriations of premodern literary forms. In the process, notions of scientific thinking were conflated with political orthodoxy in promoting public health and political campaigns, and science was dismantled as a professional institution, shifting from a rationalized bureaucratic endeavor to grassroots efforts aimed at solving pragmatic problems. Through education in what I term the "quotidian utopian"—small health and hygiene measures that had the potential to ameliorate major health challenges—these popular science genres also straddled the line between Frederic Jameson's "Utopian form and Utopian wish," between what was part utopian text and part expression of the impulse to enact utopia through changes in policy and reconfigurations of the collective body.

INTRODUCTION

This study places an emerging popular science/fiction genre—*kexue xiangsheng* (science crosstalk)—in the context of China's Mao-era field of instrumentalist art, and the

* Department of Foreign Languages and Literatures, North Carolina State University, Withers 310, Campus Box 8106, Raleigh, NC 27695-8106, USA; nkisaacs@ncsu.edu.

I wish to thank Ruth Gross, head of the NCSU Department of Foreign Languages and Literatures; Jeff Braden, dean of the College of Humanities and Social Sciences; and the National Humanities Center for their generous provision of a 2018 Summer Residency that enabled me to conduct the necessary research to bring this project to completion. I would also like to thank Wu Yan, Li Guangyi, Jia Liyuan, and the rest of the diligent organizers of the "First International Conference of Utopian and Science Fiction Studies" (Beijing, China, December 2016) for inviting me to present an early version of this article.

era's politicized definitions of science and superstition. First appearing in 1956, *kexue xiangsheng* created a new didactic genre by combining popular science with a pre-existing mode of comedic dialogue. During the early years of the Republic of China, with the establishment of the journals *Kexue* (Science, introduced January 1915) and *Xin qingnian* (New Youth, August 1915), science fiction soon "ceded its position to instrumentalist genres richer in explanatory quality, the *kexue xiaopin* (science essay) and the *kepu sanwen* (popular science essay)," eventually eclipsing the first wave of Chinese science fiction (hereafter SF).[1] *Kexue xiangsheng* is a uniquely Chinese contribution to popular science writing, one of many popular genres that emerged in twentieth-century China. Alongside the scientific essays were *kexue tonghua* (children's science stories), *kexue shi* (science poetry), and *kexue yuyan* (science fables).

Emergent genres like *kexue xiangsheng* often repurposed literary forms predating China's literary revolution for the sake of popular science writing, combining fictional settings and characters to present basic knowledge in science and technology to a broad audience. Broadly speaking, these genres appended scientific knowledge to pre-existing writing styles, deploying fictional narratives to illustrate practical aspects of science to a juvenile or amateur audience. Various definitions and descriptions further divide these genres into subgenres, occasionally blurring the lines between forms and identifying public confusion as to the differences between them. All sought to use simple, yet richly descriptive language to teach science in discrete, easily digestible units. These topics ranged from Newton's laws of motion to evolution, to the structure of atoms, and to more quotidian subjects like how fabric dyes work.

Kexue xiangsheng is a performance genre at the margins of mainstream SF that emerged in the context of instrumentalist science and fiction, and sought to influence the near future. In contrast to Stephen Hilgartner's description of a "culturally-dominant" view of popular science as a "distortion" by nonexperts,[2] midcentury Chinese popular science writing valorized amateur production and dissemination of scientific knowledge. As works of socialist realism, they narrated transformations in the consciousness of their characters as they came to understand the guiding material principles of the world around them. The evolution of political and scientific consciousness of the characters in these narratives was to be mirrored in a transformation of the consciousness of their audience, thereby advancing the scientific education of the populace and leading to practical action in the real world that would help establish a socialist utopia.

Xiangsheng, or crosstalk, is a traditional, popular performance style in China, most often featuring a comedic dialogue between two characters, though some examples feature solo monologues, or ensembles of three or more. Performances typically feature punning and allusions, and jokes centered around differences in local Chinese dialects. Acts may also feature passages of *kuaiban* narrative rhyming accompanied by a wooden clapper, singing, or musical instruments. Popular science writing from the late 1910s through the end of the Mao era was formally influenced by prose gen-

---

[1] Wu Yan, *Kehuan wenxue lungang* [Science fiction criticism] (Chongqing, 2013), 234. Unless otherwise indicated, Chinese names are rendered in *pinyin*, with family names appearing first and given names second. Translations from Chinese are my own, unless indicated otherwise. Courtesy translations of titles and keywords appearing throughout the text, and in brackets in footnotes, are largely my own, though I adopt standardized translations whenever possible.
[2] Stephen Hilgartner, "The Dominant View of Popularization: Conceptual Problems, Political Uses," *Soc. Stud. Sci.* 20 (1990): 519–39.

res that predated the rise of science fiction as a tool of modernization, while its content focused on political and scientific literacy. After the establishment of the PRC in 1949, the state made a conscious effort to incorporate *xiangsheng*, traditionally associated with street performance, into the project of socialist-realist didactic art. *Kexue xiangsheng* is a uniquely Chinese form of popular science writing and was part of the popular science network of the PRC in the 1950s. In the starkest terms, *kexue xiangsheng* is neither science nor fiction, but my examination of how the genre fit into the media landscape sheds light on the political purposes of science and art during the Mao era, and how notions of scientific thinking were conflated with political orthodoxy. For *kexue xiangsheng*, the medium was the message—the characters in the dialogues perform the democratization of science, demonstrating the significance of scientific knowledge in the world around them and its implications for national strength, and explicitly inviting the audience to participate in this process as part of the dialectical triumph of Marxist thought.

I have previously argued that the position of the SF imaginary in the polygeneric literary landscape of the late Qing and Republican era fits models of SF in particular, and literature more broadly, as part of a cultural field that recognizes literary works not in terms of adherence to a Platonic ideal, but in terms of their economic, political, and artistic valuation in comparison to other similar works and media.[3] In this study, I propose that this somewhat amorphous model allows examination of *kexue xiangsheng* as a minor discourse closely related to SF as a global selective tradition, and within China's own SF history. Through an examination of this amorphous conceptual map of thematically and politically connected works, I place *kexue xiangsheng* within the cultural field, as a close neighbor of mainstream SF. Rather than asking why there is so little SF in China in comparison to any number of other nations, we might instead ask why genres like *kexue xiangsheng* appeared only in China.

A NEW GENRE—SCIENCE CROSSTALK

Kexue xiangsheng emerged in the context of a debate regarding the appropriate language for science and education, and can be seen in evolutionary terms as part of the developmental history of the Chinese language.[4] In the pages of volumes like *Zenyang xie ziran kexue duwu* (How to write about the natural sciences), and the pages of *Renmin ribao* (well known as "People's Daily" but not published in English), popular science writing was meant to be concise and favor accessibility over potentially obscure or difficult scientific topics. These writings implicitly called for the professionalization of popular science writers as the vanguard of a program of children's scientific education. The debate over the vernacular language, while officially settled in the canonization of May Fourth and the New Culture Movement (1915–21), continued in terms of explicitly identifying the appropriate register and content of educational material. Thus, science fiction as a speculative genre of adventure and fantasy catering to an educated adult audience was remolded into a children's genre of plainspoken lan-

---

[3] For a summary of recent approaches to SF and their potential application to the history of Chinese SF, see Nathaniel Isaacson, *Celestial Empire: the Emergence of Chinese Science Fiction* (Middleton, Conn., 2017), 29–31.

[4] He Dazhang and Ma Dechong, "Zenyang bianxie ziran kexue tongsu zuopin" [How to produce works of science popularization], in *Zenyang bianxie kexue tongsu zuopin* (Beijing, 1958), 145–6.

guage and quotidian concerns. Author and communist leader Guo Moruo (1892–1978) advocated a more direct language of clarity and precision, even going so far as to suggest that authors use fewer adjectives. Guo saw concise writing as a means to bring citizens to terms with the realities of dialectical materialism.[5]

Literature now labeled *kexue huanxiang xiaoshuo* (science-fantasy fiction) saw the second high tide of original and translated works of SF in mainland China during the late 1950s, when it shared numerous characteristics with contemporaneous popular science forms. Like its nonfiction cousins, SF during the 1950s was didactic and predominately written for a juvenile audience. Paralleling the shift in the literary field at large and intensifying advocacy of the primacy of political content over entertainment value, SF writers felt an even stronger need to create work that would address immediate social issues and spur on social change.[6]

The various restrictions on the form, content, and political purpose of the arts meant that *kexue xiangsheng* was quite like contemporary works labeled as SF. Two stories published in 1956 that received attention in People's Daily as examples of SF that would foster children's interest in science are Chi Shuchang's "Gediao bizi de daxiang" (The elephant who lost its trunk),[7] and Ye Zhishan's "Shizong de gege" (The lost brother).[8] "Gediao bizi de daxiang" is set in a very near future where scientific advances lead a group of students visiting a pig farm to mistake an oversized hog for an elephant. People's Daily praised this dream of socialist plenty as a realistic possibility, noting that the launch of Sputnik and other advances signaled that such abundance was not far off. The newspaper characterized fantasies of the power of the people as "completely accurate." It went on: "A young veterinarian at the Five Star Agricultural Institute in Mingcheng Village, Panshi County in Jilin Province has injected cow's milk into the thyroid glands of living pigs weighing over 60 *jin*. Using milk to control their rate of metabolism causes them to grow faster. The fourth test resulted in one pig that gained an average of 2.48 *jin* per day. Although they are not as big as elephants, this demonstrates the science fantasy in 'Gediao bizi de daxiang' has real significance."[9]

Rudolph Wagner categorizes "'science phantasy fiction' (*kexue huanxiang xiaoshuo*)" as a subset of science belles-lettres (*kexue wenyi*), which is in turn a subcategory of popular science writing (*kexue puji*), noting that as of the early 1980s, all of the SF authors accepted to the Writers Union belonged to the category of "Children's Lit-

---

[5] Guo Moruo, "Zenyang ba wenzhang xiede zhunque, xianming, shengdong" [How to write accurate, fresh, and vivid articles], *Renmin ribao* [People's Daily], 21 March 1958, 10.

[6] Wu Yan, *Kehuan wenxue lungang* (cit. n. 1), 234–6. See also Isaacson, *Celestial Empire* (cit. n. 3), 167–75.

[7] Chi Shuchang, "Gediao bizi de daxiang" [The elephant who lost its trunk], *Zhongguo kehuan xiaoshuo jingdian* [Classics of Chinese sf], ed. Ye Yonglie (Wuhan, 2006), 122–35.

[8] Ye Zhishan, "Shizong de gege" [The lost brother], in *Zhongguo kehuan* (cit. n. 7), 136–62.

[9] Guan Yu, "Liang ben kexue huanxiang xiaoshuo" [Two science fiction stories], People's Daily, 30 May 1958, 8. One *jin* is approximately 1.1 pounds, or half a kilogram. Xie Jin's 1962 film, *Da Li, Xiao Li he Lao Li* [Big Li, little Li and old Li] also echoes the themes of Chi Shuchang's dream of socialist plenty, depicting the production team at a pig farm and drawing a direct connection between the healthy bodies of the communal work unit and the clean efficiency of their meat-packing operation. A central plot point focuses on commune members' willingness to participate in a radio calisthenics program. Anchored around the technological device of the radio, the film uses a cross-sectional set design for the shared apartment housing to establish a vision of communal space. *Da Li, Xiao Li he Lao Li*, directed by Xie Jin (1962; Guangzhou: Guangzhou qiaojiaren wenhua chuanbo youxian gongsi, 2004), DVD.

erature." This meant that both the categorical and institutional place of these works was associated with popular science and children's literature regardless of literary quality or scientific content.[10] Wagner's categorization of Chinese SF as "lobby literature," a highly politicized subset of children's literature alongside numerous popular science genres, matches prevailing definitions of SF during the 1950s. In the pages of People's Daily during the late 1950s, a turn in the discourse of *kexue huanxiang* toward an emphasis on children's literature and inspiring interest in science is apparent. Terms like science essay, science fiction (*kexue huanxiang xiaoshuo*), and science belles-lettres often appeared in the context of discussions of children's education. Calls for submissions to the newspaper appearing in the second half of 1949 asked for works that focused primarily on political leadership, while the fifth criterion on the list was children's stories and *kexue xiaopin* (science essays).

The first *kexue xiangsheng* was written by Gu Junzheng (1902–80); "Yidui hao banlü" (A good couple), appeared in February 1956 in the children's magazine *Zhong xuesheng* (Middle school student). The comedic dialogue was later reprinted in the collected volume, *Yidui hao banlü* (1963), and was performed at a convocation held by the magazine.[11] The dialogue explains the Newtonian principle of force and counterforce. Ye Yonglie (b. 1940), SF author and critic, notes that the story was translated and published in "North Korean script" (*Chaoxian wenzi*) by the Yanbian People's Press. The eponymously named press was in the ethnically Korean Yanbian autonomous region of northeastern China, and its publications were available primarily to Chinese citizens of Korean descent, though this implies that North Korean citizens were also able to access the script.[12]

The format of "Yidui hao banlü" is typical of the genre: the dialogue opens with the principal character suggesting that they perform a *xiangsheng*. When their partner—often a comedic foil—agrees, the principal suggests a topic. In this case, the principal suggests that the *xiangsheng* dialogue is a good metaphor for discussing the Newtonian principle of force and counterforce. The foil often assumes a role allowing their exploratory premise to proceed—for example, by pretending to have contracted a disease due to their ignorance of hygiene or by pretending to be a robot. At this point, the foil begins to ask questions about the topic, and the principal answers them. In "A Good Couple," the foil begins to question how a horse and cart can move if the cart pulls back on the horse as strongly as the horse pulls on it. Their repartee proceeds through a series of comedic misunderstandings and clarifications until the principal determines the subject has been adequately educated. The characters in "Yidui hao banlü" illustrate force and counterforce by pushing against a table on the stage, and then pushing one another. As a mnemonic, the characters close with a *kuaiban* rhyme reiterating the basic principles of force and counterforce. Although "Yidui hao banlü" does not make use of the device, the conclusion often also features a return to the metatextual level as the characters acknowledge that they are part of a propaganda troupe and it is their duty to disseminate this information to the masses.

[10] Rudolph Wagner, "Lobby Literature: The Archaeology and Present Functions of Science Fiction in China," in *After Mao: Chinese Literature and Society, 1978–1981*, ed. Jeffrey Kinkley (Cambridge, Mass., 1985), 19–20.

[11] Gu Junzheng, "Yidui hao banlü" [A good couple], in *Yidui hao banlü*, ed. Zhongguo shaonian ertong chubanshe [publisher] (Beijing, 1963).

[12] Ye Yonglie, "Lun kexue xiangsheng," *Quyi yishu luncong* (Beijing, 1981), 87.

THE SOVIET MODEL

Soviet authors and literary critics like Mikhail Il'in and Maxim Gorky contributed to this discourse via translation into Chinese. Chinese authors were encouraged to follow Il'in's exhortation that while science writing should be accessible to a mass audience, it could also be attractive to readers, with the potential to change both knowledge and society.[13] Il'in's developmental narrative of socialism and communism as matters of scientific fact, and of the power of science as a tool for humankind's mastery of nature, was reiterated in Chinese intellectuals' writings on the subject. In the pages of People's Daily, prolific SF and popular science author Gu Junzheng exhorted readers to "learn from Il'in," praising him as an exemplar of the genre. He noted the following: "Il'in believes that communism is built upon the foundation of science, and people study science to transform nature, and to create the people's happiness. His works not only teach one how to understand nature, they teach one how to transform nature."[14]

Soviet-modeled theories of science writing also emphasized the narration of socialism's defeat of capitalism. "Scientific prediction is the foundation of the Soviet Communist Party leadership. Our soviet society is built upon the foundation of knowledge of the laws of social development; so Soviet *belles lettres* can only narrate the direction of the development of social life based on the immutable principles of Marxist-Leninist science."[15] We might refer to Chinese SF under the soviet model as a social-science fiction—a literature based firmly in contemporary realities that tried to imagine into being new social(ist) realities based on the "scientific" principles of dialectical materialism.

Further informing the artistic and political parameters of *kexue xiangsheng* was the discourse of the transformative mission of drama previously laid out by the leadership of the Chinese Communist Party (CCP). The PRC's formative years featured a dogged insistence on the power of drama as a tool of revolution, often in the very midst of military campaigns. During the October 1927 "Sanwan Reorganization," Mao first stressed the role of drama.[16] This emphasis was officially adopted in Maoist discourse in Mao's 1942 Talks at Yan'an, at which point, "art became the 'gears and screws' of revolution."[17] Chen Duxiu (1879–1942), CCP cofounder, educator, and philosopher, argued the following: "Some are promoting social reform by writing new novels or publishing their own newspapers, but they have no impact on the illiterate. Only the theater, through reform, can excite and change the whole society—the deaf can see it and the blind can hear it. There is no better vehicle for social reform than the theater."[18] Like prose, drama drew heavily from the theory and experience of Soviet mod-

---

[13] Mikhail Il'in, "Tantan kexue" [A discussion of science], trans. Shi Ying and An Ji, in *Zenyang bianxie kexue tongsu zuopin* [How to produce popular science works] (Beijing, 1958), 77–9; originally published in *Kexue Dazhong*, September and October 1952.

[14] Gu Junzheng, "Xiang Il'in xuexi" [Learn from Il'in], People's Daily, 24 December 1953, 3.

[15] O. Chuze [O. Hujie], "Lun Sulian kexue huanxiang duwu" [A discussion of readings in Soviet science-fantasy], trans. Wang Wen, in *Lun Sulian kexue huanxiang duwu* (Beijing, 1956), 9.

[16] Brian DeMare, *Mao's Cultural Army: Drama Troupes in China's Rural Revolution* (Cambridge, UK, 2015), 28.

[17] Ying Yi and Xiaobing Tang, *Art and Artists in China since 1949* (New York, N.Y., 2017), 4.

[18] Quoted in DeMare, *Mao's Cultural Army* (cit. n. 16), 9. DeMare notes that the drama school in Jiangxi Soviet was named the Gorky Drama Academy, while the school's troupe was called the Soviet Drama Troupe by educational leader Qu Qiubai (1899–1935).

els. Despite constant attacks from the Guomindang (GMD, Chinese Nationalist Party) and the deprivations of the Long March (1934–35), drama troupes continued to perform throughout the Revolutionary War and War of Resistance against Japan (1937–45). While full-scale stage performances were rare and hampered by logistical limitations, participatory propaganda and performance was a key part of the CCP's educational strategy. "Red Drama existed not for entertainment or art, but to serve the dictates of base area political and mass work needs, which meant above all else military survival in the face of a series of 'encirclement' campaigns launched by the Guomindang."[19]

Linkages between military and stage performance established in the CCP war with the GMD, and in resistance against Japan, carried through into *kexue xiangsheng* in explicit terms. Having come under fire during the Cultural Revolution, genre historian Luo Rongshou praised CCP Chairman Hua Guofeng for rehabilitating the genre in the late 1970s: "It is [*xiangsheng*] performer's duty to promote party policy and Mao Zedong thought. Performers are unarmed soldiers. Chairman Mao said that arts and letters are 'powerful weapons for unifying and teaching the people, for striking at and eliminating the enemy.'"[20] Staged performances enacted revolutionary struggles, framing the discourse for relationships between national fate and military strength, local politics and land reform, and the personal evolution of ideological consciousness. *Kexue xiangsheng* and other popular science genres straddled the line between Frederic Jameson's "Utopian form and Utopian wish"—between what was part utopian text and part expression of the impulse to enact utopia through changes in policy and reconfigurations of the collective body.[21]

The state also encouraged efforts to create educational films dedicated to science, framing these as an educational tool that led to increased production capacity. In a report for People's Daily, Hong Lin listed four main categories in science film: sharing experiences in advanced production, popularizing scientific knowledge, educational material, and documentaries on science and technology. He went on to list three other minor film categories: science research, science "news-magazine" (*zazhi*) films, and science fiction. However, Hong noted that "whatever the category of science-educational film may be, all must be endowed with a high level of political and ideological character, rich scientific content, and lively, vivid imagery."[22] Science film, like other forms of art that communicated scientific content, operated under the imperative that ideological content and the will of the party leadership be foregrounded. Despite its recognizable significance, film was less effective than other low-tech media because many rural sites lacked electricity and other facilities for accommodating even a traveling projection team.

Another early example of *kexue xiangsheng*, appearing in the pages of People's Daily in December 1959, tied evolution to Marxist notions of historical progress through a discussion of how humans evolved from apes. The comic foil in "Xianzai de yuanhou hui bu hui bianren?" (Can modern apes turn into humans?), has written an essay in which he argues, "since humans came from apes, and there are still apes,

---

[19] Ibid., 27; see also 35, 43–51.

[20] Ibid., 5.

[21] On utopian texts and programs, see Frederic Jameson, *Archaeologies of the Future: The Desire Called Utopia and Other Science Fiction* (New York, N.Y., 2007), 3–6.

[22] Hong Lin, "Shi kexue jiaoyu dianying geng hao de wei shehui zhuyi jianshe fuwu" [Let science-educational films better serve the establishment of socialism), People's Daily, 28 April 1960, 7.

modern apes can turn into humans," and the principal speaker disabuses him of this mistaken notion, first by describing differences between modern apes and *Australopithecus*, and then by describing evolution in terms of a family tree. Explaining the role of environmental adaption in human evolution, he states that some apes remained in forests, while others "moved to the ground to live." The principal continues: "Because they stood up straight, their arms were liberated, and they used sticks and stones to defend themselves from enemies. They gradually learned to use tools and labor was born. Their feet and hands performed different tasks, and their brains developed, giving birth to consciousness, language, and society. After a long period of labor, they became modern humans. That's why Engels says, 'labor created man himself.'"[23] The principal summarizes Engels's 1876 essay, "The Part Played by Labor in the Transition from Apes to Man."[24] While giving voice to a pervasive concern of twentieth-century Chinese intellectuals regarding the soundness of evolutionary theory and the need for common citizens to understand its significance, the dialogue also illustrates how evolution was subsumed under the rubric of Marxist historiography.

SCIENCE CROSSTALK, PUBLIC HEALTH, AND THE NATIONAL BODY

Kexue xiangsheng often centered on what we might call the "quotidian utopian"—many of the pieces in *Yidui hao banlü* (1963) cover banal topics like brushing your teeth, getting a good night's sleep, and preventing the spread of disease, but we should understand these as part of the utopian project of socialist realism. *Kexue xiangsheng* also incorporated other traditional forms within its structure, particularly the use of rhymed verse as a mnemonic within the science lesson, which repurposed preexisting art forms. The content of *kexue xiangsheng* overlapped with other public health campaigns, many of which also appeared in propaganda posters.

Cui Daoyi's "Aiyo, Wode duzi teng" (Oh, my stomach hurts) features a primer on avoiding dysentery that is delivered by the principal while the comic foil interjects with questions and steps he has taken to do so. The dialogue runs through various vectors of disease transmission and steps people can take to ensure their food and drink is pathogen free. After asking how he could have contracted the disease, his partner informs him in rhyme:

> When wells and streams are near toilets,
> and sick people's feces carry pathogens,
> night soil will run into the stream or seep into the well
> whoever drinks unboiled water will be in for it.[25]

A rhyming ditty about toilet placement is likely one of the last things one might name if asked to give examples of SF, but this and other works deserve consideration as part of the enlightenment mission of the genre; they are a deeply practical iteration of the

---

[23] "Xianzai de yuanhou hui bu hui bianren?" [Can modern apes turn into humans?], People's Daily, 12 December 1959, 8.
[24] Friedrich Engels, "Laodong zai renlei laiyuan zhong de zuoyong" [The part played by labor in the transition from apes to man], trans. Lu Yiyuan, in *Makesi zhuyi de renzhong youlai shuo* [Marxist theories of the origins of man] (Shanghai, 1928), 35–46.
[25] Cui Daoyi, "Aiyo, Wode duzi teng" (Oh, my stomach hurts), in *Yidui hao banlü* (cit. n. 11), on 9.

"dialectic of enlightenment and romanticism" that Andrew Milner characterizes as the thematic core of the genre.[26]

A major part of the Communist Party's effort to create a healthier and more productive citizenry literally involved teaching people where to defecate, and how to handle the night soil afterward. This and other advice in Cui Daoyi's dialogue—exhortations to clean hands, prepare food properly, and treat drinking water for amoebic bacteria—are undeniably quotidian. At the same time, they are profoundly utopian; that is, they are intended to be part of establishing a disease-free, well-nourished, and productive society.

> A: It's not just that you'll be skinny as a monkey,
>   You also won't be able to work or study,
>   If this illness is pathogenic,
>   without medical attention, your life is over.
> B: That's terrible!
> A: Terrible indeed, so we need to be vigilant about prevention.
>   "Eliminate the Four Pests," and observe proper hygiene
>   a strong body doesn't get ill.
> B: It's not just about you,
>   you need to promote this to everyone.
> A: We are promoters after all, aren't we? Let's go, on with the show![27]

By the time Cui Daoyi's *xiangsheng* on personal hygiene was published, the notion of hygiene and its association with the strength of the national body already enjoyed a long-standing prominence.[28] In 1954, playwright Cao Yu produced a play, *Minglang de tian* (Bright skies), that features a group of cadres at an American-founded hospital in Beijing who "guide foreign-trained Chinese physicians to understand the evil nature of imperialism."[29]

The admonition to "eliminate the Four Pests" was part of a patriotic hygiene campaign initiated in 1958. The kill campaign was prefigured by one key aspect of the Patriotic Hygiene Campaign (*Aiguo weisheng yundong*)—the Five Annihilations (*Wu mie*)—that focused on the mass mobilization of Tianjin's citizens to "eliminate the Five Pests: flies, mosquitoes, mice and rats, lice, and bedbugs."[30] In the less compelling "Big Cleanup" that followed, women, domestic laborers, and university and high school students were tasked with trash removal and filling in morasses and canals that held standing or fetid water.[31] "Aiyo, Wode duzi teng" was part of a massive and ongoing effort to change people's understanding of pathogenic disease, to associate the

---

[26] Andrew Milner, *Locating Science Fiction* (Liverpool, UK, 2017), 154, 178–82.

[27] Cui, "Aiyo" (cit. n. 25), on 11.

[28] Ruth Rogaski, *Hygienic Modernity: Meanings of Health and Disease in Treaty Port China* (Berkeley, Calif., 2005).

[29] Ibid., 238–95, on 292.

[30] Rogaski argues that *weisheng* might best be rendered as "hygienic modernity," one aspect of the general condition that Tani Barlow describes as "colonial modernity." See Barlow, *Formations of Colonial Modernity in East Asia* (Durham, N.C., 1997), 1–20. Rogaski further situates the term in the context of Lydia Liu's *Translingual Practice* (Redwood City, Calif., 1995). For quotation, see Rogaski, *Hygienic Modernity* (cit. n. 28), 15.

[31] Rogaski, *Hygienic Modernity* (cit. n. 28), 296–8.

dangers of this disease and its vectors with the threat of imperialism, and to urge people at all levels of society to join in the effort for national defense.

The discursive terrain of the Great Leap Forward (1958) was dominated by calls for increased production, with officials insisting that progress and production in both agriculture and science could be sped up through concerted effort.[32] Propaganda units were organized in paramilitary fashion, with over half a million cultural workers, including professional and amateur cultural teams working in a number of media that encompassed songs, stage performances, visual displays like posters and "blackboard news," exhibitions (both touring and fixed), social clubs, and science demonstrations.[33] This was a media network for promoting party goals by sharing knowledge of physical and social science. The intensive and conscious use and reconfiguration of "symbolic resources" has been posited as one reason for the CCP's success.[34]

By mobilizing people at the county and village level through institutions like organizing committees, Communist youth leagues, women's federations, epidemic prevention stations, and hygiene associations, the party successfully generated affective engagement to change behavior and make people personally aware of hygiene issues.[35] This was done by deploying popular culture to convey stories "about a happy, disease free society." Kill campaigns used something "discernable as a proxy for the pathogen."[36] The slogan to "eliminate the Four Pests and observe hygiene" condensed a complex and abstract topic unlikely to be understood by a semiliterate population into a visible and actionable routine with clear implications for national survival. Gross notes the following: "Chairman Mao felt that science had to be enacted by the masses . . . Health campaigns seemed like a good place to jump-start scientific understanding because the science involved in counteracting diseases was concrete, pragmatic, and directly relevant to improving people's livelihood."[37] When these campaigns failed, it was often because they were not framed in terms that people understood, and the party made efforts on multiple fronts to ensure that they would be.

Opposition to the campaign demonstrates how poorly considered some policy measures were, and how they were deployed to define ideological orthodoxy as "scientific" thinking. Nearly immediately after promulgation of the plan in 1957, some scientists voiced opposition to the plan to eliminate sparrows, as the crop damage they caused was minimal and they were helpful in eliminating insects. Mao Zedong

---

[32] Lin Yunhui, *Zhonghua Renmin Gongheguo shi*, vol. 4, *Wutuobang yundong—cong da yuejin dao da jihuang* (1958–1961) [History of the Republic of China, vol. 4, Utopian movement: From the Great Leap Forward to the Great Famine (1958–1961)] (Hong Kong, 2008), 241.

[33] Ibid., 246; Miriam Gross, *Farewell to the God of Plague: Chairman Mao's Campaign to Deworm China* (Oakland, Calif., 2016), 220–1.

[34] Demare, *Mao's Cultural Army* (cit. n. 16), 3.

[35] Zeng Xuelan, "'Chu sihai, jiang weisheng' yundong zhong de jiceng shehui dongyuan—yi Hebei Lixian we zhongxin de kaocha" [Grassroots mobilization in the campaign to 'Eliminate the Four Pests and Observe Hygiene'—an investigation focusing on Li County, Hebei Province], *Hebei guangbo dianshi daxue xuebao* [Journal of Hebei Radio & TV University] 22 (25 August 2017), 87, 88. If Hebei Province is any indication, these efforts were assiduously (though likely not accurately) documented. According to Zeng Xuelan's study, for example, in the winter of 1955, 190,000 people helped to catch 10.5 million rats and 1.54 million sparrows, destroyed 1,400 *jin* of sparrow's nests, and repaired cisterns and toilets. After two years of effort, they had apparently succeeded, and the area was dubbed a "four naughts" village for having eradicated rats, sparrows, mosquitoes, and flies; see Zeng, "'Chu sihai, jiang weisheng,'" 88.

[36] Ibid., 89, 101.

[37] Gross, *Farewell* (cit. n. 33), 88.

did not respond until 1960, at which point he was reported to have dismissively quipped "forget it" (*suanle*), and then ordered that extermination efforts be turned to bedbugs instead.[38]

Zeng Xuelan notes that in the case of Yi County in Hebei Province, locals expressed various other doubts about the campaign. Cadres complained, "we are so busy with production and disaster relief, where do we have time for eliminating the Four Pests, and even if we tried, we'd never get rid of them all." Local farmers, on the other hand, argued that "rats and sparrows are bestowed from on high; eliminating them will bring disaster on us." Such opposition was characterized as conservatism and superstition. The CCP seized control of the definition of science in order to "dislodge scientific authority over what was termed science in public discourse and to define traditional culture and objects of professional science as superstition."[39] The desire for the party to assert authority over science and for the "scientific" principles of Marxism to be borne out as universal truths meant that the principal role of science was legitimation of the party line. Doctors refusing to engage in various forms of malpractice could find themselves accused of superstition, regardless of the scientific grounds of their opinions.

Locals were fond of singing the following doggerel rhyme:

> Listen to me, fellow villagers
> sparrows roost in rafters at night
> by day we track down their nests
> by night we clear them away.[40]

Parades were organized in villages with teams who would chant, "stamp out rats and exterminate sparrows, conserve grains, and support national construction."[41] Villagers were encouraged to tell their stories of the woes visited upon them by the Four Pests, and other local officials seized on stories like that of one man's son who contracted lymphatic adenitis after being bitten by a rat; these were turned into *kuaiban* rhyming songs and drum songs (*guci*), alongside *xiangsheng* performances. The campaign also deployed display culture, commissioning over fifty exhibitions of the tools and methods used to eradicate pests. Those mobilized in the effort were organized into various paramilitary "squads," including navy, infantry, air force, flamethrower squads, and young pioneers. Drawing on a script from popular culture and fantasy, these squads gave their methods and weapons names reminiscent of *fabao* (magical weaponry) and mystical techniques common to martial arts and *wuxia* fantasy fiction, coining phrases like the "demon revealing lens" (*zhao yao jing*), and "flamethrower fusillade" (*huoqiang lianhuan zhen*).[42] Popular participation in this pseudoscientific program for national defense was performed by repurposing a host of preexisting folk culture forms.

---

[38] Judith Shapiro, *Mao's War Against Nature: Politics and the Environment in Revolutionary China* (New York, N.Y., 2001), 86.

[39] Gross, *Farewell* (cit. n. 33), 184–5, 187.

[40] Zeng, "'Chu sihai, jiang weisheng'" (cit. n. 35), 90.

[41] Ibid.

[42] Ibid.

This multipronged assault on pests extended into the realm of song. He Lüting (1903–99) wrote "Saochu faxisi" (Drive out the fascists), which starts with the following lines:

> Wash clothes often, get sick less
> carefully clean the house, observe hygiene
> if we don't clear out the fascists
> there's no way to live on.

The chorus sings: "Sweep, sweep it clean, eliminate the fascists and save the people, establish a new unified government. . . ." The second verse continues: "Fascists are like flies / their harms make China ill, . . ." obviating any need for interpretation in realizing the connection between domestic hygiene and national health.[43]

The fictional science presented in the opportunity to smash fascism by swatting flies on the windowsill offered a plausible means to participate in bolstering the national health, and yielded almost immediate results. According to Zeng Xuelan's Yi County investigation, local incidence of dysentery in 1957 was down eighty-two percent compared to 1952, and gastroenteritis was reduced by fifty-two percent. Another report, documenting a visit to the model family home of one Wang Kuibin, found it to be in immaculate order, and that the residents "tidied up every day, often set their blankets out in the sun, cut their fingernails, and changed clothes at least once every ten days." The report was sure to note, "it had been a year or two since anyone had had lice."[44] *Kexue xiangsheng* and other public health propaganda campaigns functioned as a form of triage; this was a way to overcome the challenges of China's large and widely dispersed population and limited access to trained medical professionals. Though the figures recording success were likely inflated, these efforts were part of a successful public health strategy. Despite the massive death toll of the famine that followed the pseudoscientific disaster that was the Great Leap Forward, and continuing violence and turmoil through the Cultural Revolution, access to health care and many markers of life expectancy had increased markedly since the establishment of the PRC.[45]

Slogans exhorting thrifty consumption appeared in *kexue xiangsheng* as well. In "Xiangsheng de kexue" (The science of "physiognomizing sound"),[46] it is discussed whether tapping on a melon to see if it is ripe and performing chest percussions during a medical exam are similar to methods used to diagnose problems with heavy machinery and subterranean plumbing. The foil assumes that gurgling pipes must signal

[43] He Lüting, "Saochu faxisi" [Drive out the fascists], in *He Lüting quanji* [Collected works of He Lüting], ed. [Shanghai music press] (Shanghai, 1997), 85. See also Huangcao, He Lüting, and Chi Liu, *Wei minzhu ziyou er zhan* (*gequ ji*) [Fighting for democracy (song lyrics)] (Ha'erbin, 1948), 18–20.

[44] Zeng, "'Chu sihai, jiang weisheng'" (cit. n. 35), 90.

[45] Health markers in China have vastly improved over the last 60 years. Between 1965 and 2010, average life expectancy in China has gone from under 49.5 years to 76 years, and the number of doctors per 1,000 individuals has gone from 1.07 to 3.6. See "Life Expectancy at Birth, Total (years)," World Bank, Data, 2017 revision, https://data.worldbank.org/indicator/SP.DYN.LE00.IN?locations=CN-US-GB; and "Physicians (per 1,000 people)," World Bank, Data, https://data.worldbank.org/indicator/SH.MED.PHYS.ZS?locations=CN-US-GB (accessed 17 September 2018).

[46] This title involves a pun that is difficult to translate—the *xiang* in *xiangsheng* can mean to physiognomize, or otherwise appraise based on visible features. Thus, the title might also be awkwardly rendered as "the science of physiognomizing sound."

that the municipal plumbing system has bronchitis, and then argues that there is no need to fix minor leaks. The principal lectures: "You are so wasteful, so careless with national property! Do you understand practicing austerity, do you understand building the country through thrift and diligence, do you understand . . ."[47] The foil signals that he does indeed understand, and the two leave the stage to report the leaky pipe. By the time it appeared as "Xiangsheng de kexue" in 1965, the slogan "build the country through thrift and diligence" was a familiar feature of the Maoist discourse of sacrifice for the sake of revolution and development. First appearing as a keyword in Mao's anti-Nationalist war movement, and then briefly co-opted by Chiang Kai-shek and the Guomindang near the end of the civil war, the term was canonized with an entire chapter of *Quotations from Chairman Mao*, and regularly appeared in the pages of People's Daily, especially during the years of 1957 and 1958.[48]

The Anti-Rightist Campaign, the Great Leap Forward, and the Cultural Revolution all contributed to dismantling science as a professional institution by consigning scientific authority to rural cadres, barefoot doctors, and technicians who carried out medical training and work "in domains and on a scale previously inconceivable." Science shifted from a rationalized, bureaucratic endeavor focused on "understanding of unsolved phenomena based on unbiased, reproducible experimentation," to a form of "'grassroots science' focused on performing field investigations to resolve pragmatic problems."[49] This policy turn was captured by the slogan, "the lowly are the most intelligent, the elite are the most ignorant."[50] When party leaders dictated that the twenty-day medical protocol for curing schistosomiasis be condensed into a mere three days, scientists who argued this was medically unsound were criticized, and the condensed treatment protocol was touted as a triumph of scientific management, with party language emphasizing increased production and speed of development.[51]

Attitudes about the democratization and deprofessionalization of science are lampooned in Yu Zu's "Zuo shiyan" (Experimenting).[52] The foil in this *xiangsheng* tells a cautionary tale of the dangers of chemistry experiments, first burning his textbook with sulfuric acid, then burning a hole in his hat with sodium hydroxide, and finally ending up in the hospital after a mishap trying to adjust the wick on an ethanol lamp. In a rare moment of satire, the foil closes by asking: "What isn't glorious about being injured for the sake of chemistry research?"[53] Although the foil in this case is a Chinese man in a classroom, rather than a young white man in the basement of his middle-class, postwar suburban home, the dialogue mirrors the clichéd image of science as an innocent, childish, and potentially dangerous pursuit that pervaded postwar American culture.[54]

[47] Xiao Xi, "Xiangsheng de kexue," in *Kexue wenyi zuopin xuan ertong wenxue*, vol. 2 [Selected works of science belles-lettres children's literature], ed. Gao Shiqi and Zheng Wenguang (Beijing, 1980), 515–6; originally published in *Women ai kexue*, v. 13 [We love science] (Beijing, May 1965).
[48] See Mao Tse-tung, "Building Our Nation Through Diligence and Frugality," *Quotations From Chairman Mao Tse-tung*, http://en.eywedu.net/maozedong/20.htm (accessed 24 January 2019).
[49] Xiao, "Xiangsheng" (cit. n. 47), 203–4.
[50] Lin, *Zhonghua* (cit. n. 32), 244.
[51] Ibid., 260; Gross, *Farewell* (cit. n. 33), 156–60.
[52] The author's pseudonym can be read as meaning either "foolish soldier" or "[a] foolish death."
[53] Yu Zu, "Zuo shiyan" [Doing experiments], in *Yidui hao banlü* (cit. n. 11), 43.
[54] See Rebecca Onion's *Innocent Experiments: Childhood and the Culture of Popular Science in the United States* (Chapel Hill, N.C., 2016). The illustrations and depictions of experimenters as male in the *kexue xiangsheng* collection *Yidui hao banlü* (cit. n. 11) seem to be in contrast to Maoist visual culture, which often emphasized the participation of women and minorities in scientific endeavors.

A VISIT TO THE MOON

Yu Junxiong's "Fangwen yueliang" (A visit to the moon) borrows its cast of characters from premodern myths about the moon, again repurposing them for scientific and political education. In an opening address bidding farewell to the "residents of the Moon" (*yueqiu de jumin*), the principal recounts meeting the goddess Chang'e, the mythical woodcutter Wu Gang, and the Jade Rabbit. He explains that he has visited Chang'e's Palace of Vast Cold (*guang han gong*), but that because the Moon lacks an atmosphere, during the daytime it is more appropriately referred to as the "Palace of Vast Heat" (*guang shu gong*). A discussion of the Moon's other physical characteristics follows, including the different gravitational pulls, periods of rotation, and geographical features. Dark spots on the moon, explained in myth as the shadows of a massive cassia tree that Wugang is trying to cut down, are demystified as the shadows of lunar mountains and craters forming lunar seas like the Mare Tranquilitatis (*fengbao hai*) and Mare Fecunditatis (*fengfu hai*).[55]

The principal recounts being treated to a performance of the Tang ballet suite, "Nichang yuyi" (Skirts of rainbow, feather coats), a musical piece purportedly written down from memory by Tang Emperor Xuanzong (713–56) after visiting the Chang'e on the Moon in a dream.[56] This lunarian serenade is reciprocated by the principal with the couplet "Chang'e," written by Li Shangyin (813–58), and she in turn responds with a play on "Jingye si" (Thoughts on a moonlit night), composed by Li Bai (701–62), that includes "I lift my head and look at the sun, I lower my head and dream of home."[57] The foil, incredulous at the idea that the principal could have brought Chang'e, the Jade Rabbit, and Wu Gang back to Earth, given the fact that they are fictional characters, retorts:

B: Stop joking, those are people from myths.
A: Who's joking? Chang'e and Wugang got
    Married, Jade Rabbit is their baby.
B: The more you go on, the more preposterous this gets.
A: If you don't believe me, go ask the Hundred Flowers Song and Dance Troupe
    if the woman who plays Chang'e isn't married to the man who plays Wu Gang,
    and their baby isn't called Jade Rabbit.
B: Then what are they doing on the moon?
A: In order to stage "Chang'e Goes to the Moon," they went there to observe and
    learn from experience.

The *xiangsheng* closes with the conceit that the mythical inhabitants of the Moon are real in the sense that the familial relationships of the performers in a propaganda troupe match those of the mythical characters they portray on stage. Potential accusations of "superstition" and "feudal" content are precluded by situating the dialogue

---

[55] Yu Junxiong, "Fangwen yueliang" [A visit to the moon], in *Yidui hao banlü* (cit. n. 11), 44–53.
[56] See Stephen Owen, "The Reign of Emperor Xuanzong: The 'High Tang,'" in *The Cambridge History of Chinese Literature: To 1375*, ed. Kang-I Sun-chang and Stephen Owen (New York, N.Y., 2010), 309–10. For more on the complex history and provenance of the song, see Fu Junlian and Li Huilin, "'Nichang yuyi qu' xin kao" [A new examination of the "Skirts of rainbow, feather coats"], *Gansu guangbo dianshi daxue xuebao* [Journal of Gansu Radio & TV University] 25 (April 2015): 1–5.
[57] Yu, "Fangwen yueliang" (cit. n. 55), 51.

within the context and discourse of socialist ideological education and performance. Aside from a lesson on basics of lunar geography and gravity, the *xiangsheng* demonstrates how premodern poetry and myth can be adopted to serve dissemination of scientific knowledge. Rather than eclipsing premodern myth, science and myth are used to explain one another and to emphasize proper political attitudes.

RESHAPING NATURE

Positive outcomes of health campaigns contrast with the massive environmental damage brought on by the pervasive discourse of development and the human suffering wrought in the harried push to modernize. Judith Shapiro observes: "Official discourse was filled with references to a 'war against nature.' Nature was to be conquered."[58] The Great Leap Forward brought on a catastrophic famine of which we may never know the full human toll, but repercussions of overly ambitious and unscientific Mao-era policies, especially in the form of an intense focus on the fossil fuel industry, led to desertification, damage to waterways, and various other environmental and human catastrophes that continue to emerge. Kill campaigns were part of the Great Leap Forward's "targeted attack on nature . . . [in which] schoolchildren were the main participants"; these schoolchildren were soldiers in a campaign that Shapiro describes as "a highly militarized event."[59] "Both Chinese tradition and modern Western science" were rejected by Maoism. "The effort to conquer nature was highly concentrated and oppositional, motivated by utopianism to transform the face of the earth and build a socialist paradise, and characterized by coercion, mass mobilization, enormity of scale, and great human suffering."[60]

The discourse of a war on nature was also influenced by the Soviet model, and by Soviet writers discussing the content and purpose of children's literature. Mikhail Il'in's literary criticism and fiction emphasized the theme of remaking the natural world. Acknowledging the colonial undertones of Anglophone SF, Il'in argued that the purpose of transforming nature in Soviet SF went beyond bourgeois fiction's depiction of gaining individual wealth, and was meant to illustrate the value of fostering the collective good.[61] Beyond science writing specifically, Il'in stated that "struggling with nature and overcoming nature has become the subject of many of our works of science *belles-lettres*."[62] Hua Li and Nicolai Volland observe that Zheng Wenguang's late 1950s SF stories like "Cong diqiu dao huoxing" (From Earth to Mars), "Di'er ge yueliang" (The second Moon), and "Huoxing jianshezhe" (The Mars pioneers) featured themes of terraforming—reshaping a planet's environment to resemble that of Earth—and other attempts to transform nature, a tendency that was later critiqued through Zheng's 1983 novel, *Zhanshen de houyi* (Descendant of Mars).[63]

---

[58] Judith Shapiro, *Mao's War* (cit. n. 38), 3–4.
[59] Ibid., 86–7.
[60] Ibid., 8.
[61] Mikhail Il'in, *Lun ertong de kexue duwu* [A discussion of children's science writing], trans. Shao Diansheng (Beijing, 1953), 21.
[62] Ibid., 24.
[63] See Nicolai Volland, *Socialist Cosmopolitanism: The Chinese Literary Universe, 1945–1965* (New York, N.Y., 2017), 113–8. See also Li Hua, "Are We, the People From Earth So Terrible?: An Atmospheric Crisis in Zheng Wenguang's *Descendant of Mars*," *Sci. Fict. Studies* 45 (2018): 545–59; and Li Hua, "Diqiu gaizao: qihou bianhua yu Zhongguo kehuan xiaoshuo" [Terraforming: climate change and Chinese science fiction] (paper presented at IAS HUMA Joint Conference, "Science Fiction and Its Variations in The Sinophone World," Hong Kong, 30–1 May 2018).

This tendency appears in *kexue xiangsheng* as well, both in the pervasive discourse of developmentalism outlined above, and occasionally more explicitly, as in the case of Yan Hui's "Bing de gaizao" (Transforming ice). In this dialogue, the principal has published an editorial in the newspaper suggesting that ice be "transformed" so that it will no longer expand or float when frozen. In this way, ice will no longer cause damage to objects during winter freezes, and moreover, tanker ships will be able to travel unimpeded. A fevered public response ensues. First, ice skaters complain that they will have to skate on the bottom of lakes, and skating in scuba gear will be too cumbersome; then a fisherman comes to protest (*biaoshi kangyi*) on behalf of creatures who would no longer be able to survive beneath the surface of the ice in the winter if it sank. When a shipping company manager who worries that sinking ice will never thaw, resulting in permanently sealed waterways, the principal relents and decides to give ice its original properties once again.[64]

XIANGSHENG IN THE POST-MAO ERA

As with almost all other art forms, there is no record of *kexue xiangsheng* produced during China's Great Proletarian Cultural Revolution (and indeed very few *xiangsheng* dialogues at all were produced from 1966 to 1976). Almost immediately after the end of the Cultural Revolution, *xiangsheng* burst back on the cultural scene. The dialogues often continued with the scientific and political objectives that *kexue xiangsheng* aimed to achieve in the late 50s and early 60s. In 1976, a collection of *xiangsheng* on family planning expanded the number of people on the stage to five, often reciting lines as a chorus. These performances adopted the family mnemonic of "later, longer [between births], fewer" (*wan, xi, shao*), and were also aimed at countering people's preferences for a male heir, in particular the practice of continuing to have children until a male child is born. Like *xiangsheng* on public health from preceding decades, immediate and explicit connections were drawn between family planning as a means of promoting health, and the ability of the mother to contribute to the national body. This was in turn portrayed as a product of the wisdom of Chairman Mao and the party. These performances featured a number of metatextual, rhymed choruses that elucidated the role of the stage performer as a propagandist. The performances also enacted the socialist-realist emergence of politically correct, "scientific" attitudes in the characters; for example, a husband learns to reform his "feudal thinking" regarding childbirth. An entire volume of *xiangsheng* devoted to eliminating the "Four Pests" was published in 1977, substituting the Gang of Four for the various rodents and insects of previous targets of the slogan.[65]

Some *kexue xiangsheng* served as direct responses to the tumult of the Cultural Revolution. "Weiba de kexue" (The science of tails) follows a familiar pattern of using the familiar to illustrate a more abstract topic, moving from a discussion of the

---

[64] Yan Feng, "Bing de gaizao" [Transforming ice], in *Yidui hao banlü* (cit. n. 11), 54–62.

[65] Renmin weisheng chubanshe [publisher], ed., *Jihua shengyu wenyi xuanchuan cailiao huibian* [Collected family planning art-propaganda], vol. 2 (Beijing, 1976). See also *Chu 'sihai' zawen ji* [Collected writings on eliminating the "Four Pests"] (Beijing, 1977); *Chu 'sihai' (xiangsheng ji)* [Eliminate the "Four Pests" (xiangsheng collection)] (Beijing, 1977).

tails of farm animals into a paleontological explanation of how dinosaurs' tails served an evolutionary purpose. This lesson on evolution involves a disputation against Lamarckism (the notion that traits acquired by an individual could be passed on to coming generations). This disquisition on tails eventually turns to the question of the political uses of science, alluding to the "horses' tails" scene in the 1975 film, *Jue lie* (Breaking with old ideas).[66] In the film, long-winded educator Dr. Sun is ridiculed for his "capitalist educational ideology" when the third day of lectures on horse anatomy is spent explicating the function of horses' tails, and he refuses to examine a sick water buffalo brought to him by a local villager. The foil offers a full-throated denunciation of the revolutionary fervor for practical education above all else featured in the film, and the principal replies with a *kuaiban* rhyme:

> Smash the evil Gang of Four
> Science and technology flourish once more
> Erase the lingering poison of the Gang of Four
> Keep lecturing on 'the function of horses' tails!'[67]

The irony that the call for science decoupled from politics is as politically charged as the Cultural Revolution film that they decry seems to be lost on the two speakers.

Other *kexue xiangsheng* mirrored trends in the reemerging field of highbrow and pulp SF. Xiao Shengmi and Sun Ming's "Jiqiren" (Robot), published in 1979, mirrors the problematic, and arguably nonsocialist, approach toward the division of labor that Paola Iovene has identified in mainstream post-Maoist SF.[68] After comically convincing his partner that he is a robot, the principal character in "Jiqiren" alludes to Asimov's "three laws of robotics," informing the foil that robots are inferior to and must obey human beings:

> A: No, robots still have to be invented by people, and they need to obey our commands and instructions.
> B: Oyoh! People are more important, after all.
> A: Yes, people are the most important. One day, we'll send out all sorts of robots to explore all kinds of places where humans can't go.
> B: Wow!
> A: For example, when a volcano erupts, and magma is soaring through the air . . .
> B: You can send me, Robot, into the mouth of the volcano, and explore the mysteries inside of Earth.
> A: On the eve of an earthquake . . .
> B: You can have me, Robot, drill into the Earth's crust, to examine the seismic focus.
> A: One day we can have them ascend the Himalayas without the slightest effort . . .
> B: And I, Robot, can establish mines to harvest precious treasures.

---

[66] *Jue lie*, directed by Li Wenhua (1975; Guangzhou: qiaojiaren wenhua chuanbo youxian gongsi, 2008), DVD.

[67] Liu Qiusheng, "Weiba de kexue" [The science of tails], in Gao and Zheng, *Kexue wenyi zuopin* (cit. n. 47), 524; story originally published August 1978 in *Kexue huabao* [Science illustrated].

[68] Paola Iovene, *Tales of Futures Past: Anticipation and the Ends of Literature in Contemporary China* (Redwood City, Calif., 2014).

A: Great, we can also dispatch them to the ocean floor at the Paracel Islands . . .

B: And I, Robot, can explore the mysteries of life beneath the seas.

A: When I come home one day after work, and there's no one to help me cook dinner . . .

B: I, Robot, can help you lay out a bounteous feast.[69]

The piece concludes with more talk of the robot's role as a manual and effective laborer, subject to human control. Authors like Mikhail Il'in in the 1950s had suggested that science would soon eliminate the distinction between manual labor and intellectual labor, ushering in a new era distinguished by the union of science, labor, and life.[70] The "bourgeois" image of robot as manual laborer inferior to the intellectually laboring scientist, almost entirely "peculiar to *kehuan xiaoshuo* in the late 1970s,"[71] also appears in SF's close cousin, *kexue xiangsheng*. Robots as manual laborers appear again in Yang Zaijun's "Mengyou taikong cheng" (Visiting a space station in a dream) as the principal recounts being ignored by robot construction workers in a space station.[72] The dialogue uses the pursuit of a potential love interest in outer space as a way to introduce a series of technicalities regarding gravity and relativity. The dialogue borrows from the classical corpus to describe the utopian possibilities of space travel, alluding to the Tang Dynasty poem by Wang Zhihuan (688–742), "Deng guangque lou" (Climbing stork tower), to describe the magnificent views afforded in outer space.[73] Yang Zaijun's characters compare the space station to the utopia in the fable by Tao Yuanming (365?–427)—that utopia in "Taohua yuan" (Peach Blossom Springs) is separated from earthly concerns.[74] This legend of the cowherd and the weaver maid, who cross over the Milky Way once a year to reunite, is used to describe the view of space beyond the space station. The plot is as much about the pursuit of romantic interests as it is a device for introducing potential technologies of sustaining human life in outer space. The notion of heroic individual inventors-cum-scientists like Edison was critiqued directly in Il'in's treatise on scientific literature. Il'in lambasts praise for Edison as a case of "the history of scientific understanding being replaced by the history of patent rights."[75] In Yang Zaijun's 1981 *kexue xiangsheng* "Congming de mijue" (The mystery of intelligence), the image of the inventor as an individual genius has returned, signaling a further drift from the Soviet model introduced in the late 1950s that was consistent with broader trends in the era of post-Mao marketization.[76]

---

[69] Xiao Shengmi and Sun Ming, "Jiqiren" (Robot), in *Zhongguo ertong wenxue daxi*, vol. 3, *Kexue wenyi* [Science belles-lettres], ed. Fang Weiping (1979; Taiyuan, 2009), 670. Throughout the dialogue, the "robot" foil uses the phrase "wo, jiqiren," to refer to himself, a clear allusion to the Chinese title of Asimov's *I, Robot* (1950).

[70] Il'in, "Tantan kexue" (cit. n. 13), 10.

[71] Iovene, *Tales of Futures Past* (cit. n. 68), 34.

[72] Yang Zaijun, "Mengyou taikong cheng" [Visiting the space station in a dream], in *Congming de mijue* [The mystery of intelligence] (Changsha, 1981), 24–36.

[73] For a translation of Wang Zhihuan's poem, see Stephen Owen, *An Anthology of Chinese Literature: Beginnings to 1911* (New York, N.Y., 1996), 408.

[74] Tao Qian, "Peach Blossom Spring," trans. Cyril Birch, in *Anthology of Chinese Literature, vol. 1, From Early Times to the Fourteenth Century*, ed. Birch (New York, N.Y., 1965), 167–8.

[75] Il'in, "Tantan kexue" (cit. n. 13), 21.

[76] Yang Zaijun, "Congming de mijue" [The mystery of intelligence], in *Congming de mijue* (cit. n. 72), 14–23.

CONCLUSION

Kexue xiangsheng emerged in the late 1950s and is a uniquely Chinese contribution to popular science writing. The genre was a product of the Maoist use of art and science in service of party policy and was thoroughly imbricated with the political and popular culture of its era. One of the most prominent aspects of the connections between *kexue xiangsheng*, popular culture more broadly, and the mobilization of political sentiment was in the transformation of meanings of science and superstition in the late 1950s. Using *kexue xiangsheng* as the primary vector of analysis, I have sketched out some of the various connections and cross-pollinations pertaining to notions of science in China's midcentury media network. Socialist realism presented and sought to bring about a highly plausible, and very near, future. The bright future that *kexue xiangsheng* sought to establish was not a world of rockets and cold fusion, but one where scientific knowledge was democratized, and correct political attitudes illustrating the social-scientific truths of Marxism and their supremacy over other ways of knowing the world were shared enthusiastically. *Kexue xiangsheng* was utopian in that it laid out an actionable program by which the average citizen could contribute to a healthier national body. At the same time, the genre reflects a democratization of science, the production of knowledge, and participation in politics. Mao's vision of science by the people and for the people wrested control from experts, but unfortunately led into an aestheticization of politics and a breakdown of objective scientific truth during the Cultural Revolution. In the wake of the Cultural Revolution, *kexue xiangsheng* again followed broader trends in the cultural field as "bourgeois" elements reemerged. As part of China's literary tradition, *kexue xianghseng* offers a deeper understanding of the functions and forms of SF by highlighting how science pervaded discourses of social and political transformation.

Playing Games with Technology:
Fictions of Science in the *Civilization* Series

by Will Slocombe*

ABSTRACT

This article investigates the ways in which the history of technology has been modeled in "4X strategy" games, especially in a series called *Civilization* (which comprises six games and expansions introduced from 1991 to 2016). Although there have been various studies interrogating the ideological biases in strategy games' modeling of civilization and society, to date there has only been partial exploration of the ideological biases within their models of technological and scientific development involving "technology trees." Moving from discrete analysis of individual instances of technology trees within strategy games, the aim of this article is to demonstrate not only the fundamental issues behind the notion of these trees in all of the *Civilization* games, but also to demonstrate ways in which they can reveal particular historicized perceptions of technologies over the period they were developed. This investigation furthermore reveals that many players of the games may bring assumptions embedded in their sense of the history of technology, and that these present a particular problem for those who might uncritically accept the games' underlying axioms.

INTRODUCTION

Video games might be thought to be too recent, and too concerned with entertainment, to reveal appropriately historicized attitudes to technology. However, as this article will demonstrate, there is value in considering how one type of game, the "4X strategy" game, encodes particular assumptions about technology and its development. Although closely affiliated to the "God game" (a video game in which the player is given absolute, and often supernatural power over a civilization or tribe's development), 4X strategy games are defined by the phrase "eXplore, eXpand, eXploit, and eXterminate." That is, such games encourage players to "explore" the in-game environment, "expand" their civilizations, "exploit" the resources of their civilization's territory, and "exterminate" rival civilizations. While this phrasing is loaded, it provides a sense of the kinds of activities within the game, although it does not foreground the importance of technological research. Using the *Civilization* series of video games (six games and various "expansions" over the period 1991–2016) as

---

* Department of English, University of Liverpool, 19 Abercromby Square, Liverpool, L69 7ZG, UK; W.Slocombe@liverpool.ac.uk.

an exemplar, this article asserts that "technology trees," as the underpinning structural element of many such 4X strategy games, reveal not only a specific cultural attitude to technology, but also toward the history of technology. While this article will describe and examine "technology trees" later, it is important to realize at the outset that the gameplay is structured by the technology trees; they serve as a timeline of a civilization's development and facilitate effective play.

For those unfamiliar with the games in the *Civilization* series, the player takes control of a civilization and guides its development over a period of time, from its earliest origins (the virtual equivalent of a few tribespeople and huts) in 4000 BCE, through various ages, until it achieves dominance over the other civilizations—usually controlled by the game—via victory conditions set in advance, or until the default of about 2000 CE. Players control broad elements, such as how the civilization's government and economy are organized and what buildings are built in its cities; buildings can include banks, aqueducts, libraries, and even various "wonders." The last are unique structures such as the Hoover Dam, the Great Wall (of China), or the Hanging Gardens (of Babylon). Players also guide units, including aircraft, tanks, and destroyers, or warriors and cavalry in earlier stages of the game, around the map to expand a civilization's territory (fig. 1).

Depending upon the specific *Civilization* game played and how the player sets up the game, victory can be achieved by various means, ranging from the construction of a spacecraft to establishing military, cultural, or economic dominance. Issues of "winning" *Civilization* aside, the game is, at least in some respects, quite sophisticated in its modeling of cultures, and considers factors such as happiness of populations, food and resource production, and political structures and policies. But most importantly, the series emphasizes the technological development of a society.

A fair amount of research has been published in relation to the *Civilization* games, specifically on its use for the study of history, or on its cultural biases.[1] However, no

---

[1] For the use of selected *Civilization* games in history education, see, for example, Dom Ford, "'eXplore, eXpand, eXploit, eXterminate': Affective Writing of Postcolonial History and Education in *Civilization V*," *Games Studies* 16 (2016), http://gamestudies.org/1602/articles/ford; Claudio Fogu, "Digitalizing Historical Consciousness," *Hist. & Theory* 48 (2009): 103–21; A. Martin Wainwright, "Teaching Historical Theory Through Video Games," *The Hist. Teach.* 47 (2014): 579–612; and Kimberly Weir and Michael Baranowski, "Simulating History to Understand International Politics," *Simulation & Gaming* 42 (2011): 441–61. For cultural critiques and ideological analyses of selected *Civilization* games, including their approach to technology, see, for example, Christopher Douglas, "'You Have Unleashed a Horde of Barbarians!': Fighting Indians, Playing Games, Forming Disciplines," *Postmodern Culture* 13 (2002), http://pmc.iath.virginia.edu/issue.902/13.1douglas.html; Ted Friedman, "*Civilization* and its Discontents: Simulation, Subjectivity, and Space," in *On a Silver Platter: CD-ROMs and the Promises of a New Technology*, ed. G. M. Smith (New York, N.Y., 1999), 132–50; Matthew Kapell, "*Civilization* and its Discontents: American Monomythic Structure as Historical Simulacrum," *Popular Culture Review* 13 (2002): 129–35; and Kacper Poblocki, "Becoming-State. The Bio-Cultural Imperialism of Sid Meier's *Civilization*," *Focaal—European Journal of Anthropology* 39 (2002): 163–77. Diane Carr's work takes issue with such ideological readings, however, and would equally take issue with much herein, because "criticizing a simulation for being reductive is nonsensical" and because merely revealing that "fantasies [of Western chauvinism] are persistent" is not the same as revealing that players' attitudes are somehow informed by games; see Carr, "The Trouble with Civilization," in *Videogame, Player, Text*, ed. B. Atkins and T. Krzywinska (Manchester, UK, 2007), 222–36, on 234. However, implicitly reinforcing such fantasies is clearly a problem, especially as using such games for educational purposes means that the *framework* of the games must also be critiqued. That is, if they are to be used to simulate counterfactual histories for educational purposes (as they often are), it is surely as important to query the assumptions behind the simulation itself; otherwise they merely reinforce such assumptions.

Figure 1. Screenshot from Civilization IV, *showing cities, farms, and units on the game map. The game is of course in color. From Robert Zak, "Every Civ Game, Ranked from Worst to Best,"* PCGamesN, 4 November 2016, *https://www.pcgamesn.com/best-civilization-games.*

work has yet linked such studies with the games' fundamental aspect, which is a model of the history of technology that uses scientific "progress" to define the advancement of a culture. That is, there is a built-in version of the history of technology within these games and, more importantly, one cannot have such a game without such a model in place. To that end, this article offers a speculative interpretation of the *Civilization* games in which that model of the history of technology is itself revealed to reflect particular historical biases over the relatively short lifespan of the series.[2]

Thus, the purpose of this article is twofold. On one hand, it examines the intrinsic biases of the *Civilization* series in relation to the practices and assumptions of the associated histories of science and technology; that is, this technological artifact (a computer game) is studied, across its various versions, to explore a broader sense of the history of technology. This article could be said to sit in the tradition of George Basalla's seminal work *The Evolution of Technology*, were it not for the fact that in 1999 he had already dismissed "the home computer boom" as having been a fad, stating that "playing electronic games [on home computers] was an activity that soon lost its novelty, pleasure and excitement."[3] As a problem of hindsight familiar to anyone trying to be up to date when talking about the history of technology, the future caught

---

[2] There are dangers inherent to this latter point, given that this article approaches a large conceptual field through merely twenty-five years, and with little reference to other cultural touchstones. Yet, as Michael Mahoney notes, much of the emphasis in the history of computing has been on hardware. "The emphasis," he writes, "has lain on what the computer could do rather than on how the computer was made to do it"; Mahoney, "What Makes the History of Software Hard," *IEEE Annals of the History of Computing* 30 (2008): 8–18, on 12. The present article extends this awareness to consider how particular things the computer was made to do (play games) have been constructed in a particular way; software encodes the structural and cultural biases of its creators, and the *Civilization* games are no different.

[3] George Basalla, *The Evolution of Technology* (1988; repr., Cambridge, UK, 1999), 185.

Basalla out. In contrast, this article's second purpose is to show that, as a vector of cultural transmission for ideas about history, technology, and the history of technology, games such as *Civilization* have significant influence in determining common perceptions of the relationship between technology and culture. Most importantly, and worryingly for historians, such influence can be seen to reinforce overly simplistic and deterministic models of the history of technology; since its first electronic edition, the *Civilization* series has sold over 33 million copies, 8 million of which were for *Civilization V* alone,[4] which clearly attests to the popularity of the series and to the number of players who will have, to a degree, internalized the logic of its model of history.

MODELING THE HISTORY OF TECHNOLOGY

Since the first iteration of *Civilization* as a video game, one of the key elements to a civilization's success has been the "technology tree." This is a list of skills and types of knowledge possessed by the civilization that determines what possible governments it can have, what buildings can be constructed, and what kinds of units can be fielded. Indeed, the original board game of Civilization (produced in 1980) is often credited as being the first game to incorporate a technology tree. These trees have proven to be influential within the genre of strategy games and computer games more broadly. They facilitate coding by setting parameters on the availability of units, resources, or even character traits of inhabitants, based on a programmable hierarchy, and allow coders to balance the game to enable different styles and types of solution to problems; in short, they usefully simplify a larger-scale problem, such as "model global technological development," into a discrete set of steps. Thus, the easiest way to conceive of a technology tree is to imagine a genealogical tree of technologies in which earlier research provides the basis for a civilization to discover more advanced technologies. That is, each technology—also called an advance in earlier versions of the game—is discrete and leads to others. For example, to develop the Compass advance in *Civilization IV*, a civilization must have learned Sailing and Iron Working first, and it is one of the prerequisites for a civilization to discover Optics.

Each player assigns a proportion of their civilization's "work" to research, which determines how quickly an advance is realized. A player then selects a given technology on which to focus. Once that technology is "discovered," it serves as a prerequisite and others then become available. At that point, players can select another technology, with some not becoming available until still further prerequisites have been met.[5] Ever since the second *Civilization* game, in 1996, the series has also utilized the concept of "Eras," whereby particular technologies are mapped into a distinct historical epoch.

---

[4] Stephany Nunneley, "The Civilization Series Has Sold 33 Million Copies since It Debuted in 1991," *Videogames 24/7*, 19 February 2016, https://www.vg247.com/2016/02/19/the-civilization-series-has-sold-33-million-copies-since-it-debuted-in-1991/.

[5] For an example of a technology tree, see "Civilization II: Posters," Civilization Fanatics Center, https://www.civfanatics.com/civ2/downloads/reference/posters/; this web page provides links to scans of the technology tree from *Civilization II*. Within the field of history, the notion of an implicit technology tree is perhaps most obvious in the work of James Burke, who in *Connections* (1978; repr., New York, N.Y., 2007), based upon the television program of the same name, and to a lesser extent in *The Day The Universe Changed* (Boston, 1985), demonstrated the interdependence and interreliance of technological developments and the ways in which they influence how a society perceives the world. One might also consider Augustus Pitt Rivers's nineteenth-century approach to cultural evolution via technological artifacts being arranged in "order" of complexity as another instance of a historical technology tree, albeit specific to one technology; see Basalla, *Evolution* (cit. n. 3), 17–21.

Only when all the requisite technologies are discovered can a civilization proceed to the next Era (say, from ancient to classical, or industrial to modern). These Eras model something akin to Kuhnian paradigm shifts, where a civilization's view of the world and its ability to interact with it ostensibly change as the result of a specific discovery, invention, or technology. In this way, technologies within the technology tree are classified by their respective Eras. It is worth noting, however, that this "paradigm shift" is primarily cosmetic, and does not affect gameplay.

Another important element of the technology trees are the intertwined notions of prerequisites and obsolescence. Each technology after the initial few has a set of prerequisites that determine how and when it might be researched. The notion of a technological prerequisite is not that difficult to fathom; an automobile cannot be imagined without the conception of a wheel, and construction of a combustion engine is facilitated by various advances in science and technology. However, as Basalla notes, the gasoline engine versus steam or electric engine might be said to the product of specific cultural factors,[6] but *Civilization* elides these factors. As a result, the teleological structure of the technology tree means that particular "prerequisites" can seem idiosyncratic, especially as the player "knows" the outcome of the technology before it is researched. In *Civilization V*, for example, the Refrigeration advance requires a player to have researched Biology and Electricity, and Refrigeration in turn allows players to then research Penicillin, which in turn is required for a player to later research Ecology, and this in turn allows Telecommunications. This causal chain—Refrigeration is necessary before a civilization "understands" the potential of Ecology, and in turn Ecology is necessary for Telecommunications—seems unusual, but it is intended to facilitate a clear line of technological advancement that enables players to build units and enhance cities (technology is instrumental in strategy games, and particularly in the *Civilization* series).[7] Similarly, the use of technological advances to make other advances or units obsolete is intended as a game mechanic to promote gameplay balance; in that way the effects of certain structures or units are weakened over time, and players can therefore use advances to overcome deficiencies in military power. However, this in-game mechanic presents a linear model of technological development rather than a more complicated network or repertoire of available technologies and techniques that might function in different ways in different contexts.

The assumptions behind the concept of *Civilization*'s technology trees are thus worth noting. The game designers have encoded a simplified history of Western civilization—ideological, political, and technological—into a world simulator, along with all the concomitant baggage this implies.[8] It is, as has been examined, a fundamentally determin-

---

[6] Basalla, *Evolution* (cit. n. 3), 197–202.

[7] Tuur Ghys notes that, in relation to *Civilization IV*, it is impossible for a player to research Robotics without having first researched Mysticism; Ghys, "Technology Trees: Freedom and Determinism in Historical Strategy Games," *Game Studies* 12 (2012), http://www.gamestudies.org/1201/articles/tuur_ghys. It is possible to find similar examples in other *Civilization* games.

[8] In many respects, the games become more culturally aware over the series by bringing in less Western-centered perspectives on world history, such as through the inclusion of non-Classical "wonders" from non-Western cultures. Responding to the criticism that the more technologically advanced a civilization is, the more it begins to "resemble contemporary America," Sid Meier (the lead designer on many of the *Civilization* games) responded: "I can also blame the internet now. The world has become flat, we are more aware and sensitive to the globalness of the world. The early 1990s world was reflective of our thinking. China was still this mysterious hidden kingdom, Russia was the evil empire"; quoted in Kanishk Tharoor, "Playing with History: What Sid Meier's Video Game Empire Got Right and Wrong About 'Civilization,'" *Longreads*, October 2016, https://longreads.com/2016/10/26/what-sid-meiers-video-game-empire-got-right-and-wrong-about-civilization/.

istic model of technological development and thus of civilization. For example, Tuur Ghys's work examines the implications of the technological determinism at work in four strategy games, one of which is *Civilization IV*.[9] However, defending the genre's focus on technology, Ethan Watrall notes the following: "Of all the variables wrapped up in the process of culture change, technology is arguably one of the easiest to quantify and track. Technology leaves a lot of stuff behind for archaeologists like myself to find and study."[10] Watrall remains skeptical of technology trees as they are currently conceived, however, and as he later asserts, "designers desperately need to realize that for in-game technological innovation to happen, some of the direct control must be taken out of the hands of players";[11] as of 2018 this has yet to really occur, despite the verisimilitude it would promote. Determinism is also evident in the fact that the more technologies a civilization possesses, the more "advanced" it is, thereby positing a technocentric view that is measured in progress along a predefined tree.

Civilization's creator, Sid Meier, affirmed the game's relation to progress: "The tech tree represents a certain kind of optimism, the idea that we are constantly progressing . . . It's true, we don't represent the Dark Ages in the game. It's an optimist [*sic*], progress-based view."[12] *Civilization* is very much rooted in a "progressive" model of the history of technology.

A civilization that does not develop certain technologies therefore would find it almost impossible to "win" the game, as they would not survive their neighbor's advantages, however "advanced" they might be in other ways, such as in agriculture.[13] Moreover, the technologies that can be researched are, to all intents and purposes, assumed to be "researchable"; in the case of technologies like Rocketry and Computers, both of which have existed since the first iteration of the game, this is eminently justifiable, even if the extent to which they are technologies *per se* is debatable. However, in the cases of Mass Production, Communism, Philosophy, or Religion, it is debatable as to whether these are "discoverable," let alone technologies; they are categorized this way only to simplify the act of modeling. Ghys observes the following in technology trees: "We find machines (steam engine), techniques (sailing), sciences and bodies of knowledge (physics), abstract and religious ideas and rituals (polytheism, philosophy) and forms of social organisation (guilds, feudalism). The latter examples can be called 'social technologies' (no relation with recent network technology)."[14] This corresponds to W. Brian Arthur's view, whereby perceiving technology as any "purposed system" means that "musical structures, money, legal codes, institutions, and organizations" can all be viewed as technologies.[15]

---

[9] Ghys, "Technology Trees" (cit. n. 7).

[10] Ethan Watrall, "Chopping Down the Tech Tree: Perspectives of Technological Linearity in God Games, Part One," *Gamasutra*, 31 May 2000, 1–5, on 1, http://www.gamasutra.com/view/feature/131570/chopping_down_the_tech_tree_.php.

[11] Ethan Watrall, "Chopping Down the Tech Tree: Perspectives of Technological Linearity in God Games, Part Two," *Gamasutra*, 7 June 2000, 1–6, on 6, https://www.gamasutra.com/view/feature/131571/chopping_down_the_tech_tree_.php.

[12] Quoted in Tharoor, "Playing with History" (cit. n. 8).

[13] As Christopher Douglas summarizes, "A civilization further up what is termed in the strategy game genre the 'technology tree' has a competitive edge—economic, social, military—over its rivals"; Douglas, "'You Have Unleashed'" (cit. n. 1). Douglas's phrasings of "further up" and "competitive edge" also reveal the extent to which technologies facilitate one civilization being "better" than another.

[14] Ghys, "Technology Trees" (cit. n. 7).

[15] W. Brian Arthur, *The Nature of Technology: What It Is and How It Evolves* (London, 2009), 56.

The technology trees of the *Civilization* games are thus constructed from a series of fundamental axioms. Some of these are obviously real-world knowledge constraints (the games cannot include "unknown" technologies), whereas others are due to coding constraints (the need to streamline the history of technology to a simplified model). Some, however, are also implicitly based upon a particular set of assumptions about the history of technology:

1. All possible fundamental technologies (such as using fire) have been discovered already. All fundamental technologies are known to any sufficiently advanced civilization.
2. A civilization is conceived as one homogenous society working toward one goal. (A civilization can only research one advance at a time; they research a specific advance, such as Electronics, rather than an advance within a broad area, such as agriculture or the military. The player controls the outcome of the research in advance.)
3. The path through the technology tree is unidirectional, and applies to the whole civilization. (Once a technology has been discovered a civilization cannot forget it, even if that civilization is reduced to a small population in one city, and inequalities—geographic, socioeconomic, and so forth—in technological distribution do not appear within the games, unless one counts which buildings exist in which cities.)
4. Technologies have applications, whether militaristic or civil, and contribute to the civilization in some manner, even if only to open more options on the technology tree. (There is no sense of discovery "for its own sake" that might lead to a dead end, and there is no associated "risk" with a program of research.)[16]
5. There is no possible technology outside of the technology tree and its associated applications. (A player cannot use the Computers advance to create Artificial Intelligence, or Genetics to create transspecies hybrids, such as Genetically Modified Organisms. In general, all nonfundamental technologies have specific prerequisites and do not merely "appear" in a particular age.)
6. The technology tree might appear "evolutionary," as technologies are discovered based upon prerequisites, but the tree itself is static, as are the technologies themselves. (The tree can only be "modded" from outside of the game, and the technologies do not fundamentally change the game environment even if understandings of chemistry or physics can change in the real world.[17])

While it is of interest to consider what is called a technology and what is not (that is, what appears in the game as a technology), it is more important to observe that all social "advances" or "technologies," through the act of naming them as such, reconfigure society as a system in the process of ongoing refinement. That is, the *Civilization* series, through its very encoding of a technologically determined society by way of a (technological) algorithm, posits that human societies are in some ways "codable" or "systemic" in nature. To reframe the postcolonial critique that has often been lev-

---

[16] Watrall, "Chopping Down the Tech Tree, Part One" (cit. n. 10), 3.
[17] For a discussion of the role of players' "modding" activities—that is, amending the game's code to change how the game is played—in relation to *Civilization*, see Trevor Owens, "Modding the History of Science: Values at Play in Modder Discussions of Sid Meier's *Civilization*," *Simulation & Gaming* 42 (2011): 481–95.

eled at the series—that it reveals a fundamentally Western, often US-centric view of civilization at the expense of other cultural values and norms—implies that the technological discourse surrounding computer games and the act of recreating technological innovation as a "tree" have themselves colonized the view of civilization that the series uses.

WHAT DO YOU NEED TO MAKE COMPUTERS? CONTEXTUALIZING
TECHNOLOGIES IN CIVILIZATION

Given this backdrop of a technology tree, various changes have taken place that refine or amend the ideas behind what constitutes a technology in all the *Civilization* games. *Civilization IV*, for example, allowed players to research technologies that enabled them to enact civics, or rules that govern their civilizations, rather than simply suggesting a direct corollary between researching Democracy and then enacting it as a government (which could occur in the first few games of the series). One such civic is Bureaucracy, which one can utilize once the Civil Service advance has been discovered. But to what extent is Civil Service really a "technology," and to what extent does recasting such civil advances as technologies reconstitute how we understand their social functions? Similarly, corresponding loosely to Jared Diamond's work on the environmental features of technological development, *Civilization VI* incorporates research bonuses for technologies related to an environment that the player controls, such as quarries; in earlier versions of the game, potential benefits in the environment had little bearing on technological progress, despite the simulation of different environments and variable sizes of land mass.[18]

It is precisely because of the consistency of the design of technology trees in the *Civilization* series that we can move beyond the game-specific examples mentioned so far. Despite the analysis of ideological assumptions behind the series,[19] or examination of the ways in which technology trees might function in *Civilization* and other strategy games,[20] the very act of creating different trees in different versions facilitates a broader comparative analysis of the specific decisions made within each game. That is, while the structural principles of the trees are largely the same across games, the positions, prerequisites, and operations of different technologies within each tree can change. Erin MacNeil begins to approach this insight when she observes the following:

> Sequels often advance the timeline/chronology of a particular plot or set of characters and contain massive updates to the engine and mechanics of the game. Spoliated games, in contrast, actually mine the older game, often using the older game's engine, mechanics, and digital assets. *Civilization* is an iterative franchise, meaning that it continues to develop a central conceit (rule an important civilization from the dawn of time onwards) all while updating and changing its approach to core mechanics, visual style, and gaming engine.[21]

One of the spolia that remains throughout these spoliated *Civilization* games is the concept of the technology tree itself. Thus, while "each new *Civilization* game starts

---

[18] Jared Diamond, *Guns, Germs, and Steel: The Fate of Human Societies* (New York, N.Y., 1999). Diamond's work is mentioned in both Douglas, "'You Have Unleashed'" (cit. n. 1); and Wainwright "Teaching Historical Theory" (cit. n. 1), in regard to the *Civilization* games they discuss.
[19] Douglas, "'You Have Unleashed'" (cit. n. 1); Poblocki, "Becoming-State" (cit. n. 1).
[20] Ghys, "Technology Trees" (cit. n. 7); Watrall, "Chopping Down the Tech Tree, Part One" (cit. n. 10); Watrall, "Chopping Down the Tech Tree, Part Two" (cit. n. 11).
[21] Erin McNeil, "Ludic Spolia in Sid Meier's *Civilization: Beyond Earth*," *Journal of Games Criticism* 3A (2016): in section titled "Practical Spoliation," http://gamescriticism.org/articles/mcneil-3-a/.

with the desire to model history through this god-king lense, but attempts to differentiate itself from the one that came before by modifying rule sets, and adding new forms of interaction," the technology tree remains broadly static.[22]

Because of this static model, looking at how one particular advance changes across the series facilitates an examination of the ways it has been reconceptualized as a technology, and as a precursor to, or extension of, other technologies. The Computers advance is obviously something close to *Civilization*'s heart; *Civilization* is a longstanding series that relies upon computers to exist, let alone function, and its complexity (in comparison to the relatively simple board game incarnation) is only possible precisely because of computing advances. Yet, is Computers a technology in and of itself?[23] Table 1 outlines the relevant conceptions of the advance across the series.

The fact that Computers are features of different "ages" is revealing. In *Civilization II* to *IV* the advance is a feature of the "contemporary" Era—that is, the last Era that the game utilizes, and which brings the player's civilization into the present-day (Western) technological milieu. However, in relation to *Civilization V* and *VI*, Computers is conceptualized as a "historical" technology (that is, its Era is earlier than the final Era of the game's technology tree). Moreover, in *Civilization I*, mathematics and a working knowledge of electronics is required to gain the Computers advance, whereas from *Civilization IV* onward it is fundamentally linked to mass media. There is also a tension about whether a society requires Computers to consider the possibility of Miniaturization, or Miniaturization to consider the possibility of Computers. Similarly, although the Computers advance is assumed to lead to Robotics throughout most of the series, it is only explicitly linked with space exploration in the earlier versions. Although the later games do feature a space component, primarily through their continuation of the Space Race victory condition, the SETI (Search for Extraterrestrial Intelligence) program does not get featured in versions later than *Civilization III*. This seemingly corresponds with the more overt links to telecommunication and mass me-

---

[22] Ibid. The *Beyond Earth* game within *Civilization* to which MacNeil refers utilizes a "technology web" rather than a "technology tree," although the basic axioms of the tree nevertheless remain in effect. That said, MacNeil rather optimistically believes that technology in *Beyond Earth* is viewed through a "postmodern vantage point" and that "once a player has interacted with the Technology Web, the lock-step nature of *Civilization V*'s Technology Tree becomes apparent, along with its implicit Enlightenment biases"; see MacNeil, "Ludic Spolia" (cit. n. 21), in section titled "Technology Trees and Technology Webs." Using the Technology Web to view the tree facilitates critique, in essence. It rather seems that its changes are relatively minor, despite the interesting inclusion of "Affinities" deriving from the types of research a player's civilization undertakes, and it does little to ameliorate the "colonization" of human society by technocentric discourse.

[23] In an article on evolutionary paradigms in relation to technology, and in one of the earliest uses of a technology tree to model technological change, Chris De Bresson sketches a technology tree for what he calls "Micro-Computers" (thereby dating the paper technologically) in which he links the necessity of combining "semi-conductor effects" (transistor, rectifier, amplifier) into an "integrated circuit" and then linking this into conceivable "functions" of computing (input, output, operations [ROM], and memory [RAM]); he does this in order to consider how it becomes possible as a technology, thereby demonstrating how any given technology operates through a *"synthesis* of many different types of technologies"; De Bresson, "The Evolutionary Paradigm and the Economics of Technological Change," *Journal of Economic Issues* 21 (1987): 751–62, on 755. Although De Bresson does not include a conceptual history of computing (his tree is a deliberately limited diagram), the article nonetheless demonstrates the interreliance of many different fields of endeavor to any one technological advance. This might today be updated to include touchscreen technologies as well as different types of motherboard and microprocessor manufacture to factor in the emergence of the tablet computer and smart phones, demonstrating the further synthesis of available technological advances, without even considering the sociological effects of a given advance. Moreover, none of this takes into account the sociological effects emerging from the manufacture, distribution, and use of such technologies.

Table 1. The "Computers advance" shown across the Civilization series. ("Obsolete" indicates that once a player has researched this technology, it reduces the effect of other game elements.)

	Era	Prerequisites	Units / Buildings Available	Acts as Prerequisite For	Obsoletes
Civ I (1991)	n/a	Mathematics, Electronics	SETI Program*	Robotics, Space Flight	
Civ II (1996)	Modern	Mass Production, Miniturization	SETI Program, Research Lab	Robotics, Space Flight	
Civ III (2001)	Modern	[In Modern Times]	SETI Program, Research Lab, Mechanized Infantry	Laser, Miniaturization	
Civ IV (2005)	Modern	Radio [and Plastics added in expansion]	Modern Armor, Laboratory, The Internet	Genetics, Fiber Optics, Robotics	Angkor Wat, Spiral Minaret, University of Sankore
Civ V (2010)	Modern	Electronics, Mass Media	Nuclear Submarine, Mobile SAM	Robotics	
Civ VI (2016)	Atomic	Electricity, Radio		Telecommunications, Robotics	

Note. Search for Extraterrestrial Intelligence (SETI).

dia, which make computing for calculation and for network capabilities almost mutually exclusive as the series progresses.

In this manner, what is historical contingency, or perhaps more accurately, fashionable perceptions of particular technologies, becomes reified within the series as the only outlet for that technology when the technology tree is created, and this is generally retrospectively predicated upon "what actually happened." (In this case, the historical contingency involves whether computers are used in space exploration or mass/social media.) MacNeil writes the following in regard to the *Civilization V* technology tree requiring Atomic Theory to develop both The Manhattan Project and the Ecology technologies: "It makes such historical facts feel necessary and unavoidable. It reinforces the descriptive trajectory of our technological advancements and argues for that path being the only way things could have happened. We did research nuclear technologies before we really concerned ourselves with our conservationist technologies, and permuted through the logic of *Civilization V*, we *ought* to have done so."[24]

For many of the technologies in the games, what is meant by the signifying name of the technology is assumed to be its limits—its culmination or *raison d'être*. In relation to Computers, the long view (whereby we might incorporate Charles Babbage and Ada Lovelace through to modern conceptions of artificial intelligence and algorithms) is elided by its virtual incarnation as that which facilitates the next technology, which has itself shifted from data management and analysis, and the ability to

[24] MacNeil, "Ludic Spolia" (cit. n. 21), in section titled "Technology Trees and Technology Webs."

model dynamic situations in real time, toward communications and (social) networking. In this case, "what actually happened" has shifted over the duration the *Civilization* games have been in existence (as a general trend, computers became popularly used for media rather than calculation/exploration) and the technology tree has been amended to reflect the very contingency denied within the determinism of the tree itself. Clearly, the games do not refine their technology trees to promote accuracy or increase coverage, as the number of technologies in the trees remains roughly the same across the series, and it is other aspects of the games that increase in complexity.[25] By looking at the ways in which specific technologies have changed across the technology trees, it becomes apparent that these trees respond to and reflect paradigms of the current uses and applications of specific technologies.

TECHNOLOGICAL CONTINGENCY AND *CIVILIZATION*

If this is how the *Civilization* series structures historical technologies, then a similar structural bias is built into its imaginings of contemporary society. Because historically focused strategy games consider the "counterfactuals" of history, A. Martin Wainwright has used them with his students—since the games simulate the emergence of civilizations, students can manipulate variables and outcomes.[26] This can also occur, in a more limited sense, within the *Civilization* series because the outcomes of particular technological developments, and technological interactions, can also be simulated. For example, "what if," a player might ask, "a culture had focused on agrarian and pacifistic technologies at the expense of military or economic expansion?" Indeed, this simulation of a set of historical givens (say, the development of a particular technology leading to another) marks the *Civilization* games as being useful for asking questions about technological determinism versus technological contingency.[27] But, importantly, the *Civilization* series demonstrates a series of iterations in the act of modeling technological development itself. As mentioned above, this is not a refinement of the accuracy of the modeling (as simulations are generally assumed to be), but one that responds to enhancements in computer technologies themselves (such as the game's user interface shifting from top-down colored squares to isometric and animated maps), consumer markets (what sells and what does not in computer games), and in what is indeed considered "technological" at a given historical juncture. That

---

[25] Each game in the series has approximately seventy to ninety technologies, although this depends upon expansion packs. The first *Civilization* video game and *Civilization VI* have the fewest (approximately seventy), whereas *Civilization II* and *IV* have the most (approximately ninety). In this sense, the technology tree itself is spoliated throughout *Civilization*; although the technology tree could have been made less rigid (particularly in terms of how it structures the game), it has remained in a relatively static form, with a comparable number of technologies across the series.

[26] Wainwright, "Teaching Historical Theory" (cit. n. 1), 591–3.

[27] For example, taking into account Francis Bacon's assertions that printing, gunpowder, and the magnetic compass are key Western technologies, it is worthwhile noting that they "were the products of Chinese, not European, civilization," although China did not develop them in the same directions as Renaissance Europe because of specific cultural values; Basalla, *Evolution* (cit. n. 3), 169. To this list we could add optics and glass technology, which might be construed as key to Western development of the microscope and the telescope; however, these were also not developed in the societies that first discovered glass and optics because of the relative cultural value of ceramics over glass. In light of such generalizations, it would behoove us to consider the extent to which technologies that might today be perceived as "primitive" could someday be seen as the first steps to a far more advanced technology. Still, today we cannot yet draw a lineage because the technology has not yet been invented.

is, the *Civilization* series is—via its very acts of technological modeling—a set of historically contingent models about technological development; it not only illustrates a particular view of technological and scientific development, but also reveals, to a limited extent, historical shifts in the perceptions of particular technologies.

To provide a clear indication of this, one simply needs to consider the ways in which technologies have been added to or removed from the franchise over its various incarnations. Table 2 presents a simplified list of the various advances that can be researched across the *Civilization* series in the final Era(s) of the game: "Modern" in *Civilization II* and *III*, "Future" in *Civilization IV*, and the "Information" Eras in *Civilization V* and *VI*.

Although comparing the games is difficult because they utilize slightly different technology trees and variant prerequisites for units or buildings, and the like, there are clear changes here. Perhaps the most obvious difference is the way in which the Era system developed over the course of the franchise. Whereas *Civilization II* and *III* end with Modern technologies, *Civilization IV* suggests technologies for the Future Era, at least one of which actually exists (stealth technologies, such as those used for the F-117 Nighthawk or the B2 Spirit). Then the latest two versions, *Civilization V* and *VI*, divide this Future Era into the Modern, Atomic, and Information Eras. The Modern Era, thus conceived, has become something of a historical epoch that laid the ground for subsequent Eras, and the other two named—Atomic and Information—have only relatively recently gained some traction as terms for distinct historical epochs.[28]

Similarly, the contexts in which technologies are situated can change over the series. Whereas Plastics was situated as a "modern" technology until *Civilization V*, in *Civilization VI* it became an Atomic Era advance, as the boundaries on what constituted a particular level of technology changed, and it came to be perceived as more "historical" in the light of more recent innovations. The same is also true of the [Nuclear] Fission advance, which shifted between the Industrial and Atomic Eras over the franchise; this has occurred as whatever [Nuclear] Fission enables has been altered. That is, the "knowledge" encapsulated by the Atomic Theory, Fission, and Fusion advances varies according to what the particular game considers to be the "threshold" of a particular technology (usually in its realization via a real-world application). Atomic Theory, for example, can signify a classical conception of the notion of the "atom" as well as how that understanding plays out through applications of the underlying scientific principles, depending upon which edition of *Civilization* is played.

Alongside this, the appearance (and disappearance) of technologies as the series develops is also worth noting. Space Flight disappears as a discrete technology relatively early in the series; it is, for example, phased out alongside the introduction of others, such as Satellites (from *Civilization III*). Then there are some advances that only occur once (such as The Internet or Particle Physics in *Civilization V*, Smart Weapons in *Civilization III*, or Guidance Systems in *Civilization VI*), whereas others (such as The Laser) recur across the series. This feature of the series is useful for interrogating what constitutes a "technological advance" to the game designers. Perceiving The Internet as a technology, rather than a set of implementations of technologies, for example, reconstitutes its importance and

---

[28] For example, in terms of the linear model of technological Eras, Wikipedia—itself arguably an indicator of the information age—suggests that after the (second) Industrial Revolution, the atomic age, jet age, space age, digital revolution, and information age followed; see Wikipedia, s.v. "Information Age," History of Technology sidebar, https://en.wikipedia.org/wiki/Information_Age (accessed 11 February 2019).

Table 2. Summary of "Final Era" advances in the Civilization series. The original Civilization did not use the Era system and is omitted from this table.

	Civilization II (1996) [Modern Era]	Civilization III (2001) [Modern Era]	Civilization IV (2005) [Future Era]	Civilization V (2010) [Information Era]	Civilization VI (2016) [Information Era]
Advanced Ballistics	✓	[Industrial Age]	[Modern Era]	✓	[Atomic Era]
[Advanced] Flight	✓			[Modern Era]	[Atomic Era]
Amphibious Warfare	✓				
Automobile	✓				
Combined Arms	✓		[Modern Era]	[Atomic Era]	[Atomic Era]
Composites	✓		[Modern Era]		✓
Computers	✓	✓	[Modern Era]	[Atomic Era]	[Atomic Era]
Ecology		✓		[Atomic Era]	
Electronics	✓	[Industrial Age]		[Modern Era]	
Environmentalism	✓				
Espionage	✓	[Industrial Age]			
Fiber Optics			[Modern Era]		
[Nuclear] Fission	✓	✓	[Industrial Era]	[Atomic Era]	[Atomic Era]
Future Technology	✓	✓	✓	✓	✓
Genetic[s] [Engineering]	✓		[Modern Era]		
Globalization				✓	
Guerrilla Warfare	✓				
Guidance Systems					✓
Integrated Defense		✓			

170

Labor Union	✓				
[The] Laser[s]	✓	✓	[Modern Era] [in "Beyond the Sword" expansion]	✓	✓
Mass Media			[Modern Era]		
Mass Production	✓	[Industrial Age]			
Miniaturization	✓	✓			
Mobile Tactics/Warfare	✓	[Industrial Age]		✓	
Nanotechnology				✓	✓
[Nuclear] Fusion [Power]	✓		✓	✓	✓
Nuclear Power	✓	✓			
Particle Physics				✓	
Plastics	✓		[Modern Era]	[Modern Era]	[Atomic Era]
Radio	✓	[Industrial Age]	[Modern Era]	[Modern Era]	[Modern Era]
Recycling	✓	✓			
Refrigeration	✓	✓	[Modern Era]	[Modern Era]	
Robotics	✓	✓	[Modern Era]	✓	✓
Rocketry	✓	✓	[Modern Era]	[Atomic Era]	[Atomic Era]
Satellites		✓	[Modern Era]	✓	✓
Smart Weapons		✓			
Space Flight	✓	✓			
Stealth [Technology]	✓	✓	✓ [in "Beyond the Sword" expansion]	✓	✓
Superconductor[s]	✓	✓	[Modern Era]		
Synthetic Fibers/Materials		✓			[Atomic Era]
Telecommunications				✓	✓
The Internet				✓	

how it might be situated in a social context. It first appeared in *Civilization IV* as a "world project," but then became a technology in *Civilization V*, before leaving only a vestige in *Civilization VI* via the Social Media civic, which one can imagine also disappearing in later versions of the game. Though these shifts are partly due to the mechanics of the game, as the series grew and gained in popularity the internet was emerging as a term with cultural valency, meaning that it was only an interesting "talking point" during the development of *Civilization IV* and *V*; between the launch of *Civilization III* in 2001 and *Civilization V* in 2010, it became less of a "wonder" and more of a technological norm. After that, the internet became so accepted—and further built upon—that it ceased to be sufficiently conceptualized as a technological "advance" in its own right.[29]

Further, larger-scale technological shifts across the series include those concerned with genetics and genetic engineering (which disappear from *Civilization IV* and subsequent games), superconductors (which also disappear from *Civilization IV*), and nanotechnology (which only appears in *Civilization V* and *VI*). The removal of Genetic Engineering and inclusion of Nanotechnology arguably reveal the ways in which the *Civilization* series reflects the sociocultural conditions within which technological "advances" operate. Genetics, as an advance, could lead to effects and wonders such as Cure for Cancer (in *Civilization II* and *III*) and Longevity (in *Civilization III*), or to increased population health (in *Civilization IV*).[30] The removal of these from later games, incorporating politically "safer" in-game technologies such as Penicillin—an Atomic Era technology in *Civilization V*—or removing health technologies as viable advances in *Civilization VI*, speaks to a broader cultural concern about some genetic experimentation (e.g., cloning, gene splicing), at least as viable technologies to be explored.[31] Nanotechnology was arguably included in later games (post-2010) as a more "technological" technology, with more potentially interesting applications within the game mechanics, not only because it might be seen as less politically charged than Genetics, but also because it has matured—at least as a hypothetical technology—to the extent that it can be conceived as a viable technological advance.

While we cannot take one game series as anything but a general indicator of social trends and patterns, these changes suggest that, in its very attempts to be technologically aware and "model" technological developments, the *Civilization* series acts as a limited barometer of a wider public awareness of scientific and technological fashions. In relation to nuclear warfare, for example, the *Civilization* franchise demonstrates a particular view of so-called scientific progress. In the earlier games, the nuclear option is available for destroying enemy cities, and while this continues throughout the

[29] For example, a search in Google Trends (although limited in terms of evidence value) for the term and the topic "internet" suggests a decline in interest in the term since 2010; this decline is even more evident when the "United States" (where the game designers are based) is chosen as a search origin, rather than "Worldwide"; see Google Trends, s.v. internet, https://trends.google.co.uk/trends/explore ?date = all&q = internet (accessed 7 February 2019). One might also point to the use of "internet" as an adjective or modifier amending existing activities itemized by the *Oxford English Dictionary*, many of which are listed as appearing primarily in the early 2000s.

[30] Interestingly, this application of "genetics" operates in both present and futuristic contexts; that is, we have "discovered" genetics already, but the promise of the technology, at least according to the *Civilization* games, has yet to be fulfilled.

[31] However, with the advent of "CRISPR," and the use of CRISPR-cas9 techniques to edit gene sequences, we might hypothesize that a future *Civilization VII* or *VIII* may end up revivifying genetics as a legitimate "technology."

series, the type of weapons, their deployment, and how players might defend against them changes significantly. In *Civilization II*, the "SDI Defense" is a possible building (available through the Laser advance) that is positioned in each of the player's cities to protect them and their immediate environs from nuclear weaponry. Similarly, in *Civilization III*, the Integrated Defense advance allows players to construct the Strategic Missile Defense wonder (a special type of building that is built only once by a civilization), which affords a 75% chance of intercepting an enemy's ICBM (intercontinental ballistic missile). Both of these protections clearly reference the Strategic Defense Initiative (SDI)—the so-called Star Wars program initiated by the United States during the Reagan administration. Yet, although the Laser advance enables construction of the SDI Defense in *Civilization II*, it shifts position in *Civilization III* because it is a prerequisite for Smart Weapons, which is itself the direct prerequisite for Integrated Defense. In *Civilization IV* this last element is replicated through the SDI national project, which produces the same results as in *Civilization III*. However, *Civilization IV* also introduces the Tactical Nuke unit, distinct from the ICBM unit, which can avoid interception. In *Civilization V*, however, a player can only build a Bomb Shelter within each of their cities to ameliorate population loss and contamination caused by a nuclear strike. By *Civilization VI*, players can develop nuclear and thermonuclear devices with the right advances, but only the Mobile SAM unit can help defend against them. Thus, over the franchise, there is a clear movement away from the "fantastic" technologies promised by the SDI program—of national shields and lasers shooting down nuclear weapons from within cities—to merely surviving the attack, alongside a concomitant development in the range and type of nuclear weapons (from Nuclear Missile, to ICBM, to Tactical Nuke, and so forth). With these changes, the series reveals itself to be responding to particular trends in contemporary technological developments, and perceptions of them, while not significantly changing the axioms upon which those technologies are developed.

CONCLUSION

What emerges from the ways in which the *Civilization* games "play with" technology and write fictions of (the history of) science is that their ideological nature so often observed in the ways they treat civilizations—and indeed the very notion of civilization—is equally replicated through, and perhaps even emerges from, their deterministic views of technology and its progress. Their power, as models of the "history of technology," is that they clearly demonstrate the potential for games to engage with questions of technological development, and at least imply historical contingency being divorced from "what actually happened"; as examples, a player might industrialize their society in the 1200s, develop Lasers in the 1600s, or not learn Horse Riding until the 1400s.

But for all its superficial play with technology, which nonetheless encourages players to consider it as unidirectional, the series's model of the history of technology thereby serves to limit its potential to consider cultural technological relativism in favor of reifying (contemporary) ideological concerns. The games are entertainment, not history, but they demonstrate which models of technological development and fictions of scientific progress become perpetuated (and which become hidden or ignored). Thus, the legitimacy of the technology tree as a model of technological progress remains in question, filtered as these incarnations are through a US-centered, capitalist world view. While the games purport to encompass generalizable histories of human technological progress, they in fact reinforce a capitalist and deterministic view of technology; this continues to be true despite the moves the series has made toward being more "inclusive" of other cultures and norms. They are computer games, and therefore limited in what they can do both technologically and in terms of their market demand. Nevertheless, the basic assumptions of technological progress limit the series's ability to conceive of the very futures it gestures toward in its more utopian moments. Most damaging of all, at least from the perspective of this article, is one of the most pervasive fictions within the games. This is its instrumental approach to science and technology that ignores human agency and cultural decisions at work within the respective histories of each game.

As Mahoney observes, "the history of computing, especially of software, should strive to preserve human agency by structuring its narratives around people facing choices and making decisions instead of impersonal forces pushing people in a predetermined direction."[32] The decisions and choices made by the creators of *Civilization*—what Mahoney calls in relation to software "an operative representation of that portion of the world that captures what we take to be its essential features"—clearly make a virtue of modeling technology (understood in both senses) as being central to human endeavor, and ascribe to technology an almost inexorable, inevitable ability to shape human society. The assumptions behind *Civilization*, all too often shared by those who work within technical fields, will only serve to reinforce a culturally damaging and overly jingoistic view of humanity, at the expense of the "history of technology" itself, and to the detriment of those students who might be embarking on the study of technology for the first time. The *Civilization* games can provide an effective tool in "history of technology" education, but primarily through asking students to interrogate the models of technology trees, and to question their own assumptions based on what prior experience they may have of playing such games.

---

[32] Mahoney, "What Makes the History" (cit. n. 2), 10

INSPIRING SCIENCE

War and Peace in British Science Fiction Fandom, 1936–1945

by Charlotte Sleigh* and Alice White§

ABSTRACT

Fans of science fiction offer an unusual opportunity to study that rare bird—a "public" view of science in history. Of course science fiction fans are by no means representative of a "general" public, but they are a coherent, interesting, and significant group in their own right. In this article, we follow British fans from their phase of self-organization just before World War II and through their wartime experiences. We examine how they defined science and science fiction, and how they connected their interest in them with their personal ambitions and social concerns. Moreover, we show how the war clarified and altered these connections. Rather than being distracted from science fiction, fans redoubled their focus upon it during the years of conflict. The number of new fanzines published in the midcentury actually peaked during the war. In this article, we examine what science fiction fandom, developed over the previous few years, offered them in this time of national trial.

WHO WERE THE FANS?

Fans in general, and science fiction fans in particular, have recently become a subject for respectable scholarly study.[1] Readers are increasingly familiar with the pleasures, politics, and schisms of early US fandom,[2] but little academic study has been carried out for the United Kingdom. A good deal has been written about British science fiction (SF) classics from the interwar period, such as Huxley's *Brave New World* (1932), but a focus on British fans of the same period (who were rather little affected by this particular novel) adjusts the historian's viewfinder. In making this adjustment, we do not presuppose that 1930s and '40s fandom constituted a predecessor to today's SF fans;

* School of History, Rutherford College, University of Kent, Canterbury, CT2 7NX, UK; c.l.sleigh@kent.ac.uk.
§ Wellcome Collection, 183 Euston Road, London, NW1 2BE, UK; a.white@wellcome.ac.uk.

[1] Gary Westfahl, "The Popular Tradition of Science Fiction Criticism, 1926–1980," *Sci. Fict. Stud.* 26 (1999): 187–212; John Cheng, *Astounding Wonder: Imagining Science and Science Fiction in Interwar America* (Philadelphia, Penn., 2012), 215–50; Andrew Pilsch, "Self-Help Supermen: The Politics of Fan Utopias in World War II-Era Science Fiction," *Sci. Fict. Stud.* 41 (2014): 524–42; Charlotte Sleigh, "'Come on You Demented Modernists, Let's Hear From You': Science Fans As Literary Critics In The 1930s," in *Being Modern: The Cultural Impact of Science in the Early Twentieth Century*, ed. Robert Bud et al. (London, 2018), 147–65. The last of these articles addresses the question of canon building in relation to identity, which is not discussed here.
[2] See, for example, Isaac Asimov, *I, Asimov: A Memoir* (New York, N.Y., 1994), 60–3.

instead, we find a multiplicity of issues and identities specific to working- and lower-middle-class young men in industrial, war-poised Britain. Still less do we presuppose what, if anything, counted as a canonical base of literature around which science fandom coalesced. In fact, such questions faded into the background in the context of war. British fans, then, offer to tell us something about science as a malleable cultural resource, and have the potential to disrupt US-centered generalizations about the nature and trajectory of SF. This topic begins to open up space to consider SF as a transnational scientific discourse in the era just prior to transnational big science. It is well known that British and American fans were in correspondence from the 1930s; during World War II, they began to meet in person, and extended their communicational networks through the exchange of fanzines.

This article also offers historians of science the possibility of considering their field in conjunction with fan studies, arguably a natural step on from the treatment of science as essentially mediatized knowledge as pioneered by Jim Secord, Sally Shuttleworth, and others. Fan studies, as described by Gray, Sandvoss, and Harrington, may be considered in three waves that in some ways are reminiscent of waves of history of science, technology, and medicine (HSTM) scholarship.[3] A first wave of study, they assert, was constructed around a political bipolarity of absorption into, or resistance to, mass-mediated culture. This resembles early HSTM accounts of professionalization, including accounts of systematic exclusion (such as that of women). Second-wave studies, Gray et al. contend, modified this to create a consumption-oriented account of fans as mobilizing their resources simultaneously to enact and reenact cultural capital and its hierarchies. This, broadly, resembles Foucauldian and Schafferian accounts connecting science and politics. Third-wave studies, of which Gray et al. position themselves as pioneers, tend instead to consider personal and cultural/political motivations for becoming a fan; it is an account which, by their lights, makes most sense in only the very recent past, in which opportunities for fandom multiply, diversify, and fragment—and in which, as numerous commentators have observed, "we are all fans now." This article pushes some of those third wave questions into an earlier period—the early to mid-twentieth century—asking how and why an engagement with science provided fans with a cultural resource in a period dominated by the threat and then reality of conscription. The culture of autodidactic engagement with, and satire on, science-related writing provides a *longue durée* prologue to the rise of science skepticism.

"Fan" is an abbreviation—a shortening of the word "fanatic." Deployment of the word "fan" in early twentieth-century Britain suggested a reference to the fan's class, or a somewhat sneering judgement; it was applied to football-goers, to indiscriminate theatergoers, and to bystanders of street fights. It connoted also a discomfitingly American quality.[4] Fans, and their magazines, occupied a special position in the ecology of science publishing. The fans were consumers—at full price, at pulp price, or for free through networks of lending—of professional science publications. Such professional publications included science fact and science fiction. Some were books, but it was the

---

[3] Jonathan Gray, Cornel Sandvoss, and C. Lee Harrington, eds., *Fandom: Identities and Communities in a Mediated World* (New York, N.Y., 2017), 2–7.

[4] See exemplary quotations given in the *Oxford English Dictionary* for the period 1914–25, "fan, n.2," OED Online, June 2018, Oxford University Press, http://www.oed.com/view/Entry/68000 ?rskey=21zg3E&result=2 (accessed 10 July 2018).

fiction periodical that most overlapped with, and modeled and inspired, the world of fandom.

Who were the British science fans? For a start, they were young; too young to have served in the Great War, and indeed, many of them were born after it. They had not experienced the personal psychic trauma of conflict and killing, although doubtless the intergenerational effects of it redounded in their families and other networks. Nor had they suffered the disillusionment that the Great War had provoked in earlier generations of science participants (such as Olaf Stapledon, of whom more later). Active in the 1930s, they were young enough not to have family responsibilities that might have called them away from their reading, writing, and meeting.[5] Lacking dependents and the anxieties of maturity, they were relatively untouched by the economic worries of the Great Depression. They existed in a peculiarly fragile historical moment of relative hope.

A sense of the youthfulness and tightly bounded generational nature of these early fans can be gained by looking at the photograph of the fourteen attendees of the first Science Fiction Association Convention in Leeds, 1937.[6] They came from around the country—from Essex, London, Leeds, Liverpool, and the Midlands—so it was not as though they were a preexisting coterie of friends naturally bonded by their common age (as if they were, for example, fellow members of a youth club or workplace). The oldest-looking person in the picture is Eric Frank Russell (1905–78), and even he was too young to have fought in the Great War, though he would undoubtedly have had considerable memories connected with it. Russell was one of two people at the convention who went on to have a professional career in writing. Also born just before the war were Ted Carnell (1912–72) and Walter Gillings (1912–79); Gillings was then twenty-five, but had been founder of the first fan group at the age of eighteen. Leslie J. Johnson, a key advocate of scientifiction (the 1930s term for SF) and fandom, was born in 1914. George Airey (1916–2001) was born two years later. The convention's most famous product, Arthur C. Clarke (1917–2008), was born toward the war's end, as were John Michael Rosenblum (1917–78) and Maurice K. Hanson (1918–81). Douglas Mayer (1919–76), editor of the successful "fanmag" *New Worlds*, was born just afterward.

On a specifically scientific note, one might further note the presence of the wireless among this generation. As recent histories have indicated,[7] returnees from the Great War very often brought wireless expertise with them; they went on to build their own sets and to share them with their children. There was an overlap between magazines treating radio-related topics and those in nonfictional futurism and scientifiction. The publisher Hugo Gernsback was in one way or another involved with a dozen separate wireless magazines, along with other science and mechanics titles. From his point of view at least, these were overlapping and complementary markets. Many of the young scientifictionists were wireless fans, too. Arthur C. Clarke built an early set;[8]

---

[5] Walter Gillings had a child c. 1934; mention of his four-year-old is made in *Tomorrow*, Summer 1938, 13.

[6] Douglas W. F. Mayer, ed., "Official Souvenir Report of the First British Science Fiction Conference" (Leeds, UK, 1937), Science Fiction Foundation Collection, PX8721.B74 1937, Special Collections & Archives, University Library, University of Liverpool, UK.

[7] Tilly Blyth, ed., *Information Age: Six Networks That Changed Our World* (London, 2014), 67.

[8] Arthur C. Clarke, *Childhood Ends: The Earliest Writing of Arthur C. Clarke* (Rochester, Mich., 1996), 59.

Leonard Kippin, cofounder of the first fan club, was a wireless buff;[9] and Douglas Mayer's Institute of Scientific Research in Leeds incorporated two preexisting wireless clubs.[10]

Besides the general contextual features defining these young men as a generation, there was also a very specific historical contingency that precipitated their emergence as an identifiable group. Gernsback, the proprietor of *Amazing Stories* (from 1926)—the first periodical dedicated to scientifiction—decided to facilitate communication among his readers by printing their full names and addresses in the magazine's "Discussions" section. The tradition continued in other titles, and in 1930, Walter Gillings, a rookie reporter on the Ilford *Recorder*, spotted a letter in *Wonder Stories* written by a fellow fan who lived only four miles away.[11] This man, Leonard Kippin, was a commercial traveler, and had picked up a substantial number of magazines on his journeys. Together the young men agreed to set up a group for discussing their shared interests, and thanks to his position at the local newspaper in Ilford, Essex (just outside London), Gillings was able to use its pages to spread the word. On Friday, October 3, 1930, a letter entitled "Scientifiction" appeared in it, inviting interested Ilfordians to get in touch and help set up a "Science Literary Circle." Whether this letter succeeded in prompting replies from strangers, or whether due to preexisting connections (Gillings asserted the former), a group was indeed established. Its meetings were chronicled in the pages of succeeding issues of the *Recorder*; H. P. Lovecraft, time travel, light years, and future predictions were all subjects of discussion. A visit from the Australian James Morgan Walsh (1897–1952), then living in London, provoked debate concerning life on other planets.[12] In a letter to *Wonder Stories* in November 1931, Gillings surreptitiously bumped his group's title up to the "British Science Literary Association," and mentioned similar movements afoot in Manchester, Liverpool, and Blackpool. Gernsback wrote in response that he had also heard from Leslie G. Johnson in Liverpool (apparently a "hamlet" in Gernsback's mind), who was indeed a key figure in establishing British fandom.

Traveling and meeting were not always easy for these men of limited holiday time and means. Their main social forum was virtual—on paper—the fanmags that they wrote, produced, and circulated among themselves. The first major British fanmag, *Novae Terrae*, was launched in Nuneaton, Warwickshire, in 1936; only in the following year did its editors meet up with other fans from around the country—at a meeting hosted by the Leeds chapter of the American Science Fiction League (SFL). *Novae Terrae* was adopted as the official magazine of the British Science Fiction Association (SFA) that emerged from the Leeds meeting; along with other fanmags, it forms

---

[9] Rob Hansen, *Then: Science Fiction Fandom in the UK: 1930–1980* (Reading, UK, 2016), 15.

[10] Ibid., 25. On science fiction and the wireless, see Charlotte Sleigh, "'Not one voice speaking to many': E. C. Large, Wireless, and Science Fiction Fans in the Mid-twentieth Century," *Science Museum Group Journal*, Issue 8 (Autumn 2017), http://dx.doi.org/10.15180/170802.

[11] Their addresses appear in the Ilford *Recorder*, 3 October 1930. Gillings was eventually sacked by the newspaper in 1940 for failing to put a sufficiently patriotic spin on war stories; see *Futurian War Digest*, October 1940. Issues of *Futurian War Digest* published between October 1940 and March 1945 have been painstakingly transcribed and made available online by Rob Hansen. See https://efanzines.com/FWD/FWD.htm. All future references to this magazine refer to materials available on this website.

[12] *Australian Dictionary of Biography*, National Centre of Biography, Australian National University, s.v. "Walsh, James Morgan (1897–1952)," by Stephen Knight, http://adb.anu.edu.au/biography/walsh-james-morgan-8969/text15783 (accessed 28 November 2014).

the major archival source for this article.[13] Because the majority of the fans did not go on to become historical figures in any significant way, there is little else by way of an archival resource—no collected papers or letters, no workplace records—to go on. Instead, this article follows a broadly discourse-based methodology, based upon the fans' writings in their periodicals.

They and their writing constitute an example of what Michael Whitworth has called "virtual communities" in science; studying their productions enables the reconstruction of—in his words—"the local *zeitgeist* of a particular social network."[14] As such, one could examine the productions of fans in any country (and any period) to find a locally specific parascientific culture.[15] In the case of science fiction, Whitworth's model of localism can be stretched a little, in the sense that mid-century fans were rather international in their outlook and communications. Hence, an interesting combination of local and transnational features might be expected to emerge by combining a number of national studies, such as this. In the case of Great Britain, there are specific historical reasons for interest in the fans. How did these moderately educated young men locate themselves in relation to professional scientific culture, and the new general public culture of expertise that was emerging thanks to wireless broadcasting? In what ways was science a cultural resource through which to articulate their identity and space within society at large? And, of course, how did all this pivot around the anticipation and unfolding of what became the watershed experience of the Second World War?

IN WHAT WAYS WERE THE FANS CONNECTED WITH SCIENCE AND ENGINEERING?

The most significant fan groups to be founded during the 1930s were in Leeds, Liverpool, and the Midlands, all places noteworthy for their industrial and engineering heritage. Many of the fans had some peripheral involvement with engineering industries, though they were rarely highly qualified. Few of the first-generation science fans were educated at university level, though this was beginning to change just as war broke out. Richard George Medhurst (1920–71) was studying mathematics at London University, and relocated to Cambridge due to hostilities.[16] Rosenblum described Medhurst as having probably the biggest science fiction book collection "kept in two enormous bookcases."[17] Not long after Medhurst went to Cambridge, his mother wrote to tell him that a bomb had exploded close to their house, and a piece of shrapnel had embedded itself in, of all the many books in his collection, *The Shape of Things to Come* by H. G. Wells (1933).[18]

Douglas Webster reported having "received excellent education at a Grammar School," which he followed with "Varsity—science," though it is unclear where ex-

---

[13] In addition to *Futurian War Digest* (cit. n. 11), several of these magazines can be found on Rob Hansen's website, Then: The Archive, http://www.fiawol.org.uk/fanstuff/THEN%20Archive /NewWorlds/NewWo.htm.

[14] Michael H. Whitworth, *Einstein's Wake: Relativity, Metaphor, and Modernist Literature* (Oxford, UK, 2001), 18–9.

[15] Elsewhere Sleigh has characterized such participants as scientific outsiders. See Charlotte Sleigh, "Writing the Scientific Self: Samuel Butler and Charles Hoy Fort," *Journal of Literature and Science* 8 (2015): 17–35, on 18.

[16] JMR, "Introducing: Richard G. Medhurst," *Futurian War Digest*, October 1940, 2.

[17] Ibid.

[18] "WW2 People's War," BBC (website), submitted by Sylvia Hill (formerly Medhurst), archived 15 October 2014, http://www.bbc.co.uk/history/ww2peopleswar/stories/67/a4386567.shtml.

actly he studied, or what specific course.[19] After Donald A. Wollheim suggested that science fiction fans found science "distasteful" because they "work for science, not in it" and would not descend to practical levels, Peter G. Sherry was moved to write in to *Tomorrow* to argue that his reading of science fiction had increased since he started studying a BSc (Honors Chemistry) course at Glasgow University.[20] Doug Mayer (1919–76) read physics at the University of Leeds around the beginning of World War II. At the age of sixteen he had founded the "Institute of Scientific Research" and its "library" at the home of his parents; together with two others he later formed the Leeds chapter of the SFL, which by the end of 1935 had twelve members.[21] The authorities' condemnation and demolition of the house where they met gives an indication of the socioeconomic position of its members. Fortunately, one of them had an estate-agent father, who then gave them a virtually free meeting room.

Fans aspired to higher education on behalf of the next generation as well. The 1940 announcement of the birth of Alys Rita Rathbone's son James described the infant as looking "like a futureman, with a colossal head compared with his body, but when he obtains his University degree in two weeks, he may look more normal." The announcement continued: "Meanwhile he will be cutting his teeth on 'Principia Mathematica' and Hegel's 'Principles of Logic' with a little Euclid on the side."[22] Aside from the few fans who went to university, many of the first generation had connections to industry or had technical occupations. James Parkhill-Rathbone (1918–?), father of baby James, for instance, was a scientific instrument salesman. Walter Gillings, founder of the first UK fan group, studied at Leyton Technical College–though in what field is unclear. He was actually noted among fans as having an unusually low level of interest in physics, chemistry, and so on, but communicated his love of science fiction in the pages of the newspaper for which he worked.[23]

Ted Carnell was, like Gillings, in the world of type, and during the lifetime of *Novae Terrae* he was a printing apprentice. He had been turned down for his first choice, a Trade Scholarship in bookbinding, because the board disapproved of his practice of binding at home such tripe as the *Boys Magazine*, "a lurid red-covered 'blood' which ran lengthy serials on invasions by Martians."[24] Before the war, Maurice Hanson (1918–81) was a civil servant from Nuneaton; he later commuted from Kettering to Imperial College London, where he worked as a librarian in the transport library.[25]

Farther north, fans were more connected with big industry. The science fiction writer Olaf Stapledon, universally praised by the fans, although not one himself, had a strong family connection to the great Holt shipping company in Liverpool. His father was senior within the firm and was posted to Egypt during Olaf's early childhood. In early adulthood, Stapledon had temporary employment in his father's firm. Eric Frank Rus-

---

[19] "Introducing: Douglas W. L. Webster," *Futurian War Digest*, January 1941, 4.

[20] [Various], "Candid Comments," *Tomorrow*, Autumn 1938, 14–6.

[21] Rob Hansen, "The Rise and Fall of Leeds Fandom," *Relapse*, Spring 2013, 4–11, on 5.

[22] James Parkhill-Rathbone, "Birth of a Son," *Futurian War Digest*, February 1943, on 3.

[23] Cosmivox, "His Tale of Wonder: An interview with Walter H. Gillings, Editor of 'Tales of Wonder,'" *Tomorrow*, Summer 1938, 7 and 13, on 7.

[24] *Futurian War Digest*, November 1942, 5.

[25] Rob Hansen, "Forgotten Fans #6: The Sage of Nuneaton," *Relapse*, August 2014, 1–8, on 7. Imperial College still awards a Maurice Hanson Prize to the student who produces the best paper on the Advanced Course in Transport; see "Department of Civil and Environmental Engineering, Postgraduate prizes, Transport Cluster Awards," Imperial College London, http://www.imperial.ac.uk/engineering/departments/civil-engineering/prospective-students/postgraduate-prizes/ (accessed 2 June 2017).

sell (1905–78), another Merseysider, and the most prolific and successful storywriter of the early SF period, studied science and technology at college before working consecutively as a telephone operator, quantity surveyor, draughtsman, and then as a troubleshooter for the large engineering firm Frederick Braby & Company.[26] William Temple (1914–89) trained for some years as an engineer but by 1939 was working in the Stock Exchange. The "Sage of Nuneaton," D. R. Smith (1917–99), was a regular reader of *The Engineer* magazine and worked as a draughtsman.[27] Harry Turner (1920–2009) from Manchester worked as a rubber chemist.[28]

BEFORE THE WAR

During the late 1930s, the brief window of time in which the fans coalesced in clubs and printed their fanmags, the prospect of a second world war was an unavoidable backdrop to everything, science fiction included. Alexander Korda's *Things to Come*, released in February 1936, was universally praised by the fans, as it was by professional critics. As Jeffrey Richards points out, Korda's film more or less dispensed with the critique of capitalism that had underpinned Wells's book, in favor of a prophetic focus upon warfare.[29] The horrifying scenes of bombing with which the film opens connected filmgoers very directly to the likely future, as they prominently featured a cinema, such as that in which they were then sitting, amid the destruction. Wells was indeed an omnipresent figure in British science fiction fandom, a counterbalance to the trigger-happy tales that tended to feature in Gernsback's magazines. For Britons, there was a cultural memory (and anticipation) that perhaps tempered enthusiasm for Gernsback's worlds. Notwithstanding the thrilling excitement of *War of the Worlds* (1898), its accounts of the mass flight of ordinary people was harrowing and perturbing. Wells continued, in fiction and nonfiction, to inveigh against war mongering.

Another of the fans' favored authors—for many, ranking more highly than Wells—was Olaf Stapledon.[30] He crops up as a touchstone for the fans over and over again in their magazines, suggesting that he was widely read and respected. His sprawling, staggering evolutionary histories produced a terrifying and vertiginous sense of insig-

---

[26] John L. Ingham, *Into Your Tent: The Life, Work and Family Background of Eric Frank Russell* (Reading, UK, 2010).

[27] Hansen, "Forgotten Fans" (cit. n. 25), 2; D. R. Smith, "Cover and Other Things," *Novae Terrae*, August 1938, 32–3, on 32.

[28] RFV and SDS, "Harry Turner: Artist, Science Fiction Fan & Collector," http://www.htspweb.co .uk/fandf/romart/het/hetobit.htm (accessed 2 August 2017). The fans' connections with engineering help to make sense of the entrepreneurial nature of much scientifictional science, turned toward profit and practicability; see Charlotte Sleigh, "Science as Heterotopia: The British Interplanetary Society before WWII," in *Scientific Governance in Britain, 1914–79*, ed. Don Leggett and Charlotte Sleigh (Manchester, UK, 2016), 217–33, DOI:10.7228/manchester/9780719090981.003.0012.

[29] *Things to Come*, directed by William Cameron Menzies (1936; London Film Productions, Denham Studios, Denham, Buckinghamshire, UK, 2006), DVD; Jeffrey Richards, "Things to Come and Science Fiction in the 1930s," in *British Science Fiction Cinema*, ed. I. Q. Hunter (London and New York, N.Y., 1999), 16–32, on 19–20; H. G. Wells, *The Shape of Things to Come* (London, 1933).

[30] Maurice Hanson named Stapledon as author of his favorite book; Arthur C. Clarke gave one of his novels—uniquely among all his collection—the appellation "superb"; see *Novae Terrae*, August 1938, 25; and *Novae Terrae*, June 1938, 16, respectively. D. R. Smith was a little more ambivalent; he had a love-hate relationship with literature that attempted higher meaning. However, even he conceded that "Last and First Men by Dr. Olaf Stapledon [was] hailed as the best imaginative story since Wells left the field"; see *Novae Terrae*, June 1937, 5.

nificance in their readers, and one may well read the source of this horror as his front-line experiences. *Last Men in London* (1932) gave particularly grueling descriptions of the Great War—its terrors and aftereffects upon mind and society. What had seemed the sum of knowledge and experience for Stapledon was shattered by what he saw, just as it was for readers of his epic poems: "The world of breakfast, six-pences, and artificial silk/ . . . seems vast and safe"; but "minds are fledglings in a cliff-ledge nest," and when the drop is revealed—that is, the insignificance of the self—the mind reels.[31] Stapledon's correspondence and journalism reveals his im-mersion in pacifist circles; the fans, by and large, recognized his perspective.

The youth of science fiction fans took on a new significance in the light of likely conscription. In June 1937, Eric Frank Russell was moved to comment on "Arch-bishop Cant's" apparent calling of youth to military service.[32] Though Archbishop Lang's "Recall to Religion" campaign—the source to which Russell apparently al-ludes—was primarily aimed at inducing personal piety, it was understood in context by Russell as a patriotic call to armed service. Lang's articulation of his evangelistic message as "the consecration of the *king's people*" to the service of God was likely to blame.[33] Given God's alignment with war and fatherland, debate raged among the fans as to whether He was present in, or absent from, Stapledon's epics. Some claimed that his presence would give comfort in what was otherwise a bleak and in-human universe. One correspondent to *Novae Terrae* found Stapledon's god to be a Tennysonian figure, careless of the single life; and this "scientific" view, it seemed, was preferable to a god who personally desired the death of young men.[34] Others claimed that Stapledon's postwar universe was atheistic.

At the beginning of 1938, after a year in which Italy, Japan, and the United States had rattled their weapons, fans debated among themselves as to the purpose of sci-ence fiction. The fan and amateur writer D. R. Smith was taken to task by his fellow fan Albert Griffiths for maintaining a commitment to fiction for the sake of fiction. Griffiths despaired that, if it did not connect with higher aims, scientifiction would remain as a "sophisticated [kind of] fairy story" or else "sugar-coated science a la Gernsback."[35] Instead, "true science-fiction should attempt to portray the impact upon society and civilization of new scientific ideas and ideals." Griffiths recom-mended—as we might by now guess—one of Stapledon's novels by way of example.

Griffiths had got Smith somewhat wrong. True, Smith maintained a dogged inde-pendent mindedness in the face of the "'strafing' of the Left Book Club" and the "im-perialist and capitalist authors" who opposed them.[36] Yet he had, in fact, been explor-ing social questions through his series on "cosmic cases" in the very same fanmag as that in which Griffiths wrote. These subtle and amusing pieces preempted the inter-national courts and tribunals instigated in the wake of World War II, making fans their judges. Their accounts of interplanetary battles and the rights of nations satirized in-

[31] Olaf Stapledon, "Poem 1 from *The Nether Worlds*," in *An Olaf Stapledon Reader*, ed. Robert Crossley (Syracuse, N.Y., 1997), 307.
[32] Eric F. Russell, "Rationalism and Youth," *Novae Terrae*, July 1937, 13.
[33] Robert Beaken, *Cosmo Lang: Archbishop in War and Crisis* (London, 2012), 122, 133 (emphasis mine).
[34] E. Longley, "Regarding Mr. Knight's Letter (August issue)," *Novae Terrae*, September 1938, 11 and 17.
[35] Albert Griffiths, "Ideas and Ideals," *Novae Terrae*, March 1938, 12–14.
[36] Ibid.

ternational politics and the pointless destruction of war, as well as colonialism. In his article "The Drift Away from Scientific Fiction," Smith in fact condemned the fans for not reading *real* sociology, claiming that those who had done so—such as himself—required the soothing balm of fiction as counter to life's bitter realities. "The stalwarts are putting aside the toys of childhood," he concluded, "and are going forth to the war that really is to end war, and all other forms of man's purposeless suffering. Ave atque vale."[37] As ever in Smith's writing, the irony is complex. On one level it sounds like a (lone) call to war; on the other, Catullus's lament for his brother, revoiced by Tennyson, suggests that the suffering may not be worth it after all.

Notwithstanding Smith's tergiversation, questions of society and war continued to build among the fans. The issue of *Novae Terrae* following the first "cosmic case" included a peace pledge union folder in its mailing.[38] A subgroup of the SFA was formed, committed to progressive sociology.[39] They were inspired by the US fans Claire P. Beck and especially John B. Michel, whose socialist convictions (briefly flourishing as "Michelism") were propagated at the Philadelphia Convention of Science Fiction of 1936. British readers were bombarded with questionnaires to discover their opinions on matters of politics; one revealed that ninety-three percent of readers liked articles "with a sociological trend."[40] The Leeds fan Douglas Mayer, reflecting on all this, called on fans to "wake up."

> No one who has meticulously digested the contents of the past few issues of "Novae Terrae" can have escaped noticing the trend for contributors to ignore science-fiction itself, and to expound in their articles the ideas, beliefs and inspirations that science-fiction arouses.[41]

However, this was not movement away from science fiction, but rather toward its logical apotheosis: "British fans are at last beginning to realise that science-fiction is something more than a mere type of literature."[42] Mayer claimed two features specific to science fiction that allowed it to function as social and political criticism. One was the vastness of the canvas upon which its writers painted—a scale of worlds that was appropriate to global warfare. Stapledon, as ever, was the natural referent. Secondly, and seemingly even more important to Mayer, was the cultivation of "a detached standpoint" within science fiction. Fans, claimed Mayer, had cultured through their reading an ability to see terrestrial affairs as from afar. (Yet again, the technique is center stage in literal form in Stapledon's novels, where the reader is conducted through aeons and light years until Earth is less than a pinprick.) Mayer quoted the US fan Claire Beck:

> [The fan] finds himself set free, not only of the cramping bonds of conventional incident and conventional locale, but free also to criticize historically, allegorically, or how he will, the petty meanesses [*sic*] and misdirection of our petty affairs. He may sketch

---

[37] D. R. Smith, "The Drift Away from Scientific Fiction," *Novae Terrae*, January 1938, 6–7, on 7.
[38] Even the title of the journal, "new earths," suddenly acquired an oddly biblical resonance in relation to the "new earth" of Revelation 21:1–4, in which there is no more death.
[39] "Branch Reports," *Novae Terrae*, January 1938, 17–19, on 18.
[40] See special insert, *Novae Terrae*, January 1938.
[41] Douglas F. W. Mayer, "Wake Up, Fans!" *Novae Terrae*, December 1937, 10–3, on 10.
[42] Ibid.

the ideal with a clearness and directness that in any other type of literature would get him called a menace of a fool![43]

In the Leeds Branch, at least, Mayer perceived that such a change had taken place among its members, transforming them "from amateur scientists to idealistic sociologists."[44]

There are many ways of reading this science-fictional sense of mental superiority and freedom from convention. On one level it is a simple expression of late adolescent identity. Some authors—Gary Westfahl and Steve Silberman—have noted its appeal to some persons whom today we would class as being on the autism spectrum.[45] Still another reading might relate it to an emergent political, social, and educational identity among these young, working- and lower-class males, especially against a background of generational segmentation (due to the Great War) and the ascendency of mass media. The sense of an "establishment" was perhaps emerging as a foil for the "ordinary man" [sic]. One fan noted by way of self-contextualization that "the king of England himself had been thrown out for not conforming with the Establishment."[46] Finally, however, we may note that this identity is explicitly "scientific" as another fan noted: "The applications and teachings of science should be used to sweep away archaic beliefs and superstitions, and to create on Planet Three a world fit for the superior race that man would then become."[47] The scientifictional designation of Earth as "Planet Three" exemplifies the mind-set proposed by the author; that is, despite this planet being the author's home, it is stripped of its familiar and emotive name, and renamed—or renumbered—according to its position from a physically distant perspective, which is logical to a naming of the solar system as a whole.

There were limits to "scientific" distancing, however, increasingly vocalized by both scientists and fans. As war loomed over Europe, the fanmag *Tomorrow* reported that scientists had been making various statements of their ideals, proposing a scientific "'brains trust' for the world" that involved active planning and control of science by scientists, who also suggested that technocracy could remedy global ills.[48] In response, in the next issue (one of the last fanmags published before the war), Mayer wrote an editorial urging scientists to seize control of the application of their discoveries. Under the heading "Action Applauded," the editorial team expressed their hopes that scientists' "academical musings rapidly give way to determined action!"[49] Fans had also written in to express their approval of the scientists' intentions to act.[50]

For the science fiction fans, as for scientists, the possibility of war carried with it the potential for unleashing the full powers of science. Mayer invoked the metaphor of

---

[43] Ibid., 11.

[44] Ibid.

[45] Steve Silberman, *NeuroTribes: The Legacy of Autism and the Future of Neurodiversity* (New York, N.Y., 2015); see chapter "Princes of the Air" (233–60). Silberman quotes John Campbell: "Write me a creature that thinks as well as a man, or better than a man, but not like a man," implying that such may be the description of a person with autism spectrum disorder—and a fan. See also Gary Westfahl "Homo aspergerus: Evolution Stumbles Forward," Feature Review, (*Locus Magazine*), Locus Online 6 March 2006, http://www.locusmag.com/2006/Features/Westfahl_HomoAspergerus.html.

[46] Sid Birchby, quoted in Hansen, *Then* (cit. n. 9), 34.

[47] Mayer, "Wake Up, Fans!" (cit. n. 41), 12.

[48] "The Revolt of the Scientists: A Comprehensive Survey of How Scientists Are Setting Out to Save Our Civilisation," *Tomorrow*, Summer 1938, 12–3.

[49] Douglas W. F. Mayer, "Editorial Comment," *Tomorrow*, Autumn 1938, 2.

[50] [Various], "Candid Comments" (cit. n. 20), 14–6.

Icarus and Daedalus to imply that scientists had flown too close to the sun with their scientific developments.[51] Less abstractly, Geoffrey Daniels argued the following:

> In the arbitrary advance of science some branches have leapt ahead, while others have lain neglected through the centuries. Physics, chemistry and engineering have provided man with a means for the almost immediate destruction of a city, but no branch of science has yet provided sufficient deterrent from such things.[52]

Fans, so many of whom were from technical or engineering backgrounds, believed that "something more than scientific achievement is required" because the "arbitrary" advance of science was potentially dangerous. They believed that a "practical," planned, and applied science offered the best hope of a better future for humanity.

British fans, sometimes referring to "sociology" and sometimes to "science," espoused a progressive and antiestablishment mind-set—a Wellsian rejection of war and of capitalism as it was experienced. Fan Dave McIlwain characterized fellow members of his species as those "idealis[ing] certain states or concepts—Altruism, Pacifism, and so on."[53] Moreover, this social perspective was badged as inherently "scientific." Rob Hansen quoted fan Sid Birchby's recollection:

> As a fan, I felt it was quite right and proper that the fanzines . . . should print their steady diet of pep articles on "Whither Mankind?" and "Science Progress." It was the New World we were making, and the golden tool was Science. Around us [by contrast] the world was moving into the first steps of the dance of death.[54]

Some fans were overtly socialist; others remained remarkably thoughtful and relativist about the value systems they saw around them. Some were pacifist and some were not. Differences of opinion over precise political commitments, and indeed over the place of politics in science fiction, were among the reasons for the first schism within fandom in 1937.[55] These varying positions, and the relationships between their proponents, would be tested as warfare moved from the realm of fiction to fact.

WAR

When Britain entered the war, the Science Fiction Association disbanded, and J. M. Rosenblum noted that initially fandom was decimated as men were called up or left to serve, falling into what he termed the "draft abyss."[56] Maurice Hanson was the first to be drafted, being called up in July 1939.[57] War also caused the loss of many fanmags, whether due to the call-up of those who produced them or the destruction of materials

---

[51] Mayer, "Editorial Comment" (cit. n. 49). Mayer suggested that Professor Piccard might be a modern equivalent; this presumably refers to Jean Piccard, the high-altitude balloonist whose exploits the fans, who so closely overlapped with the British Interplanetary Society (BIS), would undoubtedly recall. Shortly before the war, Piccard had used a TNT charge to release a balloon, which set fire to excelsior wood-wool and destroyed his craft.

[52] Geoffrey Daniels, "The Future of Applied Psychology: A Forecast of the Applications of the Newest of the Sciences," *Tomorrow*, Summer 1938, 6.

[53] Dave McIlwain, "One Hour with a Psychiatrist," *New Worlds*, Autumn 1939, 28–30, on 28.

[54] Hansen, *Then* (cit. n. 9), 34.

[55] Ibid., 34–9.

[56] J. M. Rosenblum, "Editorial," *Futurian War Digest*, April 1943, 2–3.

[57] Hansen, *Then* (cit. n. 9), 51.

and equipment in the eventual bombings. One publication, however, was "British fandom's main unifying force . . . the main chronicler of the trials and tribulations facing fandom in these uncertain times": the *Futurian War Digest*, also known as FIDO.[58] During the war, it enabled fans scattered across the world to stay in touch and was even used to recruit new members. It is primarily via this publication that it is possible to explore the relationships, interests, and experiences of the fans.

Some fans were connected by their service as well as their fanmag. Rob Hansen has observed that most fans served in the Royal Air Force (RAF), which he attributes to both preference and their having "a number of things in common (i.e. being predominantly middle-class, literate, and interested in science) that the RAF was looking for."[59] While the Royal Navy drew only a couple of fans to its ranks, the army, and specifically the Royal Corps of Signals, attracted almost as many as the RAF. Fans themselves made a joke of the idea that their allocated roles were any indicator of ability. Signalman Eric Williams commented: "Isn't it amazing and significant the number of fans who are joining up in the Signals? Guess it must be the high level of intelligence and keeness [*sic*] needed to become a Signalman."[60] Signalmen's work ranged from maintaining radar systems to being radio operators and teletypists and dispatch riders. In fact, many fans performed similar, technical roles regardless of their force allocation; for instance, J. F. Burke was a radar mechanic in the RAF, and Donald J. Doughty was a wireless mechanic for the navy.[61] A book review in the February 1944 FIDO attests to the technical leanings of the fans; the main criticism of *On the Way to Electro War* by Kurt Doberer (1943) was that "radio enthusiasts in fandom may be disappointed at the incomplete nature of its technical explanations—and exasperated at the vagueness of many terms used."[62] Fans were interested in more than simply the storytelling possibilities of new technology. One fan in particular found that his technical understanding shaped his war experience. Doug Mayer wrote an article for *Discovery* journal on "the possibility of making use of the latent energy of the atom," which he closed by pondering whether this would lead to "a Wellsian chaos with each nation dropping bouquets of uranium bombs in a policy of encirclement."[63] George Airey and Herbert Warnes recalled how, shortly after its publication, Mayer was "whisked away to London, where he became one of the famous 'backroom boys.'"[64]

The common features of the wartime roles taken by fans reinforce the idea arising from their prewar jobs that many of them shared technical interests and skills; their interest in science fiction went beyond the stories. Moreover, if these skills were useful to the military, finding themselves in similar roles also benefited the fans by making face-to-face meetings easier even than before the war. As well as forging international connections, war enabled British fans to discover each other and to connect.

---

[58] Ibid., 57.

[59] Ibid., 51.

[60] "Overseas Mail," *Futurian War Digest*, August 1942, 2.

[61] Fans' wartime roles have been traced mainly through reports of their call-up and service included in the pages of FIDO.

[62] Julian Parr, "Book Review," *Futurian War Digest*, February 1944, 2.

[63] Douglas W. F. Mayer, "Energy from Matter," *Discovery: A Popular Journal of Knowledge*, n.s., 2 (1939): 459–60.

[64] George Airey and Herbert Warnes, [No title], *Crystal Ship* 14, February 1988, quoted in Hansen, *Then* (cit. n. 9), p. 53.

Three science fiction veterans, McIlwain, Russell, and Forster, were sent to No. 1 Signals School at the same time and were joined shortly afterward by Birchby; they held biweekly suppers and a mini-convention.[65] Fans hoped for the same sort of "flourishing fan colony" at No. 2 Signal School where Clarke was joined by Canadian fan E. A. Atwell.

Another group of fans, only discovered by FIDO readers in October 1942, were based at the Teddington Paint Research Station (PRS). The work conducted there could be described as materials science. The overall aim of the work was to improve camouflage and strength of equipment. In practice this involved developing more sophisticated chemical substances; the organization celebrated the construction of its first electron microscope in 1944.[66] The Teddington group were initially unknown even to one another until E. Frank Parker donated his collection of science fiction reading to help National Fire Service members pass the time as they sat on the roof of the building watching for fires.[67] For fandom, the PRS group (later renamed the Cosmos Club) presented exciting possibilities, because they had access to printing machines and technical printing abilities, which were generally in short supply. Perhaps unsurprisingly in light of this, the authorities at the Paint Research Station considered that science fiction was not good for science (or morals), which forced the newly bonded group into clandestine meetings.[68] Tribe making, for fans during the war, was closely connected with scientific and technical roles both because of common interests that resulted from the work, and because of the access to spaces such as cafeterias and rooftops that were shared during breaks from that work.

Fans connected internationally in an ad hoc way, too. In some cases, fans met by chance, and in others they were connected by FIDO, which acted as a hub for letters detailing locations. For instance, one fan reported the following from Egypt: "More fans are out in the Middle East. Letters will start to shuttle back and forth pretty soon, and we'll be a stf Colony."[69] In 1943, FIDO announced that "contact betwixt the respective fandoms of USA and England in person has at last been effected."[70] This transatlantic connection of fan communities continued throughout the war in the form of letters, magazine exchanges, and occasional meetings whenever they could be managed. British fans also connected with Australian and Canadian fandoms through military service, enabling the fans to compare their communities. Bert Castellari lamented that Australian fans often seemed to clash, and pondered the differences: "Perhaps it is because the whole make-up of British fandom differs from that of Australia and America in that British fans are genuinely interested in the problems of the future, in scientific development."[71] Likewise, Eric Frank Russell "(the Australian one!)" explained in 1943: "In Australia at present there is practically no science fiction active at all. There exists the Sydney Fantasy Society which has a lot of ups & downs. No fanmags are being published owing to their being stopped because of the

---

[65] "Per Ardua Ad Astounding," *Futurian War Digest*, July 1942, 3.
[66] "About PRA, PRA History," Paint Research Association, 2018, https://www.pra-world.com /about-pra.
[67] "FIDO Goes VOMish," *Futurian War Digest*, October 1942, 1.
[68] Rob Hansen, "The Cosmos Club & the 1944 Eastercon," ed. Peter Weston, *Relapse* 16 (2010), 22–7.
[69] "FIDO Goes" (cit. n. 67), 3.
[70] "Contact," *Futurian War Digest*, October 1943, 8.
[71] "FIDO Goes" (cit. n. 67), 4.

paper shortage. We receive some but not all of the reprints of Astounding and Un-known and occasionally there is a native Stf booklet."[72] The war made the world of fandom smaller and more interconnected as fans themselves spread across the globe.

Science fiction magazines also spread across the world, and even with paper ration-ing, the war surprisingly made it easier to circulate publications. Though secluded from the "ravings of Anglofandom" for a brief time due to the secrecy of an attack in which he was involved, Ted Carnell reported from Algiers: "Going on board a troopship in the harbour . . . we were amazed to find all the latest US mags on sale at the canteen. The steward had loaded up at New York prior to the trip. Even the British reprints were there."[73] Eric Williams also reported the international reach of science fiction publications from his posting on the Middle East front: "I expect you would like to know the position of SF out here, well, believe it or not, it has a terrific sale. New SF mags appear in the shops one day and the next you have to take a rifle with you if you want to make the wog [sic] shopkeeper hand over the last copy that he has been saving for his own use."[74] Rosenblum mused that it might be the challenges of accessing science fiction publications during wartime that had facili-tated the growing community:

> It has been almost impossible to obtain the magazines without some degree of skulldug-gery, but the generous help of American fans has sent over a sufficient supply of the mag-azines to enable most fans . . . to keep up the tradition. They have been circulated under various schemes . . . I might suggest, in fact, that the strongest force drawing fandom together during the war has been a desire to share in this slender supply![75]

Fans took advantage of these closer connections by arranging gatherings and outings whenever they were able. The activities they engaged with during their gatherings sometimes seem far from scientific discussions and experiments. In the summer of 1943, the Cosmos Club's junkets included attempts at water divining and extrasensory perception tests, for example, though the latter were dismissed as meaningless by a participant referred to as "Physicist Bullett."[76] In the "Introducing" feature of FIDO, where fans described themselves or were described to other fans, a reader is often more likely to discover that another speaks Esperanto, or is a vegetarian or teetotaler, than what they did for a prewar living, or what they did in the military. This is not to dismiss the technical interests that fans had in common; it was unnecessary to communicate something so obvious as to be assumed (the assumption of scientific interests can be seen in the decision to start a chain letter for science discussions because "there must be a number of our members who are very much interested in scientific devel-opments"[77]). It does highlight, however, that the fan community was built upon shared ideals and interests rather than scientific work or military service alone.

One particular shared ideal was remarkably common among fans during this pe-riod: pacifism. In Britain, approximately one percent of those drafted objected to the

[72] "Potted Paragraphs," *Futurian War Digest*, October 1943, 9.
[73] "Sands of Time – Dec/Jan issue," *Futurian War Digest*, January 1943, 2.
[74] "Overseas Mail," *Futurian War Digest*, August 1942, 2.
[75] "Fan – The Embers!" *Futurian War Digest*, August 1944, 13.
[76] E. F. Parker, "Report: Too Much Water in Cosmopolis!" *Futurian War Digest*, July 1943, 5.
[77] "Science Discussions Group," *Futurian War Digest*, January 1943, 6.

war on moral or political grounds, and of those, sixty thousand were granted some form of exemption from combatant service.[78] So far as a heuristic calculation (based on the data yielded by fanmags) can be obtained, it would appear that, among British science fiction fans, the proportion of conscientious objectors was far higher than in the British population as a whole.[79] Foreshadowing the arguments that his fellow fans would later make in tribunals for conscientious objection, Philip S. Hetherington wrote in 1938: "As regards the method of mending this so very ill world, my own view is that any attempt to mend it by force would be worse than the disease."[80]

For some, Nazism was such an extreme "disease" that it changed their views. Eric Williams explained to other fans: "For this war at least, I have given up my pacifistic ideas; they won't work. If we give in, in would come Nazism, out would go freedom, down would come darkness."[81] Similarly, Sam Youd (1922–2012) changed his stance due to the politics of the time, and argued with other fans such as Burke and Russell that fighting was necessary. He wrote to *Fan Dance* in August 1941, ridiculing the idea of combating Panzer tank divisions with "Ghandi" [*sic*] methods.[82] But many other fans held on to their pacifism.

At his tribunal for conscientious objection, Rosenblum engaged in fan-style hypothetical arguments with the chairman. When asked what he would do to protect other Jewish people from the Germans, Rosenblum countered: "Of course, this army you speak of is not a thing that would suddenly appear from nowhere."[83] Forced to consider the possibility of invasion, he replied: "I would not oppose them by force because I would be using their methods, opposing them on the plane on which they are naturally the stronger." Instead, he suggested that he would work to bring down their administration and asserted that "the method you prefer would destroy the things this country stands for." The chairman said that his argument was not rational, to which Rosenblum responded: "I am perfectly rational, and am prepared to follow that argument up in a rational way." The ideals of science and reason were deeply held by Rosenblum, to the extent that he risked prison for his convictions. The newspaper reports imply that the chairman grew increasingly irritated with Rosenblum, at one point telling him: "You would probably be put in a concentration camp and have

[78] Spencer C. Tucker, ed., *World War II: The Definitive Encyclopedia and Document Collection*, 5 vols. (Santa Barbara, Calif., 2016), 448.

[79] About 10 percent of fans whose wartime roles it is possible to trace either explicitly declared objections to combat and sought exemption from combatant service, or else chose to sign up for medical rather than combatant roles. By the end of 1939, 1.5 million men were conscripted to the British Armed Forces. During the course of the whole war, 61,000 appeared before tribunals as would-be conscientious objectors. Roger Broad, *Conscription in Britain, 1939–1964: The Militarisation of a Generation* (Abingdon, UK, 2006), 39 and 188.

[80] [Various,] "Candid Comments" (cit. n. 20).

[81] Eric Williams, writing in FIDO supplement *Fan Dance*, August 1941, as quoted in Hansen, *Then* (cit. n. 9), 59. Hansen has suggested that the Spanish Civil War led to a split between pacifist fans and socialist fans due to their difference in priorities over whether violence should be refused or fascism should be resisted by any means; Hansen, *Then* (cit. n. 9), 32.

[82] Sam Youd, *Fan Dance*, August 1941. Russell replied with a disgruntled letter to FIDO, calling Youd's comments "sheer twaddle" and asking: "Has the Home Guard introduced him [Youd] to that strange stuff called beer?"; Eric F. Russell, "A Few Words from Eric F. Russell," *Futurian War Digest*, September 1941, 3.

[83] "Jewish C.O. Before Leeds Tribunal: First of His Faith in the Area," *Yorkshire Evening Post* (West Yorkshire, England), 15 December 1939, 10.

your toes broken so that you could not walk away from it."[84] Nonetheless, he was registered for noncombatant duties.

Douglas Webster, like Rosenblum, knew that his views were not without risk; he wrote to fellow fans that he expected to finish his degree in jail due to his pacifism.[85] He refused even to facilitate the fighting, telling his tribunal that "if by taking work on a farm he allowed others to go to fight he would be unwilling to do so, but if it would help to feed people then it would be a reasonable thing to do."[86] Other fans elected to serve in other ways that better fit with their ideals: Harold Gottliffe and the "pacifistic" James Parkhill-Rathbone served with the Royal Army Medical Corps (RAMC); Osmond Robb and Arthur Busby served with the National Fire Service; and Rob Holmes worked in Air Raid Shelters and for the Pacifist Service Unit until his father's death, when he was permitted to leave and care for his family.

Jack Walter Banks was not so fortunate; he was rejected because the courts did not believe he was sincere in his objections, and was sent to prison in early 1942 for refusing to submit to medical examination.[87] Walter Gillings lost his job on the *Ilford Recorder* due to what he said was refusal to "change my attitude to suit a more war-like policy." He was then rejected in his application to become a conscientious objector at both his tribunal and appeal, and was in military service with the Royal Armoured Corps by late spring of 1941.[88] Rosenblum observed that Gillings had been forced to "succumb to his fate for the sake of his wife and child"; Rosenblum had also mentioned at his own tribunal that he knew many more who would become conscientious objectors if they could afford to, so it is possible that their lower-middle-class situations masked a potentially even wider pool of pacifist-leaning fans. Even without the potential for wider objection among fandom, for a significant number of science fiction fans, reason was a counterpoint to violence on a personal as well as an international scale, and was an ideal for which they were willing to risk humiliation and even imprisonment. Science fiction fans did not simply give voice to scientific ideals in their homemade magazines—they actively sought out roles in wartime that fit with their view that the best hope for humanity lay in planning, co-operation, and negotiation, rather than physical power. The reactions of the fans to conscription may be anecdotally compared to the British subjects from this period who were not science fiction fans, and were interviewed by oral historians for the book *Men in Reserve* (2017).[89] These men, in reserved occupations during World War II, were of the same class and generation as the fans (including, for example, a draughtsman). The book describes a similar set of identity issues, differently configured—masculinity, duty, patriotism, autonomy. Some were simply happy to continue in safety, earning well, while others articulated conscious fulfilment in "playing their part" on the home front. Others (half the sample) were desperate to join up and made multiple attempts to do so. In general, this "control set" articulated more patriotism and duty, and less political skepticism, than the fans, and were more concerned to preserve or at least reconfigure a version of masculinity in their self-identity.

---

[84] Ibid.

[85] "Introducing" (cit. n. 19), 4.

[86] "North East C.O. Tribunal," *The Scotsman* (Midlothian, Scotland), 4 July 1940, 3.

[87] J. M. Rosenblum, "Editorial," *Futurian War Digest*, January 1942, 2–3.

[88] "Forum," *Futurian War Digest*, October 1940, 3.

[89] Juliette Pattinson, Arthur McIvor, and Linsey Robb, *Men in Reserve: British Civilian Masculinities in the Second World War* (Oxford, UK, 2017).

Besides the fans themselves, women frequently crop up in the pages of FIDO. Some appear as passive; Mrs. Smith, Mrs. Fearn, and Mrs. Rosenblum were presented as long-suffering martyrs who permitted fans to colonize their homes whenever the opportunity presented, and Webster's sister allowed him to use her nail varnish as correcting fluid.[90] Others, though, were enthusiastic facilitators and fans. Stoke-on-Trent's club lost its only female member to the Land Army in early 1943, but other women became active participants in fandom precisely because of the war. Joan Lane was vital to the existence of the April 1944 issue, which she duplicated and distributed with her brother Ron.[91] Ann Gardiner was described as taking her brother Derek's place as a "fantast" on his departure for Farnborough, and the community offered "Welcome indeed to a recruit fannette."[92]

Women's engagement with fandom during the war was not considered remarkable. Female names crop up regularly in lists of subscribers and letter writers, indicating that although they may not have written so many public pieces for fanmags, they felt comfortable inhabiting the world of fandom. They were also frequent attendees at meet ups. The Cosmos Club "girls" mocked the "far from divine" antitank ditch found during the water-divining expedition, for example. Joan Temple both hosted gatherings and made trips to participate in fan gatherings, and one fan's sister wrote that, following her experience at a convention, she was "on the verge of becoming a fan MYSELF despite the dissuasion of my brother who never had a friend to warn him!!!"[93] Many of these women had relationships to men science fiction fans, but they were engaged with fandom in their own right. Marion Eadie, who worked for the Transport Department and was founder of the Junior Branch of the British Astronomical Association, married fellow fan Harry Turner because of their shared interests. Though Hilda M. Johnson contributed a comic poem on the "Despair" of being a "S.F.F.'s G.F" (science fiction fan's girlfriend), she was more than just an "SF-widow."[94] Poems such as "Inspiration" indicate that she was herself engaged with fan culture and values:

> In a world gone all awry, where the bombs and bullets fly,
> And killing is the order of the day—
> There's a steady little band of intelligentsia, who demand
> That the Password should be "culture," not "decay."
> . . .
>
> They won't use their wits for war, for it's war that they deplore:
> They're planning earth's rebirth and not its end.
> So, re the future, don't look glum, for the "shape of things to come"
> Has a scientific, idealistic trend.[95]

This echoes precisely the sort of views expressed more widely in FIDO and in fans' justifications for their conscientious objection to war.

---

[90] D. W. Webster, "Stenciller's Oar," *Futurian War Digest*, June 1944, 6.

[91] J. M. Rosenblum, "Editorial," *Futurian War Digest*, April 1944, 9.

[92] J. M. Rosenblum, "Oddments," *Futurian War Digest*, May 1943, 3.

[93] A Brother's Sister, "My Impression of Fans: 'Orrible Disclosures by An Outsider. . . . . . .," *Futurian War Digest*, February 1944, 10.

[94] Hilda M. Johnson, "'Despair'; to the S. F. F.'s G. F.," *Futurian War Digest*, August 1943, 4.

[95] Hilda M. Johnson, "Inspiration," *Futurian War Digest*, February 1944, 3–4.

The "shape of things to come" was, perhaps unsurprisingly, one of the biggest concerns of fans. In this respect, they were no different from the wider British population.[96] For science fiction fans, their hopes and fears for the future were always linked with their fan interests; while they had always speculated about how society should or would be, the war gave immediacy to their musings.

Their primary concern was not scientific or technical, and an explanation for this can be inferred from a letter from Stapledon. He had become disengaged from fandom during the war but was moved to express his views when scolded for ingratitude for the copies of FIDO he had been sent:

> You asked me about my attitude to the present condition of the world. I think we have come to one of the great turning points in the career of our species. I agree with Wells that we may crash into extinction or at least a dismal dark age, but that we may also turn the corner and create a world order far more favourable to human capacity than anything that has existed before. If the corner is successfully turned, a few generations may transform the race into something amounting almost to a new species, and in a sense a super-human species. The future, I believe, depends less on the further advancement of science (though that is important) than on the search for wisdom, and the realization of the kind of being that we ought to desire man to be. We have power, but not wisdom.[97]

Many prewar fans had believed that planning could direct science toward the improvement of humanity. This appears to still have been considered possible, but the war had caused fans to direct their attention to the *sort* of person who should be planning this future, and how active the fans should be in setting the direction of society. In writing of Eric C. Hopkins, for instance, John Burke wrote that "entry into the Forces has made him think more deeply about post-war planning and all the difficulties attendant on the abolition of war from human relationships."[98] Rather than focusing on technologies or science, fan debates centered on the question of how a future society should be governed, and in particular whether this should involve a technocratic state headed by people like themselves.

Significant amounts of ink were spilled over this topic when the "socialistic" Julian Parr triggered a discussion of moral obligations based upon something he appears to have written in *Fantast*. One reader, R. R. Johnson, acknowledged that the war, and H. G. Wells's *Phoenix*, had caused a change in his views:[99]

> I used to deny social obligation—had some lovely scraps with Julian Parr over that point, in and out of DT's [fanmag *Delirium Tremens*]—because of the worthlessness of Homo Sap(ien)s. I used to say something like this: It is the duty of every man to his fellows to take a complete interest in social, political, &c. matters; but since the majority of men are too unintelligent and apathetic to do this, any duty towards them can be disregarded, and therefore we who are intelligent enough to take notice of things have no obligations to do so.

---

[96] The Mass Observation Project highlighted such concerns; see Tom Harrisson and Charles Madge, *War Begins at Home* (London, 1940), 40.

[97] Letter from Olaf Stapledon, 7 September 1942, in "FIDO Goes" (cit. n. 67), 2–3

[98] John F. Burke, "Spirit of the New Age #4: Eric Charles Hopkins," *Futurian War Digest*, February 1944, 5.

[99] In 1943, Johnson was still a student, though he had just passed his matriculation exams and was ready to head off for university; he was attending a public school, which was unusual for a fan.

I still can't think out a reasonable reply to that argument. It simply depends on whether the present apathy of man is indeed synonymous with worthlessness—or whether there is, indeed, still some hope. I used to think the former; I now believe the latter.[100]

Like Johnson, Jack W. Banks believed that H. G. Wells's recent works had issued the challenge to fans to "'become a conscious devoted Revolutionary' . . . whether one chooses to interpret 'Revolutionary' in the usual sense or not."[101]

In the same issue in which Banks echoed Wells's challenge, Sidney Dean took a very different view, and posited the following in opposition to Parr (and Johnson):

My idea: First Oblig. is to yourself . . . be happy. Second, to the present . . . be productive enough to maintain the status quo. Third, to the future . . . contribute to posterity. For the mass of us, this third obl. means to produce more than enough, and try by our increased efficiency & production to further progress. Some of us can invent, some uncover new sociological knowledge, some of us can delve into philosophy and theoretical physics. Even as amateurs in these fields we can achieve somewhat. But first, we should earn our way, better our relations (social & industrial) with the masses, and then . . . as a hobby . . . try our hand at some world shaking ideas (if we think it will do any good.) But I cannot agree that it is imperative for us to lead & guide, point the way for, etc, poor ignorant humanity. Nor can I see why we just gotta save the world from a dastardly fate. It got along by itself some millions of years, suddenly it will all collapse if we don't go absolutely crazy & fly into a frenzied burst of activity. Well, suppose somebody tells me what dangers must steer humanity clear of? And also, what powers have we that fit us for leadership? Imagination and broadness of view is not enough, without mental and emotional stability. And most fans lack both.[102]

By 1944, Johnson was more emphatic in his views in his response to Dean, when he expressed his belief that it was the duty of "every educated, intelligent person . . . to act so as to point the way for the unthinking and unintelligent masses, who otherwise are swayed by what they are told."[103] Johnson was vehemently against capitalism, believing that efforts to overthrow it could quite easily cause "continuous war." He presented his solution (which was slightly undermined by Douglas Webster's editorial intrusion):

But if the intelligent and educated people of today start with a full knowledge of the problem, a complete awareness, and an absolutely unbiased attitude with regard to "propaganda" the removal of capitalism can be effected simply and easily. [Hey presto! And there's your rabbit, complete with spats and plus-fours, for the asking. Who is kidding who? —DW] That is a fact, and that is why Julian Parr and I are starting an intensive campaign of sponsoring political awareness.[104]

Not all fans were as earnest about their hopes and fears for the future, however. Eric Hopkins's response to the topic from Canada eventually arrived:

---

[100] R. R. Johnson, "Inspiration - Concerning Social Obligation," *Futurian War Digest*, August 1943, 8.

[101] Jack Banks, "Sociological-'The Rape of the Masses,'" *Futurian War Digest*, October 1943, 2.

[102] Sidney "Simald" Dean, "Reactions – 'Why Fans' — & — 'Moral Obligations,'" *Futurian War Digest*, October 1943, 2–5.

[103] R. R. Johnson, "Sociological – 'The Reason Why,'" *Futurian War Digest*, February 1944, 2–3.

[104] Ibid.

> Since Fido went intellectual it has become more entertaining if not amusing. R. R. Johnson's castigations of the pore old people used to tickle me immensely. His conversion to social obligation etc. is not quite so jolly but still leaves room for a rib-tickler or two. I am enormously enthusiastic for these quiet critical closely banded people who understand all and await everything. Furthermore, I admire Passive Revolutionaries; I coup at their d'etat. Lastly, I could not help but love a guy who eminently disagrees. Such insouciance. What's the beezer mean, anyway? Starting a campaign for the Awakening of Politically Unconscious Fans! Starting! Listen, R.R., rigor mortis has been setting in for years. Vote for Laney! Down with Temple and Clarke!!![105]

Rosenblum also made light of some of his fellow fans' heated discussions in the strapline (tagline) of FIDO, where he described it as "being an amateur magazine devoted to the central and moral elite of mankind known amongst ourselves as 'Fandom.'"[106]

However they envisioned the postwar future, British fans were transformed by their experiences of war itself. Their speculations about the future were given a particular intensity, their beliefs about peace were put to the test, their fascination with technological systems had new outlets, and their communities grew and became internationally connected.

CONCLUSION

To return to the questions that closed this article's first section, we find that the British fans appropriated a public culture of expertise and made science their own; it gave them a sense of participating in the conversations of, as Hilda M. Johnson put it, "the intelligentsia."[107] In a period on the brink of war, they used science as a cultural resource to define and assert their independence of thought—their freedom from what they perceived as patriotic or superstitious cant. Thus, they were, for example, more likely than average to be pacifists or conscientious objectors.

The outbreak of war gave many fans the opportunity to participate in technical activity beyond what their education in civilian life would ordinarily have made possible. There is good circumstantial evidence to suggest that wartime roles enabled the type of intelligent, technical work they had been unable to access before the war. Moreover, war enhanced fans' sense of community, though of course this was far from a unique experience in wartime. The fannish sense of being a rare voice of sanity amid the unthinking masses was, perhaps, a successful coping strategy under conscription, where unquestioning obedience to commands was expected. Being part of a tribe that rejected this mode of life and conducted a parallel life in print that was lived otherwise may well have been a source of psychological and social sustenance. Thus, World War II did not end the nascent tribe of fandom; in many ways it strengthened its emergent features. Moreover, thanks to the heightened emotion and organized communication circuits of wartime, the *Futurian War Digest* succeeded in a way that *Novae Terrae* (the "official" first fanmag) had hoped to but had not, drawing together fans around the world and recruiting new ones.

Whatever their precise opinions on politics and pacifism, fans' utopian inclinations were refocused by the war; where they had previously discussed hypothetical rede-

---

[105] Eric Hopkins, "Airgraph Dept.," *Futurian War Digest*, December 1943, 9–10.
[106] *Futurian War Digest*, August 1942, 1.
[107] Johnson, "Inspiration" (cit. n. 95).

signs of society, there was now a sense that this was a real-life project. Their reading knowledge and their cognitive skills equipped them, they believed, unusually well to the task. The fans' ruefully anticipated disappointment—that they would not in fact be recognized as the "central and moral elite" of postwar society—may suggest topics for future historical inquiry concerning the failure of the postwar elites (within universities and the media) to sustain public buy-in to their vision of democracy, ultimately resulting in the crisis of expertise of the late twentieth century.[108]

[108] On democracy, productivity, and identity see Peter Miller and Nikolas Rose, "Production, Identity, and Democracy," *Theor. Soc.* 24 (1995): 427–67; and Alice White, "From the Science of Selection to Psychologising Civvy Street: the Tavistock Group, 1939–1948" (PhD diss., Univ. of Kent, 2016).

Old Woman and the Sea:

Evolution and the Feminine Aquatic

by Erika Lorraine Milam*

ABSTRACT

A curious sympathy between second-wave feminism and evolutionary theory forged a powerful connection between women and the sea. Speculative nonfiction by Elaine Morgan rewrote humanity's evolutionary past to be more fluid and more feminist in her *Descent of Woman* (1972). Later fiction—including Kurt Vonnegut's *Galápagos* (1985) and biologist Joan Slonczewski's *A Door into Ocean* (1986)—posited alternative futures in which long association with the ocean resulted in the evolution of new forms of biological and social order. The elusive boundary between science and fiction in these narratives highlights both the moral authority of nature and the subversive connotations of the aquatic.

> Anthropology, like the best science fiction, celebrates cultural difference as productive tension between the culture-specific and the universally human.
> —Regna Darnell, 2001[1]

INTRODUCTION

It was the dawning of the Age of Aquarius.[2] Women poured into professional anthropology, as graduate students and established faculty members alike questioned the necessity of conceptualizing the evolution of humanity as driven primarily by men on the hunt.[3] Rigorous scientific work in cultural anthropology and primatology allowed "Woman the Gatherer" to take her place alongside "Man the Hunter."[4] Elaine

---

* History Department, 136 Dickinson Hall, Princeton University, Princeton, NJ 08544-1017, USA; emilam@princeton.edu.

For fruitful comments on this paper, I am grateful to Amanda Rees, Iwan Morus, Suman Seth, W. Patrick McCray, Jenne O'Brien, and the anonymous reviewers of this volume. I benefited, too, from a vibrant conversation with New York University's History of Women and Gender Colloquium. Thank you.

[1] Regna Darnell, *Invisible Genealogies: A History of American Anthropology* (Lincoln, Nebr., 2001), 322.

[2] James Rado and Gerome Ragni, lyrics; Galt MacDermot, music, *Hair: The American Tribal Love-Rock Musical* (1967). In the original off-Broadway release of *Hair*, the character Claude was a space alien.

[3] Donna Jeanne Haraway, *Primate Visions: Gender, Race, and Nature in the World of Modern Science* (New York, N.Y., 1989).

[4] Richard B. Lee and Irven DeVore, eds., *Man the Hunter* (Chicago, Ill., 1968); Sally Linton, "Woman the Gatherer: Male Bias in Anthropology," in *Women in Cross-cultural Perspective*, ed. S. E. Jacobs (Urbana, Ill., 1971), 9–21; Nancy Tanner and Adrienne Zihlman, "Women in Evolution, Part 1, Innovation and Selection in Human Origins," *Signs* 1 (1976): 585–608.

Morgan, a writer of radio shows for BBC Wales, provided one of the first pop-evolutionary alternatives to the "Tarzanist" assumptions of many tales of human evolution. She noted: "Everything depends on context. A knife is a weapon or a tool according to whether you use it for disemboweling your enemy or for chopping parsley."[5] Morgan imbued agency in humanity's female ancestors by adding a controversial aquatic phase to the narrative of human evolution.[6] Living at the water's edge, she suggested, may have protected early humans during a time of Pleistocene drought and provided them with plenty of food. Substantial time in the water, she reasoned, would have left indelible marks on our bodies, which she identified in our layer of subcutaneous fat, the swirled patterns of hair on the backs of men, or the breasts of women that provide babies with a firm grip on their mothers' bodies.

Whether based in the ocean, the desert, or the forest, ecological interpretations of human evolution played a crucial role in postwar evolutionary epics of humanity's past.[7] Scientists suggested our ancestors' transition from forest-dwelling apes to hunters on the open savannah depended crucially on this same shift in ecological environment. Forest-dwelling vegetarian gorillas and peaceful chimpanzees provided powerful alternative primate models to the hierarchical baboons that roamed far beyond the protection of the trees.[8] Cross-cultural anthropological comparisons appeared to echo assumptions that members of human communities who dwelled in the forests should be more pacific than those who lived in the open savannahs.[9] The forest, in each of these examples, provided analogical evidence for an irenic Edenic past before our ancestors' profound transformation into hunters in the crucible of the African savannah. These narratives also spoke to ecology's functional centrality to reconstructions of humanity's past after the Second World War. In the mid-1970s, this easy contrast between forest and savannah met with a crucial difficulty when Jane Goodall's research group at Gombe realized chimpanzees killed each other too. According to *National Geographic*, they were capable of war.[10]

In imagining alternative possible futures for humanity, writers of speculative science and science fiction thus found in the aquatic a realm of conceptual possibility

[5] Elaine Morgan, *The Descent of Woman* (1972; repr., London, 1985), 156. On the opening of the ocean to humanity's gaze in the 1950s, see Helen Rozwadowski, "From Danger Zone to World of Wonder: The 1950s Transformation of the Ocean's Depths," *Coriolis* 4 (2013): 1–20.

[6] Morgan first came across this idea in Desmond Morris, *The Naked Ape: A Zoologist's Study of the Human Animal* (New York, N.Y., 1967), 43–5, as Morris discussed Sir Alister Hardy's proposal that humanity might have had a heretofore unrecognized aquatic past; Alister Hardy, "Was Man More Aquatic in the Past?" *New Scientist*, 17 March 1960, 642–5. On the narrative conventions of evolutionary histories as heroic tales, see Misia Landau, "Human Evolution as Narrative," *Amer. Scient.* 72 (1984): 262–8; and Landau, *Narratives of Human Evolution* (New Haven, Conn., 1991).

[7] Nasser Zakariya, *A Final Story: Science, Myth, and Beginnings* (Chicago, Ill., 2017); Erika Lorraine Milam, *Creatures of Cain: The Hunt for Human Nature in Cold War America* (Princeton, N.J., 2019).

[8] Donna Haraway, "The Politics of Being Female: Primatology Is a Genre of Feminist Theory," part 3 in *Primate Visions* (cit. n. 3), 279–382; Susan Sperling, "The Troop Trope: Baboon Behavior as a Model System in the Postwar Period," in *Science without Laws: Model Systems, Cases, Exemplary Narratives*, ed. Angela Creager, Elizabeth Lunbeck, and Norton Wise (Durham, N.C., 2007), 73–89.

[9] Julian Steward, *Theory of Culture Change: The Methodology of Multilinear Evolution* (Urbana, Ill., 1955); Marston Bates, *The Forest and the Sea: A Look at the Economy of Nature and the Ecology of Man* (New York, N.Y., 1960). In anthropology, see, for example, Colin M. Turnbull, *The Forest People* (New York, N.Y., 1962); and Turnbull, *The Mountain People* (New York, N.Y., 1972).

[10] Jane Goodall, "Life and Death at Gombe," *National Geographic*, May 1979, 592–621.

that potentially avoided the violence of terrestrial landscapes.[11] The US Navy had certainly tried to militarize the oceans, even going so far as to call their depths "inner" space, a direct parallel to the involvement of the air force with the exploration of "outer" space. In 1963, Vice Admiral William F. Radborn suggested that in a decade, the navy would operate submarines that could conduct research at a depth of 20,000 feet and atomic-powered transponders to communicate under sea ice, and officers would work at depths of 1800 feet for a month at a time. An editor of the *Boston Globe* added, "were it not for the author . . . the description of what the United States fleet will be like 10 years from now would seem like a page out of science fiction."[12] Yet the promise of the oceans as a limitless frontier eventually paled in comparison to the vastness of space and excitement over the Apollo missions.[13]

In deep evolutionary time, our most distant ancestors had crawled out of the salty brine that spawned all planetary life, long before they encountered the arid savannah.[14] Even without the aquatic ape, oceans thus offered a reminder of humanity's epic primordial past. As Rachel Carson wrote in her immensely popular *The Sea Around Us*, "as life itself began in the sea, so each of us begins his individual life in a miniature ocean within his mother's womb, and in the stages of his embryonic development repeats the steps by which his race evolved, from gill-breathing inhabitants of a water world to creatures able to live on land."[15] Although the far distant ancestors of the human lineage had left the ocean millions of years ago, Carson noted, humanity had slowly been finding its way back to the sea. Humans "could not physically re-enter the ocean as the seals and whales had done," but using their ingenuity and reason, they could "re-enter it mentally and imaginatively."[16] Carson wrote with a sure pen about the charismatic wonders of the ocean, including the origins of life itself in the murky, wet past; "the unending darkness" of the deep abyss; and the "inverted 'timber line'" below which vegetation cannot grow.[17] Throughout the book, she also wove a message of hope and cyclical renewal. Even "the very iciness of the winter sea" promised a new spring, just as particles of material were "used over and over again, first by one creature, then by another."[18]

---

[11] On the long fascination with forests as a wild precursor to civilization in Western thought, see Robert Pogue Harrison, *Forests: The Shadow of Civilization* (Chicago, Ill., 1992). On the role of oceanic cetaceans in Cold War imagination, see D. Graham Burnett, *The Sounding of the Whale: Science and Cetaceans in the Twentieth Century* (Chicago, Ill., 2012).

[12] Vice Admiral William F. Raborn, "The Navy's Role in Nuclear Age: Exploring 'Inner' Space," *Boston Globe*, 25 February 1963, 10.

[13] Helen Rozwadowski, "Arthur C. Clarke and the Limitations of the Ocean as a Frontier," *Environ. Hist.* 17 (2012): 578–602; Neil Maher, *Apollo in the Age of Aquarius* (Cambridge, Mass., 2017); Michael J. Neufeld, ed., *Spacefarers: Images of Astronauts and Cosmonauts in the Heroic Age Era of Spaceflight* (Washington, DC, 2013); Kelly Moore, *Disrupting Science: Social Movements, American Scientists, and the Politics of the Military, 1945–1975* (Princeton, N.J., 2008).

[14] Stefan Helmreich, *Alien Ocean: Anthropological Voyages in Microbial Seas* (Berkeley, Calif., 2009). These associations continue to reverberate through recent science fiction narratives. In *Gravity*, directed by Alfonso Cuarón (2013; Burbank, Calif.: Warner Bros. Entertainment, 2014), DVD, Sandra Bullock is reborn to a new life as she staggers out of the ocean after her harrowing space incubation and splashdown into the watery depths.

[15] Rachel Carson, *The Sea Around Us* (1951; repr., London, 2014), 20.

[16] Ibid., 21.

[17] Ibid., 65, 81. See also Jacques-Yves Cousteau, with Frédéric Dumas, *The Silent World* (New York, N.Y., 1953).

[18] Carson, *The Sea* (cit. n. 15), 46, 38.

Countercultural associations of the oceans added to their draw as an alternative to normative terrestrial conceptions of intelligence and sexuality.[19] Dolphins entered the public spotlight as the aquatic equivalent of primates—an intelligent species capable of recognizing themselves in mirrors and communicating through sound.[20] Connections between women and water have a long history, from mermaids to synchronized swimming.[21] More important for Morgan may have been local folkloric transformative traditions that featured female selkies, who could take both human and seal forms.[22] Building on these tropes, Morgan crafted a feminist, aquatic account of humanity's evolutionary past as an antidote to stories about the transformative power of men on the savannah. She wrote of a past in which the key characteristics that made us human—our bipedal posture, our capacity for spoken language and facility with tools, our ability to cooperate with others—arose when our human ancestors partially adapted to life in the intertidal zone. With the added buoyancy of water, Morgan suggested, they learned to wade in search of food after the plentiful supplies of the forest vanished. They wielded rocks to crack open shellfish. With most of their bodies surrounded by water, they lost their fur, gained subcutaneous fat to protect against hypothermia, and over time developed the capacity to communicate verbally. Morgan's aquatic apes were more cooperative than the savannah beasts of anthropological tomes, as their ecology required gathering proteinaceous creatures rather than hunting them. Even if readers enjoyed her sarcastic rejoinder to life on the savannah, however, most assumed her reconstruction of humanity's past was mere conjecture. In *The Descent of Woman*, Morgan suggestively reinterpreted existing paleoanthropological data but supplied no new evidence of her own beyond appeals to readers' common sense experience of their own bodies. Practicing evolutionists deemed her aquatic theory the worst kind of pseudoscience (fig. 1).[23]

For authors of science fiction, however, oceans continued to provide an ecologically plausible alternative to humanity's own past and a fruitful space for speculation about its future. This essay examines two American science fiction novels—Kurt Vonnegut's *Galápagos* and Joan Slonczewski's *A Door into Ocean*, both published in the mid-1980s—that create a distant future for humanity defined by water.[24] Von-

---

[19] D. Graham Burnett, "A Mind in the Water," *Orion Magazine*, May–June 2010, 38–51; John C. Lilly, *Man and Dolphin* (New York, N.Y., 1961).

[20] D. Graham Burnett, "Shots Across the Bow," chap. 6 in *The Sounding* (cit. n. 11), 517–645; see also Leo Slizard, *The Voice of the Dolphins, and Other Stories* (New York, N.Y., 1961); Arthur C. Clarke, *Dolphin Island: A Story of the People of the Sea* (New York, N.Y., 1963); Douglas Adams, *The Hitchhiker's Guide to the Galaxy* (1979; repr., New York, N.Y., 1981); and Alexander Jablokov, *A Deeper Sea* (New York, N.Y., 1992). In Stanisław Lem's *Solaris* (1961; repr., New York, N.Y., 1970), the planetary ocean itself is a self-aware, intelligent entity.

[21] Astrida Neimanis, "Hydrofeminism: Or, On Becoming a Body of Water," in *Undutiful Daughters: New Directions in Feminist Thought and Practice*, ed. Henriette Gunkel, Chrysanthi Nigianni, and Fanny Söderbäck (New York, N.Y., 2012), 85–99.

[22] See, for example, Rosalie K. Fry, *Child of the Western Isles* (London, 1957), a fantasy novel for children that John Sayles would later use as the basis for his film *The Secret of Roan Inish* (1994; Culver City, Calif.: Sony Pictures Home Entertainment, 2000), DVD.

[23] Erika Lorraine Milam, "Dunking the Tarzanists: Elaine Morgan and the Aquatic Ape Theory," in *Outsider Scientists: Routes to Innovation in Biology*, ed. Oren Harman and Michael R. Dietrich (Chicago, Ill., 2013), 223–47; Jerold Lowenstein and Adrienne Zihlman, "The Wading Ape: A Watered-Down Version of Human Evolution," *Oceans* 13 (1980): 3–6; Ian Tattersall and Niles Eldredge, "Fact, Theory, and Fantasy in Human Paleontology," *Amer. Scient.* 65 (1977): 204–11, on 207.

[24] Kurt Vonnegut, *Galápagos* (1985; repr., New York, N.Y., 1999); Joan Slonczewski, *A Door into Ocean* (1985; repr., New York, N.Y., 1986).

Figure 1. *Elaine Morgan reasoned in* Descent of Woman *(1972) that extended exposure to water over generations would have caused human ancestors to lose their body hair, learn to walk upright thanks to the added buoyancy of the water, and gain a layer of subcutaneous fat to help regulate internal body temperatures. This cartoon accompanied a thoroughgoing critique of Morgan's arguments and her evidence cowritten by a paleoanthropologist and a physician: Lowenstein and Zihlman, "Watered Down Version" (cit. n. 23). (Illustration by Bill Prochnow.)*

negut and Slonczewski grounded their narratives in current biological theory, paleo-biological and microbiological, respectively. Both imagined life after an apocalypse of human causation. In Vonnegut's darkly humorous account, human evolution continues but only when people are stripped of weapons and other forms of technology. Slonczewski maintained a fierce optimism in her writing as she forged a link between women and nature to envision a future in which (some) humans learned to live in harmony with the ocean rather than seeking to conquer it. Both books served to warn readers of the hubris inherent to capitalism run amok.

Although it hardly comes as a surprise that authors of science fiction, even evolutionary sci-fi, worked to make their narratives scientifically accurate, both books would have been classified in the 1980s as "soft" (rather than "hard") science fiction—a term often accompanied by a healthy dose of derision. Science fiction author and critic Charles Platt, for example, vilified the rise of such "New Wave" science fiction, lamenting that "the body of literature I love has been doped up and defiled, draped in fake finery and turned into a flabby old hooker smelling of festering lesions and cheap perfume."[25] He lamented the loss of the brash, rebellious feel of the science fiction he had read and fallen in love with as an adolescent boy in the 1950s. In its place, he suggested, came a "new 'soft' science fiction" that verged on fantasy. Due to the popularity of Richard Adams's *Watership Down* and J. R. R. Tolkien's three-volume *Lord of the Rings*, followed by novels by Robert Heinlein, Robert Howard, Frank Herbert, Terry Brooks, and others, Platt posited that fantasy had become a genre in its own

[25] Charles Platt, "The RAPE of Science Fiction," *Science Fiction Eye*, July 1989, 44–9, on 45. See also Pamela Sargent's response in her introduction to *Women of Wonder: The Contemporary Years: Science Fiction by Women from the 1970s to the 1990s*, ed. Sargent (Orlando, Fla., 1995), 3–6.

right.[26] At the same time, women had created a new strand of science fiction with an "admirable" "concern for human values," he wrote. The success of Joan Vinge, Vonda McIntyre, Ursula Le Guin, Joanna Russ, Kate Wilhelm, and Carol Emshwiller had helped erode "science fiction's one great strength that had distinguished it from all other fantastic literature: its implicit claim that events described *could actually come true*."[27] At the root of his jeremiad was a belief that the commercial success of science fiction starting in the 1960s produced a glut of mediocre writing, all of which sold brilliantly and made it difficult to discover any true emerging talent.[28] Too often, he deemed, authors invented the impossible, and bent the known rules of the physical world to save themselves from working out rigorously scientific mechanisms. (His complaints echoed those mounted by practicing anthropologists against Morgan's aquatic ape.)

Platt's historical taxonomy of the genre mirrored the rising currency of "soft science" as a term used to describe nonlaboratory social science research, from anthropology to human biology.[29] Yet, his characterization of some science fiction as "soft," and therefore unconcerned with plausibility, elided the careful attention many of these authors paid to contemporary social science. Platt's categories thus implicitly reflected the denigration of the social sciences as *science*, rather than a lack of engagement with their precepts or analytical methods.[30] For technological optimists in the 1980s, the progress of science and technology could and should exist outside the political realm. Good science—"real" science—they believed, was apolitical, especially when it came to theories of human nature.[31] With the benefit of hindsight, we can see more easily that authors of soft science fiction in the 1980s, like Vonnegut and Slonczewski, took world building to be serious scientific work impossible to separate from moral import.[32]

In their hands, aquatic landscapes provided a powerful means of challenging traditional accounts of human evolution that correlated arid savannahs with all-male hunting groups, while still adhering to the biological principle that through natural

[26] Richard Adams, *Watership Down* (New York, N.Y., 1972); J. R. R. Tolkien, *The Fellowship of the Ring* (Boston, Mass., 1954); Tolkien, *The Two Towers* (Boston, Mass., 1954); Tolkien, *The Return of the King* (Boston, Mass., 1955).

[27] Platt, "The RAPE," (cit. n. 25), 46 (emphasis in the original); Joan Vinge, *The Snow Queen* (New York, N.Y., 1980); Vonda McIntyre, *Dreamsnake* (New York, N.Y., 1978); Ursula Le Guin, *A Wizard of Earthsea* (Berkeley, Calif., 1968); Le Guin, *Left Hand of Darkness* (New York, N.Y., 1969); Le Guin, *The Dispossessed* (New York, N.Y., 1974), Le Guin, *The Word for World is Forest* (New York, N.Y., 1976), among many others; Joanna Russ, *The Female Man* (New York, N.Y., 1975); Kate Wilhelm, *Where Late the Sweet Birds Sang* (New York, N.Y., 1976).

[28] See also Amanda Rees, "From Technician's Extravaganza to Logical Fantasy: Science and Society in John Wyndham's Postwar Fiction, 1951–1960," in this volume.

[29] Based on searches in JSTOR and Ngram Viewer, both "hard science" and "soft science" existed in earlier decades, but they increased in frequency as analytical terms in the 1960s.

[30] Mark Solovey and Hamilton Cravens, eds., *Cold War Social Science: Knowledge Production, Liberal Democracy, and Human Nature* (New York, N.Y., 2012); Joel Isaac, *Working Knowledge: Making the Human Sciences from Parsons to Kuhn* (Cambridge, Mass., 2012).

[31] Ullica Segerstråle, *Defenders of the Truth: The Battle for Science in the Sociobiology Debate and Beyond* (New York, N.Y., 2000); Aaron Panofsky, *Misbehaving Science: Controversy and the Development of Behavior Genetics* (Chicago, Ill., 2014); David Kaiser and W. Patrick McCray, eds., *Groovy Science: Knowledge, Innovation, and American Counterculture* (Chicago, Ill., 2016).

[32] Despite the left-leaning politics of many academics, some "hard social science" fiction remained quite conservative in outlook; see many of the stories collected in Leon E. Stover and Harry Harrison, eds., *Apeman, Spaceman* (New York, N.Y., 1968); and Willis E. McNelly and Leon E. Stover, eds., *Above the Human Landscape: A Social Science Fiction Anthology* (Pacific Palisades, Calif., 1972).

selection physical environments inexorably shaped the creatures inhabiting them.[33] Alternatives to traditional evolutionary narratives took a variety of forms. Vonnegut used *Galápagos* to highlight the importance of contingency to the normal operation of evolution. In *A Door into Ocean*, Slonczewski took inspiration from the resonance between new-wave science fiction and second-wave feminism that allowed authors to explore alternatives to the techno-utopian worlds imagined for women by male visioneers.[34] (Ironically, although Platt lamented the demise of techno-utopian factualism, he—like Vonnegut and Slonczewski—feared the destructive potential of rampant capitalism, albeit in its short-term capacity to ruin the core of science fiction by catering too heavily to new audiences.) Alien ecological landscapes thus provided a plausible scientific mechanism for imagining a universal humanity governed by unconventional politics, genders, and economies. The power of the feminine aquatic stemmed not from its association with a single set of affective connotations, but from its generative intellectual slipperiness.

KURT VONNEGUT'S FURRY FEMININE FUTURE

In Kurt Vonnegut's cynical guide to the evolution of humanity, *Galápagos*, 1985 marked the year when a global economic crisis would cripple the world's infrastructure. A devastating strain of bacteria that consumed human egg cells would render all but a handful of women infertile. Only a few humans would escape this fate, having been marooned on the Galápagos Islands by a series of unlikely events, beyond the reach of other people and therefore the spread of the disease. This small handful of women and a single man thus became the progenitors of all future humanity. The story is told from the perspective of the ghost of Leon Trout a million years in the future, having observed the fate of humanity in the meantime.[35] In retrospect, Trout deemed the twentieth century the "era of big brains." Our brains caused us endless amounts of trouble, he repeats throughout the book. As humans had adapted to life in the water, natural selection had shrunk our brains and transformed our hands, which had once been so dexterous, into flippers. Humans became pacific because we could neither conceptualize how to make weapons nor use them. Humanity's evolutionary fate as sleek, furry, "innocent fisherfolk," had been as much a matter of chance as of fitness.[36]

After the Second World War, Vonnegut spent five years studying anthropology at the University of Chicago. Although his anthropological musings often found their way into his fiction, he never published as an anthropologist. When Vonnegut and his second wife, Jill Krementz, visited the Galápagos in 1981, he reported being "fascinated by the island's natural life." He added: "I spent as much time there as Charles Darwin did—two weeks. We had advantages that Darwin didn't have. Our guides all had graduate

---

[33] Lee and DeVore, *Man the Hunter* (cit. n. 4); E. O. Wilson, *Sociobiology: The New Synthesis* (Cambridge, Mass., 1975); Richard Dawkins, *The Selfish Gene* (New York, N.Y., 1976).

[34] On the masculine precepts of science fiction as a genre, see N. Katherine Hayles, *How We Became Posthuman: Virtual Bodies in Cybernetics, Literature, and Informatics* (Chicago. Ill., 1999); and W. Patrick McCray, *The Visioneers: How a Group of Elite Scientists Pursued Space Colonies, Nanotechnologies, and a Limitless Future* (Princeton, N.J., 2012), 94–104.

[35] Leon Trout is the son of Kilgore Trout who makes an appearance in a variety of other Vonnegut novels, including *Breakfast of Champions, or Goodbye Blue Monday* (1973) and *Timequake* (1997).

[36] Gilbert McInnis, "Evolutionary Mythology in the Writings of Kurt Vonnegut, Jr.," *Critique: Studies in Contemporary Fiction* 46 (2005): 383–96, on 391; Sheila Pardee, "Drifting and Foundering: Evolutionary Theory in Kurt Vonnegut's *Galápagos*," *DQR Studies in Literature* 57 (2015): 249–65.

degrees in biology. We had motorboats to move us around the islands more easily than rowboats could when Darwin visited the Galapagos in the 1830s. And, most important, we knew Darwin's theory of evolution, and Darwin didn't when he was there."[37]

Soon after returning, Vonnegut gave a lecture in New York City at the Cathedral of St. John the Divine. (According to their online material, the cathedral is the length of six blue whales.) Vonnegut spoke about the strange creatures he had seen on the Galápagos Islands—especially the blue-footed boobies who in courtship iteratively and solemnly raised each beautiful, bright foot to show their prospective mates. He thought about the millions of years needed to create such natural intricacies; this was a span of time vast to us but a mere wink in nature's eye. However long it had taken for nature to craft humans, he feared we were running out of time. Death itself was old, he told his audience, but the scale of our destructive capacity threatened our very existence as a species. The previous night, Vonnegut reported, he had dreamed of meeting the descendants of humanity in a thousand years. In his dream, he asked these survivors how humanity had managed to persist for so long. Their reply? "By preferring life over death for themselves and others at every opportunity, even at the expense of being dishonored."[38]

Three years later, Vonnegut published *Galápagos*, a longer reflection on what would be required for humanity to survive for a million years—orders of magnitude longer than his earlier thought experiment. In Vonnegut's fantasy, an ill-fated celebrity cruise to the renowned islands strands a handful of lost souls on the entirely fictional Santa Rosalia. The members of this genetic bottleneck are rich, poor, likeable, insufferable, and ethnically diverse. Vonnegut took great care to establish the truly random circumstances that led each individual to a place on *Bahía de Darwin*, humanity's new ark to the future.

Vonnegut based his evolutionary theory in *Galápagos* on current trends in biological thought, especially those espoused by Stephen Jay Gould in the pages of *Natural History* magazine.[39] Evolution, for Vonnegut, was necessarily contingent and inconsistently progressive, and it changed its focus in fits and starts. As humanity regressed to a more animalistic state, we also became more peaceful, retreating to a small-brained primeval innocence.[40] (Leon Trout called this progress indeed.) The genetic bottleneck of humanity created by the small handful of survivors in turn creates substantial genetic drift, so that future generations of humans resemble those few individuals left to repopulate Earth. One of the survivors includes a young girl, the first

---

[37] Quote from Herbert Mitang, "Advantages Darwin Lacked," *New York Times*, 6 October 1985, BR7. See also, Lorrie Moore, "How Humans Got Flippers and Beaks," *New York Times*, 6 October 1985, BR7.

[38] Kurt Vonnegut, "Fates Worse than Death," *North American Review* 267 (1982): 46–9.

[39] Many of Gould's essays were then reprinted in paperback collections, also available when Vonnegut wrote: Stephen Jay Gould, *Ever Since Darwin: Reflections in Natural History* (New York, N.Y., 1977); *The Panda's Thumb: More Reflections in Natural History* (New York, N.Y., 1980); *Hen's Teeth and Horse's Toes* (New York, N.Y., 1983); and *The Flamingo's Smile: Reflections in Natural History* (New York, N.Y., 1985). On Vonnegut's reading of Gould prior to writing *Galápagos*, and his desire to make the novel "reputable scientifically," see his interview with Hank Nuwer, "A Skull Session with Kurt Vonnegut," *South Carolina Review* 19 (1987): 2–23.

[40] Leonard Mustazza, "A Darwinian Eden: Science and Myth in Kurt Vonnegut's 'Galápagos,'" *Journal of the Fantastic in the Arts* 3 (1991): 55–65; Donald Morse, "Thinking Intelligently about Science and Art: Kurt Vonnegut's *Galápagos* and *Bluebeard*," *Extrapolation* 38 (1997): 292–303. Mustazza claims this primeval innocence as ancestral, a contention most evolutionists in the 1980s would not have endorsed. In fact, most "regressions" of humanity in science fiction led to more violent behavior, not less. See David Kirby, "Darwin on the Cutting Room Floor: Evolution, Religion, and Film Censorship," in this volume.

child born on Santa Rosalia, who had a fine pelt of dark hair—a genetic consequence of her grandmother's survival of the atomic bomb in Hiroshima. Trout notes sagely that humans almost certainly would have become hairier eventually, but this happy circumstance speeded the process considerably. Contingency, as manifested in genetic drift as well as punctuated evolutionary changes, was key to paleobiologists' re-imagining of contemporary evolutionary theory in the 1980s, especially for Gould.[41]

Gould hoped to break down popular assumptions that evolutionary fitness some-how meant that the best or brightest individuals were necessarily those who left the most offspring. Sometimes, he reasoned, individuals survived and reproduced be-cause they happened to be in the right place at the right time. Selection could favor, after all, ridiculous traits. The "Irish Elk" (so named despite the fact that it is neither an elk nor exclusively Irish) constituted one of Gould's favorite examples of this phe-nomenon.[42] Vonnegut discusses the Irish elk explicitly, linking the fate of their antlers to that of human brains. The large size of human brains, according to Vonnegut, had brought humans nothing but misery and had come to imperil our very existence. Build-ing on long-standing tropes in the colloquial science literature of the day, Vonnegut sug-gested that humans' capacity to conceptualize, manufacture, and use nuclear weapons had outstripped our social savvy in maintaining peace once they existed.[43] "Can it be doubted," the ghostly narrator asks, "that three-kilogram brains were once nearly fatal defects in the evolution of the human race?"[44] He returns again and again to this theme—the very trait that assured our survival, on which humans judged their value and self-worth, was the very same trait that needed to be tamed to ensure the survival of the species. Vonnegut's satire feels especially dire in these moments. To survive bi-ologically, we would need to sacrifice all literature and Beethoven's Ninth Symphony (a point that Vonnegut repeated eight times over the course of the novel).

Yet if Vonnegut really had read Gould as closely as he claimed, he likely knew that evolutionists no longer propagated this mono-causal tale of the Irish elk's extinction. The exceedingly large Pleistocene deer exhibited giant antlers that were so big as to be functionally useless in battle. Gould had suggested they attracted female mates and so were useful in and of themselves, without having to be twisted or turned, much less bashed against the antlers of another male.[45] Despite modern folklore, Gould ar-gued that the species later went extinct because of changing climatic conditions, not the size of the antlers. He also argued vociferously that if scientists were to replay the tape of life, it would never turn out the same.[46] Chance events would intercede. Life would turn out differently. This opens the possibility that for Vonnegut, humanity might not have been as doomed as the narrator—already reconciled to humanity's fate as fisherfolk—insisted.

[41] David Sepkoski, *Rereading the Fossil Record: The Growth of Paleobiology as an Evolutionary Discipline* (Chicago, Ill., 2012); Myrna Perez Sheldon, *Darwin's Heretic: Stephen Jay Gould, 1941–2002* (unpublished).

[42] Stephen Jay Gould, "The Misnamed, Mistreated, and Misunderstood Irish Elk," *Ever Since Darwin* (1979; repr., New York, N.Y., 2007), 79–90.

[43] See, for example, Charles Osgood, *An Alternative to War or Surrender* (Urbana, Ill., 1962), 19; Konrad Lorenz, *On Aggression*, trans. Marjorie Kerr (New York, N.Y., 1956). On colloquial science, see Milam, *Creatures of Cain* (cit. n. 7).

[44] Vonnegut, *Galápagos* (cit. n. 24), 9.

[45] Stephen Jay Gould, "The Origin and Function of 'Bizarre' Structures: Antler Size and Skull Size in the 'Irish Elk,' *Megaloceros giganteus*," *Evolution* 28 (1974): 191–220.

[46] David Sepkoski, "'Replaying Life's Tape': Simulations, Metaphors, and Historicity in Stephen Jay Gould's View of Life," *Stud. Hist. Phil. Biol. Biomed. Sci.* 58 (2016): 73–81.

That humanity had any future, furry or otherwise, looked rather bleak once the hand-ful of remaining people had been marooned on Santa Rosalia. They included newborn Akiko Hiroguchi, her mother Hisako Hiroguchi, Selena MacIntosh, Mary Hepburn (age sixty-one), Captain Adolf Heist (age sixty-six), and six Kanka-bono women from the mountains of Ecuador who kept their names hidden from the rest of the group. As the captain was "a racist," Vonnegut wrote, he was "not at all drawn to Hisako or her furry daughter, and least of all to the Indian women."[47] It was Mary whose curiosity was piqued by her big brain and desire to know "whether a woman could be impregnated by another one on a desert island without any technical assistance."[48]

Leon Trout narrates the scene as if it took place in a movie: "Mary Hepburn, as though hypnotized, dips her right index finger into herself and then into an eighteen-year-old Kanka-bono woman, making her pregnant."[49] She then repeats this five more times, and all six women bore children. When Mary had first stepped foot on Santa Rosalia (150 pages earlier), she stumbled, abrading her knuckles. A small finch landed on that same finger and gently drank the droplets of blood that had appeared. In fact, this was how she had known they were marooned on Santa Rosalia—*Geospiza difficilis*, that queer, bloodsucking finch lived only there. (It seems fitting that finch speciation should play at least a bit part in any novel about evolution in the Galápagos, even if the species itself does not exist.[50])

Vonnegut's evolutionary vision could be read as darkly optimistic. Certainly, hu-manity at the end of his million-year glance into the future had managed to survive the ravages of natural selection, chance mutation, and extreme local conditions (albeit in newly aquatic form). By returning to the sea and abandoning military adventurism, humanity might yet survive enlarged, three-kilogram brains.[51] But at what cost? Hu-mans had reverted to their animalistic, primordial past. Any attempt to escape back to a terrestrial existence was met by those ravenous egg-eating bacteria.

Shortly after publication, Gould read *Galápagos* quickly, over one weekend. He wrote to Vonnegut the following Monday, praising the novel as "beautifully accurate" in its depiction of evolution's quirkiness and punctuated progress. Gould approved of the novel's emphasis on contingency as inherent to the process of evolution and would on occasion assign it in his courses. Elsewhere, Gould would suggest, "In Vonnegut's novel, the pathways of history may be broadly constrained by such gen-eral principles as natural selection, but contingency has so much maneuvering room within these boundaries that any particular outcome owes more to a quirky series of antecedent events than to channels set by nature's laws."[52] Vonnegut replied imme-diately, admitting that Gould had been constantly on his mind as he wrote.[53] Like Gould, he had sought to undermine sociobiological arguments that implied the most

[47] Vonnegut, *Galápagos* (cit. n. 24), 289.
[48] Ibid., 292.
[49] Ibid.
[50] On Charles Darwin's voyage to the Galápagos and the role of finches in his theory of natural selection, see Janet Browne, *Charles Darwin: Voyaging* (New York, N.Y., 1995).
[51] Pardee, "Drifting and Foundering" (cit. n. 36), 265.
[52] Stephen Jay Gould, *Wonderful Life: The Burgess Shale and the Nature of History* (New York, N.Y., 1989), 286.
[53] Letter from Stephen Jay Gould to Jill Krementz and Kurt Vonnegut, 7 October 1985, Box 111, Folder 6; letter from Kurt Vonnegut to Stephen Jay Gould, 10 October 1985, Box 698, Folder 3; M1437 Stephen Jay Gould Papers, 1899–2004, Department of Special Collections & University Ar-chives, Stanford University Libraries, Stanford, California.

successful members of society had attained their positions because they were smarter or more attractive.[54] In *Galápagos*, people had not survived because they were more fit than their neighbors; instead, they survived thanks to sheer chance.

Reflecting on his science fiction writing, Vonnegut later claimed he found it difficult because he needed to attend to two things at once—he sought to both "hold the reader emotionally" and "make sense scientifically." Vonnegut ensured that *Galápagos* was "responsible" in its presentation of evolution and natural selection. Good science fiction, for Vonnegut, made people rethink the capacity of science to solve some kinds of questions but not others. He added quickly, "It's a lot easier if you're not funny."[55]

JOAN SLONCZEWSKI'S SYMBIOTIC SHARERS OF SHORA

When Joan Slonczewski read Frank Herbert's *Dune*, she was inspired to write her own ecological science fiction—she believed she could do better, scientifically and politically.[56] Slonczewski trained as a microbiologist and began teaching at Kenyon College in Gambier, Ohio, in 1984. She published *A Door into Ocean* a year later, although she had been working on it for some time. The students who worked in her laboratory would come to refer to her, with affection and awe, as "The Sloncz," and appreciated her use of science fiction in her science classes. Slonczewski set *A Door into Ocean* in the watery world of Shora, home to an aquatic, decentralized, nonviolent, all-female Sharer society. Shora was the antithesis of Herbert's desert planet where harsh environmental conditions bred a race of natural warriors. Slonczewski's ecofeminist approach fit well into an established genre populated by authors like Margaret Atwood, Ursula Le Guin, and Marge Piercy, who combined second-wave feminism with environmental concerns.[57] Similar trends characterized contemporaneous scholarship in the humanities, where deep ecology and feminism combined in Carolyn Merchant's *Death of Nature*.[58] Whereas Merchant sought to document the early modern slaughter of nature in feminine guise at the hands of machines and masculine reductionism, Slonczewski resurrected feminine nature in futuristic form on an alien planet. Like Merchant, Slonczewski mobilized femininity as a foil to the extractive violence she saw as characterizing contemporary political and economic arrangements in "the West."

In her work as a microbiologist, Slonczewski explored the interior workings of the gram-negative bacteria *Escherichia coli*, commonly found in the intestines of mammals. Human digestive health symbiotically depends on some strains of *E. coli*, although other strains can cause major foodborne illnesses. Her research at the time of *A Door into Ocean*'s publication explored how these common bacteria maintain a consistent in-

---

[54] Sheldon, *Darwin's Heretic* (cit. n. 41).

[55] Zoltán Abády-Nagi, "'Serenity,' 'Courage,' 'Wisdom': A Talk with Kurt Vonnegut," *Hungarian Studies in English* 22 (1991): 23–37.

[56] Author interview with Joan Slonczewski, 8 March 2017; Frank Herbert, *Dune* (Philadelphia, Penn., 1965).

[57] See also Naomi Mitchison, *Memoirs of a Spacewoman* (London, 1962); Le Guin, The Word (cit. n. 27); and Marge Piercy, *Woman on the Edge of Time* (New York, N.Y., 1976).

[58] Carolyn Merchant, *The Death of Nature: Women, Ecology, and the Scientific Revolution* (San Francisco, Calif., 1980); Joan Cadden, "Introduction" (485–6), and editor of Focus section, "Getting Back to the *Death of Nature*: Rereading Carolyn Merchant," *Isis* 97 (2006): 485–533, including essays by Katharine Park, Gregg Mitman, Charis Thompson, and a response by Carolyn Merchant.

ternal environment, especially pH levels.[59] The idea that symbiotic relationships existed between organisms with radically different evolutionary origins was hardly new in the 1980s, but the importance of symbiosis as a process in the history of life on Earth was steadily gaining traction among both biologists and colloquial science readers.[60]

Slonczewski used *A Door into Ocean* to address questions of diversity and cooperation in the living world. Herbert had created an ecologically distinct desert planet in which to base his epic story. Yet no living ecology could exist with such paltry species diversity, she insisted. (When David Lynch directed his recounting of the *Dune* story, released in 1984, he ventured to the quartz fields of the Samalayuca Desert in Mexico, known for its spectacular sandscapes, and lying only an hour south of El Paso. Lynch deemed the area insufficiently "pristine" to represent the desolate world Herbert had envisioned and hired 300 men to remove all rocks, shrubs, and cacti from the 25 square miles where he would be filming.[61]) Slonczewski bridled, too, at Herbert's normalization of aggression in his depiction of humans as born warriors, hardened by circumstance and environment. At the other end of the political spectrum, she found herself equally disappointed by Le Guin's *Word for World is Forest*, where the peaceful, idyllic creechies, a far distant descendent of human space travelers, were able to repel their colonial invaders (another species of humans) only after adopting the violence of their oppressors.[62] Slonczewski saw in the history of colonialism a similar tragedy, where local leaders had to become westernized to emerge as national heroes.[63]

In search of an alternative to *Dune*'s dry vistas, Slonczewski chose an aquatic environment.[64] She had read Morgan's *Descent of Woman* and was impressed with the notion that fish have excellent protein composition for human brain development.[65] Ad-

---

[59] Publications contemporary with *A Door into Ocean* include J. L. Slonczewski, M. W. Wilde, and S. H. Zigmond, "Phosphorylase a Activity as an Indicator of Neutrophil Activation by Chemotactic Peptides," *Journal of Cell Biology* 101 (1985): 1191–7; J. L. Slonczewski et al., "Effects of pH and Repellent Tactic Stimuli on Protein Methylation Levels in *Escherichia coli*," *Journal of Bacteriology* 152 (1982): 384–99; and J. L. Slonczewski et al., "pH Homeostasis in *Escherichia coli*: Measurement by P31 Nuclear Magnetic Resonance of Methylphosphonate and Phosphate," *Proceedings of the National Academy of Sciences USA* 78 (1981): 6271–5.

[60] On the importance of symbiosis to the history of life on Earth, see Lynn Margulis, *The Origin of Eukaryotic Cells* (New Haven, Conn., 1970), 45–68; and James Lovelock, *Gaia: A New Look at Life on Earth* (New York, N.Y., 1979). See also Jan Sapp, *Evolution by Association: A History of Symbiosis* (New York, N.Y., 1994); Rachel Mason Dentinger, "The Nature of Defense: Coevolutionary Studies, Ecological Interaction, and the Evolution of 'Natural Insecticides,' 1959–1983" (PhD diss., Univ. of Minnesota, 2009); and Michael Ruse, *The Gaia Hypothesis: Science on a Pagan Planet* (Chicago, Ill., 2013).

[61] *Dune*, directed by David Lynch (1984; Hollywood, Calif.: Universal Studios Home Entertainment, 1984), DVD. In 2009 the Mexican government created the Área Natural Protegida Médanos de Samalayuca, although tourism still allows sand-boarding, 4 × 4 dune tours, and other activities.

[62] Le Guin, *The Word* (cit. n. 27). The progressive politics of several of Le Guin's earlier novels may account for Slonczewski's disappointment with *The Word*; see, for example, *Left Hand*, and *Dispossessed* (both cit. n. 27).

[63] In my interview with Slonczewski, she mentioned the life of Mahatma Gandhi as an example of this process, although Frantz Fanon would seem an equally natural parallel; Frantz Fanon, *Wretched of the Earth*, trans. Constance Farrington (New York, N.Y., 1963).

[64] Slonczewski provides a study guide for the book on her website and believes science fiction, when constructed well, can provide students with an introduction to biological concepts: "*A Door into Ocean* Study Guide," last updated 4 January 2001, http://biology.kenyon.edu/slonc/books/adoor_art /adoor_study.htm.

[65] See also Betty Meehan, "Hunters by the Seashore," *Journal of Human Evolution* 6 (1977): 363–70; Meehan, "Man Does Not Live by Calories Alone: The Roles of Shellfish in a Coastal Cuisine," in *Sunda and Sahul: Prehistoric Studies in Southeast Asia, Melanesia and Australia*, ed. J. Allen, J. Golson, and R. Jones (Cambridge, Mass., 1977): 493–531.

ditionally, her imagination had been sparked by the pages of *National Geographic*, where she read about the Ama of Japan, women who dove without SCUBA gear in search of mollusks.[66] In *A Door into Ocean*, readers first meet the Sharers who live on an aquatic moon called Shora, through the eyes of another culture, the Valans who occupy the terrestrial planet Valedon, around which Shora orbits. Whereas the Valans are miners, stone shapers, capitalist, and militaristic, the Sharers are aquaculturalists, nonviolent, and all female. Regarded by the Valans as "pre-stone age," the Sharers are marked, too, by their lavender or deep violet skin color. Sharers live on floating rafts that sit atop the ocean, shaping their dwelling structures from sea silk, and diving to great depths to find and cultivate food. The purple color of the Sharers' skin derives from a symbiotic relationship with "breath microbes." In the novel, Slonczewski detailed how these breath microbes possessed a color-changing molecule shaped like a "ring of dots." And "when this molecule held oxygen, it turned purple, like a light switch."[67] When it gave up its oxygen, however, it turned white. Embedded in the skin of Sharers, these molecules released oxygen at times when their host was deprived of sufficient air, allowing them to dive for up to fifteen minutes on a single breath—when they surfaced, their skin was quite pale. Initial impressions thus paint a picture of two cultures: one white, technologically advanced colonizer trading small baubles for boatloads of sea silk; and one purple, living in symbiotic harmony with their world, and in danger of extermination.

As a teenager, Slonczewski had read Margaret Mead's *Coming of Age in Samoa*, and her books reflect Mead's commitment to fluid sexual identity, especially for women.[68] She has described her characters as "pansexual." (The transgressive erotics of mermaids held greater social relevance in the early modern period, although the subversive connotations of the aquatic have never fully dissipated.[69]) The gendered and racialized tropes that open *A Door into Ocean* slowly erode over the course of the novel. Valan visitors to Shora find themselves changing color; at first the hollows of their cheeks look a little lavender, and then even in bright sunlight their skin gains a vibrant violet hue as the breath microbes integrate themselves into their physiologies. Visitors could choose to remedy this situation with a phalanx of antibiotics. Other physical differences are far more ingrained in each population. With the centuries of their separation, Sharers possess inner eyelids that act as natural goggles when they dive. They also can no longer copulate with males, reproducing instead through the fusion of ova from two women bonded as partners. These processes, the novel makes clear, had been accelerated with the guidance of "life-shapers" and provide an early clue that the Shorans' lack of technological prowess might not be as straightforward as it first appears.

---

[66] Luis Marden, "Ama: Sea Nymphs of Japan," *National Geographic*, July 1971. A grandmotherly character in the book is even named Ama.

[67] Slonczewski, *A Door* (cit. n. 24), 142.

[68] Margaret Mead, *Coming of Age in Samoa*, foreword by Franz Boas (1928; repr., New York, N.Y., 1961). Mead's classic was reprinted in the early 1960s thanks to a burgeoning interest in "intellectual paperbacks." See Hayward Cirker, "The Scientific Paperback Revolution: A Traditional Medium Assumes a New Role in Science and Education," *Science* 140 (1963): 591–4; and Melinda Gormley, "Pulp Science: Education and Communication in The Paperback Book Revolution," *Endeavour* 40 (2016): 24–37.

[69] See Tara Pedersen, *Mermaids and the Production of Knowledge in Early Modern England* (Farnham, UK, 2015). On the legacy of mermaids, see Harriet Ritvo, *The Platypus and the Mermaid and Other Figments of the Classifying Imagination* (Cambridge, Mass., 1997).

In vitro fertilization (IVF) had become more fact than dystopian speculative fiction with the 1978 birth of Louise Brown in Oldham Hospital in the United Kingdom. American researchers at the time had been hard at work on a similar process, but Doris Del-Zio had found her hopes dashed in 1973, when Dr. Vande Wiele let the test tube with her eggs and her husband's sperm stand out on a hospital counter, effectively stopping cell division and the procedure. She claimed that Vande Wiele had killed her baby and sued him, the hospital, and Columbia University, where the work had taken place; he in turn insisted he had stopped a dangerous procedure and that the concoction would have harmed and possibly killed her. Louise Brown was born just one week after the American trial began. Her birth made international news, not only for the success of the process but also because, despite earlier fears that IVF babies would likely be born in some way malformed, Louise was completely healthy. (Meanwhile, the jury found that Dr. Vande Wiele had been at fault, but awarded the Del-Zios only a small fraction of the damages they had demanded.) By September of 1982, 124 IVF babies had been born around the world.[70] In the intervening years, the media hubbub over IVF had died down considerably. Slonczewski raised the political stakes by imagining a world in which the cold metal of the laboratory was replaced by the botanical warmth of the Shorans' life-shaping chambers, where they could craft an embryo through the fusion of eggs.

Narratively, Slonczewski interwove two parallel stories throughout the novel—a love story between a Valan and a Sharer, and a chronicle of the fates of their respective cultures. Spinel and Lystra learn to respect each other's strengths and weaknesses. Valans and Sharers come to a mutually beneficial agreement regarding the resources of Shora. To counter the Valans' weapons, the Sharers offer passive resistance. Some Sharers take their own lives rather than remain captives. When the Valans imprison Sharer children, they let them remain captive. Sharer children also join their elders on hunger strikes in protest of the Valans' military presence on their planet. They get to know the Valan soldiers, successfully breaking their resolve to fire on unarmed Sharers. As events become more heated, they even offer themselves for execution, stepping in front of a firing squad bent on shooting imprisoned Sharers, and thus requiring the already hesitant Valan soldiers to kill even more people in carrying out their orders. Equally crucial to the plot, a handful of "life-sharers" alter the breath microbes rendering them resistant to antibiotics. The inevitable spread of purple through the soldiers' bodies leaves them visibly marked by their experiences in Shora. If the purple color of their skin served as an analogy for race, Slonczewski's narrative demonstrated the biological arbitrariness of any such physical marker. This physical transformation also reinforces the Valans' increasing belief that the Sharers possess far more scientific skill than they initially had thought.

A telling passage late in the book between Siderite, a Valan scientist who has lived on Shora for years to understand their life-shaping skills, and Realgar, head of the Valan military forces, explains a bit more about the history of the moon. "What sort of people are likely to develop methods of confrontation which exclude violence?" Siderite poses to Realgar, who responds: "People who have no weapons." Siderite waves his response away: "The first tools man invented were knives and arrows.

---

[70] For a thorough account of these two families and the politics of IVF in the 1970s, see Robin Marantz Henig, *Pandora's Baby: How the First Test Tube Babies Sparked the Reproductive Revolution* (Cold Spring Harbor, N.Y., 2004).

Think again. Who were the Sharers?" Realgar later replies, "A people whose weapons are too deadly to be used."[71] In this exchange, Slonczewski flips reader expectations (although to be fair, there are plenty of hints earlier in the book). The Shorans had become peaceful *because* they had been scientifically advanced.[72] In the words of one Valan who had spent considerable time on Shora, the residents were "post-metal age." For Slonczewski, radical pacifism could succeed only in a society committed to nonviolent resistance, where members refused to share a fate with others if they were unwilling to accept that fate for themselves, too. Writing during the Cold War, she lamented that Americans considered it unpatriotic to talk of peace and thought citizens too easily dismissed pacifism as a "fairy tale." She especially disapproved of narratives in which white saviors (*Dune* again) arrived to save the planet, or indeed even the universe.[73] The Shorans' strategy of passive resistance ultimately found sympathetic ears among the Valans. By the end of *A Door into Ocean*, not only had the Sharers scuttled Valan control, but key members of the Valan military had internalized their strategic disobedience.

Although the first settlers to Shora had the capacity to wreak death upon unwelcome visitors, their modern sisters no longer remember how to construct such bioweapons. For Slonczewski, then, the real power of passive resistance comes not from mutual fear but from the Valans' realization that Shora and Valedon shared the same fate. In a confrontation between Merwen, a Sharer, and her captor Realgar, she tells him, "When you come to see that your survival is inseparable and indistinguishable from mine, then we both will win." This strategy works because the Valans begin to see their own position as under the control of a much larger military power that is similar to what they have forced upon the Sharers. Their mutual survival is very much connected after all.

Slonczewski uses the watery world of Shora to demonstrate the potential sympathetic connections between pacifism, feminism, and environmentalism. Unlike Vonnegut, Slonczewski approached her Shoran subjects with sincere optimism, conceptualizing her novel as an existence proof for the kind of world she saw around her. Nonviolent resistance did work. Feminism was powerful. Thoughtful, collaborative science could lead to breakthroughs in humanity's capacity to live in greater harmony with the environment.

CONCLUSION

Read together, Vonnegut's *Galápagos* and Slonczewski's *A Door into Ocean* resist any neat separation of science from worldly affairs by pushing against narratives of biological determinism that circulated in the science and the fiction of the 1980s. Vonnegut savagely revealed the social assumptions behind an adaptationist perspective on human progress—the most "fit" either biologically or socially were not always the in-

---

[71] The idea that there are weapons, from gunpowder to nuclear arms, so deadly they can stop war has a long history; Osgood, *Alternative* (cit. n. 43).

[72] Gene Sharp, *The Politics of Nonviolent Action*, editorial assistance of Marina Finkelstein, 3 vols. (Boston, Mass., 1973).

[73] Slonczewski disapproved, too, of Herbert's invocation of a deep species-level memory and ESP among the Bene Gesserit and the Fremen; see Joan Slonczewski and Michael Levy, "Science Fiction and the Life Sciences," in *The Cambridge Companion to Science Fiction*, ed. Edward James (Cambridge, UK, 2003), 174–85.

dividuals who survived and reproduced. The future, for him, depended far more on chance than those in positions of power were liable to admit. Slonczewski built her commitment to diversity and her scientific fascination with symbiosis into her novel. In using fiction as a means of communicating both, she hoped to convince her readers that cooperation played a powerful role in shaping the social and natural worlds we inhabit.

Charles Platt's horror at "New Wave" science fiction recognized this tendency in what he deemed the social science fiction of the Cold War. In hindsight, we can see more easily that even the science fiction authors he had valorized, from Frederik Pohl to Harlan Ellison, embedded politics into their scientific speculations about the future. He wrote of falling helplessly in love with their technologically plausible futures, without recognizing that their visions secured this future for only a sliver of the modern world.[74] For a different set of readers, Vonnegut and Slonczewski embraced "rigorous plausibility" as a feature of their narratives, but swapped the inevitability of technological progress for the contingent messiness of the life and social sciences.

For Ursula Le Guin, an author Platt singled out for particular scorn, the point of speculative fiction had been to lie convincingly in order to help readers see more clearly the present in which they lived. In her 1976 introduction to *Left Hand of Darkness* she proffered this vision of what science fiction could, indeed should, be: "Science fiction is not predictive." All authors can tell you, she wrote, "is what they have seen and heard, in their time in this world, a third of it spent in sleep and dreaming, another third of it spent in telling lies." Platt's objections to social science fiction and Le Guin's embrace of the malleability of truth drew on existing tensions within the sciences that exploded in the "science wars" of the 1990s.[75]

Reading Platt's concern in this context, we can imagine his distress (even if his target was misplaced) over the blending of science fictions and facts as a new generation of authors used their narrative skills to support and popularize nonmainstream scientific theories. Peter Dickinson took inspiration for *A Bone from a Dry Sea*, for example, from Elaine Morgan's theory of the aquatic ape.[76] In the opening pages, "the child" (one of two at the center of the story) slowly comes into focus—with a "snapped off hoot," she warns her tribe of an approaching shark, and points with a "web-fingered hand." Readers also come to understand that this girl differs from the other members of her tribe; she can strategize about the future in a way they cannot, and she is aware of the difference. Dickinson avoided causal explanations of why this young girl was so much smarter than her companions, but his decision reflected an

---

[74] Platt, "RAPE" (cit. n. 25), 45. See De Witt Douglas Kilgore, "On Mars and Other Heterotopias: A Conclusion," in *Astrofuturism: Science, Race, and Visions of Utopia in Space* (Philadelphia, Penn., 2003), 222–38.

[75] Consider the controversial reception of Donna Haraway's *Primate Visions* (cit. n. 3); for example, Peter S. Rodman, "Flawed Vision: Deconstruction of Primatology and Primatologists," *Curr. Anthropol.* 31 (1990): 484–6; and the Sokal affair, starting with Alan Sokal, "Transgressing the Boundaries: Towards a Transformative Hermeneutics of Quantum Gravity," *Social Text* 46–47 (1996): 217–52; Sokal, "A Physicist Experiments with Cultural Studies," *Lingua Franca* 6 (1996): 62–4; and Jennifer Ruark, "Bait and Switch," *Chronicle of Higher Education*, 1 January 2017. See also Ursula Le Guin and Margaret Atwood's delightful conversation about writing and science fiction in Portland, Oregon, 23 September 2010, 52 min., available online: Literary Arts, "Ursula Le Guin & Margaret Atwood," https://literary-arts.org/archive/ursula-le-guin-margaret-atwood/ (accessed 5 September 2018), which includes a brief dig at Desmond Morris's conceptualization of humans as naked apes.

[76] Peter Dickinson, *A Bone from a Dry Sea* (New York, N.Y., 1999).

embrace of quantum evolutionary change—one closer to Richard Goldschmidt's hopeful monsters than Stephen Jay Gould would have found comfortable.[77] In the second chapter, the narrative focus shifts to the present day, when another young girl is accompanying her father on a paleontological expedition in Africa. Vinny, like "the child" from the first chapter, excels at thinking through complex questions. Subsequent chapters oscillate between "THEN" and "NOW," iteratively telling the tale of these two girls whose fates are intertwined. (The omniscient narrator decides to call the prehistoric child "Li" for convenience at the expense of accuracy, for none of the members of her tribe used names, readers are told, as they were only "half-way towards words."[78])

In his young adult novel, Dickinson brought to life Morgan's aquatic ape at the same time that he championed the role females played in the success of the human species. Li, whose skin is a "very dark purply brown," invents a mesh with which she can more efficiently capture shrimp, learns to hunt fish collaboratively with dolphins, and devises a splint to help heal the broken leg of a tribe member. Vinny, for her part, seeks to convince the scientists on the dig that Elaine Morgan might have had a point. Most express deep skepticism, from her father's "she's not respectable" to another paleoanthropologist's "I think she's wrong, but not crazy wrong." In the end, Vinny (it will come as no surprise) plays a crucial role in uncovering and interpreting the bones of members of Li's tribe who had died millions of years earlier. In Dickinson's hands, at least, Elaine Morgan was vindicated.[79]

Today, new links are being built between oceans and climate change, in which displaced water serves as one of the primary mechanisms by which human lives are uprooted and forever altered.[80] In *The Sea Around Us*, Rachel Carson had already noticed the ocean's role as a regulator of global temperatures. "Day by day and season by season," she wrote, "the ocean dominates the world's climate." Oceans had necessarily played a significant role in past climatic shifts. In fact, she reasoned, they were involved again. Carson warned, "Now in our own lifetime we are witnessing a startling alteration of climate." The Arctic and sub-Arctic regions were warming and changing the habits and migration patterns of the nonhuman world. "The long trend," she cautioned, "is to a warmer earth."[81] These speculative futures warn once again that human

[77] Michael Dietrich, "Reinventing Richard Goldschmidt: Reputation, Memory, and Biography," *J. Hist. Biol.* 44 (2011): 693–712. See also Greg Bear's *Darwin's Radio* (New York, N.Y., 1999) for his vision of what quantum evolution might look like in the future, where the trigger for an event is a virus; the story is part Richard Preston's *Hot Zone* (New York, N.Y., 1994) and part evolutionary drama.

[78] Dickinson, *Bone* (cit. n. 76), 24.

[79] Morgan's feminist theory also makes a cameo appearance in Naomi Alderman's *The Power* (New York, N.Y., 2016), when an anthropologist suggests that girls' new ability to produce electricity from specialized organs by their collar bones "is proof positive of the aquatic ape hypothesis—that we are naked because we came from the oceans, not the jungle, where once we terrified the deeps like the electric eel, the electric ray" (22).

[80] Brian Fagan, *Attacking Ocean: The Past, Present, and Future of Rising Sea Levels* (London, 2013); Amitav Ghosh, *The Great Derangement: Climate Change and the Unthinkable* (Chicago, Ill., 2016); Elizabeth DeLoughrey, "Submarine Futures of the Anthropocene," *Comparative Literature* 69 (2017): 32–44; Ursula Heise, *Imagining Extinction: The Cultural Meanings of Endangered Species* (Chicago, Ill., 2016). For fictional dramatizations, see Octavia Butler, *Parable of the Sower* (New York, N.Y., 1993); *Waterworld*, directed by Kevin Reynolds (1995; Universal City, Calif.: Universal Pictures Home Entertainment, 2016), DVD. More classically, see J. G. Ballard, *The Drowned World* (New York, N.Y., 1962); Ballard, *The Burning World* (New York, N.Y., 1964); and *Mad Max*, directed by George Miller (1979; Beverly Hills, Calif.: Twentieth Century Fox Home Entertainment, 2016), DVD.

[81] Carson, *The Sea* (cit. n. 15), 201, 208, 213.

survival depends on our capacity to conceptualize ourselves as part of the natural world rather than independent of it.

Even as the politics of earth and air appeared increasingly constrained, the generative openness of the water remained. Authors like Vonnegut, Slonczewski, and even Dickinson made use of this conceptual space to imagine alternatives to standard scientific narratives of our past and future. These novels suggested that humans could learn and could evolve—the future was not fixed, it was ours for the making.

Ahead of Time:

Gerald Feinberg and the Governance of Futurity

by Colin Milburn*

ABSTRACT

Looking back at his research on tachyons in the 1960s and 1970s, the physicist Gerald Feinberg recalled that he started thinking about particles that go faster than light after reading James Blish's 1954 science fiction story "Beep." While the technical conceits of Blish's tale may have stirred Feinberg's curiosity, its literary implications were yet more significant. As a story about faster-than-light messages that travel backward in time, "Beep" thematizes the capacity of speculative fictions to affect the present and reorient the future. For Feinberg, stories like "Beep" and Arthur C. Clarke's 1953 novel *Childhood's End* offered conceptual resources as well as models for practice, affirming a science fiction way of doing science. By attending to Feinberg's work on tachyons as well as his ventures in futurology, such as *The Prometheus Project*, this essay shows how Feinberg's reading of science fiction reinforced a speculative approach to knowledge and innovation, an understanding of theoretical science as intimately aligned with science fiction, and a conviction that science fiction was a vital instrument for science policy and social change.

In 1967, the physicist Gerald Feinberg published a quantum field analysis of a purely hypothetical class of faster-than-light particles that he dubbed "tachyons." This was a rather unorthodox research project, to say the least. While the possibility of objects moving at superluminal velocities had been considered since the late nineteenth century, the impact of Einstein's 1905 special theory of relativity seemed to put further study of faster-than-light speed into the realm of pure science fiction. According to special relativity, it would require infinite energy to accelerate an object with real mass up to the speed of light. In other words, as Einstein put it, "superluminary velocities have no possibility of existence."[1] Even were it possible to transmit signals faster than light, such transmissions could appear to some observers as having trav-

---

* Department of English, 273 Voorhies Hall, One Shields Ave., University of California, Davis, CA 95616, USA; cnmilburn@ucdavis.edu.

My sincere thanks to Amanda Rees, Erika Milam, Joanna Radin, Michael Gordin, W. Patrick McCray, Projit Mukharji, Victoria de Zwaan, John Fekete, Veronica Hollinger, and the anonymous reviewers for their comments and suggestions. I benefitted tremendously from participating in the "Histories of the Future" workshop at Princeton University in February 2015, organized by Erika Milam and Joanna Radin.

[1] Albert Einstein, "On the Electrodynamics of Moving Bodies," in *The Collected Papers of Albert Einstein, Volume 2: The Swiss Years: Writings, 1900–1909*, ed. John Stachel, David C. Cassidy, Jürgen Renn, and Robert Schulmann, English translation supplement, trans. Anna Beck (Princeton, N.J., 1989), 140–71, on 170.

eled backward in time, introducing the uncanny possibility of information from the future influencing past events. Not only would this violate the principles of physical causality, it could also lead to certain paradoxes, for example, if messages were to arrive before they were sent and preempt the sender from sending them. Special relativity therefore seemed to squash any fantasies of faster-than-light travel and backward communications from the future—at least in the zones of serious science.[2]

But in 1962, the physicists Oleksa-Myron Bilaniuk, V. K. Deshpande, and George Sudarshen postulated the existence of subatomic particles that always travel at superluminal velocities. Such "meta" particles would never confront the speed of light as an upper barrier because they would be already beyond it. Moreover, the problem of temporal paradoxes could be solved by relativistic reinterpretation, insofar as observers in some frames of reference would perceive such a particle "not as a weird negative energy particle traveling backward in time, but as a positive energy particle traveling forward in time, but going in the opposite direction."[3] Likewise, any apparent violations of causality could be reinterpreted as causal events, as a matter of perspective.

Feinberg, who had joined the faculty at Columbia University in 1959, was fascinated by this possibility. From the late 1940s onward, the field of high energy physics had witnessed the number of elementary particles grow by leaps and bounds—an explosion of the so-called particle zoo. Thanks to cosmic ray research and, especially, the development of increasingly powerful accelerators and detectors, more than one hundred subatomic species had been discovered by the mid-1960s.[4] Feinberg himself was already a contributor to the particle zoo, having predicted the muon neutrino in 1958. (Its existence was demonstrated in 1962 by his colleagues Leon Lederman, Melvin Schwartz, and Jack Steinberger, who later shared the Nobel Prize for their work.) He was confident that even more exotic beasts would yet emerge from the depths of the subatomic universe. But he also foresaw that, more than simply adding another strange critter to an expanding menagerie, adventurous research on bits of matter that go faster than light could potentially upend modern physics—whether these things existed or not.

So, encouraged by his previous success with the muon neutrino, Feinberg took up the question of faster-than-light particles with a bold territorial move: he named them. Rendering them as objects of relativistic quantum field theory, figuring them as excitations of a quantum field with imaginary mass, Feinberg also conjured them as objects of discourse: "One description is presented . . . for noninteracting faster than light particles, which we call *tachyons*."[5] Shortly afterward, in a *Scientific American*

---

[2] On the history of faster-than-light communication theories and time-travel scenarios, see Paul J. Nahin, *Time Machines: Time Travel in Physics, Metaphysics, and Science Fiction*, 2nd ed. (New York, N.Y., 1999).

[3] O. M. P. Bilaniuk, V. K. Deshpande, and E. C. G. Sudarshan, "'Meta' Relativity," *Amer. J. Phys.* 30 (1962): 718–23, on 719. For reflections on this foundational work, see Oleksa-Myron Bilaniuk, "Tachyons," *Journal of Physics: Conference Series* 196 (2009): 012021.

[4] On the history and practices of particle physics, see Peter Galison, *Image and Logic: A Material Culture of Microphysics* (Chicago, Ill., 1997); David Kaiser, *Drawing Theories Apart: The Dispersion of Feynman Diagrams in Postwar Physics* (Chicago, Ill., 2005); Andrew Pickering, *Constructing Quarks: A Sociological History of Particle Physics* (Chicago, Ill., 1984); and Sharon Traweek, *Beamtimes and Lifetimes: The World of High Energy Physicists* (Cambridge, Mass., 1988).

[5] Gerald Feinberg, "Possibility of Faster-Than-Light Particles," *Physical Review* 159 (1967): 1089–1105, on 1090.

article introducing tachyons to a wider audience, Feinberg highlighted the strategic nature of this maneuver: "In anticipation of the possible discovery of faster-than-light particles, I named them tachyons, from the Greek word *tachys*, meaning swift."[6] Feinberg's philological and theoretical production of these speculative objects ahead of time, I venture, represents a science fiction way of doing science, a way of staking claims on the future.

After all, Feinberg was known as much for his intense love of science fiction as for his prodigious talent as a physicist. When he passed away in 1992, the obituaries described him as a "devoted scientist and avid science-fiction fan."[7] And sure enough, his work on tachyons was to some degree indebted to his reading of science fiction. As the physicist and science fiction writer Gregory Benford recalled, "He told me years later that he had begun thinking about tachyons because he was inspired by James Blish's short story 'Beep.' In it, a faster-than-light communicator plays a crucial role in a future society . . . The communicator necessarily allows sending of signals backward in time . . . Feinberg had set out to see if such a gadget was theoretically possible."[8]

While Blish's 1954 story no doubt piqued Feinberg's curiosity, I want to suggest that its value for Feinberg's own work was less in its technical provocations than its literary ones. For "Beep" affords an understanding of theoretical science as equivalent to science fiction: not a purveyor of mere flights of fancy, but a generator of *consequential fictions* with the capacity to inform the circumstances of their own actualization. In other words, they are fictions that contribute to the conditions of possibility, the enabling contexts for discovery and innovation. From Feinberg's perspective, his reading of "Beep" reinforced a forward-looking, conjectural approach to the production of scientific knowledge. It confirmed his sense of the motivating power of fictive narratives, the robust forms of deliberation and anticipation made available by engaging in pretense. Moreover, it indicated that science fiction could become a model for science policy, shaping the trajectory of high-tech society. For Feinberg, then, it validated a lifetime of thinking about imaginary futures, because it suggested how science fiction could change the world.

READING THE FUTURE

Blish's text unfolds as a story within a story. The framing narrative focuses on the governmental organization known as the Service, which is responsible for overseeing the smooth advancement of galactic civilization. This massive interplanetary operation depends on the technology of the Dirac communicator. Any message sent through one Dirac communicator is received instantaneously by all others, preceded by a small

---

[6] Gerald Feinberg, "Particles That Go Faster Than Light," *Sci. Amer.* 223 (February 1970): 69–77, on 70.

[7] Leyla Kokmen, "Prof. Dies after Fight with Cancer," *Columbia Spectator*, 23 April 1992, 1.

[8] Gregory Benford, "Old Legends," in *New Legends*, ed. Greg Bear and Martin H. Greenberg (New York, N.Y., 1995), 270–84, on 276. See also Benford, "Time and *Timescape*," *Sci. Fict. Stud.* 20 (1993): 184–90. Linking his interests in superluminal matters to Blish's 1954 story, Feinberg implied that his research antedated Bilaniuk, Deshpande, and Sudarshan, "'Meta' Relativity" (cit. n. 3); as well as Sho Tanaka, "Theory of Matter with Super Light Velocity," *Progress of Theoretical Physics* 24 (1960): 171–200. Feinberg's 1967 *Physical Review* paper, "Possibility" (cit. n. 5), cites these earlier publications in its first footnote, along with another predecessor: "G. Feinberg (unpublished)."

beep of sound. In the nested story, which concerns the origins of the communicator, the video commentator Dana Lje discovers that the "beep is the simultaneous reception of *every one of the Dirac messages which have ever been sent, or ever will be sent.*"[9]

The narrative symbolically aligns Lje's discovery with lies and illusions—the fictive as such. Her disclosure of the beep's meaning depends on an elaborate charade where she poses as a man named J. Shelby Stevens. On the other hand, Robin Weinbaum, the director of the Service, is fully aligned with the regime of truth. He represents a government that does not tolerate fiction: "Just in case you're not aware of the fact, there are certain laws relating to giving false information to a security officer . . . plus various local laws against transvestism, pseudonymity and so on" (30). Weinbaum demands transparency. He likewise insists on stripping any obscuring "noise" from otherwise pure signals. He instructs his assistant to excise the beep: "Margaret, next time you send any Dirac tapes in here, cut that damnable *beep* off them first" (23). Weinbaum, agent of truth and law, disregards this literally preposterous noise. Its meaning only comes to light through Lje's fictive performance.

Lje's process of extracting discrete transmissions from the beep is an interpretive practice described as reading: "I can read the future in detail" (31). It apprehends the faster-than-light messages as *texts,* many of which are strange, mystifying, and unnerving: "Once you know, however, that when you use the Dirac you're dealing with time, you can coax some very strange things out of the instrument" (39). Yet these "strange things" are recognized as consistent with reality, explicable by scientific logic. Which is to say, they have the qualities of science fiction:

> She paused and smiled. "I have heard," she said conversationally, "the voice of the President of our Galaxy, in 3480, announcing the federation of the Milky Way and the Magellanic Clouds. I've heard the commander of a world-line cruiser, traveling from 8873 to 8704 along the world line of the planet Hathshepa, which circles a star on the rim of NGC 4725, calling for help across eleven million light-years—but what kind of help he was calling for, or will be calling for, is beyond my comprehension. And many other things. When you check on me, you'll hear these things too—and you'll wonder what many of them mean." (39)

These vignettes draw from a common repertoire of science fiction tropes—enormous space ships, intergalactic civilizations, time travel—and they bring forth a sense of wonder and wonderment: "Weinbaum, already feeling a little dizzy . . . wanted only scenes and voices, more and more scenes and voices from the future. They were better than aquavit" (42). Alluring and bewildering, they evade full rational comprehension even while indicating that such comprehension is possible—some day: "You'll know the future, but not what most of it means. The farther into the future you travel with the machine, the more incomprehensible the messages become, and so you're reduced to telling yourself that time will, after all, have to pass by at its own pace, before enough of the surrounding events can emerge to make those remote messages clear" (39).

---

[9] James Blish, "Beep," *Galaxy Science Fiction*, February 1954, 6–54, on 36. Further page citations appear parenthetically in the text (emphasis in the original).

The effect of these weird messages from the future is instead a new perception of the present, namely, as the historical context for incredible changes to come.[10] Viewing a beep transmission of a "green-skinned face of something that looked like an animated traffic signal with a helmet on it," Weinbaum exclaims, "And we'll be using non-humanoids there! What was that creature, anyhow?" (42). Weinbaum sees the alien as his own future, albeit radically estranged. This surprising future puts the present in new light, exposing its humanist biases ("looked like an animated traffic signal"), even while suggesting the possibility for things to be otherwise.

The beep is therefore a figure for science fiction as such. While Lje's "method of, as she calls it, reading the future" (35) understands the beep messages to represent inevitable events—evidence for a perfectly acausal universe—she nevertheless interprets them as self-fulfilling prophesies: "Since I was going to be married to you [Weinbaum] and couldn't get out of it, I set out to convince myself that I loved you. Now I do. . . . But I had no such motives at the beginning. Actually, there are never motives behind actions. All actions are fixed. What we called motives evidently are rationalizations by the helpless observing consciousness, which is intelligent enough to smell an event coming—and, since it cannot avert the event, instead cooks up reasons for wanting it to happen" (32–33). Taking the beep messages seriously, Lje "cooks up reasons" to ensure the events will take place. She invents fictions of causality ("rationalizations") that become indistinguishable from actual causes.

It is a method of reading the future that the Service then implements as *policy*. In the framing narrative, hundreds of years later, Serviceman Krasna says,

> Our interests as a government depend upon the future. We operate *as if* the future is as real as the past, and so far we haven't been disappointed: the Service is 100% successful. But that very success isn't without its warnings. What would happen if we *stopped* supervising events? We don't know, and we don't dare take the chance. Despite the evidence that the future is fixed, we have to take on the role of the caretaker of inevitability. We believe that nothing can possibly go wrong . . . but we have to act on the philosophy that history helps only those who help themselves. . . . Our obligation as Event Police is to make the events of the future possible, because those events are crucial to our society—even the smallest of them. (43, emphasis in the original)

This "as if" approach does not draw a hard line between fact and theory, truth and lies, or science and science fiction.[11] The Service accepts the beep messages as if they were reliable prophesies, but treats them in practice as speculative forecasts that might fail without a vast technoscientific infrastructure designed to "make the events of the future possible." In its role as "Event Police," the Service assembles numerous technical resources: "We have some foreknowledge, of course. . . . But we have ob-

---

[10] Fredric Jameson argues that science fiction's "mock futures serve the quite different function of transforming our own present into the determinate past of something yet to come. . . . SF thus enacts and enables a structurally unique 'method' for apprehending the present as history"; see Fredric Jameson, *Archaeologies of the Future: The Desire Called Utopia and Other Science Fictions* (New York, N.Y., 2005), 288. This function is one aspect of cognitive estrangement, described by Darko Suvin, *Metamorphoses of Science Fiction: On the Poetics and History of a Literary Genre* (New Haven, Conn., 1979).

[11] Michael Saler, *As If: Modern Enchantment and the Literary Prehistory of Virtual Reality* (Oxford, UK, 2012). See also Hans Vaihinger, *The Philosophy of "As If": A System of the Theoretical, Practical and Religious Fictions of Mankind*, trans. C. K. Ogden (New York, N.Y., 1925).

vious other advantages: genetics, for instance, and operations research, the theory of games, the Dirac transmitter—it's quite an arsenal, and of course there's a good deal of prediction involved in all those things" (11). Even with knowledge of things to come, the Service marshals the tools of various scientific disciplines to turn speculations and "as if" scenarios into lived events.

All of this represents an explicitly *constructivist* approach to truth, a model for the generation of scientific knowledge that does not excise science fiction (as Weinbaum once excised the beep), but rather upholds its crucial role for the advancement of science and society.

VERGING ON FICTION

To the degree that "Beep" represents a science fiction way of governing the future, its appeal to a "devoted scientist and avid science-fiction fan" like Feinberg seems obvious, in retrospect. The Service's decision to reinterpret causally uncertain events in a causal way, sidestepping potential paradoxes, is tantalizingly similar to the principle of reinterpretation developed in Feinberg's own work. According to Feinberg, actual instrumentation for detecting the absorption of tachyons from the future would make distinctions between emission and absorption, sending and receiving, undecidable. Inclined to causal explanations, an observer would therefore "naturally describe" the tachyon detector as if it were spontaneously sending signals forward rather than receiving them from the future.[12]

For the same reason, Feinberg suggested that a device such as the Dirac communicator, if taken literally, would not be possible.[13] The relevance of "Beep" for thinking about tachyons, however, is more figural. For the story presents a model of scientific practice and technological governance that treats physical theories and interpretations of data as fictions, but not "mere" fictions; rather, the Dirac messages become the conditions for further experimentation, triggering additional research and decisive actions. They are science fictions that enable their own materialization in the form of consequential practices.

Feinberg offered a similar perspective in his writings on tachyons: "Having convinced ourselves that the existence of faster-than-light particles does not imply any contradiction of relativity, we must nevertheless leave the determination of whether such objects really happen in nature to the experimental physicist."[14] Feinberg here presents tachyons as consequential fictions that galvanize experiments—no more,

---

[12] Feinberg writes, "Therefore, while it does appear possible to construct kinematic closed cycles using tachyons in which signals are sent back to the past, a careful examine of the methods of detection, with due regard to the interpretation of absorption of negative-energy tachyons as emission of positive-energy tachyons, leads to the conclusion that such closed cycles will not be interpreted as reciprocal signaling, but rather as uncorrelated spontaneous emission"; Feinberg, "Possibility" (cit. n. 5), 1103. Though such reinterpretation was already suggested by Bilaniuk, Deshpande, and Sudarshen, it is now often called the "Feinberg reinterpretation principle." Notably, it echoes the solution proposed by Ernst Stueckelberg and Richard Feynman to account for the negative energy of antimatter particles in quantum field theory; see R. P. Feynman, "The Theory of Positrons," *Physical Review* 76 (1949): 749–59.

[13] Feinberg writes, "A conclusion warranted by this argument is that tachyons cannot be used to send reliable signals, either forward or backward in time, in the sense that one cannot completely control the outcome of an experiment to produce or absorb them"; see Feinberg, "Possibility" (cit. n. 5), 1092.

[14] Feinberg, "Particles" (cit. n. 6), 72.

no less. Certainly, they have provoked numerous studies and experiments over the last half century. Even failure to prove their existence has led to new theoretical interpretations. For example, we could point to the "tachyonic field" concept in quantum field theory, taken to mathematically indicate field instabilities rather than real particles. The appearance of tachyons—and efforts to resolve them—have likewise been critical in the history and development of string theory.[15] We could also point to the abiding hope for faster-than-light signals—a hope projected always into the future—as suggested by recent research on quantum tunneling, or more tellingly, by the premature 2011 declaration by scientists at CERN, the European Organization for Nuclear Research in Geneva, Switzerland, that they had discovered neutrinos traveling faster than light (a claim later disproved).[16]

Yet even in generating such experimental adventures, tachyons have not ceased to be understood as science-fictional objects. Already in 1970, Feinberg conceded that experimental results seemed to suggest that tachyons are not actually real, but to him this only further demonstrated the value of such entities. After all, it may be likely that "tachyons simply do not exist . . . [but] we may not understand why it should be so until we reach a much deeper understanding of the nature of elementary particles than now exists."[17] Which is to say, as nothing other than science fictions, they indicate the degree to which extant physical theories are themselves provisional, constructed, and prone to change in the future.

In this regard, Feinberg aligned with other physicists who emphasized the science-generating capacities of theoretical entities that, even if they might not actually exist, could nevertheless defamiliarize inherited models and presage an altered way of seeing. Compare, for example, Murray Gell-Mann's proposition in 1964 that hadrons are not elementary particles but are instead built up from combinations of things he called "quarks." For Gell-Mann, the ontological status of quarks was indeterminate, but fair game for speculation. As his first paper on quarks famously concludes, "It is fun to speculate about the way quarks would behave if they were physical particles of finite mass (instead of purely mathematical entities as they would be in the limit of infinite mass)."[18] Like Feinberg's writing on tachyons, Gell-Mann's quark paper underscores the "as if," the subjunctive mood ("if they were physical particles"). It extrapolates the implications of these prospective entities into a speculative chronology, a "what if?" narrative that looks backward as well as forward. It imagines a previously undisclosed

---

[15] Dean Rickles, *A Brief History of String Theory: From Dual Models to M-Theory* (Heidelberg, Ger., 2014).

[16] Günter Nimtz, "Tunneling Confronts Special Relativity," *Found. Phys.* 41 (2011): 1193–9; E. Kapuścik and R. Orlicki, "Did Günter Nimtz Discover Tachyons?" *Ann. Physik* 523 (2011): 235–8; T. Adam et al. (OPERA collaboration), "Measurement of the Neutrino Velocity with the OPERA Detector in the CNGS Beam," preprint, submitted 17 November 2011, *arXiv:1109.4897v1*. For an account of what happened in the OPERA experiment, see Ransom Stephens, "The Data That Threatened to Break Physics," *Nautilus* 24 (2015), http://nautil.us/issue/24/Error/the-data-that-threatened-to -break-physics.

[17] Feinberg, "Particles" (cit. n. 6), 77. Bilaniuk and Sudarshan had made the same point: "We find the question of existence of the hypothetical superluminal particles to be a challenging new frontier. Regardless of the outcome of the search for tachyons, investigations in this field must invariably lead to a deeper understanding of physics. If tachyons exist, they ought to be found. If they do not exist, we ought to be able to say why not"; see Olexa-Myron Bilaniuk and E. C. George Sudarshan, "Particles Beyond the Light Barrier," *Phys. Today* 22 (1969): 43–51, on 51.

[18] M. Gell-Mann, "A Schematic Model of Baryons and Mesons," *Physics Letters* 8 (1964): 214–5, on 215.

past: "Ordinary matter near the earth's surface would be contaminated by stable quarks as a result of high energy cosmic ray events throughout the earth's history, but the contamination is estimated to be so small that it would never have been detected." It then dashes ahead, venturing upon experimental trials to come: "A search for stable quarks . . . at the highest energy accelerators would help to reassure us of the non-existence of real quarks." The conditional futurity ("would help to reassure us") allows Gell-Mann to hedge his bets, demurring on the reality of "real quarks" while remaining open to whatever might serendipitously emerge from cosmic rays or particle accelerators; the provisory formulations invite further research. As late as 1969, Gell-Mann observed, "The quark is just a notion so far. . . . It is a useful notion, but actual quarks may not exist at all."[19]

Yet even if they were to prove "only fictitious," Gell-Mann indicated it would still be feasible to speak of them in "as if" terms, for example, in the case where "hadrons act *as if* they were made up of quarks but no quarks exist."[20] Quarks were objects of theoretical conjecture, signifiers of possibility. Of course, quarks would eventually become real—affirmed retroactively by experimental data, even as the theoretical framing continued to be debated and refined.[21] The same can be said for many other wild hypotheses, prototypes, forward-looking statements, and promissory visions in the history of science and technology. As tools of the imagination, fictions to think with, they solicit experiments and other technical responses, stirring further speculations in their wake. They instantiate a way of doing science that often verges on the domain of science fiction, concerned with the articulation of cognitively estranging concepts and extrapolating the conditions under which things that seem bizarre or impossible in the present could be taken for granted in the future. It is an approach that embraces the "as if" and the exercise of pretense, serious and playful at the same time. According to Gell-Mann, who had been a reader of science fiction since his youth, it even provides a similar recreational pleasure. As he noted, "it is fun to speculate."[22]

To be sure, the discourse of science fiction regularly affirms the pleasures and playfulness of scientific speculation. In February 1967, five months before the publication of Feinberg's first tachyon paper, the science fiction writer and biochemistry professor Isaac Asimov devoted his regular science column in *The Magazine of Fantasy and Science Fiction* to the question of superluminal velocities. Irritated by an episode of the television show *It's About Time*, a 1966–1967 sitcom about time-traveling as-

---

[19] Gell-Mann quoted in "Gell-Mann of Caltech Awarded Nobel Prize," *Los Angeles Times*, 31 October 1969, 29.

[20] Murray Gell-Mann, "Quarks," *Acta Physica Austriaca, Suppl.* 9 (1972): 733–61, on 746 (emphasis mine).

[21] On the shifting reality of quarks, see Pickering, *Constructing Quarks* (cit. n. 4); and George Johnson, *Strange Beauty: Murray Gell-Mann and the Revolution in Twentieth-Century Physics* (New York, N.Y., 1999). In 1964, the physicist George Zweig independently developed an equivalent theory of hadrons as made of more fundamental particles ("aces"), but he held a stronger view about the concrete reality of such particles than Gell-Mann initially did; see G. Zweig, "Memories of Murray and the Quark Model," *International Journal of Modern Physics A* 25 (2010): 3863–77.

[22] Gell-Mann has often indicated his interest in science fiction and its concepts; see Murray Gell-Mann, "Opening Remarks to the Session on New Concepts," *Nuclear Instruments and Methods in Physics Research* A271 (1988): 165–6; and Gell-Mann, *The Quark and the Jaguar: Adventures in the Simple and the Complex* (New York, N.Y., 1994), 19, 21, 113, 165, 370. On speculative play and scientific make-believe, see Colin Milburn, *Mondo Nano: Fun and Games in the World of Digital Matter* (Durham, N.C., 2015). On fictionalism in science more generally, see Mauricio Suárez, ed., *Fictions in Science: Philosophical Essays on Modeling and Idealization* (New York, N.Y., 2009).

tronauts, Asimov diligently explained why anything moving faster than light was preposterous, forbidden by special relativity: "Impossible, That's All!"[23] A year later, the science fiction writer and futurist Arthur C. Clarke published a good-humored rejoinder in the same magazine titled "Possible, That's All!" Clarke poked fun at Asimov for being out of touch with current research:

> I am indebted, if that is the word, to Dr. Gerald Feinberg of Columbia University for this idea. His paper "On the Possibility of Superphotic Speed Particles" (privately printed; posted in plain, sealed envelopes) points out that since sudden jumps from one state to another are characteristic of quantum systems, it might be possible to hop over the "light barrier" without going through it. . . . Even if there is no way through the light-barrier, Dr. Feinberg suggests that there may be another universe on the other side of it, composed entirely of particles that cannot travel slower than the speed of light.[24]

An early recipient of Feinberg's manuscript, Clarke located himself on the cutting edge, ahead of the curve, while suggesting that Asimov had slipped behind. Clarke even jested about temporal reversal, pretending to be much younger than Asimov now, despite actually being three years older: "The galactic novels of my esteemed friend Dr. Asimov gave me such pleasure in boyhood that it is with great reluctance that I rise up to challenge some of his recent statements."[25] Caught off guard, Asimov wrote two responses, conceding the theoretical interest of tachyons but doubling down on the Einsteinian light barrier, the "luxon wall," as a fundamental limit separating our real-mass universe from an imaginary-mass universe.[26] He also addressed the history of thinking about superluminal particles: "This was done for the first time with strict adherence to relativistic principles (as opposed to mere science-fictional speculation) by Bilaniuk, Deshpande, and Sudarshan, in 1962, and such work hit the headlines at last when Gerald Feinberg published a similar discussion in 1967."[27] Asimov allowed that, while tachyons were probably imaginary in more than one sense, their abidance by the rules of modern physics would distinguish them from earlier examples of "mere science-fictional speculation." His comment nevertheless suggested how much theoretical science actually needs science fiction, if only for the sake of contrast. Certainly, in February 1969, *Time* magazine featured the dispute between Asimov and Clarke as a framing anecdote for its profile of Feinberg: "For Columbia University Physicist Gerald Feinberg, the monthly magazine *Fantasy and Science Fiction* is as compelling as any learned scientific journal. It has printed a continuing debate between authors Isaac Asimov and Arthur C. Clarke over the existence of a particle that travels faster than light. . . . Feinberg's fascination is understandable. The particle is his conception, although he is still not certain that it really exists."[28]

---

[23] Isaac Asimov, "Impossible, That's All!" *Magazine of Fantasy and Science Fiction*, February 1967, 113–23.

[24] Arthur C. Clarke, "Possible, That's All!" *Magazine of Fantasy and Science Fiction*, October 1968, 63–8, on 64.

[25] Ibid., 63.

[26] Isaac Asimov, response to Clarke, "Possible, That's All!" *Magazine of Fantasy and Science Fiction*, October 1968, 68–9; Isaac Asimov, "The Luxon Wall," *Magazine of Fantasy and Science Fiction*, December 1969, 96–105.

[27] Asimov, "Luxon Wall" (cit. n. 26), 103.

[28] "Exceeding the Speed Limit," *Time*, 14 February 1969, 50. On the prominence of Asimov, Clarke, and other fiction writers in futurological discourse, see Peter J. Bowler, *A History of the Future: Prophets of Progress from H. G. Wells to Isaac Asimov* (Cambridge, UK, 2017).

Situated against a background of science fiction, the work of theoretical physics seemed a more intense form of the same kind of fabulation, a more technical game of make-believe.

Meanwhile, tachyons triggered extended discussions among physicists—and the research publications often invoked fictive narratives and motifs. For example, the physicist Lawrence Schulman argued that the time-travel stories of Robert A. Heinlein, Michael Moorcock, Brian Aldiss, and Robert Silverberg were helpful resources for theorizing about tachyonic physics and reverse causality: "It will be seen that the results of this discussion will be in general accord with a consensus of science fiction writers who have dealt with this theme."[29] The physicist Frank Jones noted that Schulman's analysis exemplified the tendency of tachyons to elicit a certain genre of images: "The sort of images that are conjured up by this possibility is illustrated by the fact that in one recent discussion in the literature almost half of the references cited were to science fiction stories."[30]

Tachyons also proliferated in science fiction itself, entering the repertoire of standard genre tropes. Remarkably, the first novel to include tachyons as a plot device was written by James Blish. Since 1967, Blish had been adapting the *Star Trek* television series into books of short stories—a contract job that secured his financial livelihood in those years, because the *Star Trek* volumes outsold his other books by a fair margin.[31] As part of this arrangement, in 1970, he published *Spock Must Die!*, a full-length *Star Trek* novel. While it featured characters from the television series, *Spock Must Die!* was Blish's creation: an allegory about the dilemmas of originality that, like "Beep," depicts a future built from other speculative fictions. In the novel, the chief engineer of the starship *Enterprise*, Scotty, decides to redesign the ship's transporter system to utilize tachyons: "Tachyons *canna* travel any *slower* than light, and what their top speed might be has nae been determined."[32] Instead of physically beaming crew members to distant locations, the new transporter sends tachyon copies, zipping them through Hilbert space, while the original people remain safely aboard the *Enterprise*. During the first test, however, something goes wrong; the transporter creates a physical duplicate of Mr. Spock.

The two Spocks appear identical. Yet one Spock is actually the mirror image of the other, with reversed chirality all the way down to his elementary particles. Scotty discerns that the transporter's tachyon beam hit a "deflector screen" around the planet Organia, which inverted the signal. Whereas the original Spock is loyal to the *Enterprise*, the duplicate Spock is secretly an enemy, working to sabotage the mission.

Eventually, the two Spocks face off in a psychic battle behind the tachyon-deflecting screen of Organia, each trying to destroy the other through the force of imagination alone: a "combat of dreams" (89). During this climactic duel, Captain Kirk is caught

---

[29] L. S. Schulman, "Tachyon Paradoxes," *Amer. J. Phys.* 39 (1971): 481–4, on 481. The first peer-reviewed response to Feinberg was Roger G. Newton, "Causality Effects of Particles That Travel Faster Than Light," *Physical Review* 162 (1967): 1274. Newton argued that tachyons need not be fatally paradoxical; if we run an "experiment in which an effect would precede its cause," we simply need to acknowledge a weird universe that "would, in principle, make precognition experiments possible."

[30] Frank C. Jones, "Lorentz-Invariant Formulation of Cherenkov Radiation by Tachyons," *Physical Review D: Particles and Fields* 6 (1972): 2727–35, on 2727.

[31] David Ketterer, *Imprisoned in a Tesseract: The Life and Work of James Blish* (Kent, Ohio, 1987), 249–50, 358.

[32] James Blish, *Spock Must Die!* (New York, N.Y., 1970), 13. Further page citations appear parenthetically in the text.

in a torrent of dream stuff, a flood choked with creatures from science fiction and fantasy lore. He sees "the bedraggled bodies of a wide variety of small animals from a dozen planets—rabbits, chickens, skopolamanders, tribbles, unipeds, gormenghast-lies, ores, tnucipen, beademungen, escallopolyps, wogs, reepicheeps, a veritable zoo of drowned corpses, including a gradually increasing number of things so obscene that even Kirk, for all his experience in exoteratology, could not bear to look at them" (92). It is a cascade of allusions. The tribbles from *Star Trek* mix with unipeds from Stanley G. Weinbaum's "Parasite Planet" (1935); the gormenghastlies nod to Mervyn Peake's *Gormenghast* series (1946–1949); the tnucipen reference the tnuctipun of Larry Niven's *World of Ptavvs* (1966); the reepicheeps point to C. S. Lewis's *Chronicles of Narnia* (1950–1956); the wogs, laden with racist associations, refer to the lexicon of L. Ron Hubbard's Scientology (as a term for non-Scientologists); and so forth. Blish even cites his own fictions here; the skopolamanders evoke "A Hero's Life" (1966), and the beademungen come from "Common Time" (1953). As Kirk watches these imaginary casualties float by, it becomes clear that the battle between the Spocks is a conflict over speculative fiction, or the literary imagination as such.

The replicate Spock asserts his superiority by drawing an overt literary metaphor: "My existence . . . is a fortuitous revision, and a necessary one, of a highly imperfect first draft. It is the scribbled notes which should be eliminated here, not the perfected work. . . . Perhaps, crude recension though you are, you could be brought to understand that" (91). Playing along, the original Spock responds, "The true scholar . . . prizes all drafts, early and late. But your literary metaphor is far from clear, let alone convincing" (91). While the duplicate Spock derides his predecessor's "smudgy incunabular existence" (91), the original Spock summons an imaginary tornado, which thrusts the replicate into the same tachyon screen that created him in the first place. The fiction is so persuasive that the replicate literally dies: "It was a combat of illusions—and in the end, the replicate *believed* he had been driven into the screen. That was sufficient" (95). Thus ends the adventure of a "fortuitous revision," bounced from screen to text and back again.

Transparently, *Spock Must Die!* is a metaphor for its own composition, coming to terms with its relation to preexisting fictions. As Blish writes in his authorial preface, "Unlike the preceding three STAR TREK books, this one is not a set of adaptations of scripts which have already been shown on television, but an original novel built around the characters and background of the TV series conceived by Gene Roddenberry" (ix). He conjectures that whereas he had been translating the TV show into prose, *Spock Must Die!* might reverse the process: "And who knows—it might make a television episode, or several, some day" (ix). Although the novel establishes that adaptations may not have the best interests of the original at heart, it nevertheless indicates that a derivative work may make a separate claim to authenticity. As one Spock notes, "I can assure you that I *know* that I am the original—but this knowledge is not false even if I am in fact the replicate" (378). Indisputably, adaptations take on a life of their own.

By the same token, *Spock Must Die!* is also about the relationship between science and science fiction. Tellingly, the story begins with Scotty's effort to harness the untapped power of tachyons, to boldly go where no one has gone before ("particles that travel faster than light—for which nobody's ever found any use" [13].) The book highlights the poaching of scientific concepts for purposes of fiction. But the tachyons only create a simulacrum, Spock's sinister twin: a high-tech derivative of a science fiction icon. The tachyonic reproduction of Spock points to the novel's more extensive thematization of fiction as an adaptive resource for technical innovation. For example, to outwit the Klingon forces threatening the *Enterprise*, Uhura proposes to transmit a message to Starfleet Command in Eurish:

> It's the synthetic language James Joyce invented for his last novel, over two hundred years ago. It contains forty or fifty other languages, including slang . . . You know the elementary particle called the quark; well, that's a Eurish word. Joyce himself predicted nuclear fission in the novel I mentioned. I can't quote it precisely, but roughly it goes, "The abnihilisation of the etym explodotonates through Parsuralia with an ivanmorin-thorrorumble fragoromboassity amidwhiches general uttermost confussion are perceivable moletons scaping with mulicules." There's more, but I can't recall it—it has been a long time since I last read the book. (49)

Uhura draws attention to the polysemic, heteroglossic nature of literary discourse—hyperbolized in Joyce's invented Eurish that "contains forty or fifty other languages" in itself—as well as the interplay of science and fiction foregrounded by Joyce's last novel. As Uhura recalls, *Finnegan's Wake* is rife with references to modern physics; she even attributes the "prediction" of nuclear fission to Joyce, retroactively. Moreover, Uhura remembers that *Finnegan's Wake*—in particular, the line "Three quarks for Muster Mark!"—was a semiotic resource for Murray Gell-Mann. Certainly, Gell-Mann footnoted Joyce's book in his 1964 quark paper, and while he later claimed that he took only Joyce's spelling—applying it to a nonsense word he devised in his head—he maintained that *Finnegan's Wake* was a novel he knew well, having first encountered it when he was ten years old and revisiting it for "occasional perusals" thereafter.[33] Perhaps Gell-Mann did not have Uhura's nearly perfect recall of the text, but he resorted to it when needing to confine his speculative particles in a word that, while suggesting literary significance, exceeded the familiar and the intelligible. For Uhura as well, this is the virtue of Joyce's estranging language, for it allows the *Enterprise* to transmit secret information and "scientific terms" in plain sight. It is a language that performs intertextual entanglements, demanding interpretation. As Uhura says, "Nobody's ever *dead* sure of what Eurish means . . . But I can probably read more of it than the Klingons could. To them, it'll be pure gibberish" (49).

Over and again, the novel emphasizes the indebtedness of the high-tech future to literary history. Spock refers to Shakespeare's *Othello* (1603) and Milton's *Paradise Lost* (1667), Kirk gestures to Conan Doyle's Sherlock Holmes stories (1887–1927), and Scotty recalls the Hobbit concept of useless "mathoms" from Tolkien's *The Lord of the Rings* (1954–55). Blish's own oeuvre is invoked several times; for example, a Klingon remembers the Xixobrax Jewelworm, which is a nod to "This Earth of Hours" (1959). Likewise, when Scotty describes the *Enterprise*'s transporter, he notes it is a Dirac technology: "What the transporter does is analyze the energy *state* of each particle in the body and then produce a Dirac jump to an equivalent state somewhere else" (3). The transporter thus joins other Dirac technologies that populate Blish's fic-

---

[33] Gell-Mann, "Schematic" (cit. n. 18); Gell-Mann, *The Quark* (cit. n. 22), 180. See also Johnson, *Strange Beauty* (cit. n. 21), 43, 214; Peter Middleton, *Physics Envy: American Poetry and Science in the Cold War and After* (Chicago, Ill., 2015), 51–9. On Eurish (a term Joyce never used), see Anthony Burgess, *Re Joyce* (New York, N.Y., 1965). On Blish's frequent allusions to Joyce, see Grace Eckley, "*Finnegan's Wake* in the Work of James Blish," *Extrapolation* 20 (1979): 330–42; and Ketterer, *Imprisoned* (cit. n. 31), 19, 96–8.

tions. These include the spindizzy engines of the *Cities in Flight* series (1955–62), which achieve antigravity and faster-than-light speeds by virtue of the "Blackett–Dirac equations"; the faster-than-light ship in "Detour to the Stars" (1956) and "Nor Iron Bars" (1957) that behaves as a Dirac hole; and, of course, the Dirac communicator. The Dirac communicator appeared in several stories following "Beep," including the *Cities in Flight* sequence, *Midsummer Century* (1972), and *The Quincunx of Time* (1973), an expansion of "Beep" that would be Blish's final novel before his death in 1975. Offering a subtle reminder that Blish had fabulated things that go faster than light throughout his career, *Spock Must Die!* represents a reclaiming of tachyons. It recycles these particles "for which nobody's ever found any use" in a story that is literally about a struggle over creative priority, set in a future where tachyonic duplication has threatened to reverse the proper order of things.

But the point of *Spock Must Die!* is not simply that the future inherits from the past, that new scientific propositions as much as literary fictions swipe from earlier cultural resources. It also suggests how the past inherits from the future, reinterpreted in light of new developments that drive further speculation, further innovation. For instance, the original Spock does not, at first, understand how his twin created a portable warp drive to draw energy from Hilbert space. But after investigating, Spock comprehends how the trick was accomplished, and he sees a path forward to even greater feats: "Anything the replicate could do, I can probably do better" (104). In other words, the adaptation compels the original to catch up, to go further: a game of modifiable futures that both Blish and Feinberg knew well.

PROMETHEAN PROJECTIONS

Around the same time, and playing the same game, Feinberg proposed an ambitious global endeavor that he called the Prometheus Project. He sketched out this proposal in his 1968 book, *The Prometheus Project: Mankind's Search for Long-Range Goals*, and he continued to develop these ideas over the following decade, reiterating them in his 1977 book, *Consequences of Growth: The Prospects for a Limitless Future*. The Prometheus Project was to be a scheme for reshaping social organization to achieve futuristic visions. In effect, Feinberg proposed establishing an Event Police, governing the future in the image of science fiction.

In his books, Feinberg imagines a widespread democratic process, involving stakeholders from all over the world, in which a variety of long-range future scenarios are deliberated to help our species pursue desirable sociotechnical pathways and avoid catastrophic ones: "The purpose of this book is to propose a group effort, which I call the Prometheus Project (from the Greek word *prometheus*, meaning foresight), by which humanity can choose its goals. . . . In that way, humanity would move closer to becoming the shaper of its own destiny."[34]

According to Feinberg, the accelerating entanglements of globalization and technologization indicate that choices we make today might prove to be "world-shaking decisions" with "irreversible effects" (19). He points to nuclear technologies and climate effects of industrial pollution as known instances of such world-shaking decisions, claiming that even more profound eventualities will arise. "The advent of 'world-

---

[34] Gerald Feinberg, *The Prometheus Project: Mankind's Search for Long-Range Goals* (Garden City, N.Y., 1968), 13–4. Further page citations appear parenthetically in the text.

shaking decisions,'" he argues, "thus calls for some kind of long-range planning of the future of mankind. Otherwise man, having elevated himself through his technology from being the plaything of blind nature, might become the victim of blind actions on his own part" (20–21).

Such long-range planning would involve collective assessment of various "as if" scenarios, technological and scientific developments that might utterly change society, the natural environment, or human biology. It would mean committing as a global civilization to certain goals and affirmatively answering the question, "Should we, the people alive today, bind the people of the future to our own purposes, and can we do so?" (25).

Necessarily speculative, such a massive project would benefit tremendously from drawing on one of our best resources for assessing the future impacts and potentialities of technoscientific innovations: namely, science fiction. "It is unfortunate," Feinberg writes, "that these imaginative writings have not been given much serious attention, since they are not only a fertile source of suggestions for goals, but also often point out unexpected implications that the attainment of some goals might have for human life" (18).

Among the many "as if" scenarios presenting themselves with particular intensity, Feinberg notes the unprecedented possibilities for reengineering the human body and human behavior though new tools of biotechnology and cognitive science. "Human nature is therefore not something unalterable," he writes, "but can be changed" (29). He cites several novels including A. E. Van Vogt's *Slan* (1946), Olaf Stapledon's *Last and First Men* (1930), Jack Williamson's *Dragon's Island* (1951), and others as providing insightful assessments of these issues, helping to frame the scope of further developments.

But Feinberg asserts that our capacities to speculate and deliberate on "the option of changing man" (27), and to anticipate even a small fraction of the ramifications, are circumscribed in advance—limited by human nature itself. Human, all too human. For this reason, the Prometheus Project proves to be a self-reflective critical assessment of human being, the constitutive finitude of humannness that prevents us from seeing much beyond our own presentist horizon. As Feinberg writes: "The most serious fault in the human condition lies in our finitude. We are conscious beings aware of our own limitations. Two of the most important of these are the lack of power to do things we want to do, and the specter of impending death, which always threatens to put an end to all our thinking and doing" (43).

The difficulties of securing a sustainable, desirable future ultimately reduce to the limits of human agency and foresight—the blinders of short-term thinking imposed by the span of a single human lifetime—which encourage us to push unforeseen problems onto later generations. According to Feinberg, the solution is to overcome these limits, these so-called ends of man. This would first mean mobilizing our best science and technology toward the elimination of death as an obstacle, taking "steps toward indefinite life and youth" (45). But merely eliminating death as a natural limit would not solve the seemingly more intractable problem of the metaphysics of presence, or what Feinberg calls the "temporal provincialism inherent in most human thought."[35] So, we must go further to overcome not only the condition of being-toward-death, but

---

[35] Gerald Feinberg, *Consequences of Growth: The Prospects for a Limitless Future* (New York, N.Y., 1977), 94.

also being-as-presence: "Perhaps the only true solution to the problem of death is to eliminate it for those who do not desire it. . . . But even 'at death's end' men will remain finite beings in their accomplishments if not in their expectations. . . . [So,] some other kind of reconstruction of man appears called for to deal with finitude."[36] It calls for the complete reconstitution of the human, beyond humanism and its ends: "The most radical proposals along these lines would be that man should reconstruct his biological and psychological nature so that the present faults in his condition are corrected. If this road is chosen, it means that we will have accepted the fact that not only is nonhuman nature subject to man's will in how it is to function, but that man has the choice of what he himself is to be. Such an attitude is perhaps a logical extension of setting goals for humanity" (510). The goals, the ends of humanity, now exceed humanity as such.

While conceding that others may disagree, Feinberg ventures that "changing man" is a "logical extension" of projecting our foresight into the long-term future—in other words, it is inherent to the Prometheus Project. For the social necessity of engaging in widespread deliberation about whether we *ought* to reengineer ourselves—to commit ourselves to "the biological reconstruction of the human race" (16)—leads to the conclusion that we *must* do so, if we are to have any hope of grappling with the implications of our emerging technical powers. "Changing man" becomes already inevitable, recursively transposed on the present. The "as if" future destines itself, secured by the forces of the Prometheus Project and its stakeholders who will effectively become the Event Police. Or, as Feinberg puts it, "More than ever before, we have the power to *determine* the future, rather than to predict it."[37]

Yet vigilant policing would prove unnecessary, for once begun, this imagined future becomes self-reinforcing: "If mankind is transformed through conscious biological manipulation, the new men that are produced will have a different set of interests and potentialities than we do. In making this change, we will therefore have started on a road that could not easily be retraced, not because the biological manipulation could not be undone, but because the new man is not likely to wish to reintroduce the attributes of contemporary man in himself. . . . Hence biological engineering is likely to lead to some of the irreversible changes we have discussed."[38]

With its extreme vision of refashioning human biology and behavior, Feinberg's Prometheus Project contributed to an ongoing conversation about the high-tech acceleration of human evolution, a discourse situated at the intersection of science fiction and popular science. Kickstarted in the early twentieth century by works of speculative science such as J. B. S. Haldane's *Daedalus, or Science and the Future* (1924) and J. D. Bernal's *The World, the Flesh and the Devil* (1929), the discourse of evolutionary futurism became increasingly prominent through a succession of famous texts, including Manfred Clynes and Nathan Kline's "Cyborgs and Space" (1960), Robert Ettinger's *The Prospect of Immortality* (1964), and F. M. Esfandiary's *Up-Wingers* (1973), as well as later works such as K. Eric Drexler's *Engines of Creation* (1986), Hans Moravac's *Mind Children* (1988), and Ray Kurzweil's *The Singularity Is Near* (2005).[39]

---

[36] Feinberg, *Prometheus* (cit. n. 34), 47. Further page citations appear parenthetically in the text.
[37] Feinberg, *Consequences* (cit. n. 35), 8.
[38] Feinberg, *Prometheus* (cit. n. 34), 67.
[39] See J. B. S. Haldane, *Daedalus, or Science and the Future* (New York, N.Y., 1924); J. D. Bernal, *The World, the Flesh and the Devil: An Enquiry into the Future of the Three Enemies of the Rational*

In the 1970s, taking a cue from Feinberg, the literary theorist Ihab Hassan associated this entire discourse with the figure of Prometheus. In his 1973 essay "The New Gnosticism: Speculations on an Aspect of the Postmodern Mind," as well as in his 1977 essay "Prometheus as Performer: Toward a Posthumanist Culture?," Hassan reflected on the intensification of technoscientific ambition, pointing to Feinberg's Prometheus Project among other examples as indexing a growing urgency, a transcendental yearning: "The postmodern Prometheus reaches for the fire in distant stars."[40] If the ancient myths of Prometheus once allegorized the invention of the human, the *originary technicity* of human nature—Hassan would have agreed with the philosopher Bernard Stiegler on this interpretation—the resurgence of Promethean enthusiasm in postmodernity suggested a re-visioning of humanity and humanism entirely.[41] It heralded an overcoming of physical and metaphysical constraints that Hassan, with notable irony, ventured to call *posthumanism*: "We need first to understand that the human form—including human desire and all its external representations—may be changing radically, and thus must be re-visioned. We need to understand that five hundred years of humanism may be coming to an end, as humanism transforms itself into something that we must helplessly call posthumanism." Hassan was among the first critics to diagnose a posthumanist turn in Western culture, characterized by the exercise of Promethean foresight, science fiction driving scientific practice: "Because both imagination and science are agents of change, crucibles of values, modes not only of representation but also of transformation, their interplay may now be the vital performing principle in culture and consciousness—a key to posthumanism."[42]

Feinberg certainly understood the Prometheus Project in the context of a broader discourse of posthumanism.[43] Starting in the 1960s, he advocated for a variety of forward-looking endeavors that mixed science and science fiction to recalibrate the limits of human being. For example, in 1966, he published an article in *Physics Today* called "Physics and Life Prolongation" that elaborated the scientific arguments in support of Ettinger's cryonics program. While death remains a dilemma, Feinberg agreed with Ettinger that it was just a matter of time—so freezing yourself now in

---

Soul (London, 1929); Manfred Clynes and Nathan S. Kline, "Cyborgs and Space," *Astronautics*, September 1960, 26–7, 74–5; Robert C. W. Ettinger, *The Prospect of Immortality* (Garden City, N.Y., 1964); F. M. Esfandiary, *Up-Wingers: A Futurist Manifesto* (New York, N.Y., 1973); K. Eric Drexler, *Engines of Creation* (Garden City, N.Y., 1986); Hans Moravec, *Mind Children: The Future of Robot and Human Intelligence* (Boston, Mass., 1988); and Ray Kurzweil, *The Singularity Is Near* (New York, N.Y., 2005).

[40] Ihab Hassan, "The New Gnosticism: Speculations on an Aspect of the Postmodern Mind," *boundary 2* 1 (1973): 546–70, on 555; Hassan, "Prometheus as Performer: Toward a Posthumanist Culture?" *Georgia Review* 31 (1977): 830–50.

[41] Bernard Stiegler, *Technics and Time, 1: The Fault of Epimetheus*, trans. Richard Beardsworth and George Collins (Stanford, Calif., 1998).

[42] Hassan, "Prometheus" (cit. n. 40), on 843, 838. For Hassan, science fiction made visible the accelerating convergence of science and imagination, myth and technology; see Hassan, "The New Gnosticism" (cit. n. 40), 565, 567.

[43] On posthumanism in the history of science, see N. Katherine Hayles, *How We Became Posthuman: Virtual Bodies in Cybernetics, Literature, and Informatics* (Chicago, Ill., 1999); Mark B. Adams, "Last Judgment: The Visionary Biology of J. B. S. Haldane," *J. Hist. Biol.* 33 (2000): 457–9; Richard Doyle, *Wetwares: Experiments in Postvital Living* (Minneapolis, Minn., 2003); Colin Milburn, "Posthumanism," in *The Oxford Companion to Science Fiction*, ed. Rob Latham (Oxford, UK, 2014), 524–36; and Andrew Pilsch, *Transhumanism: Evolutionary Futurism and the Human Technologies of Utopia* (Minneapolis, Minn., 2017).

hopes that science might resurrect you in the future seemed a gamble worth taking. He suggested that the "results of low-temperature biology (cryobiology) open the possibility to those living of taking advantage of this progress before the problems of aging and death are solved."[44] For the cryonics community, alas, it is a tragedy that Feinberg did not actually commit his own mortal remains to the freezer when he died of cancer in 1992.

In the late 1980s, Feinberg also joined the Board of Advisors of the Foresight Institute—an organization founded in 1986 by Eric Drexler and Christine Peterson (at the time, husband and wife) to promote research in nanotechnology and develop policies for its development. Having read Drexler's *Engines of Creation*, Feinberg became convinced that nanotechnology would dramatically change human life—indeed, it was already inevitable. In line with his thinking about the Prometheus Project, Feinberg indicated that emerging governance efforts around nanotechnology were behind schedule before they started—the Event Police were now playing catch-up with the future. "We've already lost more time than we can afford," he said. To prepare society for the upheavals of nanotechnology, Feinberg worried that "it may already be getting to be too late."[45]

Feinberg's abiding concerns about time and timeliness—the sense that, when it comes to irreversible decisions, we may already be out of time—reinforced the posthuman dimensions of his thought. To be sure, his project of setting forth science-fictional scenarios that preemptively transform the present instantiates a posthumanist *temporalization*, reordering the classical relations of present and future, disorienting the path "from where man is now to what he will become." After all, this path sometimes might appear to move backward instead of forward, or rather, the temporal frame becomes undecidable. In *The Prometheus Project*, for example, Feinberg discusses the possibility of a human hive mind:

> In a sense, developmental and transcendent goals might be regarded as successive stages in a process of going from where man is now to what he will become. But this may not be the order in which things will or should be done.
>   The formulation of transcendent goals is difficult because the lack of models makes it hard to imagine what conditions would really be like in the new situation. Consider, for example, the proposal that men would be much better off if they all shared a common consciousness, perhaps through something like an artificial telepathic communication system. While it is possible to describe this in words, and perhaps even to carry out the plan, it is very difficult to feel intuitively what life would be like for those who lived under these conditions. (96–97)

As Feinberg says, while it might seem that this process is about extending a developmental program—a prosthesis of the human present—into the transcendent future,

---

[44] Gerald Feinberg, "Physics and Life Prolongation," *Phys. Today* 19 (1966): 45–8, on 45.

[45] Feinberg quoted in Dan Shafer, "Feinberg Anxious for Policy Discussions," *Foresight Update* 9, 30 June 1990, http://www.foresight.org/Updates/Update09/Update09.1.html. On Feinberg's involvement with cryonics, nanotechnology, and other speculative sciences, see David Kaiser, *How the Hippies Saved Physics: Science, Counterculture, and the Quantum Revival* (New York, N.Y., 2011); W. Patrick McCray, *The Visioneers: How a Group of Elite Scientists Pursued Space Colonies, Nanotechnologies, and a Limitless Future* (Princeton, N.J., 2012); and McCray, "Physics at the Frozen Fringe," *Leaping Robot Blog*, 3 February 2015, http://www.patrickmccray.com/2015/02/03/physics-at-the-frozen-fringe/. On the rhetoric of the "already inevitable" in nanotechnology and its posthuman implications, see Colin Milburn, *Nanovision: Engineering the Future* (Durham, N.C., 2008).

gradually turning imagination into actuality in linear historical time, the situation is quite otherwise ("this may not be the order in which things will or should be done"). While it might be possible to "carry out the plan" of creating a human hive mind, the actual experience of such collective consciousness cannot be known beforehand. Feinberg recalls that writers such as Olaf Stapledon have presented imaginative accounts "in words" of the expansion of human consciousness beyond its native boundaries, but what it would feel like remains inscrutable to merely human minds. All we can know now is that it would profoundly change us. Indeed, it would have effects that we cannot foresee even when contemplating the goal, with all its fictive qualities. As Feinberg notes, we have the power to determine this future, rather than to predict it. The Event Police, after all, can only operate through the logic of the "as if."

Yet for precisely this reason—namely, that it would be a "world-shaking decision" with irreversible consequences—the pursuit of such a goal would require us to change ourselves *already in advance*, to think beyond our own limits and preadapt ourselves to what is to come: "The time scale of a few centuries and the knowledge of where we are trying to go make it possible for us to adapt our social institutions to the situation when the new form of consciousness becomes a reality." According to Feinberg, we must preemptively reorganize society and ourselves for the future—and there's no time like the present: "One of the things we could do soon . . . would be to begin thinking about what kind of society would go with the new forms of consciousness" (153).

In this manner, the long-term future would appear not merely as the outgrowth of near-term developmental steps, but vice versa. This speculative future urgently solicits a reordering of society in such a way that the more specific goal of expanded consciousness becomes virtually inexorable; the consequential fiction of the hive mind sets the conditions for its own actualization. The long-term goal is not the end, but the beginning. Or rather, cause and effect become undecidable, open to reinterpretation. As Feinberg suggests, once we reshape our expectations in this way, actively preparing society for an imminent posthumanization, other goals emerge that were not seen in advance but prove to have been inevitable all along, including "the extension of consciousness to its logical limit, when it becomes coextensive with the universe" (104). At this limit we would be unlimited, in excess of ends—transfinite.

For Feinberg, the goals we set are not ends but preconditions for other modes of becoming that we cannot fully predict, and these retrospectively become our destination. Our self-transformation through long-range speculation becomes an opening to futurity, radical potentiality. Feinberg's project is not a visioning program that sees the future as merely a prolongation or metastasis of the present, but instead represents an eversion of the future, aspiring to bypass the failures of the human and its being-toward-limits.

We might note that this excession of ends, as an onto-epistemological transformation of the human condition, parallels the structure of Feinberg's analysis of tachyons as particles already beyond the speed limit of light, already beyond the relativistic barrier that secures the classical principle of causality—and whose propagation backward or forward in time, as either cause or effect, would be fundamentally undecidable. Hence, the reinterpretation principle allows us to see future events as affecting the past only in the sense of the "as if."

In this regard, once again, James Blish would seem to be at the root of Feinberg's posthumanist project. But it is not Blish alone. Rather, it is science fiction as such, which characteristically bodies forth a Prometheanism in advance of itself, a *preposterous* Prometheanism. As indicated by the history of the genre, from Mary Shelley's *Frankenstein; or, The Modern Prometheus* in 1818, to Ridley Scott's film *Prometheus* in 2012, science fiction presents itself as a projection of foresight that becomes the precondition for looking backward—looking inward at ourselves . . . preposterously, without end.

PREPOSTEROUS FICTIONS

Consider, for example, the work of Arthur C. Clarke. While Feinberg does not directly cite Clarke's fiction in *The Prometheus Project*, he does discuss the author's 1962 non-fiction treatise, *Profiles of the Future*, which attends to the crucial role of speculative fiction for technoscientific progress: "The only way of discovering the limits of the possible is to venture a little way past them into the impossible."[46] Moreover, there are moments in *The Prometheus Project* when Feinberg considers the possibility of alien influences fostering our posthuman evolution—a recurring concept in Clarke's fiction, beginning with his 1953 novel *Childhood's End*.[47] To be sure, *Childhood's End* was one of Feinberg's favorite novels, and he often recommended it to other scientists.[48] Feinberg's interest in this particular novel also helps to explain why he sent his tachyon manuscript to Clarke in 1967, because the main plot twist involves information traveling backward in time. In any case, the literary conceits of Clarke's novel prefigure the approach to posthumanization in *The Prometheus Project*.

When the Overlord ships arrive at the beginning of *Childhood's End*, the event signals both recollection and prolepsis, both *déjà vu* and preview: "There had been no warning when the great ships came pouring out of the unknown depths of space. Countless times this day had been described in fiction, but no one had really believed that it would ever come. Now it had dawned at last; the gleaming, silent shapes hanging over every land were the symbol of a science that man could not hope to match for centuries."[49] The arrival represents a high-tech future incongruously transposed into the human present ("the symbol of a science that man could not hope to match for centuries"), recalling an entire narrative genre conspicuously preoccupied with such events ("Countless times this day had been described in fiction"). The arrival of the alien Overlords symbolizes science fiction as such, a recursive figuration with all its generic clichés as well as its conceptual innovations. Or, to say it differently, the arrival of the Overlords takes the form of science fiction—it is a science-fictional event—even within the world of the novel. Furthermore, the discourse of science fiction has prescribed the experience of so-called first contact. One character even re-

---

[46] Arthur C. Clarke, *Profiles of the Future: An Inquiry into the Limits of the Possible* (New York, N.Y., 1962), 21.

[47] Arthur C. Clarke, *Childhood's End* (New York, N.Y., 1953). Clarke continued to explore this idea in other narratives, including *2001: A Space Odyssey* (1968) and its sequels.

[48] For example, the physicist Jeremy Bernstein recalled that Feinberg recommended *Childhood's End* to him in the early 1960s; Jeremy Bernstein, "The Grasshopper and His Space Odyssey," *Amer. Sch.* 77 (2008): 154–7. Feinberg's recommendation led to Bernstein striking up a friendship with Clarke and writing one of the earliest overviews of Clarke's career; Jeremy Bernstein, "Out of the Ego Chamber," *New Yorker*, 9 August 1969, 40–65.

[49] Clarke, *Childhood's End* (cit. n. 47); quotation here is from the Del Ray Impact edition (New York, N.Y., 2001), 13. Further citations of the Del Ray Impact edition appear parenthetically in the text.

bukes another for holding preconceived notions about the Overlords: "You . . . have been reading too much science-fiction" (21).

Even long after their arrival, the Overlords hide their true appearance from human eyes, because they recognize that the entire history of myth, fantastical literature, folkloric and religious tradition—the prehistory of science fiction in the largest sense—has prejudiced the human species against them.[50] For the Overlords look like demons: "There was no mistake. The leathery wings, the little horns, the barbed tail—all were there. The most terrible of all legends had come to life, out of the unknown past" (71).

Once the Overlords reveal their true forms, a number of human scholars theorize that some horrific situation in ancient history, some aboriginal encounter between the Overlords and our ancestors, must account for the instinctual human fear of demonic figures. Yet the reality is more shocking. As it turns out, the appearance of the alien other in the long history of folklore and mythology does not record ancestral memory. Instead, it records *futurity*—a prescription or program of events yet to come. For the Overlord's secret purpose is to shepherd the orthogenetic evolution of *Homo sapiens* into something more than human, guiding social and biological conditions toward the emergence of the posthuman.

The ancient fear of demonic figures and monstrous others, as recorded in the genealogy of fantastical literature, proves to have been a memory in reverse—a psychic adjustment to the posthuman future. The human apperception of its own closure, the end of its childhood, has therefore been narrated since time immemorial. The Overlords explain, "For that memory was not of the past, but of the *future*—of those closing years when your race knew that everything was finished. . . . And because we were there, we became identified with your race's death. Yes, even while it was ten thousand years in the future! It was as if a distorted echo had reverberated round the closed circle of time, from the future to the past" (225–26). Fables of satanic possessions, monstrous births, and alien invasions prove to have been letters from the posthuman future, though we did not recognize them in time.

Clarke's novel depicts the genealogy of fantastical stories as a series of distorted messages, misunderstood allegories of the human engagement with world-changing forces. Recursively situating itself in this literary history, *Childhood's End* represents a final, undisguised account of the end of human existence and its transmutation into something unspeakably other. But the narration also thematizes its own belatedness, the sense in which its full implications may remain indecipherable in the present. Jan, the lone human witness to the ascension of posthumanity, remains on Earth even after the Overlords themselves have fled to safety. He records his observations with a telephonic device, describing the final moments of the doomed planet in his own words: the last human narrator, reporting from the edge of apocalypse. Jan's words are transmitted to the Overlord ship, helping the aliens to imagine a posthuman event that they cannot witness directly. However, due to the time delay in relaying messages to a ship

---

[50] On science fiction's continuity with mythic and religious discourse, see David Ketterer, *New Worlds for Old: The Apocalyptic Imagination, Science Fiction, and American Literature* (Bloomington, Ind., 1974); Alexei Panshin and Cory Panshin, *The World Beyond the Hill: Science Fiction and the Quest for Transcendence* (1989; repr., Rockville, Md., 2010); Thomas M. Disch, *On SF* (Ann Arbor, Mich., 2005); Jeffrey J. Kripal, *Mutants and Mystics: Science Fiction, Superhero Comics, and the Paranormal* (Chicago, Ill., 2011); and Adam Roberts, *The History of Science Fiction*, 2nd ed. (London, 2016). The extent of science fiction's relationship with myth has been much debated. Clarke's novel imaginatively resolves the debate by rendering it into an explicit plot element.

traveling near light speed, the Overlords receive Jan's narration of the posthumanizing process long after it has concluded, long after the last man himself has been absorbed by the transformative energies: "It was strange to think that the ship of the Overlords was racing away from Earth almost as swiftly as his signal could speed after it. Almost—but not quite. It would be a long chase, but his words would catch [them]" (232).

As usual, the posthuman narrative comes too late, always playing catch-up. Even when sent early, it proves legible only after the fact. It addresses a destination that it has not yet reached, depicting a future that has already taken place. Which is to say, Jan's transmission—his words racing ahead, looking backward on what will have been—has all the characteristics of science fiction.

Childhood's End offers a parable about speculative narratives that presage and inform the circumstances of their own actualization, even if they appear a bit off schedule. So it's really no wonder that Feinberg was such a fan. His practice as a theoretical physicist and science policy advocate seems to have taken seriously Clarke's trope of literary precognition—a distorted memory of the future that, while it may appear as fantasy and make-believe, nevertheless conditions the future as it arrives. If nothing else, as a metaphor, Clarke's idea that speculative fiction might both record and distort genuine precognition—recognized only belatedly—resonates with Feinberg's endorsement of the "as if" method for scientific innovation.

Indeed, he applied the same approach when taking Clarke's metaphor literally. In 1974, Feinberg contributed a paper called "Precognition: A Memory of Things Future?" to a conference in Geneva, Switzerland on "Quantum Physics and Parapsychology." Similar to his assessment of tachyons, Feinberg granted that precognition may not be a real phenomenon, but he nevertheless offered "a very speculative model for precognition," complete with testable hypotheses that would encourage further research to better understand the model and its implications for physics in general—the mode of the "as if":

> I am suggesting that precognition, if it exists, is basically a remembrance of things future, an analogy to memory, rather than a perception of future events, an analogy to sense perceptions of the very recent past. This suggestion has at least the merit of being fairly easy to test through simple experiments . . . If it is correct, it would not directly indicate the physical mechanism for precognition, any more than the existence of memory indicates its physical mechanism. However, if it does turn out that memory can operate into the future as well as into the past, it would suggest that the symmetry of physical laws . . . is involved, and that physicists have been premature in discarding those solutions to their equations that describe reversed time order of cause and effect.[51]

Preemptively claiming stakes on any science of precognition yet to emerge, Feinberg also pointed out that, should his speculative model prove viable, the practice of excluding perplexing solutions to physical equations as well as the scientific consensus on classical causality would both need to be revised. As in his work on tachyons, Feinberg indicated how scientific speculations—whether theoretical constructs or preposterous fictions—might afford new ways of looking at the actual practices of science, rethinking received knowledge, and imagining alternatives, even if only to bet-

---

[51] Conference paper published as Gerald Feinberg, "Precognition: A Memory of Things Future," in *Quantum Physics and Parapsychology*, ed. Laura Oteri (New York, N.Y., 1975), 54–73.

ter recollect how it is that we know what we know. For Feinberg, the science fiction way of doing science was never about predicting the future, as such. Rather, it was about attentively establishing the conditions in which the future could take place, un-restrained by the present and its limits, the known and the familiar: an open solicita-tion of realizable worlds and the shape of things to come . . . ahead of time.

Environmental Futures, Now and Then:
Crisis, Systems Modeling, and Speculative Fiction

*by Lisa Garforth**

ABSTRACT

Postwar environmental concern has been powerfully shaped by projections of ecological catastrophe. Indeed, it can be said that the global environment as an object of social and political concern came into existence in part through narratives of future crisis. This article explores two successive framings of environmental crisis and the kinds of knowledges that made them up. It examines the announcement of ecological limits to economic growth in the early 1970s, the culmination of an early wave of popular green concern that modeled the future as a choice between the catastrophic continuation of business as usual and the prospect of eco-utopian alternatives. It considers the crisis logics of contemporary climate dynamics, where the power of scientific modeling leaves little room for the imagination of radically different futures. Environmental crisis now cannot perform the anticipatory and utopian functions that it once did. The "apocalyptic horizon" of limits has given way to the collapse of crisis into the present and new kinds of colonization of the future. But in both cases, environmental crisis can be read as a science-fictional object, simultaneously descriptive and speculative, scientific and fictional. Science fiction tropes were crucial to early constructions of environmental crisis, and speculative climate fiction will be a vital resource for negotiating the social-natural futures of the Anthropocene.

THE IDEA OF ENVIRONMENTAL CRISIS

For the past fifty years, from the emergence of the idea of limits to growth in the 1970s to climate change today, environmental concern has been shaped by projections of potentially catastrophic disruption to ecological and economic systems. Indeed, in many ways the idea of the global environment as an object of social and political concern does not exist without narratives concerning the "crisis about its future."[1] As Andrew

* School of Geography, Politics and Sociology, Newcastle University, NE1 7RU, UK; lisa.garforth @ncl.ac.uk.

I am grateful to the issue editors and to the anonymous reviewers for helping me clarify my terms and highlight the science fiction novels that sometimes risked getting lost in this essay. The research on contemporary climate futurities that informs the arguments here on climate models, postwar futures studies, and contemporary climate fiction was supported by the Arts and Humanities Research Council (AHRC) in relation to the Unsettling Scientific Stories Project (2016–2018) and benefited enormously from conversations with all the USS project team members.

[1] Libby Robin, Sverker Sörlin, and Paul Warde, "Introduction: Documenting Global Change," in *The Future of Nature: Documents of Global Change*, ed. Robin, Sörlin, and Warde (New Haven, Conn., 2013), 1–14, on 6–7.

Ross has argued, of all the new social movements of the 1960s and 1970s, "the ecology movement was the one most tied to an explicit set of theses about the future: how to avoid a disastrous, and generate a better, future."[2] In this article, I explore the kinds of knowledges that have made up successive ideas of environmental crisis and environmental futures. I focus in particular on the relationship between technoscientific projections and wider cultural repertoires of speculation—that is, between environmental science and green science fiction, and between predictions of ecological catastrophe and dreams of better futures with and for nature.

Environmental crisis is closely linked to claims of scientific certainty, predictive statements, and policy expertise. As Libby Robin et al. put it, "the Age of Environment has been nurtured by the Era of Prediction."[3] But both have also been deeply entangled with speculative and fictional imaginaries that raise ethical, political, and even metaphysical questions about human prospects in a changing natural world. Ideas of environmental crisis since the 1970s have been composed of varying admixtures of systematic and self-consciously objective attempts to model ecosocial future trajectories on one hand, and heuristic and affective future imaginaries on the other. In this respect, environmental concern typifies the sometimes synergistic and sometimes sterile tension between technocratic extrapolation and humanist utopianism that Jenny Andersson identifies at the core of postwar futures studies.[4] Understanding science and speculation together offers a rich picture of the history of contemporary environmental ideas and a nuanced understanding of what has changed in crisis narratives between the 1970s and the present day, both in terms of the scientific epistemologies that make up crisis, and in terms of the work that crisis does on and in the popular and political imagination.

With that in mind I work through two arguments in this article. The first is that environmental crisis now is not the same as environmental crisis then. In the 1970s, ecological crisis usually took the form of a future projection based on highly abstract models of large-scale global systems. The idea of crisis in this period introduced a shocking and novel "apocalyptic horizon"[5] into modern Western imaginaries of progress and the future, and as such it opened up debates about radically different alternatives: apocalypse or utopia, economic expansion or a no-growth community, heedless progress or mindful stability, collapse or sustainability. Today, environmental crisis is no longer a novelty. We are effectively living through what Brian Wynne calls the "predictive shadow"[6] of the first announcement of impending crisis. Frederick Buell argues that environmental apocalypse now feels something more like "a way of life"[7] than a projection. The present is saturated with slowly unfolding environmental collapse, and the future is colo-

---

[2] Andrew Ross, *Strange Weather: Culture, Science and Technology in the Age of Limits* (London, 1991), 184.

[3] Robin, Sorlin, and Warde, "Introduction" (cit. n. 1).

[4] Jenny Andersson, "Midwives of the Future: Futurism, Futures Studies and the Shaping of the Global Imagination," in *The Struggle for the Long Term in Transnational Science and Politics: Forging the Future*, ed. Andersson and Egle Rindzeviciute (London, 2015), 16–37; Andersson, "The Great Future Debate and the Struggle for the World," *Amer. Hist. Rev.* 117 (2012): 1411–30; Andersson and Sibylle Duhautois, "Futures of Mankind: The Emergence of the Global Future," in *The Politics of Globality Since 1945: Assembling the Planet*, ed. Rens Van Munster and Casper Sylvestre (New York, N.Y., 2016), 106–125; see also Ross, *Strange Weather* (cit. n. 2).

[5] John S. Dryzek, *The Politics of the Earth: Environmental Discourses* (Oxford, UK, 1997), 37.

[6] Brian Wynne, "Strange Weather, Again: Climate Change as Political Art," *Theory Cul. Soc.* 27 (2010): 289–306, on 298.

[7] Frederick Buell, *From Apocalypse to Way of Life: Environmental Crisis in the American Century* (New York, N.Y., 2003).

nized by warming mechanisms already in train. In these circumstances, the idea that the prospect of crisis might stimulate radical social change can seem much less tangible. In what follows, I focus on some of the contrasts between the projection of environmental crisis under the sign of the limits to growth in the 1970s, and the idea of climate crisis as a public, cultural, and technoscientific object after the publication of the Intergovernmental Panel on Climate Change (IPCC) Fourth Assessment Report in 2007.[8]

My second argument is that environmental crisis can productively be read as a science-fictional object—that is, as an epistemic entity composed of orientations to planetary futures that are at once descriptive and speculative, scientific and fictional. I base this argument in part on a theoretical claim about the nature of the science-fictional imagination and the distinctive position of the genre as "the literature of technoscientific societies."[9] I understand science fiction after Istvan Csicsery-Ronay as a cultural sensibility that emerged over the course of the twentieth century in response to intense scientific and technological change. Its fictional mode combines extrapolative and cognitive mapping of technoscientific innovation with sociopolitical and ethical critique of likely outcomes. This sensibility helped make environmental crisis thinkable. My argument also involves the more specific suggestion that popular science framings of environmental crisis in the 1960s and into the early 1970s drew extensively on science fiction tropes, rhetorics, icons, and narratives already in cultural circulation. In this sense, the scientific and the science fictional were constitutively entangled in making up the environmental crisis as a matter of public and political concern. As we grapple now with a present and a future being rapidly reshaped by climate dynamics, science fiction is doing vital work examining the prospects for a liveable Anthropocene, and even generating new ways of thinking about the possibility of a better one.

THE SCIENCE FICTIONALITY OF ENVIRONMENTAL CRISIS

In histories of modern environmentalism the announcement of ecological crisis in the 1960s and 1970s is seen as a critical moment.[10] In this telling, crisis is understood as political and metaphysical—as a means of making a radical intervention into the politics of unsustainability and challenging humanity to reject exploitative and destructive capitalism and pursue a more sustainable life in touch with nature. More recently, in histories of science and science and technology studies (STS), the focus has been on how systems modeling and the cybernetic sciences helped to create the idea of

---

[8] Intergovernmental Panel on Climate Change (IPCC), *Climate Change 2007: Synthesis Report. Contribution of Working Groups I, II and III to the Fourth Assessment Report of the Intergovernmental Panel on Climate Change*, ed. R. K. Pachauri and A. Reisinger (Geneva, Switzerland, 2007), https://www.ipcc.ch/report/ar4/syr/.

[9] Istvan Csicsery-Ronay Jr., *The Seven Beauties of Science Fiction* (Middletown, Conn., 2008).

[10] Selective examples include Andrew Dobson, *Green Political Thought*, 4th ed. (1990; repr., London, 2007); Robyn Eckersley, *Environmentalism and Political Theory: Toward an Ecocentric Approach* (Albany, N.Y., 1992); Samuel Hays, *Beauty, Health, and Permanence: Environmental Politics in the United States, 1955–1985* (Cambridge, UK, 1989); David Pepper, *The Roots of Modern Environmentalism* (London, 1984); Douglas Torgerson, *The Promise of Green Politics: Environmentalism and the Public Sphere* (Durham, N.C., 1999); and Donald Worster, *Nature's Economy: A History of Ecological Ideas*, 2nd ed. (Cambridge, UK, 1985). These histories also clarify that the making of modern environmental thought involved the recovery and recontextualization of earlier nineteenth- and twentieth-century writers who emphasized not crisis but conservation, living in place, and romantic accounts of nature's intrinsic value—Rachel Carson, Aldo Leopold, John Muir, and Henry David Thoreau. See, for example, Worster, *Nature's Economy*, 342–56.

planetary crisis, and how in turn the prospect of system collapse helped to construct emerging discourses of globality and planetary environmental management.[11] In this telling, environmental crisis is a product of shifts in the coproduction of scientific knowledge in the middle of the twentieth century, or appears as a crisis in objectivity revealing the hybridity of social, natural, and technological systems, and an opportunity to reimagine the politics of nature.[12]

Both environmental histories and STS accounts touch on the links between environmentalism and science fiction. Environmental histories often identify Rachel Carson's *Silent Spring* as a foundational text, especially the opening chapter with its powerfully apocalyptic image of a future American small town leached of life by chemical pesticides and pollution. Histories of environmental science often touch on some of the science fiction texts that accompanied the announcement of environmental crisis in the 1960s and 1970s—in particular the dystopian narratives of overpopulation and pollution that formed part of the trope or metaphor of "Spaceship Earth"[13] that Sabine Höhler argues was critical to the emergence of discourses of planetary management. But few studies have looked systematically at the changing relationship between science fiction and environmental crisis since the 1970s or drawn on the resources of science fiction literary criticism to do so. Accounts that *have* explored the nexus of science fiction and environmental rhetoric have focused almost exclusively on dystopian and (post)apocalyptic narratives at the expense of exploring the strands of utopian thinking that have been such an important part of modern environmental thinking.[14] To understand environmental crisis, then, we need to understand its historical emergence at the intersection of politics, fiction, and science, as well as its changing cultural functions in mobilizing debates about ethical and social alternatives—including both apocalyptic visions and intimations of better futures with nature that would be sustainable, satisfying, and equal.

Approaching the environmental crisis through the lens of science fiction is a way of exploring the history of environmental ideas across varied epistemic claims and multiple knowledge contexts. It offers new insights into how ecopolitical ideas emerged from a cultural backdrop rich in science-fictional tropes, and how environmental modeling has been taken up and elaborated in genre literature. I will sketch some of the specificities of these intertextual connections in what follows. But I also want to suggest a more fundamental and general way in which the science-fictional imagination has played a constitutive part in the very idea of environmental crisis. Csicsery-Ronay argues that over the course of the twentieth century science fiction literature and its ideas have saturated modern Western cultures. This makes possible a "mood," "habit

[11] Sabine Höhler, *Spaceship Earth in the Environmental Age, 1960–1990* (London, 2016); Fernando Elichirigoity, *Planet Management: Limits to Growth, Computer Simulation, and the Emergence of Global Spaces* (Evanston, Ill., 1999). See also Andersson and Duhautois, "Futures of Mankind" (cit. n. 4).

[12] Bruno Latour, *Politics of Nature: How to Bring the Sciences into Democracy* (Cambridge, Mass., 2004).

[13] Höhler, *Spaceship Earth* (cit. n. 11).

[14] Rachel Carson, *Silent Spring* (Boston, Mass., 1962); Buell, *From Apocalypse* (cit. n. 7); Ross, *Strange Weather* (cit. n. 2); M. Jimmie Killingsworth and Jacqueline S. Palmer, "Millennial Ecology: The Apocalyptic Narrative from Silent Spring to Global Warming," in *Green Culture: Environmental Rhetoric in Contemporary America*, ed. Carl G. Herndl and Stuart C. Brown (Madison, Wisc., 1996), 21–45; Killingsworth and Palmer, "*Silent Spring* and Science Fiction: An Essay in the History and Rhetoric of Narrative," in *And No Birds Sing: Rhetorical Analyses of Rachel Carson's Silent Spring*, ed. Craig Waddell (Edwardsville, Ill., 2000), 174–203, on 177.

of mind," or "kind of awareness"[15] that is powerfully alert to the constant emergence of technoscientific novelty and that offers a tool kit of tropes, metaphors, icons, and narratives for speculating about the futures and social forms that those innovations might bring about. The science-fictional imagination is about the pleasures of playing in and with those possible futures. It also involves a distinctive attitude of estrangement and critique. This is a way of stepping back from how things are and from the apparent certainties of prospective scientific knowledge claims in order to enact what Csicsery-Ronay calls a double hesitation—a pause in which the critical imagination is mobilized to interrogate what is plausible, and to consider what is ethical or desirable.

The very idea of environmental crisis involves a kind of science-fictional imagination. It involves being able to apprehend Earth, its ecosystems, and its socioeconomic arrangements as a single entity, viewed from outside. It involves being able to project a common future for humanity over a hundred-year timescale. Modern environmentalism has always depended on projected planetary futures to make its case, and those futures have always been in the most basic of senses fictional: imagined, created, narrated. They have been *science*-fictional insofar as they partake of science fiction's extrapolative critique of technoscience. It is not a coincidence that modern environmentalism emerged contemporaneously with a major shift in science fiction. Peaking in the early 1950s,[16] science fiction had been a genre largely celebratory of scientific futurism and technocratic models of progress. By the early 1970s, it was increasingly critical of the powers of technoscience and alienated from the supposed benefits of Western modernity. Both science fiction and popular environmentalist nonfiction texts have offered images of overcrowded cities; burning, drowning and drying worlds; or angry protagonists fighting back against despair. Dystopian science fiction from the 1950s to the 1970s, in the novels of J. G. Ballard, John Wyndham, Frank Herbert, Ursula Le Guin, Phillip K. Dick, John Brunner, and Harry Harrison, both anticipated and reflected environmentalist themes of pollution, overpopulation, limits to growth, and widespread alienation.[17]

In this period, an emergent environmentalism and a changing science fiction were operating together to map and contest the globalizing hubris of military-capitalist technoscience. They did so in the shadow of the bomb and the threat of nuclear catastrophe. Worster dates the "age of ecology" to the first atomic explosion in 1945, the consequences of which included widespread doubt about "the entire project of the domination of nature that had been at the heart of modern history," and "the moral legitimacy of science, the tumultuous pace of technology."[18] Thomas Disch, editing an early collection of science fiction short stories about environmental crisis in 1971, also makes this connection by characterizing ecological destruction as "bombs" that are "already dropping."[19] Today's crisis is shaped not only by different substantive threats but by a dif-

---

[15] Csicsery-Ronay, *Seven Beauties* (cit. n. 9), 2–3.

[16] This peak was specifically in 1953, according to Thomas M. Disch, "On Saving the World," in *The Ruins of Earth*, ed. Disch (London, 1973), 9.

[17] In this article, my primary focus is on novel-length science fiction; science fiction film and short stories would obviously add further examples and themes to this list.

[18] Worster, *Nature's Economy* (cit. n. 10), 342, 343.

[19] "In effect the bombs are already dropping—as more carbon monoxide pollutes the air . . . as mercury poisons our waters, our fish, and ourselves, and as one by one our technology extinguishes forms of life upon which our own life on this planet depends"; Disch, "On Saving" (cit. n. 16), 11. This connection is a culturally complex one. The continuities are important—but modern environmentalist discourse is distinctive in its addition of a pervasive sense of the vulnerability of nature and a focus on

ferent logic of chronic, ongoing, and well-known environmental dynamics, including the logics of climate change that are already in train. Science is now better at mapping threats and modeling likely futures; but critical and imaginative political and social thinking is more, not less, important. Science fiction continues to furnish contemporary cultures with crucial resources for imagining environmental futures.

INVENTING ENVIRONMENTAL CRISIS: LIMITS, SCIENCE, AND SCIENCE FICTION IN THE 1970S

> If the present growth trends in world population, industrializa-
> tion, pollution, food production and resource depletion continue
> unchanged, the limits to growth on this planet will be reached
> sometime within the next one hundred years. The probable re-
> sult will be a rather sudden and uncontrollable decline in popu-
> lation and industrial capacity.
>
> —*The Limits to Growth*, 1972[20]

In 1972, the Club of Rome published *The Limits to Growth: A Report for the Club of Rome on the Predicament of Mankind*. This widely publicized assessment of future global development aimed to show that over the next hundred years unconstrained economic growth would compromise and ultimately break the capacity of Earth's ecosystems to support human life. In the fifty years since its publication, the idea of limits to growth has been dismissed as economically naïve neo-Malthusianism and green doomsday pessimism. Its models and methodologies have been critically deconstructed, and its data has been dismissed. By the mid-1980s, global policy makers were keen to dismantle the notion of a single planetary limit to growth and instead frame the ecological challenge in terms of sustainable development and multiple development pathways.[21] But in the decade or so after its publication, *The Limits to Growth* "opened the public imaginary to the possibility of thinking anew about the relation between humanity and the biosphere."[22] It presented that relationship in terms of a globally interconnected system, remaking the planet as a unified object of technoscientific knowledge and a field for management and intervention.[23] It also provided new conceptual resources for a nascent environmental movement by holistically projecting dynamics of resource depletion, population growth, and industrial expansion over the next century to suggest that the trajectory of industrial capitalist growth was heading inevitably toward global collapse.

The Limits to Growth did not invent the idea of environmental crisis, nor was it unique in attaching environmental concerns to uncertain futures. But it did introduce

---

crisis in interconnecting systems; see Worster, *Nature's Economy* (cit. n. 10); and other useful discussions in Ross, *Strange Weather* (cit. n. 2); Buell, *From Apocalypse* (cit. n. 7); Killingsworth and Palmer, "Millennial Ecology" (cit. n. 14); and others.

[20] Donella H. Meadows et al., *The Limits to Growth: A Report for the Club of Rome's Project on the Predicament of Mankind* (New York, N.Y., 1972), 23, https://www.clubofrome.org/report/the-limits -to-growth/.

[21] World Commission on Environment and Development (WCED), *Our Common Future* (Oxford, UK, 1987), 43; Lisa Garforth, *Green Utopias: Environmental Hope Before and After Nature* (Cambridge, UK, 2017), 40–8.

[22] Elichirigoity, *Planet Management* (cit. n. 11), 111.

[23] Ibid., 7.

a way of projecting earthly trajectories that could claim to be scientific. The world system about which *The Limits to Growth* made its arguments was not the empirical Earth of geology and ecology or the sensory Earth of everyday experience. It was the product of new epistemic practices that allowed the globe as an object to be seen and measured from the outside via new technologies of visualization—in the photographic images of Earth enabled by human space flight in the late 1960s and 1970s,[24] and via computer simulations constructed in the logical terms of systems theory. As has been extensively documented,[25] in *The Limits to Growth* the humanistic, existential "world problematique" of the Club of Rome met the "mathematically conceptualize[d] planet" of Jay Forrester's dynamic systems models.[26] Forrester had already used his training in postwar cybernetic sciences to simulate the future behaviors of systems including factories and cities. He had developed techniques for abstracting and quantifying complex, dynamic interactions among multiple interdependent factors, in particular the positive and negative feedback loops through which systems self-regulate. In the early 1970s Forrester offered this methodology to the Club of Rome. A team led by Dennis Meadows modeled relationships among global rates of population, industrialization, pollution, food production, and resource depletion, and then generated aggregated global data for each variable. Digital algorithms enabled interactions among variables to be simulated across different "runs" over a hundred years or so.

The data on world industrial development, resource use, pollution, and so on that fed the world system model had to be conceptualized and aggregated from multiple incommensurate sources. As Elichirigoity points out, the model in effect created the global data rather than vice versa.[27] The Club of Rome's aim was to "represent real world relationships pictorially or mathematically," with the emphasis on visual simplicity and ease of understanding.[28] The graphs that illustrate the world system runs in *The Limits to Growth* use vague quantities and shifting timescales. Numerical values, the authors admit, are "only approximately known."[29] The point is not quantitative accuracy but rather demonstrating the "relations between variables."[30] The emphasis is on the "functional interdependence" of a system constituted via "conceptual and arithmetic abstraction and simplification."[31] Paul Edwards noted that "system structure and dynamics mattered far more than precise inputs."[32]

The scientific knowledge that makes up environmental crisis was then not simply a product of empirical observations, but was related to abstractions based on principles derived from systems theory. It is well known that Forrester's models foregrounded certain kinds of dynamic relationships—in particular, logics of exponential growth

[24] Wolfgang Sachs, "The Blue Planet: An Ambiguous Modern Icon," *The Ecologist* 24 (1994): 170–5, on 170. See also Höhler, *Spaceship Earth* (cit. n. 11).

[25] W. Patrick McCray, *The Visioneers: How a Group of Elite Scientists Pursued Space Colonies, Nanotechnologies, and a Limitless Future* (Princeton, N.J., 2013); Andersson, "The Great Future Debate" (cit. n. 4); Elichirigoity, *Planet Management*; Höhler, *Spaceship Earth* (both cit. n. 11).

[26] McCray, *The Visioneers* (cit. n. 25), 29–30.

[27] Elichirigoity, *Planet Management* (cit. n. 11), 90.

[28] Dennis Meadows, cited in ibid., 97.

[29] Dennis Meadows, cited in Höhler, *Spaceship Earth* (cit. n. 11), 72. Höhler shows how the graphs in *The Limits to Growth* removed the vertical scales to bring multiple variables together on a single axis showing a multifactorial planetary catastrophe.

[30] McCray, *The Visioneers* (cit. n. 25), 30.

[31] Höhler, *Spaceship Earth* (cit. n. 11), 71.

[32] Paul N. Edwards, *A Vast Machine: Computer Models, Climate Data, and the Politics of Global Warming* (Cambridge, Mass., 2010), 367.

and delayed feedback loops. The epistemological centrality of feedback and amplification in systems dynamics and cybernetics meant that the models almost always projected sudden and catastrophic change. They emphasized exponential growth rather than gradual or linear trajectories. The continuity of the current system could not be taken for granted. Crisis was built in; the model was designed from the outset to project an impending "global emergency."[33] As Edwards argues, "no matter what they were simulating, Forrester's models tended to be insensitive to changes in most parameters . . . the models offered a way to discover the few parameters and structural changes that . . . would otherwise escape anyone's attempt to control them.[34] World systems models were designed from the outset to be about crisis and intervention rather than prediction for its own sake. *The Limits to Growth* carefully framed its futures as scenarios or projections, not forecasts. The Club of Rome in fact expected that the presence of catastrophe in the limits models would underwrite a rational argument for a planned transition to a sustainable or steady-state future.

Insisting on the possibility of the future as discontinuity, the idea of limits brought into play what had previously been an "apparently unnatural and unimaginable" prospect:[35] a no-growth economy and a society in which human well-being is not predicated on consumption and expansion. In this way, environmental crisis might be said to have a science-fictional or utopian function. The idea of planetary limits drops a *novum*, a novel piece of scientific or technological knowledge, into our cognitive worlds.[36] It has the capacity to unsettle assumptions and make existing social and political arrangements seem contingent and open to change. It works through defamiliarization and the speculative imagination to distance us from everyday perceptions and understandings.[37] The progressive future in which economic growth guarantees well-being comes to seem strange. An ecologically sustainable alternative is made thinkable, even desirable. This articulation of environmental crisis depends on a kind of science fictional imaginary that has made possible the ability to understand the earth from the outside—as a technologically transformed object and in relation to timescales outside human experience.

A key ingredient of that novum was the metaphor of "Spaceship Earth" that had been emerging in environmental discourse through the 1960s. As Höhler has shown, this Earth—our own planet "newly discovered," singular, knowable, and limited[38]—is a technoscientific object. But it also depends on a science-fictional metaphor that substitutes the soft planet of nature, everyday sensory experience, and existential comfort with a much harder and stranger one composed of technologies, manufactured objects, and limited supplies. Human space flight in the 1960s itself, Höhler suggests, was as much science-fictional as real.[39] By 1968, piloted space flight had produced images of the blue planet, of Earth seen from space, as finite, fragile, and unique. But the possibilities of exploration in the solar system and beyond, which this view of Earth opened up, remained (and remain) prospective, even fantastical. The visual figure of the spaceship that framed environmental crisis discourse in the late 1960s and early

[33] The Club of Rome's Aurelio Peccei, cited in Elichirigoity, *Planet Management* (cit. n. 11), 67.
[34] Edwards, *A Vast Machine* (cit. n. 32), 367.
[35] Meadows et al., *The Limits to Growth* (cit. n. 20), 167.
[36] Darko Suvin, *Metamorphoses of Science Fiction* (New Haven, Conn., 1979).
[37] Killingsworth and Palmer, "*Silent Spring* and Science Fiction" (cit. n. 14), 183.
[38] Höhler, *Spaceship Earth* (cit. n. 11), 8.
[39] Human space flight was "already mastered . . . and yet aspired to . . ."; ibid., 47.

1970s then seemed scientific, but this image derived as much from the television show *Star Trek* and the novel *Solaris* as it did from the Apollo 8 or 11 missions.[40]

In this and many other ways, the idea of crisis articulated as science in *The Limits to Growth* clearly emerges from a cultural context rich in science fiction. Rachel Carson's *Silent Spring* was a crucial popular science text in the mobilization of apocalyptic rhetorics and the elaboration of what Jimmie Killingsworth and Jacqueline S. Palmer call "millennial ecology."[41] The "speculative" and "emotional" qualities of Carson's text, along with its formal innovation as a thought experiment about the slow end of the world through pollution, led it to be dismissed by scientists in its own time as mere science fiction—trivial and fantastical.[42] But those qualities are now celebrated as the source of its "moral complexity" and rhetorical power.[43] Killingsworth and Palmer read *Silent Spring* as a creative experiment at the edges of popular science and science fiction, borrowing and rewriting apocalyptic tropes to insist on the possibility of change and renewal.[44] In its wake, announcements and elaborations of environmental crisis proliferated in popular culture, comingling ideas and approaches from academic ecology, systems science, popular science, and science fiction. Predictions of a ticking population bomb juxtaposed ideas about Earth's carrying capacity drawn from ecosystems ecology with social and political scenarios of war, famine, and sterilization and suicide programs.

In the 1960s and 1970s, popular science often used futuristic narratives and science-fictional rhetorics to warn of impending environmental crisis. Paul Ehrlich, a prominent bioscientist, wrote narratives that mixed systems science, statistical demography, and action-thriller storytelling to depict a world on the precipice of catastrophic overpopulation.[45] His book *The Population Bomb* presented dramatic fictional projections as popular science (although a related essay, "Ecocatastrophe!" was later published in a science fiction anthology).[46] At the same time as popular science was projecting fantastical futures, science fiction writers and editors soberly claimed that their stories were about observable facts in the present, and were "not catastrophes of the imagination."[47] Near-

---

[40] The original *Star Trek* seasons 1–3 aired from 1966 to 1969 (created by Gene Roddenberry, NBC). Stanisław Lem's novel *Solaris* was published in 1961 in Warsaw, with the Soviet Tarkovsky film following in 1972 (most recent home media release was in 2011 by the Criterion Collection, Blu-ray Disc). Apollo 8 in 1968 accomplished the earliest US moon orbit and generated the "Earthrise" images of the blue planet seen from space. Apollo 11 was the mission of the first moon landing in 1969. The metaphor "spaceship Earth" dates originally to the nineteenth century but came to prominence in popular discourse when it was used by Kenneth Boulding in 1966 and Buckminster Fuller in 1968, and then became an icon of environmentalism after its use by United Nations Secretary General U Thant on Earth Day 1971. For a fuller discussion, see Höhler, *Spaceship Earth* (cit. n. 11).

[41] Killingsworth and Palmer, "Millennial Ecology"; Killingsworth and Palmer, "*Silent Spring* and Science Fiction" (both cit. n. 14), 191.

[42] Killingsworth and Palmer, "*Silent Spring* and Science Fiction" (cit. n. 14), 175.

[43] Ibid., 181.

[44] Killingsworth and Palmer, "Millennial Ecology"; Killingsworth and Palmer, "*Silent Spring* and Science Fiction" (both cit. n. 14), 177.

[45] Killingsworth and Palmer describe Ehrlich's tone in *The Population Bomb* as that of "a prophet living in the last days"; see "Millennial Ecology" (cit. n. 14), 32. Paul Ehrlich, *The Population Bomb* (New York, N.Y., 1968). It is difficult now, however, not to read the opening of Ehrlich's book as an account of privileged white revulsion in response to an unfamiliar and densely populated city in the global South as much as a scientific treatise on global population.

[46] Ehrlich, *Population Bomb* (cit. n. 45); Paul Ehrlich, "Eco-catastrophe!" *Ramparts*, September 1969, 24–8.

[47] Disch, *Ruins* (cit. n. 16), 16, on 11. Another key anthology from this time was Virginia Kidd and Roger Elwood, eds., *Saving Worlds* (New York, N.Y., 1973), which, the back cover explained, was a "collection of astonishing science fiction about the ecological *crisis*" (emphasis mine).

future dystopian narratives of systemic environmental collapse had proliferated in the 1960s (Harry Harrison's 1966 *Make Room! Make Room*, filmed as *Soylent Green* in 1973; John Brunner's 1968 *Stand on Zanzibar* and his 1972 *The Sheep Look Up*; and Philip K Dick's 1968 *Do Androids Dream of Electric Sheep?*). In the decade between *Silent Spring* (1962) and *The Limits to Growth* (1972), the idea of a future environmental crisis emerged in multiple forms across popular science, genre fiction, and environmental politics.

There was a powerfully utopian dimension to all these constructions of environmental crisis. It is often argued that ideas about limits framed the cultural mood of the 1970s in one-dimensionally "catastrophic" or "pessimistic" terms.[48] W. Patrick McCray dwells on the "bleak assessment of the future"[49] launched by the idea of ecological limits to economic growth. His analysis emphasizes the ways in which environmental critiques in this period appeared to suggest a sudden reversal of decades of progressive optimism. Instead, they offered apprehension about technology; anxiety about pollution and resource depletion; and looked ahead to a cramped, regressive future. Here the very idea of limits becomes a "shibboleth" for technological progressives who would go on to innovate a future defined against it in the name of technological expansion and enhancement.[50] Ross reads postwar futurology in terms of the replacement of the future shock of nuclear annihilation by "dark eco-futures predicated upon slow environmental deterioration and collapse."[51] The science fiction dystopias of the period envisaged crowded cities, survivalist brutality, and relentless scarcity. The concerns and the aesthetics of these futures became less urgent but more entrenched through the 1980s, not least in cyberpunk's rainy, crowded, decrepit cities; its dirty realism; and dreams of transcending the limits of the human body.[52]

But accounts of the period that dwell only on imagined futures of collapse, pessimism, and dystopia miss the element of warning and critique in dystopian science fiction and the utopian speculation that was also part of the idea of environmental crisis. A stark binary between projected catastrophe and desired transformation was a hallmark of Western environmental concern from the early 1970s to around 1990, when sustainable development became the more powerful policy narrative framing environmental futures.[53] This mood of radical green hope is often overlooked in histories of scientific ideas, but it is vividly present in environmental histories and in ecopolitical philosophy's accounts of its origins. Visions of better futures for humans and the natural world are at the heart of deep ecology, which articulates ecocentric ethics and suggests modes of human well-being that are not dependent on economic growth and materialism. The idea of limits enables an alternative vision of living in place; of a rich culture of self-expression; and freedom from consumerism, consumption, and alienation. Economic sufficiency would restore humans to nature and to "the deep pleasure and satisfaction we [might] receive from close partnerships with other forms of life."[54] Utopian

---

[48] On "catastrophic," see McCray, *The Visioneers* (cit. n. 25), 22; on "pessimistic," see Ross, *Strange Weather* (cit. n. 2), 185.

[49] McCray, *The Visioneers* (cit. n. 25), 5.

[50] Ibid., 6.

[51] Ross, *Strange Weather* (cit. n. 2), 171.

[52] Ibid.; see also Buell, *From Apocalypse* (cit. n. 7).

[53] Douglas Torgerson, "Reflexivity and Developmental Constructs: The Case of Sustainable Futures," *Journal of Environmental Policy and Planning* 20 (2018): 781–91, DOI:10.1080/1523908X.2013.817949 (published 12 July 2013), accessed online 25 March 2019; Garforth, *Green Utopias* (cit. n. 21), 40–49.

[54] Arne Naess, "The Shallow and the Deep, Long-Range Ecology Movement," *Inquiry* 16 (1973): 95–100, on 96.

speculation was also present in ecocentric political proposals in the 1970s and 1980s, with detailed descriptions of the no-growth economy, appropriate technologies, bioregional decentralization, local democracy, and a culture of communality and cooperation. In this period, environmental thought embodied a kind of apocalyptic thinking that emphasized the renewal and change implied by the prospect of an ending.[55]

Speculation about a better future within economic limits was powerfully expressed in literary science fiction from the middle of the 1970s to the early 1990s. Le Guin's *Always Coming Home*, Kim Stanley Robinson's *Pacific Edge*, and Marge Piercy's *Woman on the Edge of Time* reinvented the genre of utopian fiction to respond to environmental crisis, articulate new political subjectivities, and bring to life alternatives to capitalist expansion.[56] These novels do not show environmental crisis directly, but it haunts their fictional futures. Piercy and Robinson set their utopian worlds one- to two-hundred years in the future and describe people and lifestyles embodying and enacting values of sustainability, small-is-beautiful economics, bioregional awareness, and local democracy. But at the limit of their worlds the environmental crisis intrudes. *Woman on the Edge of Time* depicts characters fighting threats to their green, utopian society at its geographical borders. The novel also uses the science fiction device of multiple timelines to show characters going back in time to try to ensure that their own present will prevail. Connie Ramos, the novel's central present-day character, at one point crashes into the wrong future where environmental crisis has exploded in a hypercapitalist future patriarchy. In *Pacific Edge*, we go back to the 1980s with the grandfather of the protagonist who is watching environmental collapse begin to unfold, while searching for the utopian desire and political strategies to change it. In *Always Coming Home*, Le Guin presents a radically ecocentric future documented through the texts of an imagined anthropology, a society that seems to exist because it has severed all narrative and historical connections to our own time. The people of the Valley, it is said, "have always lived there."[57] But the Valley still contains the material detritus of a long, slow apocalypse; people tell stories of stumbling into a world where "the air is thick and yellow" and "the road is coated with grease and feathers."[58]

The binary between utopian and dystopian futures both within science fiction texts and across the genre was part of a wider cultural response to ecological concerns from the 1970s onward. There was one path going ahead to the future, and it could go in one of two sharply divided directions: continuity versus change, rational intervention versus catastrophe, a better future with nature versus business as usual. This dualism, I argue, is a product of the projection of possible environmental crisis at a global scale into a faraway future. The idea of environmental crisis in the early 1970s was

---

[55] Killingsworth and Palmer, "*Silent Spring* and Science Fiction" (cit. n. 14); Stefan Skrimshire, "Eternal Return of the Apocalypse," in *Future Ethics: Climate Change and Apocalyptic Imagination*, ed. Skrimshire (London, 2010), 219–41.

[56] Marge Piercy, *Woman on the Edge of Time* (1976; repr., London, 1979); Kim Stanley Robinson, *Pacific Edge* (1990; repr., London, 1995); Ursula K. Le Guin, *Always Coming Home* (London, 1986). These are not the only ecotopian fictions or utopian fictions with ecocentric elements published in this period (although they are probably the most clearly focused on sustainable futures, as well as being narratively successful). Other examples include Ernest Callenbach's widely read *Ecotopia*, first self-published in 1975; Sally Miller Gearhart's *The Wanderground* (1979); Joanna Russ's *The Female Man* (1975); and Ursula K. Le Guin's *The Dispossessed* (1974). See also Tom Moylan, *Demand the Impossible: Science Fiction and the Utopian Imagination* (1986; repr., Berlin, 2014).

[57] Le Guin, *Always* (cit. n. 56), 99.

[58] Ibid., 155–6.

dependent upon certain kinds of scientific modeling. But the science fictionality that is also part of environmental crisis is not simply a reflection on those scientific projections. The very ideas that made up environmental crisis are science-fictional, and its elaboration in (popular) science and science fiction are in many ways chronologically and formally inseparable. Insofar as the science-fictional imagination was a critical part of environmental crisis and environmental knowledges in the 1970s to the 1990s, so was a rich sense of the social, which was also predominantly conceived in terms of a clear contrast between business as usual and the possibility of radical change. This would change significantly as ideas of environmental futures became more and more dominated by climate modeling.

UNMODELING APOCALYPSE: CLIMATE SCIENCE AND SCIENCE FICTION IN THE EARLY TWENTY-FIRST CENTURY

The scientific basis of future climate scenarios is more robust and secure than previous forms of environmental modeling. But the social and even utopian elements that were so important in the first iteration of environmentalism's futures are now less radical and more fugitive. I want to argue that this is because the kinds of knowledge involved in projecting climate futures have changed, in concert with a shifting political and cultural context. Climate science and its institutions have helped to construct more plausible and detailed knowledge claims about possible and probable environmental futures. But some have argued that this consensus has been achieved at least in part by creating a gradualist map of climate change that does not formally acknowledge potentially catastrophic scenarios. The prospect of crisis has been effaced in climate policies, even as it has arguably been normalized in contemporary culture.[59] The future scenarios of contemporary climate science and policy are multiple and overlapping, rather than binary as in the limits to growth trajectory. It is increasingly difficult to represent the future as radically discontinuous with the present, and harder still to imagine desirable possibilities in the context of locked-in carbon logics. In science fiction, however, enabled by a popular culture replete with tropes of climate crisis, the political, social, and human dimensions supposedly evacuated from science and policy are vividly present. Science fiction continues to work with apocalyptic narratives to figure the ethical, metaphysical, and even utopian possibilities of a climate-changed world.

Modeling global climate systems and projecting climate futures has a complex history. Particularly important has been the emergence of GCMs (general or global circulation models) over the course of the twentieth century,[60] and the more recent adoption by the Intergovernmental Panel on Climate Change (IPCC) of standardized scenarios for developing policy responses to the climate problematic. By the 1930s, Edwards shows, GCMs had become the ideal for long-term climate modeling. It was the emergence of the circulation models that generated attempts to reanalyze historical weather data in order to map how the atmosphere and oceans maintain thermal equilibrium by receiving and reradiating solar energy.[61] By the 1970s, research programs for developing GCMs were established at many US and other Western institutions.[62] They were de-

[59] Buell, *From Apocalypse* (cit. n. 14); Erik Swyngedouw, "Apocalypse Forever? Post-Political Populism and the Spectre of Climate Change," *Theory Cult. Soc.* 27 (2010): 213–32.

[60] Edwards, *A Vast Machine* (cit. n. 32).

[61] Ibid., 143.

[62] Ibid., 167.

signed and refined to show average climate patterns over long periods, to "display predictable symmetry, stability and/or periodicity,"[63] and to filter out local specificity and short-term and anomalous weather events. By the early 1990s, under the aegis of the IPCC, a scientific consensus had emerged that was based on converging GCM modeling. The agreement was that anthropogenic climate change was real, and it required global policy attention.[64] Forecasts based on specific data inputs are not made by GCMs. They instead "replicat[e] the world in a machine,"[65] using massive computer power to "simulate their own climates."[66] Climate dynamics drawn from historical data are spun up, stabilized (to reach equilibrium),[67] and used to project possible future atmospheric trajectories under various conditions of climate forcings—the elements external to the system that drive large-scale changes.

Like Forrester's dynamic systems simulations, climate models are not about predictions. They are designed to foreground relationships and highlight "the planet as a dynamic system: interconnected, articulated, evolving, but ultimately fragile and vulnerable."[68] Climate models, too, have depended on the retrospective mobilization and aggregation of patchy and often incommensurable data.[69] Like the world systems models, climate simulations are generally set to run over fifty to a hundred years. New epistemological insights about climate depend on very high levels of abstraction, both spatial and temporal. In this loose political and conceptual sense, climate modeling did build on aspects of the world systems dynamics. But in its institutional and epistemic development, contemporary climate science rejected Forrester's methods.[70] It emerged from a very different network of tightly interlinked research centers and communities that are now stable and long-standing, and that have little connection with the earlier phase of environmental crisis modeling. Climate modeling is more sophisticated than the early days of global systems models, and climate knowledge is more tightly integrated.[71] Through the IPCC there has been a hybridization of science and public policy knowledges,[72] as well as the emergence of a distinct scientific and regulatory regime of climate governance.[73] At its center are the models themselves, serving as "technical gateways"[74] in the stabilization of climate knowledge, showing how economic activities in the past have created accumulations of carbon gases in the atmosphere, and suggesting the paths of rising concentrations of greenhouse gases (GHGs) in the atmosphere linked to rising global temperatures over the coming decades.

Within these sophisticated and tightly integrated epistemic machineries there are of course endemic and even irreducible uncertainties associated with attempts to predict

---

[63] Ibid., 141.

[64] Ibid., 396–8.

[65] Ibid., 138.

[66] Ibid., 337.

[67] Ibid., 149.

[68] Ibid., 2; see also John Urry, *Climate Change and Society* (Malden, Mass., 2011), 25.

[69] Edwards, *A Vast Machine* (cit. n. 32), 337.

[70] Ibid., 37.

[71] Edwards, *A Vast Machine* (cit. n. 32); Urry, *Climate Change and Society* (cit. n. 68).

[72] Mike Hulme, *Why We Disagree About Climate Change: Understanding Controversy, Inaction and Opportunity* (Cambridge, UK, 2009); Piero Morseletto, Frank Biermann, and Philipp Pattberg, "Governing by Targets: *Reductio ad unum* and Evolution of the Two-Degree Climate Target," *International Environmental Agreements* 17 (2017): 655–76.

[73] Angela Oels, "Rendering Climate Change Governable: From Biopower to Advanced Liberal Government?" *Journal of Environmental Policy & Planning* 7 (2005): 185–207.

[74] Edwards, *A Vast Machine* (cit. n. 32), 37.

the future of "large, complex and chaotic systems"—a future that includes the humans and social institutions whose decisions and actions might change outcomes.[75] Many suggest that the process of managing and hybridizing climate knowledge around shared models has necessitated a reduction of the complexities in climate knowledge and resulted in an unduly narrow consensus on predicted climate futures. This has been read variously as the unintended outcome of mobilizing policy and science around a simplified boundary object;[76] an unavoidable consequence of the discursive construction of climate change as a governable knowledge-object;[77] or a deliberate strategy to make climate challenges appear amenable to rational management.[78] In any case, the result has been the dominance of gradualist models of climate change in global scientific and public policy models. John Urry argues that the "reductionism" that is the source of GCM models' projective power also pushes them to converge around an orthodoxy that excludes the possibility of nonlinear and abrupt change.[79] Climate models tend to filter out the uncertainties of complex systems and regional variation, and thus underestimate the complexity of "multiple physical and social feedback mechanisms."[80] Wynne argues that climate models have tended to smooth out amplifying dimensions of positive feedback loops and underestimate uneven accelerations and instabilities.[81]

There are policy and political dimensions to the tendency of climate models to converge around linear and gradualist futures. Since the late 1980s, climate models have been increasingly important to, and underwritten by, the IPCC, whose UN mission is to provide rigorous review and filtering of climate science for decision-makers. The United Nation's policies on climate in the twentieth century have emerged from and reproduce what Michael T. Boykoff, D. Frame, and Samuel Randalls call the "myth of climate stabilization"[82] in attempts to rationally manage a situation in which emissions have long lives and cumulative effects. Greenhouse gases remain in high concentrations in the atmosphere long after the activities that produced them (may or may not) have ceased. The recent history of intensifying carbon economies means that GHG accumulations must now get worse before they can get better. Climate policies, therefore, have tried to project forward a threshold for a safe limit to GHG accumulations in the future, and then "work back to what emissions scenarios will get the world to that concentration."[83] This is perhaps the ultimate in technocratic approaches. Anticipated futures via

[75] Hulme, *Why We Disagree* (cit. n. 72), 82.

[76] Morseletto, Biermann, and Pattberg, "Governing by Targets" (cit. n. 72).

[77] Oels, "Rendering Climate" (cit. n. 73).

[78] Michael T. Boykoff, D. Frame, and Samuel Randalls, "Discursive Stability Meets Climate Instability: A Critical Exploration of the Concept of 'Climate Stabilization,'" *Global Environmental Change* 20 (2010): 53–64.

[79] Urry, *Climate Change and Society* (cit. n. 68), 29.

[80] Ibid., 164.

[81] Wynne, "Strange Weather, Again" (cit. n. 6).

[82] Boykoff, Frame, and Randalls, "Discursive Stability" (cit. n. 78), 58.

[83] Ibid. A parallel, albeit less critical, argument is made about the emergence and stabilization of the two-degree centigrade upper target for temperature rises in United Nations Framework Convention on Climate Change (UNFCCC) policy over the last twenty-five years or so; see Morseletto, Biermann, and Pattberg, "Governing by Targets" (cit. n. 72). The authors suggest that temperature targets work as a *"reduction ad unum,"* simplifying divergent issues and elements into a "synthetic" and singular number that "increase[s] the sense of governability" of climate change (ibid., 656, 657). The two-degree centigrade target, they suggest, began as an effective boundary object in climate science and policy, but since 2010 has become "disembedded" from meaningful methods and processes for implementation and hence politically ineffective (ibid., 658).

quantitative extrapolation are expected to act back on the present. Seeking to protect a threshold between safety and danger, the stabilization approach assumes that GHG accumulations can be managed in order to keep global temperature rises within a predictable range. At the high end of that range, clearly, climate emissions scenarios imply dramatic changes. But crisis itself—the prospect of a future fundamentally discontinuous with the present, of catastrophic or runaway climate change—is not part of the model. Indeed, the model is designed precisely to treat the future as linear and calculable, and hence leave crisis out.

If the dominant mode of climate modeling has brought more certainty and expert consensus to bear on environmental futures, it has at the same time depended on and tended to reproduce an expectation (what Mike Hulme calls a "delusion" [84]) of rational control. Over the past quarter century, climate change has emerged as a governable object in part through the construction of graphic emissions scenarios used by IPCC's reports on climate change in 2001 and 2007. In its *Special Report: Emissions Scenarios* of 2000, the IPCC developed four broad story lines (the "SRES" scenarios) mapping divergent futures characterized by somewhat different demographic, socioeconomic, technological, and environmental trajectories, each with an estimate linking them to levels of annual GHG emissions and rising GHG atmospheric concentrations.[85] Based on the SRES scenarios, the IPCC's Third Assessment Report (TAR) *Climate Change* 2001 and Assessment Report Four (AR4) *Climate Change 2007* modeled increasing global average surface temperatures in a range from 0.4 to 4.0 degrees centigrade.[86] The SRES scenarios show multiple futures, usually on one synthesized graph, with seven or eight fuzzy lines that each represent one-hundred years or so. The lines all rise—some gently, some steeply. But the graph shows no break or turning point. Presenting multiple scenarios and possible development paths appears to offer a rational choice between scenarios. The future becomes a realm of technocratic risk calculation and intervention.[87]

Climate crisis scenarios can seem more scientific and less science-fictional than their earlier limits to growth counterparts. Earlier apocalyptic framings of climate change have been effaced, and technical/neoliberal governance posited as the only solution.[88] The social, political, and existential concerns that were such a rich part of the mix of the 1970s environmental crisis are also contained or effaced.[89] But outside the narrow realm of IPCC-endorsed climate science and policy, catastrophe narratives have been proliferating, offering social and political explorations of a range of climate-changed futures, including nongradualist and catastrophic ones.

Science fiction quickly generated climate stories, mainly dark and dystopian. Among the earliest novels were Octavia Butler's *Parable of the Sower* in 1993, Bruce Sterling's *Heavy Weather* in 1994, and John Barnes's *Mother of Storms* of the same year. A cluster of well-known titles were published following what Hulme has called

---

[84] Mike Hulme, "Governing and Adapting to Climate: A Response to Ian Bailey's Commentary on 'Geographical Work at the Boundaries of Climate Change,'" *Trans. Inst. Brit. Geogr.* 33 (2008): 424–7, on 424.

[85] Intergovernmental Panel on Climate Change, *Special Report on Emissions Scenarios* (Cambridge, UK, 2000), https://www.ipcc.ch/site/assets/uploads/2018/03/emissions_scenarios-1.pdf.

[86] Intergovernmental Panel on Climate Change, *Climate Change 2007: Synthesis Report* (Geneva, Switz., 2007), https://www.ipcc.ch/site/assets/uploads/2018/02/ar4_syr_full_report.pdf; Intergovernmental Panel on Climate Change, *Climate Change 2001: Synthesis Report* (Cambridge, UK, 2001), https://www.ipcc.ch/site/assets/uploads/2018/05/SYR_TAR_full_report.pdf.

[87] Boykoff, Frame, and Randalls, "Discursive Stability" (cit. n. 78).

[88] Oels, "Rendering Climate" (cit. n. 73).

[89] Urry, *Climate Change and Society* (cit. n. 68).

a "turning point" in the cultural positioning of climate change as real and urgent in the mid-2000s. Hulme, *Why We Disagree* (cit n. 72), 82. These include Cormac McCarthy's *The Road* in 2006; Margaret Atwood's Maddaddam trilogy, especially *Year of the Flood* in 2009; Stephen Baxter's *Flood* in 2008; Paolo Bacigalupi's *The Windup Girl* in 2009; and widely read young adult climate dystopias such as Saci Lloyd's *The Carbon Diaries 2015*, published in 2009, and Julie Bertagna's *Exodus* in 2002. Dystopian and apocalyptic climate science fiction is now ubiquitous in popular culture.[90] Most critics position such narratives not as warnings, as in the earlier phase of crisis writing, but as pessimistic, fatalistic, and even conservative.[91] And populist framings of future climate catastrophe can of course work to obscure or close down the possibility of renewal and social change, becoming thin "millennial" visions militating against collective historical interventions and suggesting a conservative desire to hold onto the same.[92]

But if there is nothing "inherently transformative" in apocalyptic narratives,[93] there is nothing inherently conservative about them either. In a contemporary context where climate management and hyperrational framings of climate adaptation and mitigation dominate, visions of catastrophe stir up ideas that help us to resist and complicate mainstream gradualism, and keep social and political questions on the agenda.[94] In mainstream discourse there is, for Hulme, a "new climate reductionism . . . driven by the hegemony exercised by the predictive natural sciences over contingent, imaginative and humanistic accounts of social life and visions of the future."[95] Wynne suggests the need to develop a more "poetic" articulation of the climate dilemma, to enrich its sociological imagination, and to expand "tacit imaginations of human and social actors and capacities"[96] against technocratic and gradualist discourses. For many commentators, this work should be taken up in the social and human sciences and the newly christened environmental humanities.[97] But this speculative technoscientific imagination, and especially the "poetics" of social critique, is precisely the work of science fiction.

Science fiction has brought its intense narrative focus on "the transformation of human societies as a result of the innovations attending technoscientific projects"[98] to bear on climate science, and it has brought a wealth of genre resources to the exploration of possible, probable, and desirable climate futures. As Adam Trexler and Gerry Canavan argue, this work is not being done within the utopian forms characteristic of

[90] Hulme, *Why We Disagree* (cit. n. 72), 82. This climate science fiction is perhaps even more evident in film than in literature, but here I focus primarily on novel-length science fiction.

[91] See, as one example, Hulme, *Why We Disagree* (cit. n. 72), a book that is enormously nuanced in its argument for the need for a less narrowly scientific and more richly social approach to global warming, but which criticizes a "myth of climate Apocalypse" that is based on a very thin subsection of dystopian and postapocalyptic science fiction with a weak engagement with the functions and forms of genre fiction. I discuss the social science and political theory rejection of apocalyptic narratives in relation to climate change more extensively in *Green Utopias* (cit. n. 21), 106–10.

[92] Swyngedouw, "Apocalypse Forever?" (cit. n. 59), 219; Frederick Buell, "A Short History of Environmental Apocalypse," in Skrimshire, *Future Ethics* (cit. n. 55), 13–36.

[93] Sarah Amsler, "Bringing Hope to 'Crisis': Crisis Thinking, Ethical Action and Social Change," in Skrimshire, *Future Ethics* (cit. n. 55), 129–52, on 140.

[94] Skrimshire, "Eternal Return" (cit. n. 55), 237.

[95] Hulme, "Reducing the Future to Climate: A Story of Climate Determinism and Reductionism," *Osiris* 26 (2011): 245–66, on 245.

[96] Wynne, "Strange Weather, Again" (cit. n. 6), 300. See also Hulme, "Governing and Adapting" (cit. n. 84).

[97] Eva Lövbrand et al., "Who Speaks for the Future of Earth? How Critical Social Science Can Extend the Conversation on the Anthropocene," *Global Environmental Change* 32 (2015): 211–8.

[98] Csicsery-Ronay, *Seven Beauties* (cit. n. 9), 7.

the 1970s radical sociopolitical response to environmental crisis.[99] But it is not reducible to conservative catastrophism, and it includes powerful strands of utopian hope as well as clear-eyed critique. Contemporary Anthropocene fictions are reworking dystopian narratives and repurposing apocalypse to dwell in the parts of the climate projections that both science policy and sociology have avoided. Kim Stanley Robinson's Science in the Capital trilogy, for example, brings climate crisis and the case for a more radical and just science into everyday life and politics.[100] The novels are set five minutes ahead of the reader in an already mutating and multiplying present, blurring generic boundaries between speculation and realism.[101] In *Forty Signs of Rain*,[102] a chaotic weather event disrupts politics as usual when Washington, DC, floods. Domestic and urban spaces open up to strange new connections between humans and non-human subjects as individuals camp out with escaped zoo animals in former parks, and displaced people move in to extend the small nuclear Quibler family that is at the heart of the trilogy. Change in the name of sustainability comes not through heroic individual actions but from the "bureaucratic" utopianism of networked collective agencies rippling through scientific and political institutions via "boring tasks" and everyday administration.[103]

Very recent Anthropocene science fiction has become even more explicitly utopian and activist. New climate genres like solarpunk self-consciously position themselves against the closed and gradualist logics of climate policy and the apparent deadlock of environmental politics, but they also work creatively with mainstream catastrophism. They are keeping the future open to the social imagination—to think beyond both business as usual and "mere survival."[104] Solarpunk explores "the struggles of humanity in an already apocalyptic or dystopian world."[105] Here, real utopian possibilities exist not in contrast to dystopian extrapolation but in the midst of slowly and inexorably unfolding climate change and a range of iterations of its catastrophic consequences, which are often figured spatially. In Robinson's *New York 2140*,[106] partial urban flooding of New York City creates new intertidal zones, both literally and metaphorically. There are new risks—flooded and collapsing buildings in Manhattan, the intensification of "shock doctrine" capitalism. But there are also diverse communities inventing new and better social forms and ways of life together, making both material and political change from high-rise farms to financial market disruption. In Doctorow's *Walkaway*,[107] automation leads to unemployment and precarity for the masses while a caste of the super or "zotta"[108] rich flourish. Many walk away to form improvised, shifting communes at

---

[99] Adam Trexler, *Anthropocene Fictions: The Novel in a Time of Climate Change* (Charlottesville, Va., 2015); Gerry Canavan, "Introduction," in *Green Planets: Ecology and Science Fiction* (Middletown, Conn., 2014), 1–21.

[100] Trexler, *Anthropocene Fictions* (cit. n. 99), 159. Kim Stanley Robinson's Science in the Capital Trilogy comprises: *Forty Signs of Rain* (New York, N.Y., 2004); *Fifty Degrees Below* (New York, N.Y., 2005); and *Sixty Days and Counting* (New York, N.Y., 2007).

[101] Trexler, *Anthropocene Fictions* (cit. n. 99), 12–13.

[102] Robinson, *Forty Signs of Rain* (cit. n. 100).

[103] Trexler, *Anthropocene Fictions* (cit. n. 99), 146, 159.

[104] Phoebe Wagner and Brontë Christopher Wieland, "Editor's Note," in *Sunvault: Stories of Solarpunk and Eco-speculation*, ed. Wagner and Wieland (Nashville, Tenn., 2017), 9.

[105] Andrew Dincher, "Foreword: On the Origins of Solarpunk," in *Sunvault* (cit. n. 104), 7–8, on 7.

[106] Kim Stanley Robinson, *New York 2140* (New York, N.Y., 2017).

[107] Cory Doctorow, *Walkaway* (New York, N.Y., 2017).

[108] For a sense of the emergence of the idea of "zotta" in the world of the novel, see ibid., 412.

the spatial and social edges of so-called default reality, where hollowed-out cities meet a polluted countryside that in turn meets the open spaces of a super-advanced wireless cloud. In an only partly ironic spirit, the walkaways tell each other that they are enacting "the first days of a better world."[109] They mobilize a multitude of post-scarcity possibilities created by new technologies of digital fabrication and negotiate the many interlocking and unfolding consequences of global warming. These include reappropriating some of the ruined material and communal infrastructures left behind by a predatory late capitalism. This is the environmental crisis as "way of life" that Buell somberly diagnosed in the early 2000s. But it is not the fiction of mourning and loss that Buell prescribed.[110] It is a fiction bursting with both the awfulness and the energy of the Anthropocene, with ideas for new hybrid natural-social alliances, and with a speculative eye on the diverse possible social forms of livable and, perhaps in some respects, better futures.

BACK TO THE LIMITS?

Scientific modeling has been part of modern environmentalism throughout the postwar period. In contemporary times, the most prominent models are extremely sophisticated simulations of climate change trajectories. As Edwards argues, climate science now exhibits a high degree of consensus in relation to likely scenarios—not a single predictive "bright line," but a set of "shimmering," blurring, overlapping futures that offer a complex, multifaceted, and robust mapping of what might happen. They are not "predictive truth machines." But they are pretty good "reality-based social and policy heuristics."[111] By contrast, the systems dynamics models of the limits to growth were relatively simple, unintegrated, and dependent on poor-quality data.[112] I have argued, however, that it was precisely the somewhat crude temporal projection of a singular planetary crisis in *The Limits to Growth*, coupled with the blurry boundaries between popular science and science fiction, that helped 1970s environmentalism make its epistemological intervention into culture and consciousness. Crisis then made environmental futures a matter for social, political, and ethical debate, and it produced powerful dystopian and utopian fiction. Crisis now is perhaps more realistic and less futuristic; climate change models are more scientific than science-fictional. Environmental crisis in the 1970s framed the present as a "threshold" period, full of urgency and the possibility of radical change.[113] In the early twenty-first century the framing of climate crisis suggests in contrast that threshold periods for change may already be in the past. Policy makers continue to support incremental management of the climate challenge rather than radical interventions, and climate fiction suggests futures in which we will have to live with climate dynamics already in train.

In recent years, there have been demands that contemporary social and political thought should go "back" to the concept of limits to growth (if not *The Limits to Growth*). Some of these demands are made in the name of remobilizing arguments for the end of

---

[109] Ibid., 62. The first instance of this idea, explained to three "noobs" to walkaway life by Limpopo, is couched in terms of a "better world." Subsequent occurrences are more usually phrased "the first days of a better nation"; see page 377, among other instances.

[110] Buell, *From Apocalypse* (cit. n. 7).

[111] Wynne, "Strange Weather, Again" (cit. n. 6), 295.

[112] Edwards, *A Vast Machine* (cit. n. 32), 371–2.

[113] Höhler, *Spaceship Earth* (cit. n. 11), 11.

economic growth, which were given renewed urgency and relevance in the aftermath of the global financial crises of 2008 and after. Some suggest that the Club of Rome presciently announced the beginning of a period of environmental limits that we are still living through, and that there is "unsettling evidence that society is still following the 'standard run' of the original study."[114] Others base arguments for a return to considering limits to growth on something more like its science fictionality: the novelty and shock of the environmental crisis of the 1970s, its promise to contest "the self-satisfied and self-regarding assurance of Western industrialism,"[115] and radically reimagine the future.[116]

I have tried to argue, however, that the conditions in which a book like the *Limits to Growth* could announce a shattering crisis and in which cultural narratives could play out stark political alternatives are no longer in place. Limits and climate crises belong to two distinct periods of environmental politics and consciousness—the first characterized by a forceful and radical challenge to dominant ideologies of progress and economic growth, and the second unfolding in the context of the internalization and normalization of that challenge and the new challenge of climate change.[117] A projected crisis can no longer perform the old anticipatory or utopian function—in particular as a way of thinking against dominant assumptions of unlimited economic growth or the association of economic development with progress, justice, and human well-being. As we have seen in relation to contemporary climate science fiction, crisis now has to be figured in terms of multiple futures that often shrink into the present. The binary certainties of the anticipated limits to growth have ceded to the many emergent possibilities of climate scenarios. Utopian desire and social alternatives must be thought of as growing within, not in opposition to, unfolding environmental catastrophe.

What Andersson characterizes as "the great future debate and the struggle for the world" in the 1960s and 1970s was between technocratic expertise and quantitative predictions on one side, and humanist reflection and utopian desire on the other.[118] In relation to environmental futures, by the 1990s and 2000s that struggle had been complicated by everyday apocalypse, resistance, indifference, and new kinds of uncertainty in relation to climate and environmental crisis. But, as I have indicated in relation to the emerging fictions of climate and the Anthropocene, science fictionality remains a critical part of how we can come to understand environmental futures and environmental challenges. If mainstream climate governance and politics have written crisis out of their projected futures, science fiction has continued to use and

---

[114] Tim Jackson and Robin Webster, *Limits Revisited: A Review of the Limits to Growth Debate* (London, 2016), http://limits2growth.org.uk/wp-content/uploads/2016/04/Jackson-and-Webster-2016-Limits-Revisited.pdf, 3. Others who have either continued to measure the variables included in the original models or who have returned to reexamine *Limits* data include Ugo Bardi, *The Limits to Growth Revisited* (New York, N.Y., 2011); Roberto Pasqualino et al., "Understanding Global Systems Today: A Calibration of the World3-03 Model between 1995 and 2012," *Sustainability* 7 (2015): 9864–89; and Donella H. Meadows, Jorgen Randers, and Dennis L. Meadows, *Limits to Growth: The 30 Year Update* (White River Junction, Vt., 2004).
[115] Wynne, "Strange Weather, Again" (cit. n. 6), 298.
[116] Urry, *Climate Change and Society* (cit. n. 68), 166.
[117] This internalization and normalization have been discussed under a number of signs, including sustainable development, ecological modernization, the politics of unsustainability, and others. See, for example, discussions in Torgerson, "Reflexivity" (cit. n. 53); Urry, *Climate Change and Society* (cit. n. 68); Buell, *From Apocalypse* (cit. n. 7); Swyngedouw, "Apocalypse Forever?" (cit. n. 59); Ingulfur Blühdorn, "Sustaining the Unsustainable: Symbolic Politics and the Politics of Simulation," *Environmental Politics* 16 (2007): 251–75.
[118] Andersson, "The Great Future" (cit. n. 4), 1411.

transform the trope of crisis in order to both anticipate and challenge likely ecofutures. Science fiction is already doing the work that social scientists suggest is the proper job of sociologists and science studies: talking about discontinuous futures, resocializing climate science, arguing about what matters ethically and ontologically as we move into the Anthropocene, and working in between the speculative and the real.

APPLYING (THE HISTORY OF) SCIENCE

Sleeping Science-Fictionally:
Nineteenth-Century Utopian Fictions
and Contemporary Sleep Research

by Martin Willis*

ABSTRACT

In this article, I examine historical representations of sleep found in both medical and fictional narratives of the second half of the nineteenth century. I draw primarily on medical cases constructed as narratives for specialist medical periodicals, on the one hand, and on utopian fictions (or utopian science fictions, as they might also be called), on the other. I place these narratives in dialogue with my own ethnographic writing of experiences within a contemporary sleep laboratory. The aim of this unusual conflation of past and present, and of employing different methodological approaches to the study of a specific subject, is to understand sleep better, in the first instance, but also ultimately to examine how an interrogation of science fiction might be repurposed as an interrogation of the methodology of science fiction. Science fiction is a genre that draws upon the past to imagine a future. My article considers how reimagining such temporal disjunctions as critical practice might allow for new insights, both for future methodologies bridging the sciences and the humanities, and for specific objects of study, such as pathologies of sleep, or any other that has social, cultural, and scientific purchase.

This is an article about both histories of sleep and how science fiction inspires new ways of studying those histories. The article has, therefore, twinned objects of study, and both take their moments of primacy. It can certainly be read as an article on the scientific, medical, and literary history of sleep underwritten by new methodological innovation. However, I frame it principally as an examination of a unique mode and practice inspired by science fiction that takes as its case study a range of scientific, medical, and fictional representations of sleep. Science fiction (SF) has long been considered a genre of imaginative writing that is centrally concerned to articulate an understanding of the present state of science's role within society, and does so by anticipating plausible futures. In their introduction to this volume, Amanda Rees and Iwan Rhys Morus set out some of the ways writers of SF have attempted to do this, and the critical response their work has received.[1] What has not been considered—

* School of English, Cardiff University, Colum Drive, Cardiff, CF10 3EU, UK; willism8@cardiff .ac.uk.

My thanks to Professor Penny Lewis, Director of the Neuroscience and Psychology of Sleep Laboratory, located at the Cardiff University Brain Research Imaging Centre.

[1] Amanda Rees and Iwan Rhys Morus, "Introduction: Presenting Futures Past," in this volume.

either by historians of science or scholars of literature—is how the modus operandi of SF's future-orientated examination might be repurposed, or more correctly, reimagined, as a critical methodology. What SF does, and does freely because it is not bound to any notion of objective praxis, is to enable temporalities to collide; most often these are a recent present with a future. Indeed, because SF draws on rules and builds narrative infrastructure from its own long genre history, it actually combines three temporalities: past, present, and future. What would a similar productive muddling of telos in a scholarly framework look like? One possible answer is the work I do here, in which I place nineteenth-century narratives of sleep (fictional, scientific, and medical) in dialogue with an ethnographic study of a contemporary sleep laboratory.

This seemingly undisciplined methodological innovation is a response, too, to recent scholarship articulating new epistemologies of scientific knowledge–making in both the history of science and science studies. Leading figures in both fields have offered distinct reconceptualizations of how we approach the study of science in the twenty-first century. For George Levine, epistemes of scientific truth need to reach beyond positivist understandings of objectivity to embrace, in the historic materials, "the human role in the constitution of truth." For Levine, this position could be defended by arguing that "the ideal of truth ought to inflect human behaviour."[2] One way to do this, he contends, is not to be "coolly indifferent to the conditions either of investigator or investigated" when studying past science.[3] Science studies scholar Harry Collins, by contrast, finds areas of concern not in human involvement but rather in the temporal disjunctions of knowledge acquisition. Scholars in science and technology studies (STS) who involve themselves in various forms of "participant comprehension" with science, Collins claims, understand the different roles played by the actors (scientists) and analysts (scholars studying them).[4] What remains problematic is the historian who works only "with archive material" because they cannot encompass "a real time analysis where one's interpretations are open to critical review by those one is interpreting."[5] What emerges here is a clear disjunction between the different approaches to knowledge production of historians of science and science studies experts. Nevertheless, there are some who articulate a desire to circumvent the epistemic confusion of temporal shifts and subjectivity. Historian of medicine John Pickstone, for example, argues that "it is naïve to pretend that we should or can forget our present categories," and "to do justice to the past and to use it in the present, we need broad frames in which to think comparisons."[6] My own work in this article offers a methodology that constructs a new node at the intersection of these different understandings of how truth is revealed and disseminated. By dovetailing sleep's historical narratives with a contemporary ethnographic story of sleep research, I shall show how the production of knowledge at different temporal moments need not remain confined within its own telos but can migrate between them if properly marshalled by the self-conscious subjectivity of the investigator.

---

[2] George Levine, *Dying to Know: Scientific Epistemology and Narrative in Victorian England* (Chicago, Ill., 2002), 31.
[3] Ibid., 42.
[4] Harry Collins, *Gravity's Ghost and Big Dog: Scientific Discovery and Social Analysis in the Twenty-First Century* (Chicago, Ill., 2011), 320, 341.
[5] Ibid., 321n.
[6] John V. Pickstone, *Ways of Knowing: A New History of Science, Technology and Medicine* (Manchester, UK, 2000), 5–6.

While SF's interweaving of different temporalities was key to the development of this methodology, it was additionally the interrogation of future-oriented fictions within the field of future studies that reinforced its potential. This field is ably discussed by Peter Bowler elsewhere in this volume.[7] Future studies scholarship has been hugely enriched in recent years with the work of researchers from multiple disciplines. As a community, their work has illuminated some of the conceptual territory where temporal difference can be usefully repositioned to allow for interesting new conjunctions between past, present, and future. This was first articulated as early as 1989, when literary scholar W. Warren Wagar, a writer of SF himself, argued that imagining the future was a way of understanding its "historicity"—that is, how past developments were already part of a "line of development" leading through the present to plausible future scenarios.[8] Since Wagar's ground-breaking study, Arjun Appadurai has extended his range of connections between past and future by noting that many imagined futures are not only anticipations but also aspirations built upon an understanding of the trajectory of past into present.[9] More recently, Barbara Adam has extended this insight to the work of historical scholarship. She has argued that historians should be looking both to the future as well as to the past, in recognition that the historical objects they study are already examples of futures set in motion, albeit they are still in flux.[10] The productive connections between past, present, and future envisaged by future studies scholarship point once again to the importance of developing a methodology that can take some account of the complex interrelation of past and future science, and begin to interrogate how these new temporalities inflect the production and dissemination of knowledge both within the science under analysis and for the researcher hoping to understand it more clearly.

In the sections to follow, then, I take as my key objects of study a series of Victorian utopian fictions.[11] Victorian utopias are ideal objects in this context; they are themselves examples of future-oriented SF, and they explicitly connect their own present with imagined futures. They also offer the scholar additional and productive complexity in this regard. After all, what are present and future to the writer of the Victorian utopia are the past and (often) present to the scholar. In the first section, I highlight the crucial role of sleep in these utopian fictions, and compare that to its representations in science and medicine in the same period. In the second section, I turn to the first of my ethnographic accounts of a contemporary sleep laboratory to reveal how that work influenced my understanding of the Victorian materials. Section three provides an extended examination of the historic materials in light of the ethnography. The fourth section returns to the ethnographic narrative, while section five reexamines the Victorian utopias in response. I conclude by reflecting on what my methodology

---

[7] Peter Bowler, "Parallel Prophecies: Science Fiction and Futurology in the Twentieth Century," in this volume.

[8] W. Warren Wagar, *A Short History of the Future* (Chicago, Ill., 1989), x.

[9] Arjun Appadurai, *The Future as a Cultural Fact: Essays on the Global Condition* (New York, N.Y., 2013), 286.

[10] Barbara Adam, "History of the Future: Paradoxes and Challenges," *Rethink. Hist.* 14 (2010): 361–78, on 374.

[11] The fictions that begin with a form of sleeping and that I shall focus on in this article are W. H. Hudson, *A Crystal Age* (1887; repr., New York, N.Y., 1917); Edward Bellamy, *Looking Backward* (1888; repr., London, 1986); William Morris, *News from Nowhere* (1890; repr., London, 1987); and H. G. Wells, *When The Sleeper Wakes* (1897; repr., London, 1994).

has achieved for my study of sleep, and on what emerges as useful for future research applying the same technique to other studies.

VICTORIAN UTOPIAS AND SLEEP STUDIES

Victorian utopian fictions often employ a form of sleep as the equivalent of the pulp SF space rocket—to transport characters somewhere else at speed. William Morris's 1890 novel, *News from Nowhere*, is a good example. The novel opens with a group of friends "up at the [Socialist] League" enthusiastically advancing "their views on the future of the fully developed new society."[12] From this debate, the soon-to-be uto-pian sleeper, William Guest, returns home on the underground railway, and in that "vapour-bath of hurried and discontented humanity" he hopes to be able to "but see a day" of the potential future just discussed.[13] Once at his riverside home, and relaxed by the natural landscape of the Thames, he goes to bed. Morris then describes his som-nolent performance:

> In this mood he tumbled into bed, and fell asleep after his wont, in two minutes' time; but (contrary to his wont) woke up again not long after in that curiously wide-awake condi-tion which sometimes surprises even good sleepers; a condition under which we feel all our wits preternaturally sharpened, while all the miserable muddles we have ever got into, all the disgraces, and losses of our lives, will insist on thrusting themselves forward for the consideration of those sharpened wits. In this state he lay (says our friend) till he had almost begun to enjoy it: till the tale of his stupidities amused him, and the entan-glements before him, which he saw so clearly, began to shape themselves into an amus-ing story for him. He heard one o'clock strike, then two, and then three; after which he fell asleep again. Our friend says that from that sleep he awoke once more, and after-wards went through such surprising adventures that he thinks that they should be told to our comrades, and indeed to the public in general, and therefore proposes to tell them now.[14]

The nonnormative sleep depicted here would be commonly recognized by Victorian physiologists and today's sleep researchers. Today's researchers would likely ask whether stimulants such as coffee or alcohol had been consumed at the evening meet-ing, which might account for the sleep disruption, as well as note the fact that REM sleep occurs more often later in sleep (and hence after three o'clock).[15] It is also clear, however, that Morris here represents a sleep pathology often catalogued in nineteenth-century medicine. Robert MacNish, the Glasgow physician and much-cited author of *The Philosophy of Sleep*, writes, for example, that "mental emotions, of every descrip-tion, are unfavourable to repose."[16] MacNish found the following:

> If a man, as soon as he lays his head upon the pillow, can banish thinking, he is morally certain to fall asleep. There are many individuals so constituted, that they can do this without effort, and the consequence is, they are excellent sleepers. It is very different with those whose minds are oppressed by care, or over-stimulated by excessive study . . . It is the

---

[12] Morris, *News* (cit. n. 11), 4.
[13] Ibid.
[14] Ibid., 4–5.
[15] "Sleep Laboratory Ethnography Notebooks," Notebook 1, 2017. (As described later in this paper, I conducted a participant observation of selected experimental nights at the Sleep Laboratory at Cardiff University from May 2017, and kept notebooks of my observations.)
[16] Robert MacNish, *The Philosophy of Sleep* (Glasgow, UK, 1834), 196.

same with the man of vivid imagination. His fancy, instead of being subdued by the spell of sleep, becomes more active than ever. Thoughts in a thousand forms—myriads of waking dreams—pass through his mind, whose excessive activity spurns at repose, and mocks all his endeavours to reduce it to quiescence.[17]

Similar perspectives on sleeplessness were reached by numerous physicians and physiologists across the middle and late nineteenth century, so that Morris's sleeper comes to look very much like an archetypal pathological sleeper, and the narrative of his unruly sleep comparable to many a medical study.[18]

The second half of the nineteenth century was rich with narratives and representations of sleep. Large data searches reveal a rise in the incidence of the word sleep from the beginning of the nineteenth century, reaching a peak in the 1890s, before falling off again in the first half of the twentieth century. Both Google's Ngram Viewer and the UK Medical Heritage Library depict this rising and falling pattern.[19] Medical publications give a more pronounced rise in the incidence of sleep, with fewer than 20 individual works citing the word in 1800, to 160 by 1850, and over 280 by the end of the century.[20] This then fell back to approximately 150 in the first years of the twentieth century.

The prolonged sleeping that gave a starting point to utopian fictions in the 1880s and 1890s was commonly debated by medical practitioners in the pages of leading medical journals. The *Lancet*, for example, included a letter to the editor on 3 July 1880, under the title "Sleeping Girls," which drew readers' attention to the "Sleeping Girl of Turville," who, the physician John Gay tells us, "has, it appears, been lying on her side with her hand under her face, and without apparent adequate means of sustenance, since March, 1871."[21] Her image resurfaces again in 1890, remediated through the work of Edward Burne-Jones in his Sleeping Beauty–inspired *Briar Rose* series.[22]

These narrative interweavings suggest the wide-ranging perspectives and varied commentaries on sleep available to later nineteenth-century audiences and the close ties between scientific analyses of disordered sleeping and their reimaginings. Sleep research in the 1890s bears this out. For example, one of the most celebrated of theatrical presentations of prolonged somnolence was produced at the Royal Aquarium in Westminster at the start of 1895. For over a week visitors could see "the sleeper" Henry Nolan, as the *Times* reported it, "in an easy recumbent position in a glass case without a lid." Nolan's well-being was attended to by the physician Forbes Winslow,

---

[17] Ibid., 196–7.

[18] See, for example, William A. Hammond, *Sleep and its Derangements* (Philadelphia, Penn., 1869); Marie de Manaceine, *Sleep: Its Physiology, Pathology, Hygiene, and Psychology* (London, 1897); J. Mortimer-Granville, *Sleep and Sleeplessness* (London, 1879).

[19] Google Ngram Viewer gives the incidence of sleep as 0.00395% in 1800, rising to 0.00560% in 1869, to 0.00561% in 1881, and 0.00570% in 1899. This had fallen substantially to 0.00403% in 1950 and reached a low of 0.00364% in 1965. Interestingly, the incidence has risen sharply since the middle of the 1990s and in the 2010s far exceeds the figures for 1899. For Google's Ngram Viewer, see https://books.google.com/ngrams.

[20] Jisc UK Medical Heritage Library, Historical Texts, Visualising Medical History search tool, https://ukmhl.historicaltexts.jisc.ac.uk/results?terms = sleep&date = 1799-1905&undated = exclude.

[21] John Gay, "Sleeping Girls," letter to the editor, *Lancet* 116 (3 July 1880): 31. Narcolepsy—one of a number of sleep-related conditions—was first defined around this time.

[22] See, for example, Edward Burne-Jones, *The Rose Bower* (1870–1890). The four original paintings of the *Legend of Briar Rose*, as well as additional panels, can be found at Buxton Park, Faringdon, UK.

who had earlier "pronounced Nolan to be absolutely unconscious."[23] The *Morning Post*, which described the glass case as "coffin-shaped," noted that Winslow "examined the man's pulse . . . and declared it to be weak but regular, 78 beats being recorded."[24] This glass-encased sleeper, monitored by medicine, is surely the precursor to, and potential inspiration for, H. G. Wells's own utopian sleeper. Wells was certainly living in London in February 1895 and would have had the opportunity to see advertisements or press reports of the sleeper, although whether he visited the Royal Aquarium is now impossible to know. The third chapter of *When the Sleeper Wakes* certainly provides a striking parallel to the experiment conducted there. As Graham, the sleeper of the title, awakens, he describes how he came slowly to realize "he was not in a bed at all as he understood the word, but lying naked on a soft and yielding mattress, in a trough of dark glass . . . In the corner of the [glass] case was a stand of glittering and delicately made apparatus, for the most part quite strange appliances, though a maximum and minimum thermometer was recognisable."[25] Here is the same glass case and the same physiological assessment as was displayed at the Royal Aquarium. It is a reminder of the close parallels that existed in the period between medical narratives of sleep pathologies, cultural representations of sleeping, and fictional interrogations of sleep's extreme states. The particular link between sleep and utopianism is enhanced by contemporary sleep research, an argument I shall now develop.

INTERLUDE: THE UTOPIA OF THE CONTEMPORARY SLEEP LAB

The contemporary sleep researchers in whose laboratory I undertake my participant ethnography have much in common with their Victorian predecessors.[26] They employ instruments and techniques from neuroscience to study the physiology of their subjects, but ask questions that are primarily psychological. Similarly, late Victorian experts on sleep such as William A. Hammond and Henry Charlton Bastian combined their work as physician or physiologist with the then emerging field of neurology to ask questions about sleep's relation to ailments of body and mind.

Just as striking, however, is the science-fictional nature of the laboratory. The Cardiff sleep lab has a patina of SF to it even before entering. It is housed within the Cardiff University Brain Research Imaging Centre, whose entrance signage is rendered acronymically as CUBRIC.[27] Although its spelling diverges a little from Stanley Kubrick, its sonorous resemblance is clearly not by chance. The facility is designed to appear,

---

[23] "An Experiment in Hypnotism," *Times* (London), 2 February 1895, 10.

[24] "Royal Aquarium, Westminster," *Morning Post* (London), 2 February 1895, 6. For the medical establishment's perspective on this entertainment and Winslow's involvement, see "Death and His Brother Sleep," *Lancet* 145 (9 February 1895): 361.

[25] Wells, *Sleeper* (cit. n. 11), 20.

[26] I conducted a participant observation of selected experimental nights at the Sleep Laboratory at Cardiff University from May 2017 (cit. n. 15). I was invited by two researchers—Joe and Alison (not their real names)—to attend part or all of their memory experiments, which included one or two participants at a time. I spent either a few hours or an entire night with the researchers in the laboratory. My participation was limited to a support role in attaching electrodes to participants (which I was trained to do by the researchers) and a viewing role of the monitoring processes and diagnostic tests. I recorded my own notes in field notebooks and conducted interviews with the sleep researchers outside of the experimental periods. All the subsequent material in this section is drawn from the field notebooks I kept during the summer of 2017. These I have titled "Sleep Laboratory Ethnography Notebooks." The majority of the data here described can be found in Notebook 1.

[27] The Cardiff University Brain Research Imaging Centre (CUBRIC) website can be viewed at http://sites.cardiff.ac.uk/cubric/.

and to be, at the cutting edge of scientific research. The sleep laboratory I have been invited to attend has its own acronym to bear. It is one part of a network of sleep research facilities led by the Neuroscience and Psychology of Sleep Project, or NAPS for short.[28] Although I have been invited to view a particular experiment on the role of sleep in memory, which requires playing sounds to sleeping participants throughout the night and at specific moments in the sleep cycle, it is the culture of the sleep laboratory itself that interests me. It is quickly apparent that the majority of one night's experimenting for the sleep researcher is spent at the computer screen watching multiple EEG signals trace leftward across the screen. These refer directly to the numerous electrodes attached to the head and face of the participants. The memory experiment requires twenty electrode attachments. Each of these is placed individually at the correct position on the head and its long wire traced back to a specific plug point on a paperback book sized router that flashes green and red. As more electrodes are attached each participant comes increasingly to look like a medusa. Sleeping must inevitably be difficult when encumbered with such awkward attachments. Indeed, it becomes clear, as I am shown a template EEG, that the ideal participant is someone who falls asleep quickly, who is able to ignore the wiring and the unusual nature of the laboratory rituals, and who can get to sleep, thus enabling experiments to begin at the earliest possible moment.

Sleep researchers, it also becomes clear, are overwhelmingly frustrated by the participant who thinks they are unlikely to be able to sleep, or sleep well. This is something several of the participants say when they first enter the laboratory. It strikes me as natural nervousness—both to feel this way and to say so—but the sleep researcher is inordinately consumed by the potential for wakefulness. This strikes me as strange until I realize that the sleeping or nonsleeping participant finds a parallel in the sleeping or nonsleeping researcher. The researchers, who have to watch the EEG continually, cannot indulge themselves with a short sleep if their participant remains awake. They must wait for the correct sleep cycle in order to execute the correct computer commands at the required moment. This cannot be automated. No sleeping participant, no sleep cycle, no sleeping researcher. Sleep deprivation is extraordinarily high among sleep researchers. Their conversations with me, and with each other, reveal an obsession with sleep patterns and quantities that has nothing to do with their experiments and everything to do with their own pathological sleeping. One researcher reports to me of another: "Joe's sleep pattern is a disaster. He can never get to sleep until 3 or 4 am now. Their sleep is . . . [dismissive hand gesture]." Joe shows me his own EEG. This is designed to teach me something of sleep scoring. For hours, though, I am watching the EEG of someone who is awake. When we reach a point on the EEG trace where sleep's recognizable elements begin to appear—when Joe can spot a K-complex and I attempt to find a spindle—Joe says: "at last, I'm asleep here."

Sleep is the obsession of sleep researchers, but not as I would have imagined. Their own aberrant wakefulness and their desire for participants to sleep well constructs sleep as utopian. Sleep is something they daydream of reaching. It is envied in others, and longed for in themselves. Sleep is also located; it has a significant spatiality re-

---

[28] Further information on the Cardiff arm of NAPS can be found at http://sites.cardiff.ac.uk/cubric/research-2-2/neuroscience-and-psychology-lab-naps/.

lated to their experimental practice. Sleep happens in the bedrooms of the laboratory and on its computer screens. It is not experiential for the researcher, but rather geographic and temporal. One must go to the computer room to find sleep, and one must arrive there at specific times to witness it. Sleeping is happening somewhere else, somewhere unseen, and usually unreachable.[29] Sleep is situated as both a utopian experience and a site of utopia. It can be characterized, as utopias often are, as a better state. It is also desired; like many utopias it is imagined as ideal before it is reached. Sleeping, for the sleep researcher, is to work fruitfully, but also to emerge from that work refreshed and energized. Fruitful labor that enhances ease is the keynote of nineteenth-century utopian fictions. It is my time in the sleep laboratory that has shown me that sleep is not just the subject of utopian fictions but an aspect of it, even if that ends up in more complex articulations that reach beyond the adjective of good (and its opposite, bad, as I shall discuss later). For my purposes, the reduction to the adjectival sense of good is useful as a way of reinterrogating the historical materials. It is to these materials that I now return.

THE GOOD PLACES OF SLEEP

Hudson's *The Crystal Age*, Bellamy's *Looking Backward*, and Wells's *When the Sleeper Wakes* all begin with forms of pathological sleeping that seem to position sleep at some distance from utopia. Hudson's Smith, who has a fall that renders him unconscious before awaking in the crystal age, suffers from what physicians studying sleep in the period would have called *coma somnolentum*.[30] Bellamy's sleeper, Julian West, is an insomniac who uses largely discredited mesmeric treatments to try to sleep, and Graham, Wells's protagonist, is similarly sleepless and subject to trances. Such discouraging starting points give way to a series of representations of sleep that link it much more closely with the imagined utopian society that each sleeper experiences. Their sleep in particular does not end with entry into a utopian future but becomes a de facto signifier of the state of that utopia. There are therefore periods where sleep not only represents the good place but is actualized as the good place—it becomes understood spatially as the utopian space that the sleepers desire. Graham feels this intuitively as he struggles to come to terms with the future into which he has awoken: "He wrestled with the facts in vain. It became an inextricable tangle . . . An old persuasion came out of the dark recesses of his memory. 'I must sleep,' he said. It appeared as a delightful relief from this mental distress and from the growing pain and heaviness of his limbs."[31] For Graham, sleep is both a state, of ease and pleasure, and a material location, where his embodied experience can be transformed. Similarly, Julian West also struggles to understand the futuristic Boston he has woken into and is advised to sleep in order to cope with the mental unrest it has caused: "I should strongly advise you to sleep if you can tonight, Mr. West," his host Doctor Leete tells him, "in the trying experience you are just now passing through, sleep is a nerve tonic for which there is no substitute."[32] Hudson uses sleep in a more abstract way than either Bellamy

---

[29] Sleep researchers are extremely reluctant to enter the bedroom of a participant at any point during a night's experimenting, both for experimental and ethical reasons.

[30] See W. Travers Cox, "Morbidly Protracted Sleep," *Lancet* 24 (18 June 1835): 410–1; and William T. Gairdner, "Remarks on a Case of Abnormal Disposition to Sleep, Alternated with Choreic Movements," *British Medical Journal* 774 (30 October 1875): 547–50.

[31] Wells, *Sleeper* (cit. n. 11), 70.

[32] Bellamy, *Backward* (cit. n. 11), 114.

or Wells, but likewise locates it as part of a utopian geography; utopia is, as Smith narrates, a "slumberous valley" where he had been brought by the "swift black current" of extended sleeping.[33] In fact, for Hudson, sleep acts as the natural entrance to utopic space. To reach the apogee of utopian experience in the crystal age, when "nature reveals herself to us in all her beauty," certain diurnal rhythms must be followed: "At night we sleep; in the morning we bathe; we eat when we are hungry, converse when we feel inclined, and on most days labour a certain number of hours." The first step toward utopia is sleep, and the activity of sleeping is recognized as the gateway of that journey. Sleeping, in these fictions, has a spatial economy directly related to the utopia on offer. Put simply, the place of sleep is itself utopian.

Nineteenth-century medical writing on sleep also characterizes it as a specific location within the extended experience of the body that acts to address bodily fragility or breakdown. From the 1830s, sleep was eulogized as a place of relief from the hazards of life. It is William A. Hammond, one of the most often-cited physicians writing on sleep across the English-speaking medical world in the later nineteenth century, who provides perhaps the most memorable example of sleep as a utopian space. In his book *On Wakefulness*, he cites a case first detailed by the psychologist Forbes Winslow that tells of a prisoner "sentenced to die by being deprived of sleep."[34] By the end of a week "he implored the authorities to grant him the blessed opportunity of being strangled, guillotined, burned to death, drowned, garrotted, shot, quartered, blown up with gunpowder, or put to death in any conceivable way their humanity or ferocity could invent." For Hammond, this highlights "the horrors of . . . want of sleep."[35] Like the traitor tortured by French soldiers, which is Hammond's next example, it is the "deprivation of sleep" that is "the greatest of all his torments." Sleep here acts as a utopian place to which the unfortunate prisoners are denied access. It is a place where, it is imagined at least, the pains of the body will be nullified and mental repose enjoyed.

With these examples, Hammond builds toward the keynote of his argument that it is the modern world (of the 1860s) that is causing poor sleep habits and thus keeping everyone from the utopian world of good sleep.[36] Increasing "civilisation" and "refinement" combined with poor "hygienic management" of the self had placed sleep in jeopardy.[37] One of the common ways to try to treat the nervous conditions that were seen as a product of this modernity was to institute sleep therapies that mitigated those symptoms. Henry Charlton Bastian, physician at University College Hospital in London and consultant to the national hospital for the paralyzed and epileptic, showed how chorea—a condition whose symptoms are uncontrollable jerking movements of the body—could be almost entirely cured by sleep.[38] Placing his patients into extended sleep states of several weeks' duration, Bastian finds their choreal movements much reduced when they awaken, and sometimes "gone, except very slight

---

[33] Hudson, *Crystal* (cit. n. 11), 249.

[34] William A. Hammond, *On Wakefulness* (Philadelphia, Penn., 1866), 14.

[35] Ibid., 14.

[36] Ibid., 39–40. Hammond's perspective is not singular. Across the later nineteenth century, and in Britain as well as America, this reading of modernity was common. For an alternative version, in a very different print publication, see also "Sleep and Sleeplessness," *Belfast Newsletter*, 9 April 1887, np.

[37] Hammond, *Wakefulness* (cit. n. 34), 39–40.

[38] H. Charlton Bastian, "Clinical Lecture on a Case of Protracted and Severe Chorea Treated by Prolonged Sleep," *Lancet* 134 (13 July 1889): 55–7.

tremors now and then."[39] Bastian, like William Gairdner before him, wrote that the periods of extending sleep seem rather like placing the patient somewhere else as much as in a different form of consciousness. Bastian uses phrases such as "passing out of this state" and into another, suggestive of locomotion as well as mental transformation, while Gairdner also sees the entry into sleep, or the exit from it, as a journey from one state to another.[40] "Many times in the course of an hour," he writes of a choreal patient who suffers from a sleep pathology we would now view as narcolepsy, "the patient may pass from the most perfect wakefulness into the most profound sleep."[41] So certain is Gairdner that this sleeping takes the patient to some other place that he is astonished when she awakens that she can return to a conversation she held previously: "If she be allowed to fall asleep, and then again suddenly reawakened . . . [she] resumes con-versation, or her ordinary occupations, exactly at the point at which she left off."[42] These back and forward movements between the noumenal world of the nineteenth century and the phenomenal world of sleep, and how they are characterized, match to an extraordinary degree the shifts that occur between the utopian novel's present and its utopic future.[43]

First, utopian fictions also regard the modern world as the distressing dystopian pres-ent against which their utopia will be set. Their views on modernity parallel those of the medical profession. Bellamy sees 1880s Boston defined by "squalor and malodourousness" and beset by "stunning clamour" that turns Julian West into a "Madman."[44] Similarly, Wells's sleeper, Graham, feels as though he has "no part" in the Victorian world, that its "thousand distractions" have led to the destruction of his "nervous system."[45] Further, in the continual evocation of memories of the nine-teenth century there is the same oscillation between waking distress and repose as found in the medical case notes of sleep therapy. *Looking Backward* is the most explicit example of this; West recalls the nineteenth century throughout the novel, and in one striking scene he seems to awake not into "the glorious new Boston" of the future, but his own "familiar" nineteenth-century city.[46] Luckily, for West, this proves to be a dream, which is an interesting differentiation, for the critic at least, between the un-helpful dream state and the more therapeutic sleep state that readers have already wit-nessed, leaving West both "greatly refreshed" and "in a dozing state, enjoying the sen-sation of bodily comfort."[47] *A Crystal Age*, of all the utopias of the period, shows the therapeutic effects of good sleep, and contrasts that with the counternarrative of the Victorian present that acts upon Smith "like the memory of a repulsive dream."[48] Sleeping, unlike dreaming, offers utopian ease. Smith finds this early in the novel when "merciful sleep laid her quieting hands on the strings of my brain, and hushed

---

[39] Ibid., 56.

[40] Ibid., 57.

[41] Gairdner, "Abnormal Disposition" (cit. n. 30), 548.

[42] Ibid.

[43] Jean-Luc Nancy also argues that sleep is a location as much as an experience. In sleep, he states, "the body, for its part, abandons itself paradoxically to the very location of the mind: it is no longer actually exposed in space but implicitly or virtually withdrawn into a nonplace where it anaesthetizes itself and separates itself from the world." Jean-Luc Nancy, *The Fall of Sleep*, trans. Charlotte Mandell (New York, N.Y., 2009), 35. The location of mind is where Kant would place the phenomenal.

[44] Bellamy, *Backward* (cit. n. 11), 218–9, 229.

[45] Wells, *Sleeper* (cit. n. 11), 5–6.

[46] Bellamy, *Backward* (cit. n. 11), 215–6.

[47] Ibid., 77.

[48] Hudson, *Crystal* (cit. n. 11), 149.

their weary jangling."[49] It is, however, not in offering repose that sleep has its ultimate power, but in its ability to salve the health of the mother of the house in which Smith resides. When Smith discovers that her "malady had suddenly become aggravated" so that she "could not sleep for torturing pains in her head," he is invited to lay his hands on her forehead.[50] Although Smith thinks his actions provide a "not very promising remedy," they actually induce sleep; "sleep," as Smith's companion explains, "that would save her."[51] Utopian fictions provide an interesting parallel to medical sleep studies, replicating their recognition of the powerful therapeutic effects of sleep while recognizing that beyond its experiential qualities, sleep also has spatial boundaries. Yet utopian fictions of the nineteenth century rarely remain secure within the boundaries of their own good places. Many provide potential reversals of that space, turning utopia into dystopia. The modern sleep lab also suggests the possibility of that oscillation between good and bad sleep, especially in its allocation of spaces that are reminiscent of Foucauldian enclosures of dystopic surveillance.

INTERLUDE: BAD SPACES IN THE CONTEMPORARY SLEEP LAB

The Cardiff sleep lab is a place of inhibiting spaces.[52] Fitted out like a hospital ward, with muted paints and antiseptic linoleum, it is configured around a short L-shaped corridor from which a number of bedrooms radiate. These rooms vary in size, from moderate (for participants) to cell-like (for the researchers). There is also a computer room, where all the monitoring takes place. This, too, is tiny; around four feet by eight feet, it has two old chairs and two even older computers. On the walls are pinned curling pieces of paper with lists of emergency numbers and lab instructions. A cartoon featuring Batman tells you to turn out the lights when you leave, or else. At night, when the lab conducts its experiments, its spaces are quiet and cold. The temperature is kept relatively low in order to enable good sleeping, and the lab is soundproofed for the same reason. This may be rational in the abstract, but it contributes to an atmosphere of unnatural, even spooky, isolation.

The lab's architecture of security enhances its sensory alienation. Every door is opened either by a numerical code entered on a keypad or with a swipe of an electronic card. Communication between the participants and the researchers is conducted through the use of alarms; one alarm tells the researcher that a participant has finished a test, and another that they require some help with the equipment in their room. Everyone, myself included, is accompanied everywhere. Nobody moves around by themselves, and no one goes anywhere without invitation. You do not feel that the spaces of the lab beyond your own room are for you to inhabit. Like the vampire, you wait to be invited to transgress the boundary of thresholds. Practically speaking, of course, this is so that doors can be opened for those of us without cards and codes. It feels rather different. The experience is of a severe restriction on movement. Participants fitted with electrodes look particularly encumbered as they are chaperoned from one room to another. To me, they look like shackled prisoners whose jailors are escorting them back to their cells.

[49] Ibid., 108.
[50] Ibid., 262–3.
[51] Ibid., 266.
[52] The commentary in this section is again drawn from my "Sleep Laboratory Ethnography Notebooks" (cit. n. 26). The impressions described can be found in Notebook 1.

Electrode fitting, as I hinted at earlier, produces a strange cyborg hybrid of human participant and technology. The process of fitting the electrodes is, however, awkwardly intimate. Alison, one of the researchers, first uses this word—intimate. It is not just the uninitiated like me who feel this way. As I am trained to attach the electrodes for myself, I come to understand this intimacy firsthand. To attach a single electrode, you must first mark the skull of the participant with a red crayon, finding a way through hair to bare scalp and then marking firmly. Then, to clean the scalp of oils that might lead to the electrode falling off, you deploy a cotton bud dipped in a cleanser. To clean effectively you need to rub hard. Next, the small circle of metal at the end of an electrode is covered liberally with a grayish sticky gel and stuck firmly against the head. Holding this in place, you tape over the top of it with a strip of plaster, again moving the hair to find scalp to stick to. Then you do all of these a further nineteen times, over the head, around the eyes, behind the ears, and on the chin. During this long process of around an hour, it is impossible to forget you are doing all of this to a complete stranger. To do it well—and nobody wants an electrode to fall off halfway through the night and ruin an experiment that has taken weeks to set up— you must get your hands onto this stranger, into their hair, around their neck, and down behind their ears. It is an acutely self-conscious tactile experience, a form of bodily monitoring and control that foreshadows the study of the brain's electricity that is to follow. I ask the participant on whom I am working if she is comfortable. "Yes," she says, unconvincingly.

SLEEP'S DYSTOPIAS

How do Victorian utopian fictions stage this move from good to bad sleep, in the same adjectival sense of these terms? The most dystopian elements of Bellamy's *Looking Backward* emerge from pathologies of sleep: insomnia, disrupted or restless sleep, and premature awakening. Julian West's horror at his arrival in the far future, for example, is accompanied by a rather abrupt awakening from his "somewhat too long and profound sleep" that leaves him "feeling partially dazed."[53] His fears that the Boston of the year 2000 may prove to be a dystopian rather than utopian future continue to haunt him as he contemplates the ending of his first day there. Importantly, this concern is cast in the language of sleeplessness: "Now I had been looking forward all the evening with some dread to the time when I should be alone, on retiring for the night. . . . I had had glimpses, vivid as lightning flashes, of the horror of strangeness that was waiting to be faced when I could no longer command diversion. I knew I could not sleep that night, and as for lying awake and thinking, it argues no cowardice, I am sure, to confess that I was afraid of it."[54] West connects his inability to sleep well to a fear of contemplating his horrific position as a man displaced into a world he does not understand and in which he is entirely isolated. Later, he awakens too quickly to recall that he is living in the far future, and this abrupt exit from sleep causes "mental confusion [that] was so intense as to produce actual nausea."[55] West's "mental torture" is matched only at one other moment in the novel—when he dreams of having awoken again in late nineteenth-century Boston and believes that the future

---

[53] Bellamy, *Backward* (cit. n. 11), 51.
[54] Ibid., 60.
[55] Ibid., 80.

utopia was itself the dream vision rather than the reality.[56] This vision is excited into existence by pathological sleeping that presents as sleeplessness.

Examples of the dystopian nature of wakefulness during hours when sleep is sought are manifold in the scientific and medical literature of the period. For many writers on sleep pathologies, to be awake at incorrect times is to enter into potential dystopian space or to create dystopian sites within the body. Louis King, a late Victorian writer on health and a keen reader of the work of sleep researchers, argues that hard work "cannot fail to induce healthy and sound sleep," while idleness often leads to sleeplessness and, in turn, "stores up in the system products of disease which must ultimately lead to . . . destruction."[57] For King, extended sleep pathologies would eventually upset the entire balance of the body, "and no effort on our part will then save us from ending our days as useless cripples or hopeless imbeciles."[58] In America, too, sleep researchers envisaged dystopian results for poor sleepers. Hammond, in his 1869 book *Sleep and Its Derangements*, provided an extended case study of a young sleepless woman whose mental disturbance is an uncanny parallel to Bellamy's Julian West. Cited in the chapter titled "Pathology of Wakefulness," Hammond's patient has her mind "filled with the most grotesque images which it was possible for the imagination to conceive, and with trains of ideas of the most exaggerated and improbable character."[59] Like West, the young woman suffers most from "the horror with which I look forward to the long rows of too familiar phantoms and thoughts which I know will visit me before morning."[60]

While Hammond viewed the pathological sleeping of this patient with sympathy, many poor sleepers were blamed for their own inability to sleep properly. Indeed, Hammond himself fell into this in his own work, giving the example of "an intimate friend" whose sleeplessness (caused by overwork) led to "inflammation of the brain" and ultimately "acute insanity."[61] This, Hammond warns his readers, was because "he deprived himself of the amount of sleep" he required.[62] Late nineteenth-century utopian fictions investigate this culture of assumed responsibility for one's own sleeping. Hudson's protagonist Smith finds himself sleeping poorly and suffers the consequence of his own inability to access suitable rest. Smith's first night in the crystal age is one of nervous wakefulness. Troubled by having to find his way around what he describes as "sleeping-cells" in the dark, Smith finds that he does not have the bodily repose for proper rest: "I laid me down, but not to sleep. The misery of it! For although my body was warm—too warm, in fact—the wind blew on my face and bare legs and feet, and made it impossible to sleep."[63] This sleeplessness is a warning sign of Smith's later mental distress, which brings about a similar, but more profound, wakefulness. After falling unconscious from overwork, Smith is brought to an interview with the father of the house where he is interrogated about his sleeping: "'You went to your task in the woods almost fasting, and probably after spending a restless night. Tell me if this is not so?'" Smith's response is to admit he "'did not sleep that night.'" As a result, the father orders Smith punished:

---

[56] Ibid., 78.
[57] Louis King, *Work of Body, Work of Brain, Recreation, Sleep* (London, 1885), 59.
[58] Ibid., 62.
[59] Hammond, *Sleep* (cit. n. 18), 222–39, on 236–7.
[60] Ibid., 237.
[61] Ibid., 228.
[62] Ibid.
[63] Hudson, *Crystal* (cit. n. 11), 100–1.

"Unrefreshed by sleep and with lessened strength," he continued, "you went to the woods, and in order to allay that excitement in your mind, you laboured with such energy that by noon you had accomplished a task which, in another calmer condition of mind and body, would have occupied you more than one day. In thus acting you had already been guilty of a serious offense against yourself."[64]

For this sleeplessness Smith is sentenced to imprisonment "for the space of thirteen days."[65] This reading of wakefulness as a breach of self-discipline—as a form of self-styled dystopia that should be countered by personal attention to dutiful sleeping—is a common view in nineteenth-century sleep research. The American physician Lydia Fowler concluded that leading a life of sobriety, propriety, and Godliness was essential, so that "when sleeping we may enjoy that sleep which is peaceful, and gives us rest to prepare us for our daily duties."[66] Similarly the English physician Joseph Mortimer-Granville noted in his 1879 book on sleep that the body requires "wise management" in order to sleep properly.[67]

One other failure of self-management which dominates sleep studies of the period, and which is specifically interrogated in Wells's *When the Sleeper Wakes*, is deceit. Numerous studies of extended sleeping focus attention on the potential for the phenomenon of extended sleep to be fraudulent. In the case of the Sleeping Girl of Turville, John Gay noted that the medical men who had visited the girl, "without any exception" viewed her extended sleep "with more or less scepticism."[68] While Gay himself noted that he had no reason to find her a fraud, his treatment suggests that he was, at least, determined to discover whether he had been hoodwinked:

I had a grain of tartar emetic [a toxic drug used to induce vomiting] placed on the back of the tongue. The girl began to show signs of awakening to the dismal consequences of the salt's activity . . . When these effects subsided, she relapsed into her former condition, when, after an interval of twenty-four hours, another dose was administered, and was followed by like effects. . . . A third dose was ordered, in her hearing, when she roused herself, and once and for all refused to submit. The remedy had done its work.[69]

Such was the concern over fake sleepers that doctors like Gay resorted to extreme treatments, either to rule out fraudulence or malingering, or to confirm it. There were a number of medical cases that showed the merits of this approach; for example, William Gairdner's well-known case of choreic movement treated with sleep was revealed a few years later to be a case of fakery.[70] Fears over deceitfulness placed the pathological sleeper in a distinctly dystopian medical milieu. Awakened often into violent sickness, as the Sleeping Girl of Turville experienced, or into a prison house of other medical interventions designed to manipulate and deceive the patient, extended sleepers often found themselves imprisoned not only by their pathological sleeping but also by doctors claiming to attend to their needs.

---

[64] Ibid., 238.

[65] Ibid., 239.

[66] Lydia Fowler, "How, When and Where to Sleep: A Lecture," in *Lectures by Mrs. Fowler* (London, 1865), 32.

[67] J. Mortimer-Granville, *Sleep and Sleeplessness* (London, 1879), 75.

[68] Gay, "Sleeping Girls" (cit. n. 21), 31.

[69] Ibid.

[70] William T. Gairdner, "Sequel of a Case of Abnormal Disposition to Sleep, Alternated with Choreic Movements," *British Medical Journal* 905 (4 May 1878): 635–7.

Wells's sleeper, Graham, experiences a similarly manipulative and controlling environment when he is woken up. Suffering from "the spiritless anaemia of his first awakening," Graham is unable at first to understand the control exerted upon him.[71] Wells, however, makes that control clear to the reader not only through the novel's plotting, but also in the constant relation of wakefulness to dystopia; in, for instance, connecting the untrustworthy Ostrog to the sleeper's new wakefulness. Most interesting, however, is that Wells also constructs a period in the novel when the sleeper is perceived as a fraud. In an encounter with an old man, Graham discovers that he believes the real sleeper "died years ago" and was replaced with a "poor, drugged insensible creature."[72] As they speak, the old man provides, in effect, a case history of the sleeper that leads Graham to reveal himself as the subject of their conversation:

> "I *am* the sleeper." He had to repeat it.
> There was a brief pause.
> "There's a silly thing to say, sir, if you'll excuse me."[73]
> Time and again Graham asserts that he is the sleeper, each time to be rejected by the old man:
> "Call yourself the Sleeper if it pleases you. Tis a foolish trick—."[74]

Graham, unlike the other fictional sleepers, is (at least momentarily) considered to be a fake. The medical and scientific analysis of sleep and its fraudulent abusers reveals why. It is that Wells's novel is also the only one that properly approaches and addresses its own vision of dystopia. Unlike *A Crystal Age*, *Looking Backward*, or *News from Nowhere*, Wells's other place is utopian only temporarily. Far more keenly it anticipates a dystopian future. That one method of representing this was to call into question the authenticity of sleep illuminates its importance to Wells, and indeed to all of these narratives.

CONCLUSIONS

Sleep itself begins in a fiction. In a discussion in the Cardiff sleep lab with sleep researcher Alison, she noted that sleep always begins with pretending to sleep—with lying down, closing one's eyes, and relaxing the body.[75] Sleep is therefore imagined before it becomes real; it is phenomenal before it is noumenal. This recognition of sleep as both real and imagined is one of the things that has become apparent through the specific methodology of placing past, present, and future in the new alignments above. What has been the benefit of this methodology? Examining the present has proved a creative way to access the past—not the past of records and narratives, but an experiential past that can in part be accessed by getting as close as possible to some of the same concerns and activities, even some of the same bodily actions, as the history and literature that I have investigated. Asking questions of historical materials that were formed in today's world of sleep research opened up an entirely new way of considering nineteenth-century sleep. Making use of current science,

---

[71] Wells, *Sleeper* (cit. n. 11), 217.
[72] Ibid., 87.
[73] Ibid., 94.
[74] Ibid., 95.
[75] "Sleep Notebooks" (cit. n. 26), Notebook 1.

then, offers an intriguing and different way to enter into a kind of ecotone (to use an environmental term meaning mixed habitat) between science, literature, and history. It also enables a further historicizing of present-day sleep research, and potentially gives an opportunity for the investigator to propose new ways of knowing that will prove productive in thinking about sleep's futures. That is, this temporally disjunctive method can tell us what it is in our own present and future that we should have in better focus. To that extent, it is an interventionist telos that gives the opportunity for action. It additionally gives some active part in the present-day back to the historian of science, for it is principally the scholar who is able to see the different teloi at work together.

One proof of its success here is in what new knowledge might be conceived of through the sleep of the late nineteenth century and today. Possibly the most influential work on sleep in the nineteenth century is Roger Ekirch's brilliant essay on segmented sleeping. Ekirch convincingly showed, in his 2001 article, that in the Western world up until industrialization, sleep was taken in two parts rather than in the single (consolidated) fashion that now seems the norm. Ekirch concluded that the loss of segmented sleeping meant a comparable loss for our unconscious, which no longer has "an expanded avenue to the waking world."[76] Unable to awake and reflect upon our own dreaming after a first short sleep, or even to remember those dreams we have had, Ekirch explained, we have "continued to lose touch with our dreams" and in turn have also lost "a better understanding of our deepest drives and emotions."[77] This might well seem true when looking back at the rise of industrialization and the concomitant loss of segmented sleep. But by the later nineteenth century, the scientific, medical, and literary narratives I have examined here suggest that the Western individual wanted something else from sleep and recognized that good sleeping could provide it. This was to be, for a time, apart from our selves.

Wells saw this disassembly of the self during sleep as its most wonderful characteristic, and he articulated this just before the sleeper awakes: "What a wonderfully complex thing! This simple seeming unity—the self! Who can trace its reintegration as morning after morning we awaken, the flux and confluence of its countless factors interweaving, rebuilding, the dim first stirrings of the soul, the growth and synthesis of the unconscious to the subconscious, the subconscious to the dawning consciousness, until at last we recognise ourselves again."[78] This is a direct statement of sleep as utopia *because* it offers a period of time where one is absent from the self. Wells's metaphor additionally hints at how sleep might offer the self a mode of refurbishment that enables smooth running.

Late nineteenth-century sleeping is not about dreaming or recollecting. It is about being able to break off from the self to reinvigorate it. Sleeping is essential to our social wakefulness. What utopian fictions and sleep research show is that to achieve this we need to have periods where we forget ourselves. Having a sense of relief from the place of ourselves is a way of gaining the repose we actually need. Utopia, then, is time away from ourselves, time that then allows us to be ourselves, and to inhabit our own selfhood without fatigue or fervor. This is what utopian fictions imagine, and what today's sleepers aspire to. It is late Victorian sleep. It is also our own.

[76] A. Roger Ekirch, "Sleep We Have Lost: Pre-industrial Slumber in the British Isles," *Amer. Hist. Rev.* 106 (2001): 342–86, on 344.
[77] Ibid., 385.
[78] Wells, *Sleeper* (cit. n. 11), 19.

From Technician's Extravaganza to Logical Fantasy:
Science and Society in John Wyndham's Postwar Fiction, 1951–1960

by Amanda Rees*

ABSTRACT

This article argues that John Wyndham's postwar novels represent a sustained attempt to analyze and problematize the relationship between knowledge, expertise, and society. Wyndham held vigorous opinions on the critical role that science fiction (SF) could and should play in modernity, even as his novels dissected the ultimate unsustainability of industrial urban democracies. He believed that it was the role of the SF writer to show that the pace and path of scientific and technological developments were not predetermined, but potentially subject to collective decision-making regarding the human future. This article explores Wyndham's depiction of "experts" and "amateurs" in the context of his deployment of scientific and social-scientific concepts as he examined prospective futures, and argues that aspects of SF can be understood as practical STS or applied history of science that are firmly situated in an affective moral context.

John Wyndham is best known today for the novels that he published between 1951 and 1960. These books—*Day of the Triffids* (1951), *The Kraken Wakes* (1953), *The Chrysalids* (1955), *The Midwich Cuckoos* (1957), and *The Trouble with Lichen* (1960)—are today marketed as science fiction, a category vehemently rejected by Wyndham in his lifetime as a "repellent label" and a "repulsive term."[1] The settings

* Department of Sociology, University of York, Heslington, YO10 5DD, UK; amanda.rees@york.ac.uk.

I gratefully acknowledge the financial support of the UK's Arts and Humanities Research Council in the preparation of this piece. I would also like to acknowledge the many colleagues who read versions of this article, including Rowland Atkinson, David Beer, David Kirby, Graham McCann, Joanna Radin, Sam Robinson, and Mike Savage, as well as W. Patrick McCray, Suman Seth, and two anonymous reviewers.

[1] John Wyndham, "Roar of Rockets," *John O'London's Weekly*, 2 April 1954; Wyndham, "Notes for a talk to the SF Luncheon Club," 27 September 1955, Wyndham 5/4/10, Special Collections & Archive: John Wyndham Archive, University Library, University of Liverpool, UK (**hereafter** cited as JWA). Wyndham's novels have gone through numerous reprints—*Triffids*, for example, was reprinted by the publisher more than forty times, and four times in 1981 alone. In this article, references are to the following editions, all brought out by Penguin Books: *The Day of the Triffids* (Middlesex, UK, 1987), *The Kraken Wakes* (Middlesex, 1987), *The Chrysalids* (Middlesex, 1987), *The Midwich Cuckoos* (Middlesex, 1960), and *The Trouble with Lichen* (Middlesex, 1963).

of his novels range from a typical rural English village of the 1950s to a postapocalyptic North American landscape, and his characters include such diverse figures as professional ingénues, feminist biologists, army colonels, working-class intellectuals, and professors of geography. But what his stories all have in common is their ultimate concern with the role that knowledge and expertise should play in the shaping, maintenance, and (potential) dissolution of society.

Wyndham published *The Day of the Triffids* in 1951, a dozen years before prospective prime minister Harold Wilson told Britain that her future prosperity depended on forging a new nation in the "white heat" of a new "scientific revolution."[2] But in his work, Wyndham had already been tracing out the likely consequences of this emergence of science into public life, the impact that this immensely powerful way of engaging with and manipulating the natural world would have on society, and its implications for political and cultural governance. What is particularly interesting about what Wyndham was doing, however, is the way in which he was not just analyzing and subverting the (often contradictory) attitudes of different classes of British citizens to knowledge and expertise, but also actively enlisting these attitudes in support of his imagined futures.[3] Long before there were professors of sociology at British universities, Wyndham, a professional writer of fiction, was doing critical social science, using his amateur status to lend weight to his implicit and explicit critique of the tragic flaws at the heart of the society and culture in which he lived. This article will examine Wyndham's work from two different angles. First, it will consider how images of the expert, the scientist, or the intellectual are deployed in his novels. Second, it will look at how these notions of expertise, together with his understanding of biological science and evolutionary change, relate to his examination of the modern, industrial, consumerist society, and to his analysis of its tragic flaws. It will show how Wyndham deploys the figure of the amateur—with all the commitment of disinterested passion this evokes—as his significant agent, before reflecting on the implications that this valorization of amateur status has for his own position as an exponent of applied history of science.[4]

[2] Richard Coopey, "The White Heat of Scientific Revolution," *Contemporary Record* 5 (1991): 115–27, https://www.tandfonline.com/doi/abs/10.1080/13619469108581161?journalCode=fcbh19.
[3] On the role of the intellectual in Britain, see Perry Anderson, "Origins of the Present Crisis," *New Left Review* 1 (1964): 26–53; Thomas William Heyck, "Myths and Meanings of Intellectuals in Twentieth Century British Identity," *J. Brit. Stud.* 37 (1998): 192–221; Stefan Collini, *Absent Minds: Intellectuals in Britain* (Oxford, 2006). For analysis of the emergence of professional identities rooted in technoscientific skill and expertise, see Jon Agar, *The Government Machine: A Revolutionary History of the Computer* (Boston, Mass., 2003); David Edgerton, *Warfare State: Britain, 1920–1970* (Cambridge, UK, 2005), and *Britain's War Machine: Weapons, Resources and Experts in the Second World War* (London, 2012); Mike Savage, *Identities and Social Change in Britain since 1940: The Politics of Method* (Oxford, 2010); Melissa Smith, "Architects of Armageddon: the Home Office Scientific Advisers' Branch and Civil Defence in Britain, 1945–68," *Brit. J. Hist. Sci.* 43 (2010): 149–80; and "British Nuclear Culture," ed. Jonathan Hogg and Christopher Laucht, special issue, *Brit. J. Hist. Sci.* 45 (2012).
[4] Social science fiction is a term often used to refer to a subgenre that is less dependent on technological innovation. See Darko Suvin, *Metamorphoses of Science Fiction: On the Poetics and History of a Literary Genre* (New Haven, Conn., 1979). The subgenre derives inspiration from anthropology and even sociology (examples of its authors include Ursula Le Guin, Robert Heinlein, or C. J. Cherryh). It is also increasingly the focus of study in its own right; see "Social Science Fiction," ed. Neil Gerlach, Sheryl N. Hamilton, and Rob Latham, special issue, *Sci. Fict. Stud.* 30 (2003).

JOHN WYNDHAM, SCIENCE FICTION, AND LOGICAL FANTASY

John Wyndham, or to give him his full name, John Wyndham Parkes Lucas Beynon Harris (1903–69), was born in Warwickshire in 1903 and by the early 1930s was an established contributor to American pulp magazines.[5] Spending the war initially as a Ministry of Information censor—like Sleigh and White's SF fans—before joining the Royal Corps of Signals, he returned to writing in peacetime.[6] He used the pen name of John Wyndham for the first time with *Triffids*. Eventually, John Wyndham was to be responsible for nine novels, two appearing posthumously. Oddly, Wyndham's work, regardless of what name it appeared under, has received relatively little critical attention, and it is often surprisingly cursory or patronizingly dismissive.[7] It is not easy to explain this. *The Day of the Triffids* has never been out of print, and has been adapted numerous times for radio, television, and film, as have the other three of his most well-known postwar books—*Kraken*, *Chrysalids*, and *Cuckoos*.[8] Perhaps the popularity of his work itself partly explains the relative dearth of sustained critical attention.[9] But—and this is also partly due to the enthusiasm with which his work has been adapted for other media—there are serious discrepancies between Wyndham's public image and

[5] Wyndham used various combinations of these names throughout his writing career, enabling him to distinguish between the persona who wrote magazine stories (often John Beynon, or John Beynon Harris) and the ones who wrote novels. On at least one occasion he produced a novel—*The Outward Urge* (Middlesex, UK, 1959)—that he cowrote with himself: "John Wyndham and Lucas Parkes." In private life, he used the name Beynon Harris; see E. F. Bleiler, "Luncheon with John Wyndham," *Extrapolation* 25 (1984): 314–7.

[6] Alice White and Charlotte Sleigh, "War and Peace in British Science Fiction Fandom, 1936–1945," in this volume.

[7] Most famously, Brian Aldiss dismissed Wyndham's work as "cozy catastrophe," where "the hero should have a pretty good time (a girl, free suites at the Savoy, automobiles for the taking) while everybody else is dying off." See Brian Aldiss, *Trillion Year Spree* (1986; repr., Thirsk, UK, 2001), on 280. Few critics have gone quite as far as this, and Aldiss has been directly challenged in L. J. Hurst, "We Are the Dead: *Day of the Triffids* and *1984*," *Vector* 113 (1986): 4–5. For further commentary, see David Ketterer, "John Wyndham and the Sins of His Father," *Extrapolation* 46 (2005): 163–88; Ketterer, "John Wyndham and the Searing Anguishes of Childhood," *Extrapolation* 41 (2000): 87–103; Andy Sawyer, "John Wyndham and the Fantastic," *Wormwood* 3 (2004): 51–62; C. N. Manlove, "Everything Slipping Away: John Wyndham's *Day of the Triffids*," *Journal of the Fantastic in the Arts* 4 (1991): 29–53; Owen Webster, "John Wyndham as a Novelist of Ideas," *SF Commentary* 44/45 (1975): 39–58; William Temple, "Plagiarism in SF," *Vector* 36 (1965): 3–8; Matthew Moore, "Utopian Ambivalences in Wyndham's *Web*," *Foundation: The International Review of Science Fiction* 32 (2003): 47–56; Jo Walton, "Who Survives the Cosy Catastrophe?" *Foundation: The International Review of Science Fiction* 34 (2005): 34–9; Nick Hubble, "Five English Disaster Novels," *Foundation: The International Review of Science Fiction* 34 (2005): 89–103. Andy Sawyer's "John Wyndham" (see above), and Thomas D. Clareson and Alice S. Clareson's, "The Neglected Fiction of John Wyndham: 'Consider Her Ways,' *Trouble with Lichen* and *Web*," in *Science Fiction Roots and Branches*, ed. Rhys Garnett and R. J. Ellis (London: 1990), 88–103, are unusual in that they combine an analysis of Wyndham as novelist with an examination of his skills as a social analyst.

[8] *Triffids* was adapted for radio early in the 1950s, and was most recently broadcast in the UK on BBC Radio 4 in 2008. In 1962, a film version was made starring Howard Keel. Perhaps the most influential adaptation was the BBC's 1980s serial, starring John Duttine; it caused many children (and adults) of that period to view the local vegetation with an extremely wary eye. A new version was broadcast by the BBC in 2009. The books *Kraken*, *Cuckoos*, and *Chrysalids* were all adapted for radio, as was his novel *Chocky* (Middlesex, UK, 1968). *Chocky* also became a children's TV series for the British ITV in the mid-1980s, with several spin-offs, including *Chocky's Children*. *Cuckoos* was filmed twice as *Village of the Damned*, once in 1960 and again in 1995, with the latter starring Kirstie Alley and Mark Hamill.

[9] As Brian Aldiss puts it, both *Triffids* and *Kraken* were "totally devoid of ideas but read smoothly and thus reached a maximum audience, who enjoyed cozy disasters"; see his *Trillion Year Spree* (cit. n. 7), 279.

the actual contents of his novels. Consider this: the name Wyndham conjures, for most people, lurid images of interplanetary invasion—lurching triffids, lurking sea-monster xenobathites, or cloned "Children" with Svengali-like powers of telepathic control.[10] These are staples, one would have thought, of the worst kind of pulp science fiction— in fact, the kind of hackneyed space opera that Wyndham himself condemned so vigorously, as we will shortly see.[11] But Wyndham intended his stories not to terrify or to titillate his audiences, but to make them thoughtful. He aimed, in short, to illustrate the logical and predictable consequences of the political, social, and scientific trends that he identified in the world of the 1950s.

Wyndham was deeply concerned with science fiction's social responsibilities, revisiting the topic in numerous talks and articles in which he discussed the form, meaning, and function of the genre. Frequently, he used a definition that he attributed to Edmund Crispin: "a science fiction story is one which presupposes a technology, or an effect of technology, or a disturbance of the natural order, such as humanity, up to the time of writing, has not in actual fact experienced."[12] But that was just the beginning of Wyndham's broader analysis of his preferred mode of literary expression. He usually divided SF into four categories. There was the "technician's extravaganza," primarily written by scientists for scientists, deriving initially from Verne, often using impenetrable language, and wholly unconcerned with the social implications of the technologies described. There were "cowboys in space," produced by hacks who had realized that ray guns could easily replace six shooters, and were running amok, with "all the joy of settlers perceiving a site for a new dustbowl."[13] Even worse, there were the Hollywood "B-movies," dismissed by Wyndham as "slipshod stuff" that ignored rationality and reason in favor of schlock and horror, demonstrating nothing but "contempt for [their] audience."[14] Of most significance, though, were the "speculative stories" exemplified by the early work of H. G. Wells, but including authors like George Bernard Shaw and Nevile Shute, as well as George Orwell and Aldous Huxley. The focus of these stories was the exploration of the impact that new technologies had on the lives lived by ordinary people.[15] These four types, Wyndham argued, united only by setting and stage props, should never have been placed in the same category. "Science fiction" had, he insisted, emerged in 1920 as a "trade term coined by publishers' distributors" to aid them in "sorting bundles of paper."[16] As such, it posed no problems. But when it was used as a literary category to describe material "as diverse as the first chapter of Genesis, the works of Edgar Rice Burroughs, the Book of Revelation, Quartermass, the fantasies of J.R.R. Tolkien, 1984 and the adventures of Superman," it became a positive hindrance in public life.[17]

---

[10] "Xenobath" (or xenobathite) is the neologism that the protagonists of *Kraken* use to describe the invading intelligences from outer space who can live only at great pressure (five tons per square inch)— that is, at the bottom of the sea. "Children" is the name used for the alien invaders in *Midwich Cuckoos*, and is capitalized by Wyndham to distinguish them from the human children they physically resemble. *Chrysalids* also includes a small group of telepathic human "children."

[11] See, for example, Wyndham's "Science-Fiction and Space-Opera," 5/3/7, JWA.

[12] Wyndham "Notes for a talk to the National Library Club," 31 March 1955, 5/4/9, JWA.

[13] Wyndham, "Review of Stanley Weinbaum's *The Black Flame*," n.d., 5/1/24, JWA.

[14] Wyndham, "Notes . . . National" (cit. n. 12).

[15] See Wyndham, "Notes . . . SF"; and Wyndham, "Roar" (both cit. n. 1).

[16] Wyndham, "The Future of Science Fiction," *Radio Times*, 30 January 1969, 5/3/12 and 9/2, JWA.

[17] Ibid.

Fundamentally, he felt that the term science fiction was so hopelessly contaminated by the dross contained in space opera and B movies that writers conscious of their professional reputation had been driven away to work in other literary fields.[18] But for Wyndham, writing science fiction was actually the socially responsible thing to do. Positioning himself explicitly in the Wellsian tradition of "speculative stories," he homed in on the fact that his predecessor—like himself—had been born into a world undergoing rapid social and technological change. Wells, he stressed, saw the "need for decisions in order to avoid the traps and dangers he could perceive waiting ahead [so that] people should not be surprised into accepting every novelty at its face value nor in its first form."[19] That is to say, "by working out logically the *likely* results of new inventions and the *likely* results of current trends," the readers of speculative stories would be empowered to make decisions about how technology and society could be mutually accommodated. Novels, he argued, were the best format in which to do this, both for reasons of space and publishing practice. Magazines, with their need to maintain circulation, were vulnerable to causing offence. Imagine, he suggested, the fate of a writer who speculated that the horseless carriage might lead to less church-going. They, and the magazine in which they publish, "immediately achieve the reputation of being . . . dangerously anti-religious" and become a potential target for moral panics and campaigns.[20] Magazine SF, for Wyndham, thus tended to retain its Verne-like fascination with gadgets and gangsters. In contrast, a novelist was far freer to speculate on the wider social and moral consequences of scientific discoveries and technological innovations.[21]

For this reason as well, Wyndham was relatively cavalier about the need for scientific accuracy—as opposed to consistency—in his novels. In regard to two nonfiction pieces written in 1954, he acknowledged the need for a "slight smattering of elementary science (which we ought to have from our schooldays)," but counterposed this basic requirement with the following question: "If you were a publisher anxious to publish a first-class detective story, would you commission it from an eminent policeman—or from an accomplished writer?"[22] Speculative stories, for Wyndham, primarily needed "careful, conscientious deduction and attention to logical probability": they were the product of "speculative induction . . . carefully moulded with scientific reasoning, logical deductions, extrapolations, forecasts, prophecies and probabilities."[23] As "an exercise of the imagination within known limits," they had to play by the rules, and while one could not fly in the face of scientific fact, one could, and should, extrapolate imaginatively on the basis of different interpretations of these facts.[24] It absolutely did not matter whether a writer got their science, or indeed, their predictions, right. Instead, what mattered was *the act of making them*, of forcing people to confront the fact

---

[18] Wyndham, "But Why This Science Fiction?" n.d., 5/3/13, JWA.

[19] Wyndham, "H G Wells—Prophet," n.d., 5/2/6, JWA.

[20] John Beynon [John Wyndham], "Not So Simple," *Authentic Science Fiction*, no. 30, February 1953, 5/31/1, JWA (emphasis in the original).

[21] But see also Mike Ashley's account of the difficulties that British writers had in accessing magazine outlets: *Transformations: The Story of the Science Fiction Magazines 1950–1970* (Liverpool, 2017).

[22] Wyndham, "Stuff for the N W book", n.d., 5/3/5; "N. B. L. Talk," 5/3/2; both in JWA.

[23] Wyndham, "Science Fiction" (cit. n. 11); Wyndham, "H G Wells" (cit. n. 19).

[24] Wyndham, "Science Fiction" (cit. n. 11).

that they lived in a world that was both underpinned and undermined by technological innovations.

Critically important, however, was the fact that as the product of an age that took "man's power to alter his environment" as "axiomatic," science fiction had responsibilities that went beyond mere scientific accuracy.[25] On several occasions, both in his fiction and in his nonfiction, Wyndham made trenchant criticisms of the social and ethical recklessness of scientists. Writing as John Beynon in 1939, he wondered sarcastically why anyone would ever "need to use that hoary old standby, the mad scientist, . . . when the reputedly sane scientists are quite efficiently getting on with the job of world destruction before our eyes."[26] In another piece, he angrily described the twentieth century as "built by the scientists' brains, to be blown to bits by their morals."[27] This sentiment is present—in a more controlled fashion—throughout Wyndham's novels, and is directed not only at scientists, but also at experts more generally. His "logical fantasies," exploring the predictable consequences of the political, social, and scientific trends that he identifies in the postwar world, contain a sustained critique of the incapacity of acknowledged experts and recognized intellectuals to either identify or respond to threats to human survival. Ordinary people are forced to do the best they can with the skills they have, and if success comes at all, it is achieved only where they are able to apply their knowledge and ability in a different kind of context where individuals, even if they are professionals, are acting as amateurs. This is, of course, exactly what Wyndham himself is doing, actively applying and deploying his own amateur understanding of expert scientific and sociological knowledge as he—a professional writer—constructs his imagined futures.

Crucially, and despite his reputation as the master of the alien invasion story, his books avoid the exotic or the spectacular.[28] Wyndham's horror—for these "cozy catastrophes" *are* horrifying in the way they calmly lay out the minutiae of the apocalypse of a mundane world—lies not in the promise of monsters, but in the inability of modern society to react appropriately to threats to its security, or to consider the extent to which unchecked scientific and economic developments endanger its survival. The triffids, to take the example for which Wyndham is probably most famous, are clearly the unanticipated, though not unpredictable, products of unregulated and unrestricted attempts to manipulate our biological and physical environment in an era of political instability. But what makes Wyndham interesting, both as a writer in his own right and as an example of the complex interplay between knowledge, literature, and society, is that the ultimate sources of catastrophes and social collapse are not alien, but entirely human. Moreover, they are not the result of the actions of an individual mad (or sane) scientist but are fundamentally social in their origins.

---

[25] Wyndham, "But Why?" (cit. n. 18).

[26] John Beynon [John Wyndham], "Meet the Author," July 1939, 5/4/1, JWA. This article appeared alongside his short story "Judson's Annihilator" in the October 1939 issue of *Amazing Stories*.

[27] John Wyndham, "V of the Magi," July 1939, 5/3/16, JWA.

[28] Of the true alien invaders, the *Cuckoo's* Children are nearly identical with the human form, while the audience never sees the xenobathites (other than in a state of ooze-like decomposition)—and the lichen in *The Trouble with Lichen* looks like any other scrap of adhering, green growth. The closest Wyndham comes to creating a monster is in *Triffids*, but despite the best efforts of the book's TV and cinema adaptors, his own description of the individual plants is deliberately low key; see *Triffids* (cit. n. 1), 28–9.

In each of his cozy catastrophes, an industrialized, consumerist civilization creates the conditions for its own downfall. War, pestilence, famine, and floods may decimate the human population, but these are merely acute symptoms of an underlying and chronic malaise. They are the logical consequences of human self-confidence and an unimaginative certainty that "bad things can't happen to us," a certainty that derives in no small part from one of the key themes uniting Wyndham's novels. This theme is the prevalent and persisting assumption that knowledge—or civilization, signified by its most powerful creation, the city—has defeated nature. In none of these novels will the downfall of humanity arise ultimately from an alien threat. In each, the danger is to be found in viewing "man's power to alter his environment" as axiomatic—that is, in presuming that biological and environmental processes are under expert control.[29] This is brought home sharply in John Wyndham's very first, and perhaps most famous, account of how modernity creates its own conclusions.

"WE BROUGHT THIS LOT DOWN ON OURSELVES"

The abiding image that most people have of *The Day of the Triffids* is one of invading plants from outer space that conquer humanity by first blinding them and then—ultimate humiliation for creatures who consider themselves to be the peak of the food chain—eating them. But as Bill Masen, the book's narrator, crisply points out, this is "nonsense."[30] Triffids were, instead, bred behind the Iron Curtain. The biologists of the USSR, as desperate as the capitalists to turn infertile soil to agriculture and to maximize the production of oil—in particular, edible oils—had, "under a man called Lysenko," broken with the methods and theories of the West. While the paths they had taken "were unknown, and thought to be unsound . . . it was anybody's guess whether very successful, very silly or very queer things were happening there."[31] A box of triffid seeds is smuggled out of the USSR by a man in the pay of a capitalist oil company. As an act of industrial, rather than political espionage, it was necessitated by the fact that triffid oil will drastically undercut vested and invested financial interests. However, the seeds never arrive at their destination. Masen's conclusion is that the plane must have been blown to bits in the upper atmosphere, high over the Pacific, since that is the best explanation of how a plant "intended to be kept secret, could come, quite suddenly, to be found in almost every part of the world."[32] Triffids did not just appear. They were selectively bred to feed the world's hunger for oil and food in the context of fierce economic, scientific—and military—competition between two great superpowers.[33]

The military context here is important because the triffids, of course, are only half the story. As the novel opens, Bill Masen is waking in the hospital, eyes bandaged in the hope of saving his sight from a triffid sting—and although an astonishingly high proportion of triffid victims have been blinded by triffid poison, Masen's sight is saved. Not so the rest of the world, who wake on that morning to find themselves blind. But

---

[29] Wyndham, "But Why?" (cit. n. 18).
[30] Wyndham, *Triffids* (cit. n. 1), 26.
[31] Ibid., 32.
[32] Ibid., 36–7.
[33] Wyndham was, of course, writing some years before genetic modification became either possible or publicly problematic, and was equally in advance of the launch of Sputnik in 1957.

despite the triffids' predilection for attacking human eyes, it was not the triffids who were responsible for the mass blinding of the global population: "You'll find in the records that on Tuesday, 7 May, the Earth's orbit passed through a cloud of comet debris. You can even believe it, if you like—millions did. Maybe it was so . . . yet, until the thing actually began, nobody had ever heard a word about this supposed comet, or its debris."[34] Years after the event, Masen's speculations on the origins of the near-universal blinding lead him to realize that "up there, there were—and maybe there still are—unknown numbers of satellite weapons circling round and round the earth. . . . [So,] suppose that one type happened to have been constructed especially to emit radiations that our eyes would not stand?"[35] His conclusion? "We brought this lot down on ourselves."[36]

On this reading, the day of the triffids dawns as the result of the unanticipated consequences of human ecological manipulation. Industrial competition, the state of international geopolitics, and technological expertise combine to create an environment in which human civilization cannot survive, but that triffids are uniquely suited to exploit. They had been bred to be adaptable and able to exist in a wide range of habitats. Notably, these are characteristics that mirror those of their only predator—humanity. The ability of humans to farm triffids, however, did not depend on their much-vaunted intelligence—the popular parables of human evolution that use the size of the brain to define humanity—but primarily on the brain's ability to use the eyes to engage directly with the physical world. In a later section, this article will consider the relevance of this point to Wyndham's understanding and deployment of evolutionary theory in his work. But for now, it is enough to note that, perhaps more so than any of Wyndham's other novels, the *Day of the Triffids* depends on the depiction of the complex interdependencies not just within industrial society, but between that society and the ecosystem of which it forms a crucial—but not omnipotent, and not even omniscient—part. As for science, and foreshadowing the post-Chernobyl discussions of science and risk, Wyndham here portrays it as Janus faced. It is both the ultimate source of social catastrophe (the triffids and the satellites), and the only hope for civilization's survival.

But the way this is described is telling. Masen's role at the close of the novel is to "get some research going into a method of knocking off triffids scientifically."[37] It is not clear exactly what his job was before the book opens. We are told that for his father, "the world was divided sharply into desk-men who worked with their brains and nondesk-men who didn't and got dirty." He is clearly worried by his son's lack of mathematical ability but fails to note a talent for biology, which Bill Masen parlays into a career with a company that farms triffids.[38] His position there is vague, but seems more managerial than biological: we are told that he must do a lot of traveling to see how other companies handle their plants. Most tellingly, he is shown consistently deferring to the knowledge of a self-taught colleague, Walter Lucknow, whose triffid expertise is not due to qualifications, but to the fact that "he had a kind of inspired knack with them."[39] It is Lucknow's intuition that leads him to speculate on both triffid intelli-

[34] Wyndham, *Triffids* (cit. n.1), 12.
[35] Ibid., 247.
[36] Ibid.
[37] Ibid., 257–8.
[38] Ibid., 26.
[39] Ibid., 46.

gence and the role of intelligence in evolution. Prior to the global blinding that followed the comet/satellite disaster, a high proportion of individual triffid victims had been blinded by being stung across their eyes. Lucknow argues that the triffids do this deliberately: "They know [what is] the surest way to put a man out of action. . . . Take away our vision, and [our] superiority is gone. Worse than that our position becomes inferior to theirs because they are adapted to a sightless existence, and we are not."[40] Thus, the man without formal qualifications approaches the triffids with an open mind and correctly identifies the nature of the threat they pose. In contrast, Masen, with acknowledged credentials, studies how other companies farm their triffids, presumably in the hope of improving profitability. At the end, however, it is Masen's long-unused skills on which the hope of scientifically eradicating the triffids lies. Nevertheless, he explicitly regrets the loss of Lucknow's amateur passions, and the deliberate imprecision with which the prospective work is described ("knocking off") connotes the amateur far more than the professional.

THE BOFFIN'S LAMENT: OR, THE LAY OF THE BAFFLED BOFFIN

More generally, there are two very distinct ways in which professional knowledge and expertise are presented in Wyndham's work, which in some ways resemble the distinction between the technically skilled "expert" and the conceptually oriented "intellectual."[41] Wyndham presents his readers with two types of scientist—the technician-engineer and the theoretician-philosopher.[42] When it comes to the first type, the "boffins" of World War II slang, he shows their ingenuity to be the source of immediate salvation even as it creates the conditions for ultimate disaster. In *The Midwich Cuckoos*, for example, an army colonel complains that he can't do anything about the potential UFO threat, precisely because of the contribution that scientists were now making to the art of war: "Trouble is, for all we know it may be some little trick of our own gone wrong . . . All those scientist fellers in back rooms ruining the profession . . . Soldiering'll soon be nothing but wizards and wires."[43] The pace of technological development, coupled with the need to maintain national security, is shown not just as potentially outstripping the ability of the military to make beneficial use of novelty, but as a danger to national security.[44] Wyndham's interest here lies in exploring the extent to which such resourceful expertise can be both enthusiastic and creative—but still misguided, and often eventually sterile. In particular, his critique is directed at figures who are unable to transcend their training. Incapable of realizing that the paradigm

---

[40] Ibid., 48.

[41] One could argue that his presentation of these two aspects is a reflection of the clearer class distinctions of the 1950s—as with Bill Masen's father, the world is divided into those who use their brains and work in an office, and those who work with their hands and get dirty.

[42] Of course, as historians and sociologists of science have repeatedly demonstrated, the distinction between technicians (who run the experiments) and theoreticians (who tell you why the experiment is important and what it means) is one that dates to the origin of modern science and is still present in laboratories today: Steven Shapin and Simon Schaffer, *Leviathan and the Air Pump: Hobbes, Boyle, and the Experimental Life* (Princeton, N.J., 1985); Shapin, *A Social History of Truth: Civility and Science in Seventeenth-Century England* (Chicago, Ill., 1994); Shapin, "The Invisible Technician," *American Scientist* 77 (1989): 554–63.

[43] Wyndham, *Cuckoos* (cit. n. 1), 35. It is intriguing to compare this to the arguments put forward in Edgerton, *Warfare State* (cit. n. 3).

[44] [Wyndham], "Meet the Author" (cit. n. 26).

has shifted and the world has changed, they continue to approach problems in the way that they were originally taught, using methods and concepts that have, unfortunately, been superseded by events.

See, for example, Wyndham's discussion of the boffins' activities in *The Kraken Wakes*. Quite kindly, and with a good deal of wry humor, Wyndham describes the attempts of the boys in the back room to outwit the invading and invisible underwater xenobathites. These creatures have succeeded in banishing humanity from the greater portion of the oceans by repeatedly sinking ships that attempt to cross deep water. Once a defensive weapon has been developed, the navy sends a rather nervous ship to test it. The events that follow are described verbatim by a young lieutenant. After "the boffins had finished tearing around and testing everything in sight," they head off to cross a known danger spot with their mobile defense system (a "dolphin") scouting ahead of them. Two successive dolphin missiles successfully destroy potential threats—and then the ship begins to vibrate wildly. The boffin standing beside the lieutenant "gave a grunt, and whipped back like a streak into a kind of float-off instrument room they had rigged up on deck." Suddenly, "an excited boffin bounced out of the instrument room and ordered the depth-charge-thrower to work." While this had no visible effect, it was nevertheless immediately followed by the "noise of uproarious boffins slapping one another on the back in the instrument room." After the fourth dolphin explodes, "the boffins, all of them pretty tight by this time, tumbled out on deck to cheer and sing *Steamboat Bill*, and that was about the end of it."[45] The boffins announce success, and in Churchillian language, the prime minister tells Parliament that the "Battle of the Deeps" has been won. But within a month, it becomes evident that the underwater xenobathites are prepared to wage war on a far broader front than had been hitherto imagined, attacking human civilization on land as well as at sea. Their sea-tanks begin "shrimping" for people in coastal villages around the world, while at the same time inundating the land by melting the polar ice caps. Wyndham shows that the boffins' and the politicians' overwhelming focus on solving the immediate problem lethally distracts attention, effort, and finance from the real issues at stake. They have, in the classic sense, won the battle, but lost the war.

Part of the problem, as Wyndham describes it, was that the postwar public expected scientists to solve any problems that the world faced—but to do so while causing a minimum of fuss and hardship for the population in general, and for big business in particular. In fact, a significant portion of *Kraken* is devoted to the difficulties experienced by the boffin in "a scientifically minded age," when the public ascribes astonishing degrees of omnipotence to the "boys in the back room," while still refusing to bow to their judgements.[46] If one set of answers produces unwelcome change, whether conceptual or empirical, the "boys" could always be sent back to produce another set of solutions. *Kraken*'s narrator ruefully reflects: "That the boffins would come through with a complete answer one day was not to be doubted . . . From what I had been hearing, the general faith in boffins was now somewhat greater than the boffins' faith in themselves."[47] The trouble is, as Wyndham shows, that both scientists and the public are incapable of realizing the scale of the problems that confront them. Wyndham (through the geographer-prophet Bocker in *Kraken*) argues that at the heart of this difficulty lies

---

[45] Wyndham, *Kraken* (cit. n. 1), 109–10.
[46] Ibid., 159.
[47] Ibid.

the impossibility of escaping one's engrained disciplinary expectations, and of getting outside of one's world view. Human warfare, Bocker points out, has traditionally depended on the "ability to deliver or resist missiles of one kind or another—whereas [the xenobathites] don't seem to be interested in missiles at all."[48] His complaint is echoed by the sociologist-philosopher Zellaby in *Cuckoos*, who reflects on the inability, not just of scientists, but of most writers—even H. G. Wells—to adequately visualize interplanetary war. Instead, what they produce are just variations on the same sort of scenario situation: "Something descends, and something comes out of it. Within ten minutes . . . there is coast-to-coast panic, and all highways out of all cites are crammed, in all lanes, by the fleeing populace . . . while . . . a hitherto ignored professor and his daughter, with their rugged young assistant strive like demented midwives to assist the birth of the *dea ex laboritoria* which will save the world at the last moment, minus one."[49] Wyndham repeatedly shows that this is just human warfare writ large, and represents a profound failure of imagination. To have better science—to protect humanity's future—we need better forms of imagination.

INTELLECT, EXPERTISE, AND THE INDIVIDUAL

In many ways, a decade before Kuhn's *Structure of Scientific Revolutions* appeared, Wyndham seemed to be distinguishing between those individuals who were continuing to try to do "normal science" and those who were capable of realizing that there was an entirely different—perhaps even incommensurable—way of apprehending the problem. Wyndham treated the figure of the boffin, the within-paradigm tinkerer, and the problem solver, not just with a degree of ironic amusement, but also as a key symptom of the underlying social malaise. That malaise was the belief that all problems are solvable from within the dominant world view, whether that outlook is seen through the lens of the economic, the social, or the scientific. Like the boffins, the characters in Wyndham's work who are capable of both realizing the full nature of the threat and dealing with it are all also in the business of knowledge production and scientific research. Where they differ is in their capacity to think outside their trained disciplinary expectations and to apply their knowledge and skill in fields not necessarily their own.

It is *Kraken*'s geographer, Bocker, who revolutionizes the problem of the "Deeps" by demonstrating that positing the existence of undersea intelligence not only explains all the anomalous events recorded, but enables the prediction of the xenobathites' next attack. It is also Bocker, rather than any of the military scientist-engineers, or the experts sitting in "Technical Committees," who will ultimately figure out a way to defeat the invaders and save the (remnants of the) human race.[50] In *Cuckoos*, the figure of Gordon Zellaby dominates, and it is in Zellaby that Wyndham comes closest to describing the classic professorial intellectual. He must be reminded what he does and does not like to eat, speaks with a manner "frequently allusive and often elusive [and is] given to mentioning things *en passant* [so that] you don't know whether he followed them up with serious deductions, or was simply playing with hypotheses."[51]

---

[48] Ibid., 205.
[49] Wyndham, *Cuckoos* (cit. n. 1), 187.
[50] Wyndham, *Kraken* (cit. n. 1), 98–107.
[51] Wyndham, *Cuckoos* (cit. n. 1), 101.

He contrasts sharply with the official, government-sanctioned, experts on the alien Children. These experts were a "desiccated-looking couple called Freeman [who] were continually lurking and peering about the village, often insinuating themselves into the cottages [and were] generally resented and referred to as the Noseys."[52] The Freemans' expert investigations of the Children fail entirely. Instead, it is the vague, intellectual Zellaby who is the first to posit the existence of the Children's telepathy, the first to link this with their capacity for collective action, and if not the first to realize the deadly peril they pose to the human race, certainly the person who can face up to the necessary corollary. It is Zellaby's altruistic suicide (setting off a bomb in the room in which he is instructing and entertaining the Children) that, in this novel, saves civilization, at least for the moment.

But Wyndham's ambivalence toward expertise can also be applied to this otherwise praiseworthy ability to think about one's beliefs, as well as to allow beliefs to direct one's thinking. To return to *Triffids* for the moment, this article has already discussed the importance of Walter Lucknow, the self-taught expert on speculative triffid biology who, because he has not been "trained" to ignore what is in front of his eyes, is willing to posit the existence of triffid intelligence.[53] But also worthy of note in this novel is Dr. Vorless, a professor of sociology who tries to convince the surviving sighted that customs are contingent, and where context changes, so must custom. Since most of them were "accustomed, when [they] encountered this kind of thing to turn the radio off at once,"[54] his speech—the longest in the book—is received with hostility. The radical proposals that he advocates are not convincing to most of his audience, and create a schism in the surviving community that damages their chances for survival. In regard to both Vorless and *Kraken's* Bocker, Wyndham seems to be suggesting that thinking the unthinkable, while absolutely necessary, also needs to be carefully managed in relation to pragmatic possibilities.

But there is also an important connection between this notion of the philosopher-scientist and their willingness to take the extraordinary seriously, and the broader use that Wyndham is making of expert knowledge even as he critiques it. Essentially, Wyndham's critique of boffins hinges on one key point: boffins can only work within the paradigm with which they are familiar. When the world changes, they cannot adapt their world view. This question of change and the elemental importance of adaptability within an altered environment are fundamental to Wyndham's use of science—and particularly evolutionary science—in the plots of his novels. Consistently, Wyndham argues against teleological views of evolution, notions that were as prevalent in the 1950s as they remain today, with popular accounts in particular implying that the "aim" of evolution was to produce intelligent life, or even more specifically, human life. However, as Wyndham repeatedly emphasized, evolution has no purpose, and evolutionary success is ultimately contingent only on "fitness," the adaptability of the individual to the local environment. Regardless of how successful an individual is within a given environment, if that environment changes, then the organism must adapt or die—whether it be a dinosaur or a human being. Wyndham's critique of 1950s society and its potential futures hinges on the fact that knowledge and expertise have

---

[52] Ibid., 116–7.
[53] Wyndham, *Triffids* (cit. n.1), 46–50.
[54] Ibid., 118.

made it possible for most members of urban, industrial communities to remain insulated from the global changes in the environment that industrial capitalism is itself making, with all the unforeseen and—globally—tragic consequences that Wyndham spells out.

ONE-AT-A-TIMES, OR THINK-TOGETHERS?

But before looking in detail at Wyndham's account of the ultimate tragedy of urbanity, it is essential to bear in mind this insistence on putting social matters in their ecological context when considering the wider role that scientific expertise plays in his work. Biological theory and evolutionary theory are absolutely vital to the development both of Wyndham's plots and his wider significance.[55] Wyndham, writing at a time when the neo-Darwinian "modern synthesis" had only recently entered the public domain, and many years prior to the dominance of the "selfish gene" and the sociobiological paradigm in biology, allowed his work to reflect some of the key debates that characterized the period's biological thought. These involved, in particular, the relationship between the individual and the group, and the significance of the balance between cooperation and competition in the process of evolutionary change.[56] In the 1950s, group selection—the notion that the group rather than the individual might be the unit of evolutionary change—still retained a good deal of scientific credibility.[57] Additionally, the interwar work of the Chicago School in the United States, as well as the popular writings of biologists such as David Starr Jordan and Vernon Kellogg in the United States, and Charles Elton and Alister Hardy in the United Kingdom, emphasized the importance of cooperation, rather than competition, in evolution. The biologist W. C. Allee had experimentally demonstrated that when placed in a hostile environment, creatures that cooperated had greater success than those that competed.[58] Consistently—and increasingly insistently as time went on—Wyndham used this theme of domination through cooperation as the driving device behind his plots. It is there in a protoform in *Triffids*, but emerges onto center stage in *Chrysalids* and *Cuckoos*.

In each case, but from a slightly different perspective, Wyndham is emphasizing that cooperation is more important than intelligence when it comes to biological suc-

---

[55] In this context, it is worth noting that Wyndham was writing at a time when physics still largely remained the queen of the sciences, both in terms of funding and in relation to the public perception of science; see Iwan Rhys Morus, *When Physics Became King* (Chicago, Ill., 2005). Francis Saxover, the chief male protagonist of *Lichen*, at one point states as a "self evident fact . . . that the dominant figure of yesterday was the engineer; of today, the physicist; of tomorrow, the biologist"; see Wyndham, *Lichen* (cit. n. 1), 20. The second half of the twentieth century did indeed see biology—in the form of genetics and evolutionary biology, as well as the biochemistry with which Saxover was concerned—move toward the top of the public and academic agenda.

[56] See Julian Huxley, *Evolution: The Modern Synthesis* (London, 1942). Discussion of the history of the synthesis can be found in V. Betty Smocovitis, *Unifying Biology: The Evolutionary Synthesis and Evolutionary Biology* (Princeton, N.J.,1995); Ernst Mayer and William Provine, *The Evolutionary Synthesis: Perspectives on the Unification of Biology* (Cambridge, Mass., 1998); Ullica Segerstrale, *Defenders of the Truth: The Sociobiology Debates* (Oxford, UK, 2001); and Peter Bowler, *Evolution: the History of an Idea* (Berkeley, Calif., 2009).

[57] George C. Williams, *Adaptation and Natural Selection* (Princeton, N.J., 1966); V. C. Wynne-Edwards, *Animal Dispersal in Relation to Social Behaviour* (London, 1962).

[58] See W. Clyde Allee, *The Social Life of Animals* (Boston, Mass., 1938); and his "Social Biology of Subhuman Groups," *Sociometry* 8 (1944): 21–9. See also Stevi Jackson and Amanda Rees, "The Appalling Appeal of Nature," *Sociology* 41 (2007): 917–30; and Gregg Mitman, *The State of Nature: Ecology, Community, and American Social Thought* (Chicago, 1992).

cess. It is the sheer number of triffids, not just the blinding of the human population, that enables them to dominate, with key moments in the novel marked by the collective action of the plants. The plots of both *Chrysalids* and *Cuckoos* turn on the threat posed to the rest of society by a small group of c/Children. In both novels, the nature of the threat lies in the capacity of both groups to cooperate far more closely than ordinary human beings are able to, since they are forever isolated by their individual self-interest.[59] Similarly, in the posthumous novel *Web*, the global threat posed by the postnuclear testing spiders is not due to their physiology, but to their capacity to act in unison. It might also be noted that the eponymous *Lichen* is (obviously!) the product of a symbiosis between a fungus and an alga. But in each novel, the danger for ordinary, selfish, self-interested human beings lies in the development of this talent for the intensification of cooperation on the part of another group. It is this talent that renders the human brain, usually considered to be the apex of evolutionary complexity and success, irrelevant.

"AS A SECURELY DOMINANT SPECIES . . ."

But Wyndham also draws a significant distinction between the biological capacity for change and the intransigent unwieldiness of human institutions and customs. In particular (and ironically, given the characterization of Wyndham as the creator of cozy catastrophes), the contrast is drawn over the injunction not to commit murder versus a willingness to kill in a species' self-interest. For example, in *Chrysalids*, the children are being hunted by "normal" humans who have recognized them as mutants, when they are rescued in the nick of time by a team of telepaths from the other side of the world. To the children's horror, these rescuers then calmly execute their hunters, who include members of the children's own families. The woman from Sealand reproves them for their unthinking hypocrisy: "To pretend that one can live without [killing] is self-deception. . . . We have to preserve our species against other species that wish to destroy it – or else fail in our trust."[60] In *Cuckoos*, where (for once) there has been a genuine alien invasion, Wyndham, through Zellaby, is again clear in his identification of our institutions and expectations as the ultimate source of the threat to human life. While musing generally on the role of social expectations in the production of "instinctive" mother love, Zellaby reaches an abrupt conclusion: "Each species must strive to survive, and that it will do, by every means in its power, however foul, unless the instinct to survive is weakened by the conflict with another instinct . . . odd, don't you think? We could drown a litter of kittens that is no threat to us—but these creatures we shall carefully rear."[61] From the alien side, the Children are also quite clear on this point. As they tell Bernard, the government representative sent to deal with them, for humans, killing the Children "is a biological obligation": "You cannot afford *not* to kill us, for if you don't, you are finished."[62] But they are—rather smugly—aware that English politicians will have great difficulty in convincing their electorate of this elemental need.

---

[59] The savior of the children in *Chrysalids*, it might be noted, is never even given an individual name, although she does merit the definite article—she is always referred to as "the woman from Sealand."

[60] Wyndham, *Chrysalids* (cit. n. 1), 195.

[61] Wyndham, *Cuckoos* (cit. n. 1), 113.

[62] Ibid., 197.

"You don't appear to think very highly of our institutions," Bernard put in.
The girl shrugged. "As a securely dominant species, you could afford to lose touch with reality and amuse yourself with abstractions."[63]

In many ways, this statement encapsulates the broader argument at the heart of Wyndham's fiction. Systematically, although subtly, he questions the nature of expertise and knowledge in a society ostensibly based on rational analysis and planning, basing his critique on his own (amateur) understanding of scientific and social theories and their intersection. Repeatedly, and ironically, given the persistent popular assumptions about evolution that are still characteristic of the early twenty-first century, Wyndham reverses another persistent assumption that what is natural is inevitable, while what is cultural is contingent.

Consistently, as this article has shown, the importance and nature of evolutionary change is stressed in his novels. Although to a casual examination, the triffids look like any other plant, telepathic children appear normal at first glance, spiders look like ordinary arachnids, and there is little that distinguishes the lichen, the behavior of all of these groups has changed fundamentally, and therein lies the difference and the danger. But while the plasticity of nature is stressed—as the environment changes, so must the organism—the immutability of social institutions and their fatal inability to adapt to changed circumstances is continually hammered home. And despite Wyndham's gentle humor in dealing with his boffins, his account of the failure of human minds and culture to adapt to rapidly changing ecological conditions has a far darker tone, which makes use of critical social, not just scientific, theory.

Wyndham's critique of expert knowledge is oriented toward the form of instrumental reason that the theorists of the Frankfurt school condemned as characteristic of industrial society. This is the form of reason that treated the object as a means to an end, rather than an end in itself. And here Wyndham shows, as did the Frankfurt school, human beings as subject to such reason and such treatment—not by aliens, but by other humans. *Triffids* showcases satellite weaponry and the triffids themselves as the classic products of instrumental reason: they were the results of asking not "should this be done?" but "can this be done?" *Kraken* demonstrates the consequences of applying such reasoning to political technologies. Both of the main protagonists—Mike and Phyllis Watson—are journalists, and as such, are deeply involved with both monitoring and manipulating the public mood as it relates to potential threat from the xenobathites. Consistently, they find that what they think they ought to do is not what they are permitted to do. Truth is a casualty of war, especially when it is not in the interests of powerful groups for the state of war to be acknowledged. Fear of public panic combined with entrenched commercial interests are first thrown behind the effort to keep the xenobathites off the front page—as when Mike complains at one point, "for the last few months now, not a word about those things down there has gone out from any of our transmitters; the sponsors don't like it."[64] The "organised scoffing" that the establishment has been at pains to build up, however, soon has to be broken down when the public concludes that it is Soviets, rather than space invaders, who are sinking ships. Since war would, in the short term, be even more disruptive to commerce than alien invasion, the Watsons are then called upon to "put in [their]

---

[63] Ibid., 199.
[64] Wyndham, *Kraken* (cit. n. 1), 87.

own pennyworth towards stopping the atom bombs falling."[65] Mike and Phyllis are deeply humane and admirable characters, but they are caught up in a situation in which previous political inaction swiftly transforms their job from "the task of persuading the public of the reality of an unseen, indescribable menace [into] one of keeping up morale in the face of a menace which everyone now accepted to the point of panic."[66] Ordinary people are irrelevant when balanced against wider commercial and political interests. When these are threatened, the role of the media is to ensure that, whatever happens, the public sees those interests as identical with their own. Only thus, it appears, can civilization survive, but it may be a very specific kind of civilization that thrives in the changed circumstances.

"WE MUST HAVE THE TEACHER, THE DOCTOR, THE LEADER . . ."

Consistently, Wyndham's work vividly conveys the contradictions inherent in the specialized division of labor on which civilization—or at least, industrial democracy—depends. The consequence, after all, of living in a society dependent on expert systems is that the majority of the population are ignorant of how these systems work.[67] Bill Masen in *Triffids* finds it hard to convey to his posterity what his earlier life had looked like: "It is not easy to think oneself back to the outlook of those days. We have to be more self-reliant now. But then there was so much routine, things were so interlinked . . . Our life had become a complexity of specialists all attending to their own jobs with more or less efficiency, and expecting others to do the same."[68] Other novels explore the same theme: a critique of a division of labor so complex that individuals are ignorant of matters critical to their own survival. But at the same time, the need to maintain civilization, which is variously opposed to savagery, feudalism, and tribalism, is recognized in relation to the size of the community required to maintain it. Wyndham again pokes gentle fun at the way in which bureaucratic habits—such as the collection of names and addresses required for purposes of "system, organisation and relatives"—are still maintained as the sighted gather at the university in *Triffids*. But in the same novel, he uses the organic intellectual Coker to point out the hard facts: "What we are on now is a road that will take us back and back and back . . . to savagery. . . . The short cut to save us starting where our ancestors did [is knowledge]. . . . We must have the teacher, the doctor and the leader, and we must be able to support them."[69] The complex division of labor and expertise is thus the basis of both instability and security. Its breakdown enables Masen to be his "own master, and no longer a cog," but means that his children will be deprived of the chance to make their own choices.[70] Consistently, Wyndham suggests that the only safe way forward is to live in small communities that are large enough to enable some specialization, but are characterized by a kind of Durkheimian mechanical solidarity. Even these can be unstable,

---

[65] Ibid., 90.

[66] Ibid., 100.

[67] Increasingly, the subject of ignorance is being actively explored by sociologists and cultural studies scholars; see, for example, Matthias Gross, *Ignorance and Surprise: Science, Society and Ecological Decisions* (Chicago, Ill., 2010); Robert Proctor and Londa Schiebinger, *Agnotology: The Making and Unmaking of Ignorance* (Redwood City, Calif., 2008).

[68] Wyndham, *Triffids* (cit. n. 1), 16.

[69] Ibid., 104, 203–4.

[70] Ibid., 60.

as *Chrysalids* shows, should their inhabitants persist in trying to master, rather than adapt to, their environment.

As suggested earlier, each of these novels is ultimately an exploration of civilization's relationship with nature mediated through the ambivalence of scientific expertise, whether that be through the ecological hubris of *Triffids*,[71] the inexorable rise of the floods and the "attitude of polite patronage towards Nature," or the presumption that civilized society has evolved beyond matters of biology in *Cuckoos*.[72] As Zellaby in *Cuckoos* demands, "I wonder if a sillier or more ignorant catachresis than 'Mother Nature' was ever perpetrated? It is because Nature is ruthless, hideous and cruel beyond belief that it was necessary to invent civilisation."[73] In all three novels—and as part of this process of exploration—the tendency to assume that the laws of civilization had replaced the laws of nature is relentlessly attacked. Zellaby talks of returning to the laws of the jungle. Masen looks back on the vanished time when "it was easy to mistake habit and custom for natural law."[74] Watson in *Chrysalids* describes how as London flooded, officials mistakenly assumed the city was "so cellularly constructed that as the water flowed into each cell it would be abandoned while the rest carried on much as usual."[75] In reality, of course, when the inhabitants of high ground "turned in dismay to save themselves by fighting off the hungry, dispossessed" flood victims, the law of the jungle again prevailed.[76]

"FIND A NICE, SELF-SUFFICIENT HILL-TOP, AND FORTIFY IT"

Fundamentally, it is the loss of the cities that encapsulates the collapse of civilization and its experts, and marks the triumph of the natural world.[77] At the end of his first day as a seeing man in the world of the blind, Masen looks out at the London skyline, forcing himself to realize consciously that the city is doomed to the "long, slow, inevitable course of decay and collapse."[78] This "necrosis of a great modern city" is unimaginable to him, because so ingrained was the belief that "one's own little time and place was beyond cataclysm," even though cities have fallen in the past: "And now it *was* happening here. Unless there should be some miracle I was looking on the beginning of the end of London."[79] While the triffids attack humankind directly, the more insidious forces of nature reclaim the context of civilization. Soon, almost "every building was beginning to wear a green wig beneath which its roofs would damply rot," and from being unable to imagine the city's downfall, Masen becomes incapable of remembering what it once was.[80] Mike Watson's "drawn-out story of decay" in London is kept deliberately clipped, with brief mentions of visits to Trafalgar Square to watch

---

[71] Barry Langford, introduction to *Day of the Triffids*, by John Wyndham (London, 1999), vii–xvii.
[72] Wyndham, *Kraken* (cit. n. 1), 196.
[73] Wyndham, *Cuckoos* (cit. n. 1), 112.
[74] Wyndham, *Triffids* (cit. n. 1), 16.
[75] Wyndham, *Kraken* (cit. n. 1), 219.
[76] Ibid., 216.
[77] See also Patrick Parrinder, "From Mary Shelley to the *War of the Worlds*: The Thames Valley Catastrophe," in *Anticipations: Essays on Early Science Fiction and its Precursors*, ed. David Seed (Liverpool, 1995), 58–74.
[78] Wyndham, *Triffids* (cit. n. 1), 86.
[79] Ibid.
[80] Ibid., 231.

"the water washing around Landseer's lions." His secondhand descriptions of New York City again are worth quoting:

> A man and a woman on the Empire State Building were describing the scene. The picture they evoked of the towers of Manhattan standing like frozen sentinels in the moonlight while the glittering water lapped at their lower walls was masterly, almost lyrically beautiful—nevertheless, it failed in its purpose. In our minds, we could see those shining towers—they were not sentinels, they were tombstones.[81]

The death of the cities symbolizes the ultimate failures of industrial urban civilization to overcome threat: it demonstrates the critical instability that underlies its apparent vigor.

This shrinkage of the possibilities inherent in human communities is echoed by the narrowing of each book's horizons. For both *Triffids* and *Kraken*, the disaster that befalls is a global one, as evidenced by the description of the responses. Specialized weapons for triffid decapitation are developed in the tropics, where triffids are far less domesticated than in temperate climes. These guns swiftly become essential for individual survival in the West after the mass blindings. The melting of the polar caps is a global peril, though the "chief difference was that in the more developed country all available earth-shifting machinery worked day and night, while in the more backward it was sweating thousands of men and women who toiled to raise great levees and walls."[82] But in both of these examples, the sense of being part of a global effort swiftly diminishes. The horizon contracts with agonizing abruptness, as the quondam city of London becomes a sudden series of territories. As the seas rose, those who had taken Bocker's advice to "find a nice, self-sufficient hill-top and fortify it" found themselves overwhelmed by increasingly desperate and ever-diminishing bands of survivors.[83] And in the closing lines of both novels, the narrators hope not for the defeat of the global threat, but that the island, the Britain that has been lost, can eventually be reclaimed.[84]

CONCLUSION: GO TO THE ANT, THOU SLUGGARD, AND BE WISE

John Wyndham's novels are firmly of their time and place, but they have a much wider resonance for those engaged with the study of science, technology, and society. They were written at the midpoint of the twentieth century in the aftermath of a devastating world war, and in the immediate context of a broadly accepted political commitment to rational, expert planning in the distribution and utilization of national resources. They are focused on the imagined future of urban, industrialized societies based on an intricate division of labor that in turn rested on the interrelationships of expert (technical and bureaucratic) systems. They are directed, in particular, at the form of Western scientific rationality that treats nature as something that is capable of being conquered or controlled, and assumes that the consequences of human intervention in complex systems can be predicted and managed. Drawing on established cultural ambivalence to-

---

[81] Wyndham, *Kraken* (cit. n. 1), 220–4.
[82] Ibid., 211.
[83] Ibid., 206.
[84] On the "island" in fiction, see Paul Kincaid, "Islomania? Insularity? The Myth of the Island in British Science Fiction," *Extrapolation* 48 (2007): 462–71.

ward the role and influence of knowledge, whether purveyed by "intellectuals" or "experts," in the making of political, social, and economic decisions, Wyndham's novels vivisect these presumptions of omniscience to expose them as the Achilles' heel of industrial urbanity. In his analysis, they lead to an attitude of instrumental rationality applied to all resources, human or otherwise, coupled with a frightening inability to realize the inherently limited nature of the human capacity to apprehend the full complexity of either the natural or the social world.

In using the figure of the amateur, or the nonspecialist, as the source of pragmatic salvation, Wyndham is also subverting that ambivalence even as he is himself using it to lend force to the critique he is developing. His critique of industrial modernity is based, consciously or not, on his understanding and apprehension of key aspects of scientific and social scientific theory: on the theory of evolution and its application to human society; on the nature of community and industrial life; and on the relationship between expertise, particularly scientific expertise, and society. It is the techniques and theories of the sciences and social sciences through which he justifies his accounts. And, of course, his understanding of these works is inevitably that of an amateur. Wyndham tried a variety of careers after leaving school, but sustained himself primarily through his writings, together with a private income. The figures that he shows to be the salvation of his disintegrating worlds are reflections of his own position in relation to the expert systems and intellectual theorizing that his work critiques.

Additionally, the novels are practical demonstrations of what Wyndham saw as the prime functions of science fiction—the capacity to demonstrate the mutual interpenetration of science and society, and to show that by anticipating the future, it might be both desirable and possible to change it. There was, for Wyndham, nothing preordained about the present prospect—although for a man who didn't care whether he got his futures "right," he was remarkably prescient. His account of the relationship between science and society is strongly reminiscent of the kind of sociologically informed history of science and technology derived from the Edinburgh School since the early 1970s. His analysis of the structures of modernity more generally, such as the role of the city as emblematic of civilization and the nature of the relationship between culture and nature, speak directly to some of the most important debates that still dominate our public discourse. These range from the introduction of GMOs to the biosphere, to the role of the media in creating (not just reflecting) public opinion, to the role of governmental and corporate agencies in that process, and also to the relationship between science and commerce. Most critically of all, his analysis speaks to the inability of our social institutions to respond to a series of fundamental threats to the ecosystems within which those institutions are embedded. It is also worth noting that his work has never been out of print and that his writings continue to appeal to a broad public. The sheer popularity of his work suggests that his imagined futures continue to resonate strongly with people's understandings of their past and present.[85]

Are there lessons that the science studies community could learn from Wyndham? He certainly believed that other writers could learn from science fiction. Ruefully, he noted that the

[85] Two of his novels were, in fact, published posthumously; see *Web* (London, 1979), which was published ten years after his death; and, thirty years later, *Plan for Chaos* (London, 2009). Both testify to Wyndham's enduring popularity.

whole form has been so deep in the wilderness for so long that writers of ability have worked in more financially and reputably profitable fields, but by degrees they will discover its fictility, its capacity to carry ideas that could be handled in no other medium, its potentialities for expressing not only satire and warning, but also wit, wisdom, home and, sometimes, even a kind of near poetry.[86]

It is perhaps unlikely that historians of science, in their own professional practice, could be brought to regard science fiction as a mode in which they could write, although Naomi Oreskes and Eric Conway have made a notable stride in this direction.[87] But historians of science and science fiction writers do share, at least in part, some crucial goals, not least of which is the exploration of how these categories of science and society interact and intersect. The role of narrative in this pursuit is one that we are asked to interrogate from our very first explorations in historiography, and the real-world impact of fiction can be seen in the consequences for human lives of invented traditions and imagined communities.

In fact, what may most significantly differentiate Wyndham and other science fiction writers from historians and sociologists of science is the willingness of the former group to situate their accounts of the co-construction of sciences and societies within a profoundly moral universe, where the differential distribution of power, whether intellectual, political, or economic, is lived through the emotional and empathetic impact it has on individuals. In other words, science fiction can show its audiences the *affect* as well as the *effect* that the sciences have on our apprehension of what it is to be human. Maybe that's why they achieve much greater audiences, and sales, than we do.

[86] Wyndham, "But Why" (cit. n. 18).
[87] Naomi Oreskes and Eric Conway, *The Collapse of Western Civilization: A View from the Future* (New York, N.Y., 2014).

The Speculative Present:
How Michael Crichton Colonized the Future of Science and Technology

by Joanna Radin*

ABSTRACT

I argue for the importance of considering author, director, and producer Michael Crichton (1942–2008) as a critic and student of Cold War cultures of expertise. Though best known for his blockbuster fiction, he shared a sensibility with academics in North America and the United Kingdom who were concerned with scientists' unchecked authority. These scholars created a new field that would later become known as Science and Technology Studies or STS. Like his contemporaries in STS, and often in anticipation of them, Crichton's novels relied on a blend of history, sociology, and anthropology to lift back the curtain to reveal the specialized worlds in which scientists worked, and to devote specific attention to the practices, instruments, and values that animated their knowledge-production enterprise. In doing so, his fiction both popularized STS and shaped concerns and fears about the future of science and technology. The "speculative present" is useful for making visible the ironies exploited by Crichton, including his own engagement with ideas about speculation as both a form of conjecture and a form of prospecting value. Inspired by Donna Haraway's early work on the cyborg, the speculative present helps to pry open forms of world making obscured by adherence to binaries of fact and fiction, nature and culture, with the goal of cultivating a broader array of visions about emerging science and technology.

> We live in a culture of relentless, round-the-clock boosterism for science and technology. With each new discovery and invention, the virtues are always oversold, the drawbacks understated . . . But everyone knows science and technology are inevitably a mixed blessing. How then will the fears, the concerns, the downside of technology be expressed? Because it has to appear somewhere. So it appears in movies, in stories—which I would argue is a good place for it to appear.
> —Michael Crichton, 1999[1]

* Section for the History of Medicine, 333 Cedar Street, L132, Sterling Hall of Medicine, Yale University, New Haven, CT 06520-8324, USA; Joanna.radin@yale.edu.
Beans Velocci and Elizabeth Karron were excellent research assistants. I am especially grateful for the early encouragement of Erika Milam and thoughtful suggestions of Deanna Day, Dan Bouk, Amanda Rees, and Lisa Messeri as well as the anonymous reviewers. I am also indebted to Sheila Ann Dean for her heroic copyediting.
[1] Michael Crichton, "Ritual Abuse, Hot Air, and Missed Opportunities: Science Views Media," *Science* 238 (25 January 1999): 1461–3, on 1462.

It has been said that "the golden age of science fiction is twelve."[2] That is how old I was when I first encountered Michael Crichton. As a tween in the early 1990s, I was only vaguely aware that the publication of *Jurassic Park* also marked the beginning of twinned booms in biotechnology and information technology. What I did know was that *Jurassic Park* terrified me. Not even because it depicted a world in which dinosaurs terrorized humans, but because the humans seemed utterly stunned to realize that their creations had minds of their own. I couldn't get enough.

Nearly thirty years after the blockbuster success of *Jurassic Park* and the eponymous movie directed by Steven Spielberg, Michael Crichton is still relevant as science and technology's most popular Cassandra. In the fall of 2016, *Westworld*, the HBO series based on his 1973 movie of the same name, premiered to critical acclaim.[3] The setting, like *Jurassic Park*, is a theme park, and the thrill for viewers came from watching creatures inhabiting the park—a simulacrum of the "real" world—turn on their human creators. This time, however, the creations are androids whose artificial intelligence surpasses the expectations of those who seek to dominate them. Throughout his prolific career— Crichton wrote over thirty books (and not all of them fiction), directed movies, designed video games, and served as creator and executive producer of the hit TV medical drama *ER*—Crichton returned to the same theme: the ways in which humans react, often ineptly, to unintended consequences of the science and technology they themselves bring into being.

What was it about Crichton, who died of cancer in 2008, that enabled him to deploy this theme so consistently—and so profitably—to capture the imaginations of millions around the world? While it is possible to situate him in a long and venerable tradition of writing that blurs the boundaries between fact and fiction, which I address below, I argue for the importance of considering Crichton as a critic and student of Cold War cultures of expertise. He shared a sensibility with academics in North America and the United Kingdom who were concerned with scientists' unchecked authority. These scholars created a new field that would later become known as Science and Technology Studies or STS.[4]

Like his contemporaries in STS, and often in anticipation of them, Crichton used a blend of history, sociology, and anthropology to lift back the curtain to reveal the specialized worlds in which scientists worked, and to devote specific attention to the practices, instruments, and values that animated their knowledge-production enterprise. Specifically, Crichton's fiction demonstrated what Bruno Latour would come to call "science in the making," or behind-the-sciences accounts of the construction of durable facts.[5]

---

[2] David Hartwell, "The Golden Age of Science Fiction is Twelve," in *Speculations on Speculation: Theories of Science Fiction*, ed. James Gunn and Matthew Gandelaria (Lanham, Md., 2005), 269–88, on 272. Hartwell argues: "Teenagers are not fully integrated into the tedium of adult life and tend to view such everyday life with healthy suspicion. Quite logical."

[3] Michael Crichton, *Jurassic Park* (New York, N.Y., 1990). For the movie, see *Jurassic Park*, directed by Steven Spielberg (1993; Los Angeles, Calif., Universal Pictures Home Entertainment, 2018), Blu-ray DVD. *Westworld* (the movie), directed by Michael Crichton (1973; Burbank, Calif., Warner Home Video, 2008), DVD.

[4] Those who might be considered peers include Langdon Winner, *Autonomous Technology: Technics-out-of-Control as a Theme in Political Thought* (Cambridge, Mass., 1977); Donald A. MacKenzie and Judy Wajcman, *The Social Shaping of Technology: How the Refrigerator Got Its Hum* (Philadelphia, Penn., 1985); David F. Noble, *America by Design: Science, Technology, and the Rise of Corporate Capitalism* (New York, N.Y., 1977). See also David Edge and other founding members of the Edinburgh Science Studies Unit and the Bath school of the sociology of scientific knowledge.

[5] Bruno Latour, *Science in Action: How to Follow Scientists and Engineers through Society* (Philadelphia, Penn., 1987).

If, as STS scholars have argued, science is what scientists do, Crichton's writing often sought to reveal how scientists, corrupted by political and commercial interests, could come to produce compromised science. His stories were driven by depictions of such interests, effectively serving as applied experiments in STS—offering readers demonstrations of potential unintended consequences of innovation. As a reviewer of his 1969 novel *Andromeda Strain* observed, Crichton had "constructed less a novel than a kind of Herman Kahn think-tank scenario, hypothesizing the most likely way experiments in germ warfare might turn against the experimenters."[6] Crichton's familiar, if fabricated, worlds were those in which knowledge was most profitable to those least accountable—scientists operating under the cloak of secrecy afforded to them by dint of their claims to expertise.[7]

This homology with STS is no coincidence. Crichton not only read academic STS, but his novels, especially those written after 2000, even feature detailed bibliographies that cite STS and STS-adjacent scholars including Paul Edwards and Dorothy Nelkin; sociologists such as Ulrich Beck; anthropologists Adriana Petryna and Andrew Lakoff; and historians of medicine Keith Wailoo, Stephen Pemberton, and Wendy Kline. In addition to his aptitude as an autodidact, digesting a huge range of published scholarship, Crichton took advantage of proximity to high profile social theorists and historians through his alumni connections and residencies. A decade before Bruno Latour and Steve Woolgar conducted their ethnography of scientific practice at the Salk Institute for Biological Studies in La Jolla, California, Crichton held a fellowship there; he participated in some of the first National Science Foundation workshops on biology and society, where he rubbed elbows with the scientific elite.[8] In the late 1980s, for instance, he held the prestigious Knight Science Writing Fellowship at MIT, home of a founding STS program, where he taught what administrators described as a "highly successful" course in the art of *non*fiction writing.[9]

Irrespective of whether one is a fan of his work or despises it—and there are sure to be members of both camps reading this essay—Crichton's critiques were received by many, scientists included, as especially plausible and even predictive.[10] His use of authenticating detail ripped from leading academic scholarship in both science and STS

[6] Melvin Maddox, "New Note: The Novel as Sci-Non-Fi," *Life* 30, May 1969, 15. On the shared history of Herman Kahn's scenario planning and STS, see Sharon Ghamari-Tabrizi, *The Worlds of Herman Kahn: The Intuitive Science of Thermonuclear War* (Cambridge, Mass., 2009); and Bretton Fosbrook, "How Scenarios Became Corporate Strategies: Alternative Futures and Uncertainty in Strategic Management" (PhD diss., York University, 2017). Fosbrook shows how these scenario planners, like Crichton, drew inspiration from STS.

[7] To give but one example, as a young man, Crichton was so troubled by the secrecy that characterized militarized science, he wrote an op-ed condemning it; Michael Crichton, "A Scientist Can Always Say 'No' To Secrecy," *New York Times*, 29 October 1971, 41. Secrecy has also preoccupied Cold War social scientists and historians; see Robert K. Merton, "A Note on Science and Democracy," *Journal of Legal and Political Sociology* 1 (1942): 115–26, on 115; Loren R. Graham, *Science in Russia and the Soviet Union: A Short History* (Cambridge, UK, 1993). See also Ernan McMullin, "Openness and Secrecy in Science: Some Notes on Early History," *Sci. Tech. Hum. Val.* 10 (1985): 14–22; Michael Aaron Dennis, "Secrecy and Science Revisited: From Politics to Historical Practice and Back," *Secrecy and Knowledge Production* 23 (1999): 1–16; Peter Galison, "Removing Knowledge," *Crit. Inq.* 31 (2004): 229–43.

[8] Bruno Latour and Steve Woolgar, *Laboratory Life: The Construction of Scientific Facts* (Beverly Hills, Calif., 1979).

[9] MIT Reports to the President, 1987–1988, Institute Archives & Special Collections, MIT Libraries, https://libraries.mit.edu/archives/mithistory/presidents-reports/1988.pdf.

[10] For a resonant discussion of the broader significance of plausibility—in particular the way that speculative writers have achieved and subverted it—see Peter J. Bowler, "Parallel Prophecies: Science

enabled him to speculate about speculation itself in ways that have had material conse-quences for science practice and science policy.[11] Taking these aspects of Crichton's world building seriously—viewing him as a representer of science-in-action as well as its potential consequences—enables an assessment of the impact of STS beyond the realm of academia. It also shows how in taking fears and concerns about cutting-edge science and technology as an imaginative domain for his own creative uses, he made them real. Crichton did not predict the future of science and technology, he colonized aspects of science and technology that practitioners and policy-makers themselves felt ill prepared to address.

I examine Crichton's own claims to credibility as arbiter of the downsides of science and technology by examining the multiple valences of the concept of speculation. Spec-ulative fiction is a genre that often encompasses work also classified as science fiction. This is because speculation can be used to refer to abstract processes of theorizing or con-jecturing, as well as, in a material sense, to prospecting for sources of investment. Indeed, a good deal of science fiction is animated by accounts of efforts to settle other planets in search of new resources. Those are stories that often take place explicitly either in the future or in worlds different from Earth.

In recent years, certain STS scholars have found speculative fiction to be an impor-tant crucible for contesting how science and technology have been used to produce exclusionary systems of oppression.[12] And novelists such as Ursula K. Le Guin and Octavia Butler—to name just a few whose careers overlapped with Crichton—crafted sophisticated speculative worlds that led readers to different conclusions than Crich-ton and his fans about the potential for relationships between humans and their ma-chines. I will return to this potential in the final section, but to get there I use a concept I am referring to as the "speculative present." This is a strategy for considering the re-lationship between fictional worlds depicted *in* speculative works and the real-world building for which those speculative visions are used beyond them.[13]

Fiction and Futurology in the Twentieth Century," in this volume. Science fiction author and critic Thomas Disch also singles out Crichton for the plausibility of his accounts in *The Dreams Our Stuff is Made of: How Science Fiction Conquered the World* (New York, N.Y., 1998).

[11] It is impossible to fully do justice to this phenomenon here. For example, consider books like Rob DeSalle and David Lindley's *The Science of Jurassic Park and the Lost World* (New York, N.Y., 1997); and Kevin R. Grazier's *The Science of Michael Crichton* (Dallas, Tex. 2008). See also com-mentaries on his books in leading journals like *Science* and *Nature*, such as a 2007 roundtable in *Na-ture* titled "The Biologists Strike Back," in which microbiologist and science fiction author Joan Slonczewski goes so far as to characterize the creation of a molecular biology program at the college where she teaches as a response to the interests of what her colleagues call the "Jurassic Park gener-ation"; Slonczewski, "The Biologists Strike Back," *Nature* 448 (2007): 18–21. (Slonczewski is a sub-ject of Erika Lorraine Milam, "Old Woman and the Sea: Evolution and the Feminine Aquatic," in this volume.) See also Elizabeth Jones's dissertation about the impact of *Jurassic Park* on the explosion of interest in the science of ancient DNA; Elizabeth Dobson Jones, "The Search for Ancient DNA in the Media Limelight: A Case Study of Celebrity Science" (PhD diss., University College London, 2017).

[12] This phenomenon is documented in feminist STS and Afrofuturist scholarship, including Maria Puig de la Bellacasa, *Matters of Care: Speculative Ethics in More than Human Worlds* (Minneapolis, Minn., 2017); Ruha Benjamin, "Racial Fictions, Biological Facts: Expanding the Sociological Imag-ination through Speculative Methods," *Catalyst: Feminism, Theory, Technoscience* 2 (2016): 1–28; DeWitt Kilgore, *Astrofuturism: Science, Race, and Visions of Utopia in Space* (Philadelphia, Penn., 2003); and Donna J. Haraway, "SF: Science Fiction, Speculative Fabulation, String Figures, So Far," *Ada: A Journal of Gender, New Media, and Technology* 3 (2013), doi:10.7264/N3KH0K81.

[13] While I use the phrase here to shift attention away from prognostication, "speculative present" has also been invoked by artists to hail the visionary potential of Le Guin and Butler. See, for example, Janice Lee, ed., "On the Speculative Present," *Verse*, https://verse.press/playlist/on-the-speculative

Crichton's work was grounded firmly and consequentially on this planet, and his novels were often set in the same time period in which he was writing them. Speculation was his subject *and* his method. I show this through accounts of two of his most successful novels, *The Andromeda Strain* (1969) and *Jurassic Park* (1990), wherein practices of bioprospecting—a strategy by which life forms are themselves mined for future knowledge and profit—are front and center.[14] In describing the production and reception of these works, I explain how Crichton also offered conjecture on the negative consequences of bioprospecting. His claims appeared credible because they were often, at least partially, based in fact. His academic training enabled him to treat scientific and social scientific publications as reservoirs to be mined for authenticating details that would make his narratives appear persuasive to nonexperts and, in some cases, even to experts; it inspired him to write stories about contemporaneous efforts to, as Francis Bacon centuries earlier declared of the role of science, torture the secrets of nature. Yet, unlike science or even STS, Crichton's depictions of his subjects were not tested in the realm of peer review, as they might have been had he stayed in academia, but rather in the court of public opinion.[15] As a writer of fiction, he could employ facts however he liked, even as his stories often warned against the dangers of their misuse.

In a fitting plot twist, Crichton's speculations turned out to be fabulously profitable, garnering him fantastic wealth and, later, political influence.[16] This plot twist also stands to reveal certain ironies that have come to characterize academic STS—in particular the field's relationship to politics and perpetual vulnerability to charges of relativism.[17] A concept of the speculative present, then, serves to pry open a powerful realm of world making obscured by adherence to binaries of fact and fiction, scholarly and popular, content and form, inside and outside, or realism and relativism. The speculative present is useful for making visible the ironies exploited by Crichton—an individual who has been credited with creating "an industry" based on a relentless critique of the very institutions essential to his success—with the goal of cultivating a broader array of visions about what science and technology can be.[18]

-present-2716870952890542721 (accessed 19 August 2018); and Madeline Lane-McKinley, "Notes on the Speculative Present," *Blind Field: A Journal of Cultural Inquiry*, 1 January 2017, https://blindfieldjournal.com/2017/01/01/notes-on-the-speculative-present/ (accessed 2 January 2019).

[14] Michael Crichton, *The Andromeda Strain* (New York, N.Y., 1969); Crichton, *Jurassic Park* (cit. n. 3). For a history of bioprospecting that summarizes the critical literature on the subject covering the period between the writing of these two novels, see Joanna Radin, *Life on Ice: A History of New Uses for Cold Blood* (Chicago, Ill., 2017).

[15] This is not to say that the system of peer review is itself a guarantor of objectivity. See, for example, Alex Csiszar, "Troubled from the Start: Pivotal Moments in the History of Academic Refereeing Have Occurred at Times When the Public Status of Science Was Being Renegotiated," *Nature* 532 (2016): 306–9. And see especially Melinda Baldwin, "Scientific Autonomy, Public Accountability and the Rise of 'Peer Review' in the Cold War United States," *Isis* 109 (2018): 538–58.

[16] To give an example, following the publication of *State of Fear* (New York, N.Y., 2004) Crichton was invited to testify before Congress about the relationship between science and politics. He also met with Karl Rove, advisor to President George W. Bush, to discuss climate science.

[17] This point was made in Paul R. Gross and Norman Levitt, *Higher Superstition: The Academic Left and Its Quarrels with Science* (Baltimore, Md., 1997); and more recently in Steven Pinker, "The Intellectual War on Science," *Chronicle of Higher Education*, 13 February 2018, https://www.chronicle.com/article/The-Intellectual-War-on/242538.

[18] Michael Crichton, television interview by Charlie Rose, *Charlie Rose*, PBS, 26 December 1996; published in Robert Golla, ed., *Conversations With Michael Crichton* (Jackson, Miss., 2011), 119; Golla writes that "the interviews herein have not been edited from the form of their initial publication," x–xi.

KNOWLEDGE FICTION

This would be a good moment to provide some biographical details. John Michael Crichton was born in 1942 in Chicago and raised in Roslyn, New York. The son of a leader in advertising—his father was the President of the American Association of Advertising Agencies and the editor of the journal *Advertising Age*—and a home-maker, Michael Crichton grew big along with American science.[19] Nearly every interview conducted with Crichton mentioned his height—by age thirteen he was already nearly six feet nine—as though it could account for his outsized productivity in literature, film, and TV.

Crichton enrolled at Harvard as an English major in the early 1960s. When he got a B- on a paper (which he later admitted was written by George Orwell but submitted as his own work as a jab at his professor), he switched to anthropology and studied with William White Howells.[20] Howells, a leading figure in the reinvigoration of post-war biological anthropology, is now recognized for his early investments in the use of multivariate statistics to legitimate craniometry—the measurement of skulls—as a postracial science.[21] When he was Crichton's professor, Howells was perhaps better known for the trade books he wrote about human origins, with titles like *Mankind So Far* and *Back of History: The Story Of Our Own Origins*.[22] Later in his career, Crichton would credit Howells with sparking his interest in information technology, a recurring element in many of his novels and the subject of a nonfiction book he wrote in 1983 called *Electronic Life: How to Think about Computers*.[23]

After he graduated, Crichton taught an introduction to anthropology course at Cambridge University and published scholarly articles based on research using skulls stored at the British Museum in London. He then enrolled in medical school at Harvard where he earned his MD, though he never actually practiced as a physician.[24] He had helped finance his education by writing pulpy mysteries under the pseudonym John Lange, which helped him master conventions of narrative and plot. As a medical student, he began to use these skills to take on the profession and what he saw as its deep flaws. He could express his critical opinions in fiction in ways that simply were not permissible in scientific publications. He wrote the novel *A Case of Need* (1968) under the alias Jeffrey Hudson while wrestling with the decision to leave medicine.[25]

A Case of Need was published to rave reviews, winning the Edgar Award (named in honor of Edgar Allen Poe) of the Mystery Writers of America. Crichton was emboldened and soon shed his pseudonym to establish his beat as a reporter of the speculative

---

[19] Derek de Solla Price, *Little Science, Big Science* (New York, N.Y., 1963).

[20] Michael Crichton, *Travels* (New York, N.Y., 1988), 4.

[21] Paul T. Baker, "Adventures in Human Population Biology," *Annu. Rev. Anthropol.* 25 (1996): 1–18.

[22] Henry M. McHenry and Eric Delson, "Obituary: William White Howells (1908–2005)," *Am. J. Phys. Anthropol.* 135 (2008): 249. William White Howells, *Mankind Evolving* (Garden City, N.Y., 1944); Howells, *Back of History: The Story Of Our Own Origins* (Garden City, N.Y., 1963).

[23] Michael Crichton, *Electronic Life: How To Think About Computers* (New York, N.Y., 1983).

[24] J. M. Crichton, "A Multiple Discriminant Analysis of Egyptian and African Negro Crania," in William White Howells and J. M. Crichton, *Craniometry and Multivariate Analysis*, vol. 57 of *Peabody Museum Papers* (Cambridge, Mass., 1966), 47–67. This paper has been cited sixty-seven times, and in nearly all cases for the scientific content, not the author; it was most recently cited in 2015.

[25] Michael Crichton, *A Case of Need* (New York, N.Y., 1968).

present—a rearrangement of social reality and its technological possibilities that relied upon his own scientific literacy and institutional connections.[26] In doing so, he connected with readers who were aware of the increasing ubiquity of the products of science and technology in everyday life, but whose feelings about their impact were similarly ambivalent—a jumble of wonder and fear. His breakout publication under his real name was the biomedical thriller *The Andromeda Strain* (1969), which he wrote while still in medical school.[27]

As Crichton continued to publish and broadened into movie directing, one reviewer dubbed him an author of "knowledge fiction" in apparent recognition that his work represented a road less traveled in science fiction—one that traded in realism and took expertise itself as its subject.[28] As Crichton explained to Charlie Rose in a 1999 interview, "I try and stay within the boundaries of what's not understood to be impossible, even though there's dispute about it."[29] Much like writers ranging from Isaac Asimov to John Wyndham, he claimed his best ideas came from the realm of fact; these were facts that he culled from cutting-edge scholarship across the natural and social sciences.[30]

Crichton himself disdained being labeled a writer of science fiction, choosing instead to see himself as an "intellectual."[31] Today, Crichton has come to be most closely associated with the genre known as the "technothriller," a male-dominated field (think John Grisham's legal and Tom Clancy's military-science spy novels) distinguished by its emphasis on technical details about elite institutions.[32] Though Crichton's work was similarly set in the present, he articulated a stronger kinship with authors long dead who have since been claimed for science fiction.[33] He saw himself as heir to Edgar Allen Poe, who studied engineering at West Point, and Arthur Conan Doyle, who had worked briefly as a physician. Like both men, Crichton's primary use for the facts he collected was to situate them (not always but very often) in fictional scenarios grounded in the social world contemporaneous with his own.

"Fantasy," it has been argued of the likes of Poe and Conan Doyle, "helps us accept that the real world is to some degree imaginary, relying on contingent narratives that are subject to challenge and change."[34] Crichton's novels do the same—they present readers

---

[26] This point is in harmony with Peter Bowler's arguments about the shared foundations between science fiction and popular science writing. See Bowler, "Parallel Prophecies" (cit. n. 10); and Bowler, *A History of the Future: Prophets of Progress from H. G. Wells to Isaac Asimov* (Cambridge, UK, 2017).

[27] Crichton, *Andromeda* (cit. n. 14).

[28] Patrick McGilligan, "Ready When You Are, Dr. Crichton," *American Film*, March 1979; republished in Golla, *Conversations* (cit. n. 18), 16.

[29] Michael Crichton, television interview by Charlie Rose, *Charlie Rose*, PBS, 16 November 1999; republished in Golla, *Conversations* (cit. n. 18), 168.

[30] Amanda Rees's treatment of British author John Wyndham demonstrates striking parallels with Crichton's self-fashioning and concerns about the misuse of scientific expertise; Amanda Rees, "From Technician's Extravaganza to Logical Fantasy: Science and Society in John Wyndham's Post-War Fiction (1951–1960)," in this volume.

[31] Barbara Rose, "Hollywood Gets a New Man," *Vogue*, September 1973; republished in Golla, *Conversations* (cit. n. 18), 10.

[32] In his introduction to *Conversations* [(cit. n. 18), ix], Golla notes that Crichton was delighted by "the monikers given by two friends: Tom Clancy famously called him 'Father of the Techno-Thriller,' and Steven Spielberg described him as 'The high priest of high concept.'"

[33] John Tresch, "Extra! Extra! Poe Invents Science Fiction!" in *The Cambridge Companion to Edgard Allen Poe*, ed. Kevin J. Hayes (Cambridge, UK, 2002), 113–32; Edward Lauterbach, "Conan Doyle's Science Fiction," *English Literature in Transition, 1880–1920* 40 (1997): 481–3.

[34] Michael Saler, *As If: Modern Enchantment and the Literary Prehistory of Virtual Reality* (New York, N.Y., 2012), 21.

with the problems and possibilities of the very idea of reality. In doing so, they serve as a means of considering, in "real" time, how science and technology change the status of truth and those entitled to be reliable knowers of its nature. As Peter Bowler argues, the skillful ability to blur the line between science fiction and popular science was also used as a powerful strategy by authors like Aldous Huxley and H. G. Wells to disabuse readers of an overly celebratory notion of innovation.[35]

As a matter of genre, each of these authors also exploited the limits of the scientific publication or the medical case report and appropriated mass communication to disseminate their ideas. Crichton embraced mass market fiction—in his case, these were airport novels, blockbuster cinema, and TV—as the right tools for the job of revealing Pandora's black boxes of science and technology.[36] His fiction is littered with footnotes, graphs, charts, computer readouts, and even bibliographies, "the authenticating effect" of which was described by one early reviewer as "drolly sinister."[37] Unlike the academic journal article or monograph, mass market fiction provided an ideal channel for introducing readers to cutting-edge forms of science and technology while colonizing readers' attitudes about their potential risks.[38] In narrating a present that was esoteric to many of his readers, and depicting the otherwise hazy category of unintended consequences that the present might produce, he achieved a prophetic reputation.

As the author of "some of the most lucrative prose in literary history," Crichton was often hailed for his prescience—his seeming ability to predict the future.[39] Drawing on the knowledge and credibility he gained from earning Harvard degrees in anthropology and medicine, Crichton invested his energy in crafting vivid depictions of moral crises that afflicted American subcultures held hostage by the products of their expertise.[40] These scenarios blurred the boundary between fact and fiction, between the present and the near future, to demonstrate how humans' best efforts at technological ingenuity would fail without similar investments in social and moral innovation. He instructed his readers about how to *feel* about science and technology, highlighting certain potential risks yet always admixing them with wonder.[41]

Yet, as his reputation for prescience grew, he became an outspoken critic of mass media, charging in a 2002 speech titled "Why Speculate?" that the media—of which

[35] Bowler, "Parallel Prophecies" (cit. n. 10).

[36] Crichton's intent was in a parallel tradition to work of STS scholars such as Adele Clarke and Joan H. Fujimura, *The Right Tools for the Job: At Work in Twentieth-Century Life Sciences* (Princeton, N.J., 1992); Latour, *Science in Action* (cit. n. 5).

[37] Melvin Maddox, "New Note" (cit. n. 6), 15.

[38] My use of the concept of colonialism is not metaphorical. For a succinct discussion of the stakes of this specific formulation, see Greg Gerwell, "Colonizing the Universe: Science Fictions Then, Now, and in the (Imagined) Future," *Rocky Mt. Rev. Lang.* 55 (2001): 25–47. On the role of the science fiction industry "to dramatize our incapacity to imagine the future," see Frederic Jameson, "Progress Versus Utopia; Or Can We Imagine the Future?" *Sci. Fict. Stud.* 9 (1982): 147–58, on 153.

[39] Zoe Heller, "The Admirable Crichton," *Vanity Fair*, January 1994; republished in Golla, *Conversations* (cit. n. 18), 64.

[40] In an interview about his book *Rising Sun* (New York, N.Y., 1992), a fictional account of the threat of Japanese economic domination in technology, Crichton stated, "My interest is in America, and my whole focus is on how America is responding and behaving in the contemporary world"; see Michael Crichton, "Rising Sun Author Taps Darkest Fears of America's Psyche," interview by T. Jefferson Parker, *Los Angeles Times*, 5 July 1992; republished in Golla, *Conversations* (cit. n. 18), 55.

[41] In this sense, Crichton's work presents an archive for the history of emotions and affect. Crichton may well have been familiar with Raymond Williams, "Structures of Feeling," *Marxism and Literature* 1 (1977): 128–35.

he himself was a contributing member—"carries with it a credibility which is totally undeserved."[42] To read his personal views on the dubious power of speculation or the failures of punditry is to consider that Crichton was alarmed by his own influence. It is also to entertain a darker possibility that such a performance of concern—of claiming to stand outside the mob mentality even as he stoked (and profited from) it—was meant to excuse him from accountability for the unintended consequences of the ways his work was used in the name of science itself. Last but not least, it is to beg the question of why *should* anyone have taken Michael Crichton at his word. What makes him such an important, and as yet underappreciated, figure for the history of science and STS is just how many people, from scientists to policy makers, did and continue to do just that.

CRAFTING THE SPECULATIVE PRESENT: *THE ANDROMEDA STRAIN*

At a time when many were still fixated on the threat of nuclear crises, *The Andromeda Strain* (1969) was a story of biological crisis—the introduction of a mysterious and deadly viral agent upon the return of a space voyage.[43] In the novel, a military satellite returns to earth, landing in northern Arizona. The team sent to recover the satellite breaks radio contact. Because the military suspects an infection, a new team of scientists is dispatched to investigate only to find that everybody in the nearest small town is dead (except for an old man and a screaming infant). The two survivors are taken to an underground laboratory where the culprit, an organism named the "Andromeda Strain," is identified; it then nearly kills all of humankind. American citizens never learn of this brush with apocalypse, or that the entire crisis is the product of its government's top-secret germ warfare research.

Published weeks before the lunar landing of Apollo 11, *The Andromeda Strain* played on already circulating Cold War anxiety at two levels: fear that astronauts would bring new germs back to earth, and concern that the government would prove inept at managing this threat. Crichton subsequently attributed the idea for the plot to the eminent paleontologist George Gaylord Simpson. As he recalled, while studying anthropology at Harvard, his mentor Howells tasked Crichton with reporting on Simpson's *The Major Features of Evolution* (1953). He claimed to have been struck by a footnote in which the famed evolutionary theorist—himself an author of a volume of science fiction—observed that microbes in the upper atmosphere had never been used as the subject of a science fiction story.[44] By the time Crichton would have encountered Simpson's work, the evolutionary biologist had become embroiled in debates with molecular biologist and

---

[42] Michael Crichton, "Why Speculate?" lecture, International Leadership Forum, La Jolla, Calif., 26 April 2002. A transcription of this lecture is one of three that accompanies the e-book version as an "E-book extra" at the end of his 2004 novel, *State of Fear* (cit. n. 16), which was released in 2009 for Kindle. Crichton also published the following nearly a decade earlier: "Jurassic Journalists: Is the News Media Giving Us the Full Story, or Are They Just Giving It to Us?" *Equity* 12 (1994): 50–5. Consider also the distorting role of censorship imposed by the Motion Picture Association in direct response to the realism of cinematic depictions of science, as described by David Kirby, "Darwin on the Cutting Room Floor: Evolution, Religion, and Film Censorship," in this volume.

[43] Crichton, *Andromeda* (cit. n. 14).

[44] Crichton, *Travels* (cit. n. 20). The question of whether or not Earthly microbes could develop adaptations that would allow them to occupy the upper atmosphere was ultimately to be filed by Simpson "under the heading of idle speculation." In the note that amended that statement, Simpson continued, "Or would be a large source of material for science fiction, which has so far been singularly unimaginative and uninstructed in dealing with possible future evolutionary development here on earth." George Gaylord Simpson, *The Major Features of Evolution* (New York, N.Y., 1953), 212n.

Nobel laureate Joshua Lederberg over the subject of life in outer space, a science known as exobiology.[45]

Crichton's editor, Bob Gottlieb (who "discovered" Joseph Heller and published his *Catch-22* in 1961) also saw potential in such a story and guided the book to publication at Knopf, the venerable New York City publishing house. He encouraged Crichton to write the novel in the journalistic style of a *New Yorker* profile, and aim for a cool, detached voice that blurred the line between fiction and reality.[46] In the novel, Crichton highlighted technologies that were then new, like remote surveillance, voice-activated systems, computer imaging and diagnosis, handprint identification, and biosafety procedures. The hazmat suit, which is now a commonplace visual signifier of biocontamination, made its public debut in the 1971 film based on the novel. Crichton appeared on screen, taking a small nonspeaking role.

The precise depiction of unfamiliar technical innovations made it difficult for readers to disentangle the actual from the merely possible, which heightened the thrill. The plausibility of Crichton's account, similar to Orson Welles's infamous *War of the Worlds* radio broadcast, was meant to leave readers in a perpetual state of suspense, a nightmare-like trance that might not be that different from being awake. Gottlieb later noted his author's skill: "Michael had a keen eye, or nose, for cutting-edge areas of science—and, later, sociology—that could be used as material for thrillers while cleverly popularizing the hard stuff for the general public. You got a lesson while you were being scared."[47]

The hero of *The Andromeda Strain* is a young physician named Dr. Mark Hall, whose attention to the two human survivors of the mysterious Andromeda—the young child and old man—results in revealing the cause of its lethality. Hall is the only member of the team who had been previously unaware of the purpose of Project Wildfire, the name of the top-secret program. He serves not only as the story's protagonist but as an avatar for the reader (and perhaps Crichton himself) who is amazed and horrified by the biomedical future being formulated in the underground laboratory. As part of his initiation, Hall endures a medical exam conducted entirely by a talking computer that seems to signal his own potential obsolescence as a diagnostician. Hall's disgust with his fellow team members' willingness to serve as what STS scholar Chandra Mukerji would call a "reserve labor force" for militarized science provides the moral punchline of the book. Having been kept out of the loop, Hall is the "odd man," and the only member of the team who is distressed that Project Wildfire is associated with government-sponsored germ warfare.[48] The novel concludes with a Nobel Prize–winning scientist's untempered hubris in the face of near disaster: "But things are now under control . . . We have the organism, and can continue to study it."[49]

The novel quickly went viral and ultimately served as the nail in the coffin for Crichton's medical career. Yet, he was not ready to break with biomedicine altogether. Fresh off the heels of the successful publication of *The Andromeda Strain*, in 1969 Crichton took up residency as a fellow at the Salk Institute in La Jolla, California. In an interview with Charlie Rose, Crichton recalled the following events: "Salk [an inventor of the polio vaccine] had an idea that along with the biomedical people who were primarily

[45] Audra Wolfe, "Germs in Space: Joshua Lederberg, Exobiology, and the Public Imagination, 1958–1964," *Isis* 93 (2002): 183–205.
[46] Robert Gottlieb, *Avid Reader: A Life* (New York, N.Y., 2016), 109–10.
[47] Ibid., 109.
[48] Chandra Mukerji, *A Fragile Power: Scientists and the State* (Princeton, N.J., 1989).
[49] Crichton, *Andromeda* (cit. n. 14), 289.

staffing the place, he was going to have writers and artists in residence. So I sent him a letter and said, 'I'm a writer.' He said, 'Come.' So, I went. It was great."[50]

Historian of science Elena Aronova has discussed the creation in 1968 of this program at the Salk Institute. The Council for Biology in Human Affairs was a strategy for bridging what C. P. Snow had notoriously diagnosed as the "two cultures" problem in higher education, whereby the sciences operated in isolation from the humanities.[51] "The goal of the Council," Aronova writes, "was to coordinate public opinion on the societal implications of molecular biology by communicating ideas through the web of formal and informal contacts between leading molecular biologists" on the one hand, and through "the national legislative and executive community, and the leadership in business, industry and labor, law, education, communication and foundations" on the other.[52] A related goal, she notes, was to consider how universities could develop new programs to address the ways biology was poised to remake ideas about the human species. Toward that end, there was an explicit mandate to allocate funds to support humanists. Crichton, who had a foot in both worlds, was an ideal mediator.

Crichton participated in the first meeting of the Council for Biology in Human Affairs at Salk, along with linguist Roman Jakobson, philosopher and bioethicist Daniel Callahan, Jonas Salk, and institute staff members Salvador Luria and John Hunt. A second meeting was hosted by the Harvard Club on the East Coast and included the historian of physics Thomas Kuhn among its participants.[53] Despite words of enthusiasm from the participants, external funding did not follow for this forerunner of the Ethical, Legal and Social Implications initiatives of the NSF and NIH.[54]

However, the social policy views of British mathematician and founding Salk Institute Faculty Fellow Jacob Bronowski made an impression on Crichton. In a 1971 *New York Times* editorial about the national security risks of secrecy in science, Crichton invoked Bronowski, considering the elder man's view that there should be alternatives to government funding for those who wanted to pursue science. "Perhaps the Government should provide research funds without also deciding research priorities," Crichton posited, but noted that leaving decisions about funding solely up to the scientists was "highly unlikely from a practical standpoint and it is not guaranteed to make the world a better place to live." Chrichton noted: "Scientists already bicker over research money; and if control over dispersal were entirely in their hands, some very strange decisions might emerge. . . . In the long run, science should, after all, benefit society—and some representatives of the society should have a voice in the decision-making process."[55]

By the time sociologists Bruno Latour and Steve Woolgar made Salk the site of ethnographic research (between 1975 and 1977) for their book *Laboratory Life: The Construction of Scientific Facts*, humanistic and philosophical inquiry had become detached

---

[50] Michael Crichton, "An Interview with Michael Crichton," interview by Charlie Rose, *Charlie Rose*, PBS, 14 January 1994; transcribed in Golla, *Conversations* (cit. n. 18), 91.

[51] C. P. Snow, *The Two Cultures and the Scientific Revolution* (New York, N.Y., 1959).

[52] Elena Aronova, "Studies of Science before 'Science Studies': Cold War and the Politics of Science in the U.S, U.K., and U.S.S.R., 1950s–1970s," (PhD diss., Univ. of California, San Diego, 2012), 189–90.

[53] Ibid., 192.

[54] The creation of the Ethical, Legal and Social Implications (ELSI) arm of the Human Genome Project is described in Daniel J. Kevles and Leroy E. Hood, *The Code of Codes: Scientific and Social Issues in the Human Genome Project* (Cambridge, Mass, 1992).

[55] Crichton, "A Scientist Can Always," (cit. n. 7).

from the practice of science.[56] In what is often regarded as one of the first anthropological accounts of science, Latour and Woolgar critiqued the Salk Institute for conceptualizing science as separate from society, rather than as a part of culture that needed to be examined socially. They charged the Salk Institute with casting the social study of science as a luxury—something epiphenomenal to the practice of science itself.[57]

In this light, it becomes possible to argue that Crichton capitalized on the way that social studies of science were made peripheral to science, turning the former into a ravenously consumed cultural product that would be marginalized, ironically, as too popular to be funded as a form of science. Crichton realized that as a best-selling novelist—as a writer of speculative scenarios about scientific facts—he could perform a role that science and its government funders had failed to embrace; what was more, he could mine biomedicine for authenticating details, even as he condemned the culture of the enterprise. Those who purchased his books were, whether they realized it or not, voting with their wallets, demonstrating concern about how they had been excluded from the decision-making process.

At the time of its publication, a reviewer for *The Harvard Crimson*, the newspaper of Crichton's alma mater, characterized *Andromeda Strain* as an exposé of a new era of government secrecy.[58] Crichton's personal website describes how the book became a flashpoint for Cold War concerns, ones that were shared by many early STS practitioners, including the notion that science was right wing, elite, and a handmaiden to the military. Others worried that the book's depiction of a viral threat might actually inspire the creation of bioweapons.[59]

Joshua Lederberg was the inspiration for the character of Jeremy Stone, the Nobel laureate scientist and political insider who leads the US government's top-secret plan to respond to the potential for germ warfare depicted in *The Andromeda Strain*. Lederberg read the novel with extreme interest. In June of 1969, he wrote to the publisher of Knopf, "None of my friends or acquaintances has any doubt whatever about the identification of the character in 'The Andromeda Strain,' and I find it hard to believe that the parallel is entirely fortuitous."[60] In his correspondence, Lederberg also intimated that the novel advanced a perspective aligned with his own vocal concerns about germ warfare—even if his avatar was depicted as believing otherwise. He had already written to the publisher, "The book is an interesting one, and deserves to have a wide sale."[61]

Ever the political operator, Lederberg recognized Crichton's power in shaping public opinion about the ethics of militarized biomedical innovation. In subsequent communications over the next year he sent him, via Knopf, reprints of his own congressional testimony about the dangers of germ warfare and at least a dozen of his columns in *The Washington Post*.[62] There is no record in the archive of a response from Crichton, but

---

[56] Latour and Woolgar, *Laboratory Life* (cit. n. 8).

[57] Aronova, "Studies of Science" (cit. n. 52).

[58] Gregg J. Kilday, "Infectious," 12 August 1969, *Harvard Crimson*, https://www.thecrimson.com/article/1969/8/12/infectious-pbtbhe-andromeda-strain-by-michael/ (accessed 25 August 2018).

[59] Such responses are mentioned on Crichton's personal website, http://www.michaelcrichton.com/the-andromeda-strain/ (accessed 18 August 2017). See note 7 for discussions of STS and secrecy.

[60] Letter from Joshua Lederberg to William Koshland (president of Knopf), 26 June 1969, Box/Folder 874.3 Knopf—Editors—Koshland, William A., 1970-C 2 of 2, The Knopf Records, Harry Ransom Center, University of Texas at Austin (**hereafter** cited as TKR).

[61] Letter from Joshua Lederberg to William Koshland, 9 June 1969, TKR.

[62] Letter from Joshua Lederberg to William Koshland, 25 June 1969; letter from Joshua Lederberg to William Koshland, 20 July 1970, TKR.

his subsequent work made it clear that he had found his métier—a critique that sought to demonstrate potential forms of social harm generated by the unintended consequences of science and technology. In the years that followed, the phrase "Andromeda Strain" came to be used in biomedical publications to describe emerging infectious diseases, including AIDS, Marburg, and Ebola, quietly carrying with it the baggage of its origins in Crichton's speculations.[63]

GIVE ME A THEME PARK AND I WILL THRILL THE WORLD: *JURASSIC PARK*

By the 1990s, Crichton's early fascination with blurring boundaries had morphed into what he articulated as an "uneasiness with the distinction between fact and fiction."[64] He described a feeling of vertigo when he started reading the history of the American Revolution to research a novel. He told an interviewer, "I was continuously shocked at the speculations that historians make, which as a novelist I would never dare to make."[65] Crichton had added history to his list of disciplines to be used for mining authenticating details.[66] The most successful of his historical experiments was his 1990 novel, *Jurassic Park*, which takes place in a theme park where the deep evolutionary past has been conjured in the biotechnological present.

For Crichton, the theme park served as a sort of virtual reality lab for scrambling past, present, and future. This kind of immersive simulation had long captivated Crichton. He had first begun to examine its potential in *The Andromeda Strain*, which is set in a contained underground laboratory. In its original trailer, director Robert Wise described the underground laboratory as the 1971 film's "main character."[67] In retrospect, this focus on the culture of the laboratory in *Andromeda* also presaged Latour and Woolgar's 1979 *Laboratory Life*, the latter of which helped inaugurate the STS emphasis on studying the practices and places of scientific inquiry. In the history of science and STS, this turn to place and practice, if not ethics, followed on Thomas Kuhn's influential 1962 *Structure of Scientific Revolutions*, a book that, among other things, argued against science as a teleological accretion of facts.[68] It is a book that is also often cited as an origin point for STS.[69]

The first of Crichton's works to be set explicitly in a theme park was his 1973 film *Westworld*.[70] The movie's setting is a vacation destination comprised of three historically specific worlds, including the Wild West of the United States, with cowboys;

---

[63] Priscilla Wald, *Contagious: Cultures, Carriers, and the Outbreak Narrative* (Durham, N.C., 2008); Kirsten Ostherr, *Cinematic Prophylaxis: Globalization and Contagion in the Discourse of World Health* (Durham, N.C., 2005).

[64] Michael Crichton, "Rising Sun," radio interview by Don Swaim, *Book Beat*, CBS Radio, 6 February 1992; published in Golla, *Conversations* (cit. n. 18), 43.

[65] Ibid., 44.

[66] Crichton's 1999 novel *Timeline* (New York, N.Y.) hits especially close to home, featuring historians of science and technology at Yale.

[67] *The Andromeda Strain*, directed by Robert Wise (1971; Los Angeles, Calif., Universal Pictures Home Entertainment, 2015), Blu-ray DVD. Original trailer, https://www.youtube.com/watch?v=8qEsqjJAY-k (accessed 2 January 2019).

[68] Thomas Kuhn, *The Structure of Scientific Revolutions* (Chicago, Ill., 1962).

[69] Robert J. Richards and Lorraine Daston, eds., *Kuhn's 'Structure of Scientific Revolutions' at Fifty: Reflections on a Science Classic* (Chicago, Ill., 2016).

[70] A book version was published the same year as a tie-in for the film, but Crichton conceived the project as a screenplay. The film (cit. n. 3) has since been acknowledged as innovative not only in its plot but in its use of special effects. David A. Price. "How Michael Crichton's 'Westworld' Pioneered Modern Special Effects," *New Yorker*, 14 May 2013, https://www.newyorker.com/tech/annals-of-technology/how-michael-crichtons-westworld-pioneered-modern-special-effects.

the Middle Ages, with knights; and ancient Rome, with gladiators. Primarily upper-middle-class and white people visit, as they would any all-inclusive resort. In an early scene in the film, masses of androids—battered and bruised by humans who abused them as a way to prove their own vitality—lie splayed out on gurneys awaiting repair by staff in white coats. The viewer is confronted with what Crichton later summarized as his critique of scientific medicine; he believed that it had cast "the physician as technician and the patient as a biological machine that was broken."[71] The movie was a speculative present—a fictional critique of a dismal technological future that had already arrived.

Westworld was far from a flop, but it took *Jurassic Park* for Crichton to leverage this speculative present into a bona fide cultural sensation, earning him the designation "The Hit Man" on the cover of *Time* magazine.[72] *Jurassic Park*, about which there is much more to be said than can be accommodated here, is the product of Crichton's having mined an impressive amount of scholarship from the then emerging genomic and computer sciences as well as STS. The bounds of reality were stretched not only within the novel and subsequent film (1993), but achieved a social influence akin to that of *The Andromeda Strain*; for example, *Jurassic Park* became shorthand for innovation gone amok. It also became an entertainment empire. As historian of science David Kirby has argued, "One need only to look at *Jurassic Park* in its incarnations in novels, films, comic books, and computer games as well as its incorporation into television documentaries and news articles to see the high degree of intertextuality in science-based media."[73] It is also possible to see how Crichton was engaging with STS.

In an exchange toward the end of the novel, the author emphasizes how science and technology rarely succeed in creating the kinds of social transformations they promise; he also paraphrases work by early and influential feminist STS scholars, Carolyn Merchant and Ruth Schwartz Cowan. Crichton expertly deploys Merchant's gender critiques of the destructive practices of science and technology in mining the earth for resources, resulting in what she called the "death of nature." Chaos theoretician Ian Malcolm, the novel's moral conscience (often regarded as the voice of Crichton), discusses scientists' motivation with paleobotanist Ellie Sattler: "[They] have an elaborate line of bullshit about how they are seeking to know the truth about nature. Which is true, but that's not what drives them. Nobody is driven by abstractions like 'seeking truth.' . . . Scientists are actually preoccupied with accomplishment . . . [and] discovery is always a rape of the natural world. Always."[74] He also invokes Cowan's illumination of what she aptly called "the ironies of household technology," which increased rather than reduced domestic labor for women.[75] When Sattler protests that diminishing scientists' power would jeopardize advances, Malcolm explodes, asking what advances she means: "Why does it still

[71] Michael Crichton, "Playboy Interview: Michael Crichton," interview by John Rezek and David Sheff, *Playboy*, January 1999; republished in Golla, *Conversations* (cit. n. 18), 154.

[72] Gregory Jaynes, "Mr. Wizard," *Time*, 25 September 1995.

[73] David Kirby, *Lab Coats in Hollywood: Science, Scientists, and Cinema* (Cambridge, Mass., 2011), 228; *Jurassic Park*, the movie (cit. n. 3).

[74] Crichton, *Jurassic Park* (cit. n. 3), 318. The conflation between Malcolm's views and Crichton's is made in David Crow, "Westworld Was the First Draft of Jurassic Park," *Den of Geek*, 21 June 2018, https://www.denofgeek.com/us/movies/westworld/258761/westworld-movie-jurassic-park-first-draft.

[75] Carolyn Merchant, *The Death of Nature: Women, Ecology, and the Scientific Revolution* (San Francisco, Calif., 1980); Ruth Schwartz Cowan, *More Work for Mother: The Ironies of Household Technology from the Open Hearth to the Microwave* (New York, N.Y., 1983).

take as long to clean the house as it did in 1930?"[76] Given his residency at MIT, it is not hard to imagine how Crichton learned of these works. It is also not sufficient, as I will return to in the conclusion, to regard Crichton's use of them as evidence of his own feminism.

To be sure, Malcolm is far better remembered for uttering a now iconic statement: "Scientists . . . are focused on whether they can do something. They never stop to ask if they *should* do something."[77] Later in the novel, as Malcom hovers on the brink of death, Crichton seems to suggest the futility of a certain approach to STS. An injured Malcom, contemplating the revolutionary potential of biotechnology, invokes Kuhn's famous concept of "paradigm shifts," which is described to readers as a change in "world view." Yet, as Malcolm starts "rapidly slipping into terminal delirium," he casts aspersions on the utility of the paradigm shift concept, suggesting with an ironic smile that the idea has become obsolete.[78] As he hovers between life and death, and past and present—chasms that science had breached through its wanton act of resurrection—he rasps that he no longer cares about anything, "because . . . everything looks different . . . on the other side."[79] He achieves a privileged perspective about science as well as the nature of life only as he begins to die.[80]

While Pasteur recognized that the French countryside needed to be conceived of as a laboratory in order to convince farmers of the validity of bacteriology, Crichton realized that casting theme parks as laboratories could demonstrate less favorable consequences of biotechnology. (And Crichton, not unlike Pasteur, became a celebrity in the process of using the laboratory to raise the world.[81]) Crichton's diagnosis of the problems of science and technology's present, however, were often mistaken for prescriptions about the future cure. If it was not Malcolm's or Crichton's responsibility to make decisions about the organization of society, whose should it be?

TOWARD AN ALTERNATIVE HISTORY OF THE FUTURES OF SCIENCE AND TECHNOLOGY

At the moment Crichton's star was reaching its acme in the mid-1990s, the publication of a piece of fiction in an academic journal put STS on the defensive.[82] The author, mathematician Alan Sokal, claimed he had wanted to warn that the dangers of anything-goes relativism could lead to a devaluing of facts and of science itself.[83] The Sokal hoax is often cited as the opening gambit in an ugly cultural moment

---

[76] Crichton, *Jurassic Park* (cit. n. 3), 319.

[77] Ibid., 318.

[78] Ibid., 430.

[79] Ibid., 431.

[80] This is what Hannah Arendt referred to as the "Archimedean point" in her 1957 *The Human Condition*, when she observed the need to take science fiction seriously in the context of Cold War politics. Did anyone take notice when she wrote the following in the introduction: "What is new is only that one of this country's most respectable newspapers finally brought to its front page what up to then had been buried in the highly non-respectable literature of science fiction (to which, unfortunately, nobody yet has paid the attention it deserves as a vehicle of mass sentiments and mass desires)"? Hannah Arendt, *The Human Condition* (Chicago, Ill., 1957), 2.

[81] Bruno Latour, "Give Me a Laboratory and I Will Raise the World," in *Science Observed*, ed. Michael Mulkay and K. Knorr-Cetina (London, 1983), 141–70.

[82] Alan D. Sokal, "Transgressing the Boundaries: Toward a Transformative Hermeneutics of Quantum Gravity, "*Social Text* 46–47 (1996): 217–52.

[83] Alan D. Sokal, ed., *The Sokal Hoax: The Sham that Shook the Academy* (Lincoln, Nebr., 2000).

known as the Science Wars.[84] Science and Technology Studies, a field that had emerged from a desire to make knowledge producers more accountable to their facts, became associated with enabling the opposite.

It is tempting to wonder about what might have happened if Crichton had stayed in academia and decided to take up a permanent post at MIT to publish his critiques in *Social Studies of Science* or *Technology and Culture*, rather than as pulp fiction. Would he have become lauded as a founder of STS? By attending to his self-conscious creation of a speculative present and his sustained engagement with the concerns of STS, I have shown that such forms of conjecture may be moot as they obscure how the field's practices have traveled far beyond academia. For better, and in some cases, for worse, Crichton's work has popularized STS-style critiques of science and technology. In this sense, Crichton—in crafting and speaking from a speculative present—has kept STS in circulation, and even amplified its central concerns, to feed a deeply felt cultural desire to question the authority of science.[85]

This is not to say that ideas Crichton popularized exhaust the best potentials of what STS can achieve in an era when facts appear more plastic than ever and when the future has become a subject of history. Sheila Jasanoff and Sang-Hyun Kim, for instance, have argued for the importance of attending to "sociotechnical imaginaries," or the ways that modernity colonizes the future.[86] The worlds presented in Crichton's fictions are no less real than the forms of imagination that have been undertaken by technocratic state builders. This is because the term "speculative present" does not, in fact, reveal or predict the future. Crichton's work has been powerful precisely because it operated in its present, drawing on science and technology and social science that, if not yet mainstream, was current—already flowing and often leaking beyond the circumstances of its origin. To continue with the fluid metaphor, the mass appeal of Crichton's mode of dissemination may even have been what allowed those currents to crest into waves of reality, shaping cultural conversations about emerging science and technology and their consequences.

Jasanoff and Kim's intervention is helpful for articulating a strategy for navigating these currents—ways of actually reimagining society—especially across different national contexts and political differences. It is also indebted to work in a genealogy of STS that has often been obscured by one that privileges the likes of Kuhn and Latour—that of feminist practitioners. Indeed, it is feminist scholars who have been most attentive to the epistemological and political consequences of Crichton's work.[87] While Crichton's fiction

---

[84] Andrew Ross, ed., *Science Wars* (Durham, N.C., 1996); Bruno Latour, "Do You Believe in Reality? News from the Trenches of the Science Wars," in *Philosophy of Technology: The Technological Condition*, ed. Robert C. Scharff and Val Dusek (Oxford, UK, 2003), 126–37. See also Milam, "Old Woman" (cit. n. 11).

[85] I describe Crichton's connections to more recent post-truth debates in Joanna Radin, "Alternative Facts and States of Fear: Reality and STS in an Age of Climate Fiction," *Minerva* (2019, forthcoming). "Alternative Facts and States of Fear: Reality and STS in an Age of Climate Fictions," Minerva, Published online 19 April 2019. https://doi.org/10.1007/s11024-019-09374-5

[86] Sheila Jasanoff and Sang-Hyun Kim, eds., *Dreamscapes of Modernity: Sociotechnical Imaginaries and the Fabrication of Power* (Chicago, Ill., 2015).

[87] In the case of *Jurassic Park* alone, critiques of the reproductive politics of the novel have been made by leading feminist STS scholars, including Sarah Franklin's "Life Itself: Global Nature and the Genetic Imaginary," in *Global Nature, Global Culture*, ed. Franklin, Celia Lury, and Jackie Stacy (London, 2000): 188–227; Donna Jeanne Haraway, *Modest_Witness@Second_Millennium.Femaleman©Meets_Oncomouse: Feminism and Technoscience* (New York, N.Y., 1997); Laura Briggs and Jodi I. Kelber-Kaye, " 'There is no Unauthorized Breeding in Jurassic Park': Gender and the Uses of Genetics, *NWSA Journal* 12 (2000): 92–113.

suggests a form of fealty to what Jasanoff elsewhere characterizes as a patriarchal tradition in STS, its method at times appears to wink with awareness of the matriarchal one.[88]

Dorothy Nelkin, whom Crichton cites, was an early and influential advocate of the importance of attending to scientific controversies and science in its popular guises.[89] Donna Haraway was more radical still than Nelkin, and in the 1980s embraced science fiction as a potent means of imagining alternative futures; her interest was rooted in a desire to literally remediate the most pressing problems of the present. Haraway, who advanced what might be considered a therapeutic or reparative stance—what she has more recently dubbed "speculative fabulation"—has not been content to merely diagnose how science and technology so often enact visions of the future that do violence to women, people of color, and others who are often marginalized, but to also demonstrate how the products of science and technology could be directed toward their uplift.[90]

It was from reading Haraway as an adult that I and many of my peers, weaned on Crichton as tweens, first learned about the speculative work of writers such as Ursula K. Le Guin and Octavia Butler. They were two of his contemporaries, and they claimed very different genealogies than the likes of Poe or Conan Doyle. Le Guin, the daughter of cultural anthropologist Alfred Kroeber, was as steeped in anthropology as Crichton, if not more so, but she chose to craft worlds where nonpatriarchal configurations of gender and power could be used to transform humans' relationships to science and technology.[91] Butler, whose work is often associated with a genre called Afrofuturism, drew inspiration from biology and her own experiences as a black woman to reimagine human potential itself.[92]

Haraway's early work, along with publications by these and other feminist, indigenous, and Afrofuturist writers, provides the inspiration for my use of the "speculative present." This is recourse to a provincialization of place and time in order to expose how Crichton achieved the illusion that he was speaking with special access to the future, rather than from a very specific set of influential institutions of world making: East Coast elite academia and Hollywood. In a 1988 essay that has yet to exhaust its relevance, Haraway discussed what she referred to as a "God's-eye" view, whereby the scientist could supposedly step outside society to see things as they truly were. Such an ability, she argued, was one of the greatest fictions produced by postwar science.[93]

[88] Sheila Jasanoff, "Genealogies of STS," *Soc. Stud. Sci.* 42 (2012): 435–41.

[89] Nelkin summarizes her own contribution to the study of controversies in STS in Dorothy Nelkin, "Science Controversies: The Dynamics of Public Disputes in the United States," in *The Handbook of Science and Technology Studies*, ed. Sheila Jasanoff et al. (Thousand Oaks, Calif., 1995): 444–56. Crichton cites Nelkin's later work with lawyer Lori Andrews in a bibliography he included at the end of his novel *Next* (New York, N.Y., 2006). Lori Andrews and Dorothy Nelkin, *Body Bazaar: The Market for Human Tissue in the Biotechnology Age* (New York, N.Y., 2001).

[90] Donna Haraway, "Speculative Fabulations for Technoculture's Generations: Taking Care of Unexpected Country," *Australian Humanities Review* 50 (2011): 1–18.

[91] These values are eloquently summarized in an early piece of her own literary criticism first published in 1982 and reprinted as Ursula K. Le Guin, "A Non-Euclidean View of California as a Cold Place to Be," in *Dancing at the Edge of the World: Thoughts on Words, Women, Places* (London, 1989), 80–100.

[92] On Butler and Afrofuturism, see Gerry Canavan, *Octavia E. Butler* (Champaign, Ill., 2016); and Lisa Yaszek, "Afrofuturism, Science Fiction, and the History of the Future," *Socialism and Democracy* 20 (2006): 41–60;

[93] Donna Jeanne Haraway, "Situated Knowledges: The Science Question in Feminism and the Privilege of Partial Perspective," *Feminist Stud.* 14 (1988): 575–99.

It was one that Crichton, if unaware of Haraway, nonetheless was able to exploit. His ability to mine contemporary science to provide authenticating details for his fictions, and his simultaneous lack of accountability to science itself, made his novels appear oracular even if he, like Haraway, insisted it was impossible to predict the future.[94] Nonetheless, the fiction of the "God's eye" view achieved a kind of ironic reality in Crichton's novels to the extent that their plausible depictions of science and technology appealed to the imagination of those readers who were in a position to bring the particular visions into being, including scientists like Joshua Lederberg (responding to *The Andromeda Strain*), or the science administrators who conceived of the Ethical, Legal, and Social Implication programs for genomics (inspired in part by scenarios depicted in *Jurassic Park*).[95]

In her "Cyborg Manifesto," published several years before *Jurassic Park*, Haraway argued that a persistent desire to see nature and culture, as well as fact and fiction, as separate domains makes us blind to the ways they are anything but. "The boundary between science fiction and social reality," she declares, "is an optical illusion."[96] Crichton's speculative presents were successful at colonizing the future of science and technology because they traded on the persistent power of this optical illusion. This irony—that illusion is essential to reality—is not a glitch but a feature of the speculative present, as Haraway also argues: "Irony is about . . . the tension of holding incompatible things together because both or all are necessary and true. Irony is about humor and serious play. It is also a rhetorical strategy and a political method."[97] The importance of learning how to identify and attend to the speculative present at work is that it denaturalizes notions of scientific progress that make certain visions of the future appear more plausible and desirable than others. More fundamentally, it highlights the importance of speculation itself in creating opportunities to reimagine rather than reinforce the status quo.

Crichton made a fortune by insisting on the reality of a boundary between science fiction and social reality in his novels even as he spectacularly effaced it. Like building one of Haraway's cyborgs (an entity created out of parts that have been mined from nature and culture, fact and fiction), Crichton skillfully yanked facts and theories from academia to create fictions that made speculations about science and technology and its consequences seem real enough to become so. The primary difference between Crichton and Haraway is not that one wrote novels and the other published scholarship. It is that Crichton—and those who read him—took for granted that his concerns about science and technology were the right and most important concerns.[98]

This essay has sought to demonstrate that looking at the recent history of science and speculative fiction can reveal new ways of thinking about the practice of STS. Those who are anxious about the current fate of science and technology have much to gain by pursuing forms of speculation that do not conceal the circumstances of their origins

[94] Crichton, "Why Speculate?" (cit. n. 42).

[95] Also relevant but beyond the scope of this paper are claims about Crichton's influence on nanotechnology policy and climate science. For nanotechnology, see, Geoff Hunt, "NISE Net: Making a Macro Impact with Nanoscience," *ASBMBToday*, 8 August 2016, http://www.asbmb.org/asbmbtoday /201608/Outreach/. For climate science, see Joanna Radin, "Alternative Facts and States of Fear" (cit. n. 85).

[96] Donna Jeanne Haraway, "A Manifesto for Cyborgs: Science, Technology, and Socialist Feminism in the 1980s," *Australian Feminist Studies* 2 (1987): 1–42, on 2.

[97] Ibid.

[98] Inspired by Haraway, Latour came around to realizing hybrids themselves are products of an effort to enforce a distinction between modern and nonmodern; Bruno Latour, *We Have Never Been Modern* (Cambridge, Mass., 1993).

and that express the concerns of a wider array of knowers. This means paying more attention to fiction as resource for understanding facts and the social worlds in which they acquire power. It means embracing many different forms of inspiration for forging relationships with our machines, other life forms, and each other.[99] Doing so may be the most powerful therapy for a society seeking to regain its grip on reality.

---

[99] The idea of repair has been recently advanced by feminist STS scholars. See, for example, Dana Simmons, "Repair Work," *Engaging Science, Technology, and Society* 2 (2016): 237–41, and other articles in that volume.

Gaming the Apocalypse in the Time of Antibiotic Resistance

by Lorenzo Servitje*

ABSTRACT

Since the turn of the twenty-first century, there has been an increasing number of articles in the popular press on antibiotic resistance, a great many of which present the phenomenon in the science-fictional language of the "post-antibiotic apocalypse." In an effort to increase public awareness, the Longitude Prize, funded by the United Kingdom's National Endowment for Science, Technology, and the Arts (NESTA), created the mobile video game *Superbugs* in 2016. This essay shows how the mobile gaming medium responds to and performs the science fictionality of the antibiotic apocalypse in a way that has consequences for the history of science. *Superbugs* mediates a shift in the conceptual metaphor of a war between humans and bacteria to an interrelational model of coexistence. By defamiliarizing both the history and current representations of antibiotic resistance, the game gives us a unique way to reflect on the conditions that fostered this survival facility in the bacteria genome.

Imagine a world in which even the slightest scratch could be lethal. Cancer treatments, including chemotherapy, and organ transplants are no longer possible. Even simple surgery is too risky to contemplate, while epidemics triggered by deadly bacteria have left our health services helpless.

It is science fiction, of course—but only just.
—Robin McKie, 2016

A simple cut to your finger could leave you fighting for your life. Luck will play a bigger role in your future than any doctor could. The most basic operations—getting an appendix removed or a replacement—could become deadly. Cancer treatments and organ transplants could kill you. Childbirth could once again become a deadly moment in a woman's life. It's a future without antibiotics.

This might read like the plot of a science fiction novel—but there is genuine fear that the world is heading into a post-antibiotic era.
—James Gallagher, 2015[1]

* Department of English, and Health, Medicine, & Society Program Lehigh University, Drown Hall, 35 Sayre Dr., Bethlehem, PA 18015, USA; los317@lehigh.edu.

This essay would not have been possible without the help, feedback, and suggestions of a number of individuals, especially Sherryl Vint, Gillian Andrews, Colin Milburn, Will Tattersdill, and Lisa Shlegel. I am grateful to Amanda Rees, Iwan Rhys Morus, the anonymous reviewers and editors of *Osiris*, and Sheila Ann Dean.

So much depends upon the antibiotic. A foundational technology of modern medicine, antibiotics not only treat infectious diseases like tuberculosis, cholera, and typhus, which before the mid-twentieth century asymptotically delimited longevity; but they also allow many routine medical interventions to be carried out with a low chance of fatal infection. Yet, as these brief excerpts from the news suggest, antibiotics, once the "magic bullets" of the doctor's armamentarium,[2] might soon "cast [us] back into the dark ages of medicine," as former British prime minister David Cameron warned in 2014.[3]

Countless articles follow the rhetorical structure that Rob McKie and James Gallagher deploy in the two epigraphs. These include images of a technological and civil regression due to a mass catastrophe that connote the apocalyptic science-fiction narratives, distilled into the failure of the most basic or routine medical treatments; a leading assumption that this sounds "like science fiction"; and a caveat suggesting the scenario is or soon will be more "science fact" than "science fiction." This recurrent language of science fiction and apocalypse is equally present in many journalistic accounts of the looming failure of antibiotics: "Peer into the Post-Antibiotic Apocalyptic Future of Antimicrobial Resistance," "Why the Post-Antibiotic World is the Real-Life Version of the Zombie Apocalypse," and "The Plan to Avert Our Post-Antibiotic Apocalypse."[4]

What is the history of this "science-fictional" convention, and what cultural work does it do in the current moment? In the news media, it serves the practical purpose of using tantalizing rhetoric to sensationalize what might be a bitter pill for readers to swallow. As a concept, people often find the subject unbelievable, unintelligible, or uninteresting, according to studies and focus groups, such as a frequently cited consumer study by the Wellcome Trust.[5] But the use of apocalyptic language has been criticized for fostering these reactions. The words of one respondent are an example: "I don't believe it. That's a lot of people, just from bacteria. There's been a health boom in our society. That sounds dramatic."[6]

---

[1] Robin McKie, "Antibiotic Abuse: The Nightmare Scenario," *Guardian*, 12 Febuary 2017, https://www.theguardian.com/science/2017/feb/12/antibiotics-resistance-medicine-robin-mckie-radio-drama-val-mcdermid; James Gallagher, "Analysis: Antibiotic Apocalypse," 15 November 2015, BBC, http://www.bbc.com/news/health-21702647.

[2] "Magic bullet" refers to the theory of early pharmacologist Paul Ehrlich that involved developing a synthetic compound with an affinity for, and ability to, neutralize a specific pathogen without harming healthy cells of the body. Ehrlich helped develop the first chemotherapeutic agents against microbes, beginning with organoarsenicals such as atoxyl in the early 1900s. The term "magic bullet" was appropriated into the antibiotic zeitgeist of the mid-nineteenth century as part of the discourse of "miracle drugs." See John E. Lesch, *The First Miracle Drugs: How the Sulfa Drugs Transformed Medicine* (Oxford, UK, 2007).

[3] Fergus Walsh, "Antibiotic Resistance: Cameron Warns of Medical 'Dark Ages,'" *BBC News*, 2 July 2014, http://www.bbc.com/news/health-28098838.

[4] Michael T. Osterholm and Mark Olshaker, "Peer into the Post-Antibiotic Apocalyptic Future of Antimicrobial Resistance," *Wired*, 18 March 2017, https://www.wired.com/2017/03/peer-post-apocalyptic-future-antimicrobial-resistance/; John Aziz, "Why the Post-Antibiotic World Is the Real-Life Version of the Zombie Apocalypse," *Week*, 26 November 2013, http://theweek.com/articles/455518/why-postantibiotic-world-reallife-version-zombie-apocalypse; Ed Yong, "The Plan to Avert Our Post-Antibiotic Apocalypse," *Atlantic*, 19 May 2016, https://www.theatlantic.com/science/archive/2016/05/the-ten-part-plan-to-avert-our-post-antibiotic-apocalypse/483360/.

[5] M. Mendelson et al., "Antibiotic Resistance Has a Language Problem," *Nature* 545 (2017): 23–5, on 24; Cliodna A. M. McNulty et al., "Don't Wear Me Out—The Public's Knowledge of and Attitudes to Antibiotic Use," *Journal of Antimicrobial Chemotherapy* 59 (2007): 727–38; Wellcome Trust, *Exploring the Consumer Perspective on Antimicrobial Resistance*, June 2015, https://wellcome.ac.uk/files/exploring-consumer-perspective-antimicrobial-resistance-jun15pdf.

[6] Wellcome, *Exploring* (cit. n. 5).

In response to the rising rates of morbidity and mortality attributed to antimicrobial resistance (AMR), and to the lack of public engagement despite the topic's saturation of news and social media,[7] the Longitude Prize, an initiative of the United Kingdom's National Endowment for Science, Technology, and the Arts (NESTA), created *Superbugs* (2016). NESTA claims the following: "Recognising the urgent need to raise awareness and understanding about antibiotic resistance, today we're launching Superbugs, Nesta's first ever mobile game. It is aimed at 11- to 16-year-olds, and the goal is for players to survive as long as they can against superbugs, by wisely utilising existing and new antibiotics."[8] Despite being a simple game with an explicit pedagogical and even disciplinary purpose, *Superbugs* is a product of the mutually constitutive relationship between science and science fiction, working with the same tropes and metaphors portending doomsday as many of the news articles I described. Though the game does not use apocalyptic visuals or end-of-the-world, barren-wasteland narratives,[9] it lets users play out the medical-scientific histories that inform these journalistic accounts. In this way, it helps players rethink not only the present, but the past and future of antibiotics, both for themselves and for Western culture more broadly.

In this essay, I explore the science fictionality of the antibiotic apocalypse both in terms of it being framed by popular journalism as an inexorable catastrophe, as well as the possibility of *Superbugs* "gaming it" by educating the public.[10] By *antibiotic apocalypse*, I mean the discourse that conveys the science and anxieties of AMR through the language of science fiction or the apocalypse, and often both. This discourse occurs mostly in the news media, but also in some scientific publications and a small contingent of fictional works.[11] Scholars in communication studies, the social sciences, and occasionally medical history (such as a brief discussion by Scott Poldosky) have addressed the emergence of apocalyptic language in journalistic, popular, and biomedical accounts of AMR.[12] Although substantial arguments have been made for contagion's cultural work in literature

---

[7] It is important to note that while AMR is most commonly referenced in terms of bacteria, it also includes other pathogenic resistances, such as those against antivirals, antifungals, and antiparasitics.

[8] Tamar Ghosh, "Our New Game: How Long Can You Survive against the Superbugs?" *NESTA*, 5 July 2016, http://www.nesta.org.uk/blog/our-new-game-how-long-can-you-survive-against-superbugs. For the game itself, see *Superbugs: The Mobile Game*, by NESTA, version 1.03, https://longitudeprize.org/superbugs (accessed 1 September 2018).

[9] This is different than more elaborate games like *The Walking Dead* (2012, 2013, 2015) or *The Last of Us* (2013), which both have branching narratives, alternate endings, extended cut scenes, and an array of game mechanics.

[10] Science fictionality has been described as a "mode of response that frames and tests experiences as if they were aspects of science fiction"; see Istvan Csicsery-Ronay, *The Seven Beauties of Science Fiction* (Middletown, Conn., 2008), 2.

[11] I have found three recent and explicitly SF texts that take the antibiotic apocalypse as a central thematic; these include NESTA's edited collection *Infectious Futures* (2015); Sarah Kinney's graphic novel *Surgeon X* (2017); and "Resistance," the first episode of the BBC Radio program *Dangerous Visions* (2016).

[12] For this work in the social sciences, see Nik Brown and Sarah Nettleton, "'There Is Worse to Come': The Biopolitics of Traumatism in Antimicrobial Resistance (AMR)," *Sociological Review* 65 (2017): 493–508, Online First, 27 February 2017, https://doi.org/10.1111%2F1467-954X.12446; Brigitte Nerlich, "'The Post-Antibiotic Apocalypse' and the 'War on Superbugs': Catastrophe Discourse in Microbiology, Its Rhetorical Form and Political Function," *Public Underst. Sci.* 18 (2009): 574–88, discussion on 588–90; B. Brown and P. Crawford, "'Post Antibiotic Apocalypse': Discourses of Mutation in Narratives of MRSA," *Sociol. Health Ill.* 31 (2009): 508–24, DOI:10.1111/j.1467

and film,[13] few scholars have taken into account the way video games represent infectious diseases (with the exception of the popular iPhone game *Plague Inc.*[14]). This is not surprising, because digital games are still an overlooked medium in the study of cultural representation of infection and pandemics, although they are increasingly of interest to science fiction (SF) studies.[15] Although news media related to popular technology, and other news media generally, have responded to *Superbugs*, it has not yet been the subject of an academic inquiry by SF and games scholars.

I examine how the mobile gaming medium uniquely responds to and performs the science fictionality of the antibiotic apocalypse in a way that has consequences for the history of science. To do this, I examine the role of SF in shaping the AMR narrative in the popular imagination. Using the language of SF, researchers and journalists have presented a foreboding future in two respects, with each reflecting a different meaning of the term apocalypse. First, the term is used to suggest a catastrophic event that marks the end of an epoch or way of life—here, this involves the end of the viability of antibiotics as a tenet of modern medical practice. Second, apocalypse is used to mean a revelation or demystification; in this case, it's an apocalypse of the cultural logics, conceptual metaphors, and history that led us to declare a full scale "war" on bacteria. The war on bacteria, which began with utopian aspirations, has increasingly been moving toward more dystopian consequences. Thinking about these two meanings in concert should prompt us to ask how the antibiotic apocalypse works, not only as a set of conventions for thinking about a future without antibiotics, but as an epistemic mode of SF. I argue that the way *Superbugs* works as an interactive game vis-à-vis SF affords us a unique opportunity to observe the convergence of popular culture with the history and future of medical science. *Superbugs* uses the capacities of the mobile gaming medium to mediate a change in the metaphor that structures how medicine is perceived, known, and practiced, from the binary logic of the conceptual metaphor of war to the interrelational model of coexistence.[16] The game defamiliarizes both the history and current representations of AMR, giving us a way to reflect on the conditions that fostered the emergence of microbial abilities to develop drug resistance.[17]

-9566.2008.01147.x; Scott Poldosky, *The Antibiotic Era: Reform, Resistance, and the Pursuit of a Rational Therapeutics* (Baltimore, Md., 2015), 177–8, 81.

[13] Among others, see Priscilla Wald, *Contagious: Cultures, Carriers, and the Outbreak Narrative* (Durham, N.C., 2008).

[14] See Scott Mitchell and Sheryl N. Hamilton, "Playing at Apocalypse: Reading Plague Inc. in Pandemic Culture," *Convergence: The International Journal of New Media Technologies* 24 (2017): 587–606; Lorenzo Servitje, "H5n1 for Angry Birds: Plague Inc., Mobile Games, and the Biopolitics of Outbreak Narratives," *Sci. Fict. Stud.* 43 (2016): 85–103.

[15] See, for example, the recent special issue, "Digital Science Fiction," ed. Pawel Frelik, of *Science Fiction Studies* 43 (2016). See also Pawel Frelik, "Video Games," in *Oxford Handbook of Science Fiction*, ed. Rob Latham (Oxford, UK, 2015), 226–38; and Patrick Jagoda, "Digital Games and Science Fiction," in *The Cambridge Companion to American Science Fiction*, ed. Gerry Canavan and Erik Link (Cambridge, UK, 2015), 139–52.

[16] On the metaphor of war in this instance, see Paul Hodgkin, "Medicine Is War: And Other Medical Metaphors," *BMJ* 291 (1985): 1820–1; and Giovanni De Grandis, "On the Analogy between Infectious Diseases and War: How to Use It and Not to Use It," *Public Health Ethics* 4 (2011): 70–83.

[17] I recognize AMR is a complex phenomenon that is not solely a function of allopathic medicine as such, but is in fact largely due to agricultural use of antibiotics for both the purposes of hypertrophy in livestock and increased volume produced in unsanitary conditions. Therefore, given *Superbugs's* scope of "everyday practices" and its focus on antibiotic prescriptions, I will delimit my argument to antibiotic discourse in medical practice.

I begin in the first section by outlining the current iterations of the antibiotic apocalypse in journalism to show how writers use SF tropes not just as genre conventions, but as ways of thinking about science and culture. In the second section, I discuss how *Superbugs'* science fictionally mediates the discourse of AMR in medical science for transfer to the public sphere. I show how the game achieves this through its narrative, iconographic, and ludic modes. Although I draw on the medical history of antibiotics throughout, in the first section I integrate SF criticism with biomedical research and the study of science communications, and in the second I approach *Superbugs* through recent work in game studies. Beyond offering a specific reading of *Superbugs* and the dialectical work of the antibiotic apocalypse, my inquiry suggests that through SF, video games provide a unique way to think about the history of science within the temporal play the medium affords.

THE GENEALOGY OF THE POST-ANTIBIOTIC APOCALYPSE

There are two interrelated modes in which the apocalypse relates to SF. The first, drawing on the more common denotation of apocalypse as a widespread disaster, is a set of generic conventions, tropes, or themes involving a doomsday scenario, of natural or technological origin, that ends or radically changes the world. This mode often shifts quickly to "postapocalyptic" as it begins detailing survival after the disaster, or a return to a premodern society and technology. In the second mode, drawing on the original sense of a revelation or unveiling, the apocalyptic becomes revelatory via science fictionality; it is an epistemic manner of thinking about the relationship between society, science, and technology that imagines a relationship between historical reality and its possible unfolding in the future.[18]

Both of these apocalyptic modes in SF are informed by the broader cultural discourse of the apocalyptic, as understood through a diverse range of disciplinary approaches. What ties much of this scholarship together as a critical context for AMR is meaning making; when we are talking about the apocalypse, we are talking about making sense (in the present) of the past and future, and this includes imposing an order and coherence for various, often convergent, ends—aesthetic, rhetorical, political, and religious.[19] In Frank Kermode's well-known study on fiction and the apocalypse, he suggests the apocalyptic reflects the human imaginative investment in concordant patterns, which provide an ending to produce "satisfying consonance between the origin and the middle."[20]

---

[18] Csicsery-Ronay, *Seven Beauties* (cit. n. 10), 4. On the apocalyptic and science fictionality in political critique, see Peter Yoonsuk Paik, *From Utopia to Apocalypse: Science Fiction and the Politics of Catastrophe* (Minneapolis, Minn., 2010).

[19] Scholarship on the apocalypse is too vast to be responsibly surveyed here, but notable examples that follow the meaning-making imperative include Paul S. Boyer, who contends biblical prophecy is central to American public attitudes on domestic and international issues, ranging from the Cold War to the environment; Boyer, *When Time Shall Be No More: Prophecy Belief in Modern American Culture* (Cambridge, Mass., 1992). As he traces the history of millennialism in western civilization, Damian Thompson argues that few apocalypses result in total ends but rather usher in new "golden eras," and he cites examples from early Christianity through to more recent cult followings, notably Waco and Jonestown; Thompson, *Longing for the End: A History of Millennialism in Western Civilization* (New York, N.Y., 1999). Turning to visual culture, Morton Paley rethinks Romanticism in her study of English preoccupation with apocalyptic myth reflected in artwork; Paley, *The Apocalyptic Sublime* (New Haven, Conn., 1986.).

[20] Frank Kermode, *The Sense of an Ending: Studies in the Theory of Fiction* (New York, N.Y., 2000), 17.

Thinking about apocalypse as argument, Stephen O'Leary contends that it functions as a rhetorical tool for community building, and as a mythic identification (often of course in religious iterations) by discursively reconstructing time. Apocalyptic rhetorics succeed or fail with their intended audiences, not because they are valid or have come to fruition, but rather because they persuade audiences of their context "within the particular historical pattern of temporal fulfillment represented in [their] mythic imagery." In other words, they survive repeated "disconfirmation" by a "discursive reformulation that continually ties apocalypse to the present by reconceiving the relationship of past and future."[21] The AMR apocalypse serves this argumentative function, especially in terms of the media-specific rhetoric surrounding *Superbugs*, but it draws on the science-fictional imaginary (rather than a religious mythic tradition) to make sense of present problems. The game does this by reconstructing the history of AMR—over and over again—making the player a part of and responsible for that past.

Although AMR was recognized as a problem as soon as antibiotics were discovered, its framing through SF discourse and the apocalypse did not explicitly appear in biomedical or news publications until the mid-1990s, and only gained traction in the early 2000s, as Bridgette Nerlich has outlined.[22] These anxiogenic predictions developed out of groundwork laid by popular science books, such as Laurie Garrett's *The Coming Plague: Newly Emerging Diseases in a World Out of Balance* (1994), and Geoffrey Cannon's *Superbug: Nature's Revenge* (1995).[23] Medical publications such as Caroline Ash's "Antibiotic Resistance; The New Apocalypse?" followed for the rest of that decade.[24] By 2003, a number of articles with similar titles and anxieties were being circulated by periodicals ranging from the *Atlantic* to *Wired*. The now popular term *post-antibiotic apocalypse*, referring to how life will look after the "antibiotic era,"[25] was coined by Richard James, the director of the Centre for Healthcare Associated Infections at Nottingham University; he first used it in "Battling Bacteria" (2005), an article for the university's *Vision* magazine.[26] The term was picked up by a number of outlets, such as the *Sentinel*, the *Guardian*, and the *Observer*. They often quoted an interview with James in which the popular image of the lethally cut finger first appeared: "Quite frankly, the impending crisis can be called the post-antibiotic apocalypse. We are facing a future where there will be no antibiotics and hospital will be the last place to be if you want to avoid picking up a dangerous bacterial infection. In effect, cut your finger on Monday and you'll be dead by Friday if there's nothing to prevent it."[27]

---

[21] Stephen O'Leary, *Arguing the Apocalypse: A Theory of Millennial Rhetoric* (New York, N.Y., 1998), 13.

[22] Nerlich, "'Post-Antibiotic'" (cit. n. 12), 577–80.

[23] Laurie Garrett, *The Coming Plague: Newly Emerging Diseases in a World Out of Balance* (New York, N.Y., 1994); Geoffrey Cannon, *Superbug: Nature's Revenge* (London, 1995).

[24] Caroline Ash, "Antibiotic Resistance: The New Apocalypse?" *Trends in Microbiology* 4 (1996): 371–2.

[25] By the antibiotic era I mean the period approximately from the late 1940s to the early 1970s, during which the production rate of new antibiotics and new antibiotic classes was highest. These antibiotics were a large part of what has been characterized as the "medical golden age," where infectious diseases that once topped the charts of mortality statistics were surpassed by cancer, heart disease, and other chronic conditions. This shift was primarily due to antibiotics, but of course eradications of viral diseases via vaccines such as smallpox and polio were also important. The period is often bookended around the last antibiotic class being discovered and the emergence of HIV.

[26] Richard James, "Battling Bacteria," *Vision Magazine*, University of Nottingham, Spring 2005.

[27] Richard James quoted in Nerlich, "'Post-Antibiotic'" (cit. n. 12), 580.

To better understand the science fictionality of AMR's cultural work in these accounts, we must consider the genealogy of fictional representations of world-changing disasters and the historical moments at which they emerged—in terms both of fiction and of science. Some of the earliest works of SF from the nineteenth century describe apocalyptic scenarios in which disease and "monstrous" technologies play major roles, marking them as a significant part of the SF "megatext" that the AMR apocalypse draws from.[28] This megatext notably includes what has been called the first work of apocalyptic SF, Mary Shelley's *The Last Man* (1826), which details the end of the human race in a global pandemic. Partly a response to the 1817 cholera pandemic, the novel challenges humanity's supposed superiority to nature by debunking the ideological claim that through technology and inevitable progress, civilization could ultimately resist the seemingly supernatural might of disease—an ethos evident in the utopian aspirations one character gives voice to before the arrival of the plague. This kind of check on human hubris offers the same critique of Enlightenment fantasies of progress as Shelley's *Frankenstein* (1818), in which Victor aims to "banish disease from the human frame, and render man invulnerable to any but a violent death," a project that ultimately threatens the human race.[29]

In the antibiotic era, the utopian hopes of doctors, journalists, and popular science writers evoked the same science-fictional aspirations and technological utopianism that Shelley's novels challenge. The doctor and researcher Boris Sokoloff described this in *The Miracle Drugs* (1949):

> The goal is simply to live in a world without menacing microbes; to have all disease-producing microbes rendered harmless and domesticated; *to see infectious disease vanish from the earth*, or at least be easily controlled; to make the planet free from the dangers of death from infectious disease, so that the common cold, pneumonia, meningitis, and other ailments may be as rare a phenomena as dangerous wild beast are. . . . Will such a world exist? We believe so.[30]

With similar optimism, Sokoloff declared penicillin "a weapon of such power and magnitude that it may fulfill the dreams of generations of scientists for years to come."[31] His utopian dreams were echoed in the now mythologized declaration by Surgeon General William H. Stewart in 1968 that it was "time to close the book on infectious diseases and declare the war against pestilence won" and to shift resources to problems like cancer and heart disease. Although its attribution to Stewart is unsupported,[32] this quotation is still cited as emblematic of the shortsighted utopianism of the antibiotic era. The irony now lies in the direct relation between time and antibiotic development; this progress, doomed from the start, lasted a little more than two decades. Sokoloff's dream materialized as a nightmare, and not after generations of scientists, but after generations of antibiotics. This kind of future projection was based on a conceptual metaphor of medicine as

---

[28] This megatext is the large body of works that audiences and authors consider to form the subcultural lexicon of SF. Csicsery-Ronay, *Seven Beauties* (cit. n. 10), 275, 276.

[29] Mary Shelley, *Frankenstein; Or, The Modern Prometheus* (New York, N.Y., 1823), 55.

[30] Boris Sokoloff, *The Miracle Drugs* (Chicago, Ill., 1949), 254 (emphasis mine).

[31] Ibid., 94.

[32] Brad Spellberg, "Dr. William H. Stewart: Mistaken or Maligned?" *Clinical Infectious Diseases* 47 (2008): 294; Spellberg and Bonnie Taylor-Blake, "On the Exoneration of Dr. William H. Stewart: Debunking an Urban Legend," *Infectious Diseases of Poverty* 2 (2013): 3.

war, operating under an imperative of the total eradication of microbes. It changed the future, but not at all in the way Sokoloff hoped. Coeval with Sokoloff's prediction, anxieties about another technological innovation were reflected in the SF of the period. The midcentury was as much a period of fear as of hope regarding science and technology; at the same time that the history-altering antibiotic was developed, the potential to destroy human life on a mass scale arrived in the atom bomb. After World War II, the apocalypse was increasingly envisioned as a consequence of human action.[33] Postapocalyptic SF reemerged at the forefront of the genre, but with its focus turned from natural, cosmic, and alien-invasion disasters to dystopian outcomes of technological progress, especially fixating on nuclear war, the core of much of the genre for the next two decades.[34] This period gave rise to films like *The Day the World Ended* (1955) and novels such as Nevil Shute's *On the Beach* (1959). There were exceptions involving infectious disease; for example, George R. Steward returned to the world-ending plague in his 1949 novel *Earth Abides*, in which a contagious disease decimates the United States in a few days and the remnants of the population try to rebuild in the years that follow. Richard Matheson's novel *I Am Legend* (1954) merges both catastrophe types when radiation from nuclear war causes bacteria to mutate and turn people into vampire-like creatures. This convergent logic that draws on both biological and technological etiologies subtends the contemporary consideration of antibiotic apocalypse, in which antibiotic technology has shaped bacterial life to respond. In both the real history of AMR and *I Am Legend*, the misuse of medical technology puts a selective pressure on bacterial life that catalyzes a resurgence against unchecked, anthropocentric, and shortsighted human intervention. The intertextual traces of both utopian dreams and dystopian nightmares in contemporary projections of the post-antibiotic world signal the longer history of SF's portrayal of technological progress in works like those of Shelley and Matheson. This connection between the historical changes of the nuclear and antibiotic ages is made explicit in one journalistic narrative of the antibiotic apocalyptic. In a 2013 article for the *Daily Telegraph*, "Is This the Antibiotic Apocalypse?" Michael Hanlon writes the following: "In the Forties, Fifties and Sixties, it seemed that the germs had no answers to penicillin and other wonder drugs . . . the hardiest bacteria, such as E. coli, held their ground until a new class of antibiotics, the carbapenems, was developed in the 1980s. These were the hydrogen bombs of the antibiotic world, able to outwit the cleverest evolutionary and molecular tricks of our bacterial foes."[35] Though Hanlon begins with the SF "imagine a world" convention, in his metaphorical framing of medicine the apocalypse is implicit rather than explicit. Beyond monstrous technologies, be they nuclear weapons or the selective pressures that lead to antibiotic resistance, the postapocalypse genre uses regression as a central trope to drive conflict and narrative action—a return to the past that involves the loss of either modern technologies or the civilities of modern life.

[33] Lee Quinby, "Apocalyptic Security: Biopower and the Changing Skin of Historic Germination," in *Apocalyptic Discourse in Contemporary Culture: Post-Millennial Perspectives on the End of the World*, ed. Monica Germana and Aris Mousoutzanis (New York, N.Y., 2014), 17–30, on 26.

[34] On the nuclear apocalypses, see Paul Brians, *Nuclear Holocausts: Atomic War in Fiction, 1895–1984* (Kent, Ohio, 1987).

[35] Michael Hanlon, "Is This the Antibiotic Apocalypse?" *Daily Telegraph*, 12 March 2013, http://www.telegraph.co.uk/news/science/9923954/Is-this-antibiotic-apocalypse.html.

Postapocalyptic narratives tend to be framed around an event that destroys political, legal, and material infrastructures, with mass death as both a causal factor and a consequence. The loss of these elements of the modern world inevitably leads to technological and civil regression to premodern standards that form the narrative conflicts. In Shelley's novel *The Last Man*, people regress to their "animal functions"[36]; much the same condition is physically manifested in the degenerative mutations of *I Am Legend*. In our own moment, this theme is omnipresent in biomedical zombie fiction, perhaps most famously in the television series *The Walking Dead*, where the devolution of medical care is a recurrent obstacle for Rick and his band of survivors. In the episode "Infected" (2013),[37] for example, the group holds a secure prison block that promises relative safety from both zombies and rival survivors. Unfortunately, a highly contagious disease begins to kill the group off; with the effective absence of medical care, the contagion comes to pose as great a threat as the undead.[38] On the advice of their medical professionals, the group risks going into the open to seek a supply of antibiotics from a nearby veterinary school.[39] The disease in this episode stands in for the loss of "modern medicine," much of which depends on antibiotics. When people are shot, suffer broken bones, need amputations, or sustain any injuries in their daily routine of survival, even a mild bacterial infection can herald a character's death knell. The survivors resort to aggressive quarantines and murder. These measures, along with violent conflicts over basic resources, reveal the close relationship between the degeneration of technology and the decay of the social order.

Science fictionality is a way to think about the hesitation between the imaginative-conceptual and historical reality unfolding in the future.[40] This mutually constitutive relationship between science and science fiction often emerges in public understandings of science after being filtered through journalistic accounts that treat the SF genre as a source of, and shorthand for, ideas about the implications of new technologies and discoveries. These accounts are often framed as if reality has "caught up" with science fiction,[41] as in the AMR cases, whether as a consequence of an innovation or as an example of science fiction qua "bad science."[42] The revelatory potential of apocalyptic narratives in a science-fictional capacity operates by means of cognitive estrangement—an oscil-

---

[36] Mary Shelley, *The Last Man* (1826; repr., Ware, UK, 2004), 256.

[37] *The Walking Dead*, "Infected," season 4, episode 2, directed by Guy Ferland, aired 6 October 2013, on AMC.

[38] There is a recursive relationship between the disease causing zombiism and the disease causing fevers and respiratory distress in this scenario: in the world of *The Walking Dead*, people become zombies only after death (whether or not they are bitten by another zombie, since everyone is already infected); consequently, when people are unsupervised in the prison during the epidemic and die, they turn and become zombies creating a dual threat within.

[39] *The Walking Dead*, "Isolation," season 4, episode 3, directed by Dan Sackheim, aired 27 October 2013, on AMC. The disease is referred to as possibly being "pneumococcal" (consequently bacterial), but in more instances is called "an aggressive flu," mimicking the lethality of the Spanish Influenza of the early twentieth century or fictionalized viral narratives as in the film *Contagion* (2013). A viral infection would, of course, not respond to antibiotics. That said, as in the case of Spanish Influenza, secondary bacterial infections are compounding factors for morbidity and mortality, and thus often require antibiotic intervention.

[40] Csicsery-Ronay, *Seven Beauties* (cit. n. 10), 2–4.

[41] Sherryl Vint, "The Culture of Science," in Latham, *Oxford Handbook of Science Fiction* (cit. n. 15), 306–15, on 306.

[42] Sheryl N. Hamilton, "Traces of the Future: Biotechnology, Science Fiction, and the Media," *Sci. Fict. Stud.* 30 (2003): 267–82, on 273.

lating critical distance that lets the fictive text reveal the ideological regimes that underlie reality, thereby creating the possibility for social and political critique.[43]

Thinking about the implications of future possibilities on the basis of estrangement of the present prompts us to consider antibiotics as a technoscientific innovation, the conditions that set this in motion, and how those conditions are manifest today. Popular accounts of the history of science, as well as those within the discipline of science, include comments like "we used to think . . . and now we know. . . ,"[44] although any historian of science would challenge such a reduction. In these progressive narratives, science develops by trying to change the truth of today into an error of yesterday; it works recursively with respect to the past from which it departs.[45] The case of AMR, as Hannah Landecker suggests in language evocative of SF, prompts us to think about this question differently: "'We used to think a certain way about antibiosis and pathogens. *And then we changed the future.*' What we thought we knew became the biology under study: the solution has become the problem."[46] But, Landecker argues, the recursive nature of science has not yet had to account for the scale on which AMR has changed the material world;[47] SF is one mode through which science can make sense of such a scale.[48]

The relationship between the extreme outcomes of the loss of medical infrastructure in zombie narratives, and the more "realistic" problems wrought by AMR, work in a dialectical relationship between SF and the public understanding of science. Although the overt claim is that the AMR apocalypse is no longer SF, the act of invoking the discourse of SF and its subgenre of postapocalyptic narratives reinforces our thinking the unthinkable in SF terms. Hypothetical journalistic accounts hyperbolize scenarios like those of zombie fiction—lethal contagion, or multiple injuries leading to fatal sepsis—by drastically reducing the severity of the mechanism of injury; instead of a gunshot wound, something as mundane as a cut hand or a scraped knee does one in.[49] Thus, the title of a 2013 article runs as "Why the Post-Antibiotic World is the Real-Life Version of the Zombie Apocalypse."[50] The zombie apocalypse begins to seem less frightening than the post-antibiotic apocalypse that still looks more like our everyday life. This is why, unlike zombie stories and the contagion fiction they draw on, "real-life" AMR apocalypse narratives don't tend to focus on a single pathogen threatening the human race; it is not a very contagious strain of vancomycin-resistant staphylococcus aureus (VRSA) that will decimate humankind.[51]

Like its more everyday medical consequences, AMR has social consequences that are less apocalyptic. These are explored in a number of the fictional stories in the NESTA-sponsored collection *Infectious Futures* (2015), which offer a contrast to the martial quarantines and desolate wastelands of the more familiar apocalyptic narratives. In her intro-

---

[43] Csicsery-Ronay, *Seven Beauties* (cit. n. 10), 50–51.

[44] Hannah Landecker, "Antibiotic Resistance and the Biology of History," *Body & Society* 22 (2015): 19–52, on 22.

[45] Hans-Jörg Rheinberger and David Fernbach, *On Historicizing Epistemology: An Essay* (Stanford, Calif., 2010), 22.

[46] Hannah Landecker, "Antibiotic Resistance" (cit. n. 44), 5 (emphasis mine).

[47] Ibid., 20.

[48] While focusing less on AMR, Nicholas King has addressed the phenomena of emergent disease through scalar narratives; King, "The Scale Politics of Emerging Diseases," *Osiris* 19 (2004): 62–76.

[49] Sarah Boseley, "Superbug 'Apocalypse' Warning," *Guardian*, 6 January 2007, https://www.theguardian.com/society/2007/jan/06/health.medicineandhealth.

[50] Aziz, "Why the Post-Antibiotic" (cit. n. 4).

[51] An exception to this multiple contagion scenario is Val McDermid's *Resistance* (BBC Radio 4 full-cast drama, 2017 audiobook), CD.

duction to this volume, STS scholar Brigitte Nerlich writes the following: "Some of the stories focus on subtler narratives. Once-trivial diseases become chronic and degrading. Something as simple as a urinary tract infection becomes a major disability rather than a mere annoyance as it would be today."[52] The extrapolation of the prosaic in medical terms extends to the way many of the stories reflect, in a social science fiction mode, on preexisting social trends and problematics as exacerbated by AMR. As examples, anxieties about children and digital media become a future in which children interact only online and are afraid to ride a bike lest they scrape their knee; concerns over the social determinants of health create a world in which hyperhygienic services and commodification further divide people by class; state surveillance and security are emblematized in spy drones and biological detectors of infectious disease in public spaces; and isolationism and xenophobia are represented by a society in which epidemics are always linked to immigrants, and people avoid contact with anyone outside their small social and familial circles.[53] *Infectious Futures* is unique in being the only purely textual representation of AMR in fiction. However, NESTA's more recent public engagement, the game *Superbugs*, though not immediately recognizable as apocalyptic SF, crystalizes the revelatory aspect of science fictionality by virtue of its specific media form.

"HOW LONG CAN YOU HOLD OUT?"

Superbugs provides a unique way to think about the history of science; players create their own history of antibiotic use and development, prompting them to think through the logics that catalyzed AMR. In contrast to a medical history or journalistic account, either of which *tells*, or an extended textual, graphic, and audio fiction, which *show*, the mobile game *performs* the revelatory aspect of the apocalypse by making players interactively work out the relationship between the science fact and the science fiction of AMR. That is, the players actualize a multimodal mapping of simulated experiences that mutate the game's fictional AMR to fit its real-life history through narrative, visual, and ludic mechanisms. To understand how the game represents and shapes the discourse of antibiotic resistance within the science-fictional context of the antibiotic apocalypse, we must take a holistic approach from game studies, one that accounts for the narrative, iconographic, and ludic qualities of the game.[54] We must attend to the story the game tells, including the visual and textual representations, and the mechanics—the rules and interfaces the player interacts with.

Superbugs's narrative materializes through visual and textual representation of bacteria growing in a petri dish. Players are tasked with preventing bacteria from occupying a lethal amount of the dish by judiciously applying antibiotics (fig. 1). This is a fairly easy task and operates, in narratological terms, as the exposition and rising actions. The con-

---

[52] Brigitte Nerlich and Matthew Clarke, "Resistance and Escapism: The Risks and Rewards of Disaster Stories," in *Infectious Futures: Stories of the Post-Antibiotic Apocalypse*, ed. Joshua Ryan Sasha and Lydia Nicholas (London, 2015), 9–12, on 10. The editors in their own "Introduction" make a similar point (on 7): "Authors were asked to consider not an action hero in the explosive beginning or empty aftermath of a plague, but the subtler stories of living in a world where our antibiotics are failing: the shifts in everyday life and society, the impact on families, relationships, politics and work." For a link to the book, see https://longitudeprize.org/antimicrobial-resistance/infectious-futures-stories-post-antibiotic-apocalypse.

[53] Nerlich, "Resistance and Escapism" (cit. n. 52), 10.

[54] In the previous decade, the field of Games Studies was fraught by a divide between reading games either narratologically or ludically. See Jan Simmons, "Narrative, Games, and Theory. What Ball to Play?" *Game Studies* 7 (2007), http://gamestudies.org/07010701/articles/simons.

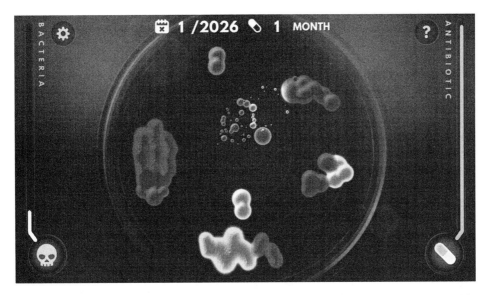

Figure 1. The narrative conflict between players and bacteria. Bacteria are represented by different colors such as yellow, purple, and green. In the edited screen captures (grayscaled), the darkened bacteria represent antibiotic resistance spreading. The leftmost bacterium, for example, is completely resistant (all red), while the bacterium toward the bottom center of the dish has begun to acquire resistance in the rightmost portion (red).

flict arises when the bacteria develop resistance, which is visually represented by their turning red.

The story progresses as an oscillating tension between bacterial growth, resistance, and elimination, and remains active only as long as players sustain the narrative conflict by forestalling the bacteria's occupation of the entire petri dish (fig. 2). The game always concludes with a graph and the statement, "You survived: x years, y months" (fig. 3).

The game's closure takes the form of a visual revelation after the player's death or after their tenure in a virtual golden age of antibiotics. Narratively, the game emerged from the cultural context of the antibiotic apocalypse; the story it tells is in part a re-mediated form of news media that deploys the language of SF and the apocalypse.[55]

Although the game's "story" is easily recognizable, its protagonist—in ludic terms, the avatar—is not, putting into question how the player is interpellated. There is no visual avatar, and the only clue to the subject position the player occupies is the con-cluding text addressing the player in the second person—"*you* survived"—implying that the protagonist is an individual infected with a drug-resistant disease. At the same time, the petri dish that becomes colonized with bacteria suggests the purview of a researcher or practitioner applying newly developed antibiotics. This ambiguous subjectivity encourages the player to consider at least these three actors in the anti-biotic network as a way of highlighting the complexity of AMR in medical science.[56]

[55] The Longitude Prize's official page on the game's website includes the following statement: "The story of this casual, real-time puzzle game comes straight from today's headlines: The rise of superbugs, resistant bacteria which can't be killed by the drugs we rely on to fight infection." *Superbugs* (cit. n. 8).

[56] I have made a similar argument about player interpellation in Servitje, "H5N1 For Angry Birds" (cit. n. 14), 98.

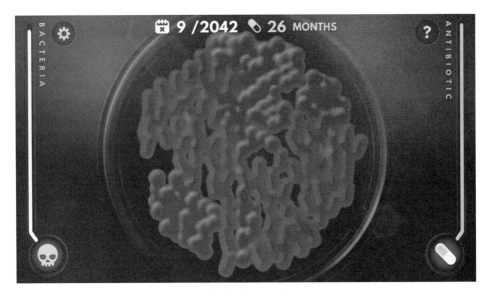

Figure 2. *Near death; all bacteria in the dish are red.*

Patients want to be free of bacterial colonization; physicians must weigh the desires and needs of their patients regarding immediate material pathogens against the less immediate and more utilitarian problem of AMR; and for researchers, antibiotics are a natural resource that become more scarce with time, and more difficult and expensive to develop. While the game cannot unpack the full overdetermination of

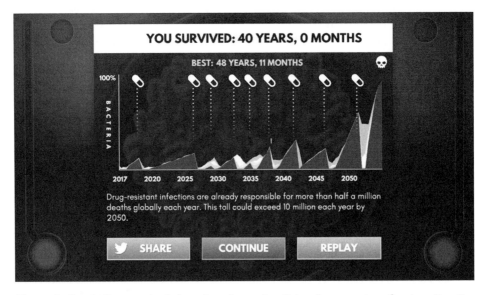

Figure 3. *Concluding graph of player's actions, visualizing the narrative of a given iteration of the game. The darker coloring is red in the game, reflecting the AMR acquisition, the lighter colors, again, being normal varieties of bacterial growth.*

AMR and the scope of the stakeholders involved,[57] it prompts players to think beyond a single dimension of the problem.[58] That is, though it might inculcate the fear of personally acquiring a drug resistant infection, the visual environment connects the individual player to the broader problem that affects the entire human race. The game thus draws on both micro- and macroimplications of AMR; it focuses on the player—the individual life—and at the same time alludes to the more widely circulated apocalyptic narrative of mass death and biomedical regression writ large. In short, the game suggests AMR is everyone's problem.

The graph is central to the apocalyptic tenets of the game (fig. 3). It makes visible the feedback loop between the game's algorithmic processes and the user's responses. The x-axis represents time, while the y-axis shows the volume of bacterial growth. Each color (here a shade of gray) represents a different strain of bacteria (presumably a species). The color red (the darkest and dominant shade in fig. 3), consistent with its associations with biohazards and danger more broadly,[59] signals drug-resistant bacterial growth. The y-axis's representation of time corresponds to the oscillation of bacterial growth; the trend, however, is a positive slope. The rising slope inculcates a sense of urgency toward the catastrophic future, one that is heightened by the skull icon, indicating the point of the player-avatar's "death," or the doomsday apocalyptic scenario for the population more broadly. This is reinforced with the addition of a number of textual facts presented below the graph, as in figure 3's predication: "By the year 2050, deaths due to drug-resistant infections could exceed 10 million per year." The pill icon marks the point at which the player deployed the most advanced generation of antibiotic. This visual argument unveils the causal, rather than simply correlative, connection between the progress of AMR over time and the development of new drugs; it prompts users to consider the relationship between antibiotic deployment and AMR.

Thinking about this visual semiotically, we can look beyond what it denotes about the events of the game to what it connotes in the larger cultural system of antibiotic discourse. The graph makes the game more "scientific," in a sense, representing the narrative and data of each iteration of play in an objective visual form; however, like other seemingly objective forms of visual representation, graphs are cultural artifacts and are embedded in and constructed through cultural and historical values.[60] This graph follows the same pattern as those described in a 2013 *Business Insider* article, "Three Scary Charts on the Post Antibiotic Era."[61] Indeed, the visualized narrative closure—the player's own history of AMR—is meant to be one of these "scary charts." These kinds of graphs belong to a recognizable genre, even to a player who hasn't seen an actual graph about AMR.

---

[57] This overdetermination and wide range includes the economics of the pharmaceutical industry, practices of the agriculture industry, social determinants of global health, and the nuances of doctor-patient interaction, just to name a few. See B. Brown and P. Crawford, "'Post Antibiotic Apocalypse': Discourses of Mutation in Narratives of MRSA," *Sociology of Health & Illness* 31 (2009): 508–24.

[58] This is in fact one of the problems with the public understanding of AMR; the most common misconception is that an individual body develops resistance—somewhat analogous to the idea of pharmacological tolerance or drug receptor downregulation—so that a person becomes "immune" to antibiotics. See Wellcome Trust, *Consumer Perspective* (cit. n. 5).

[59] On the colors and icons of "pandemic culture," see Mitchell and Hamilton, "Playing at Apocalypse" (cit. n. 14).

[60] Julie Doyle, *Mediating Climate Change* (Farmham, UK, 2011), 36.

[61] Max Nisen, "Three Scary Charts on the Post-Antibiotic Era," *Buisness Insider*, 8 January 2013, http://www.businessinsider.com/massive-risk-of-antibiotic-resistant-bacteria-in-three-charts-2013-1.

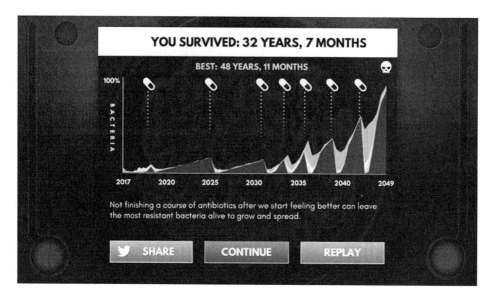

Figure 4. *The cataclysmic positive slope.*

This concluding graph's (fig. 3) visual rhetoric indexes the way AMR follows much the same pattern as visual representations of climate change and other cataclysmic processes when they are mapped onto a time axis by scientists and popular media. The central cue is the positive slope, which shows that as time progresses, the terrible apocalyptic potential also increases—as signified in *Superbugs* by the red coloring of the drug-resistant bacteria (figs. 3, 4). This visually captures the way the antibiotic apocalypse has drawn from the same kind of "catastrophe discourse" as climate change, making it seem worthy of attention.[62] Such a form of representation is rhetorically effective, drawing the player's attention by virtue of their authorship of a simulated history of antibiotic resistance, and suggesting the existence of real-world correlatives with similar projections. Beyond its connotations and intertextual associations, the concluding graph is also essential to the ludic aspects of the game. It influences the player's future strategy in the instances of gameplay that follow.

Like the narrative and iconography, the rules that govern the player's activity are simple. Bacteria grow spontaneously at a steady rate, and the goal is to prevent them from filling the petri dish. The player's primary mechanic is tapping an area of the petri dish, and holding to expand the area of application (fig. 5). Longer application of this antibiotic spray increases the likelihood and number of bacteria being killed, evoking the imperative to use the full, prescribed course of antibiotics; however, the amount one can use (represented by a light blue "Antibiotic" bar on the right) is then temporarily reduced. The antibiotic is most effective when the targeted bacteria is centered as precisely as possible within the purple application circle, a visual and ludic metaphor for the antibiotic spectrum, what has been called "the right drug for the right bug."[63] A secondary mechanic is activating a new antibiotic so that one can kill bacteria already resistant to another drug.

---

[62] Nerlich, "Post-Antibiotic" (cit. n. 12), 575–6.
[63] Podolsky, *Antibiotic* (cit. n. 12), 40. By spectrum, I mean the range of bacterial species for which a given antibiotic is effective. Given the scope of the game, I have omitted Podolsky's more nuanced

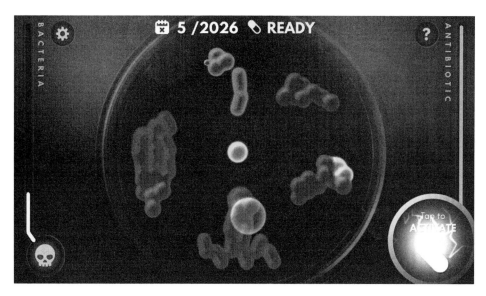

Figure 5. The small purple (central bottom and semitransparent) circle shows the primary mechanic, the application of antibiotic via the touchscreen. The secondary mechanic is deploying the next-generation antibiotic once it has been developed, by holding the circle on the lower right, outside the petri dish.

These mechanics quickly instruct players in the different causes of AMR (the three mechanisms by which bacteria develop resistance), which are random mutation, survival of exposure to antibiotics, and horizontal gene transfer. Random mutation is a product of stochasticity without selective pressure, both in real cell reproduction and in the game. This does happen occasionally, but not frequently in comparison to antibiotic-induced AMR. The second mechanism is a direct response to how the player applies antibiotics; that is, if bacteria are exposed to antibiotics directly, but the quantity is too low, they grow larger but do not die.[64] If they are in the petri dish but are not touched by the antibiotics within the dish, they will develop resistance by virtue of exposure. Finally, non-drug-resistant bacteria that grow nearby and touch resistant bacteria then adopt the trait, reflecting horizontal gene transfer. In that way, bacteria of the same generation can share genetic material and consequently have the same traits. As resistance mounts, so does the probability of failing to "survive."

LOSING, SO YOU CAN LOSE BETTER

Superbugs wants you to fail, so that you can fail better the next time. Jesper Juul has suggested that failure is an essential component of video games; failure in a game really means we are in some respect inadequate, but unlike real life, "games give us a fair

---

characterization of "rational therapeutic" imperatives (8), which he calls the right drug for the right bug for the right patient for the right price at the right time.

[64] The way in which the antibiotic spray causes bacteria to pop and disappear reflects the mechanisms of action of many antibiotics, such as the β-lactams (including penicillin), inhibiting the growth of the cell wall and causing cells to rupture from internal pressure.

Figure 6. *The results of deploying new antibiotics as soon as they become available.*

chance to redeem ourselves."[65] The pedagogical function of failure is essential to the science-fictional function in that it reveals the player's way of thinking about humans, disease, and drugs. The game encourages and highlights the option of immediately using a new drug when it becomes available, thereby reflecting one of the cultural tendencies that facilitated AMR; this was the assumption that there are unlimited resources (both conceptually and in terms of natural sources, such as soil from which so many antibiotics were derived) for creating new antibiotics.[66]

The only way for players to overcome drug-resistant bacteria is to develop and deploy the next generation of antibiotics, represented at the bottom right of the screen by a pill animated with lightning bolts and the encouragement, "Tap to ACTIVATE." At first, most players will exercise this option as soon as it is available, merely because they can, or because drug-resistant bacteria are already present in their dish. But they soon encounter a caveat: each new generation takes longer to develop. The graph in figure 6 shows the research lag in the antibiotic pipeline between bacterial generations. This process mirrors the history of the antibiotic era, when bacteria quickly developed resistance to early antibiotics like penicillin and streptomycin—often as soon as testing began—and new antibiotics, both natural and synthetic, were introduced. In the game, the first few generations of antibiotics develop quickly—18 months, 20 months, 24 months—but progress rapidly slows to impossible time frames for their use. This shift reflects the rapid discovery of new antibiotics between 1940 and 1970, in contrast to the last 30 years, which have yielded no new antibiotic classes.[67] This is one of the messages that can ap-

[65] Jesper Juul, *The Art of Failure: An Essay on the Pain of Playing Video Games* (Cambridge, Mass., 2013), 7.

[66] There have been diminishing returns in the discovery of new antibiotic compounds in the soil. By the 1960s, it has been suggested, soil-screened antibiotic compounds were "overmined"; Losee L. Ling et al., "A New Antibiotic Kills Pathogens without Detectable Resistance," *Nature* 517 (2015): 455–9, on 455.

[67] Between 1938 and 1968, eight classes of antibiotics were introduced. (There is often a significant gap between discovery and introduction, with penicillin an emblematic example; discovered in 1928 and introduced in 1938, it was not mass produced to be widely available until 1945.) Between 1968 and 2013,

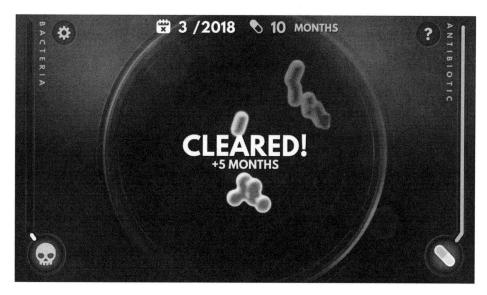

Figure 7. *Clearing. All (red) AMR bacteria have been cleared, prompting the growth of "normal" (not red) bacteria.*

pear under the concluding graph. Rather than merely seeing a visual representation of alarming trends in antibiotic development, players experience the research lag in the antibiotic pipeline by becoming agents and authors of their own "scary" graphs.

In much the same way, *Superbugs* tempts users to activate the latest drug despite not needing to—it encourages them to fail in a second way. From the bacteria's first appearance, most players will aim to kill all of them in sight. And they get a time bonus (artificially adding to their length of survival) every time they clear the dish of bacteria (fig. 7).

The catch is that immediately after the dish is cleared, the number and size of bacteria that emerge on the blank slate increase significantly. Similar to the relationship of diminishing returns between new drugs and the time to discovery, each full clearing increases both the amount of bonus time given and the initial volume of the spontaneous colonization that follows. This process reflects the overuse of antibiotics in the drive to clear out all infection and even colonization.

In one specific sense, this rapid growth following a clearing corresponds to the "scorched-earth" effect of many broad-spectrum antibiotics, as Laurie Garrett suggests in *The Coming Plague: Newly Emerging Diseases in a World Out of Balance.*[68] Deriving its conceptual metaphor from war, scorched-earth pharmacology creates an imbalance in the microbiome, catalyzing an environment where species such as

---

only five classes were advanced. It is widely cited that there have been no new antibiotics introduced in the last 30 years. This claim should be qualified, as it references the "last clinically useful" new drug, daptomycin, which was developed in 1986—but not approved for use until 2003. There have been new antibiotics since then, but they are not widely applicable due to their very narrow spectrum; the Diarylquinolines are an example, and they are effective only against TB. See Kim Lewis, "Platforms for Antibiotic Discovery," *Nature Reviews Drug Discovery* 12 (2013): 371–87, on 371. In 2015, teixobactin made headlines by possibly representing a new class of antibiotics that inhibit cell wall synthesis through a different mechanism than other drug types (β-lactams for example). See ibid.

[68] Garrett, *Coming Plague* (cit. n. 23), 618–9.

C. diff can emerge and become pathogenic because they are no longer kept in check by the other gut flora. More broadly, this excessive clearing reflects the utopian "dream of hygienic containment,"[69] which is the kind of fantastical closure to the total war on bacteria evident in the science writing of researchers like Solokoff. In these two misdirections that appeal to common cultural attitudes, the game initially fosters a stratagem along the logic of the "martial metaphor" that reflects the "arms race" between medical science and bacteria and the desire for hygienic containment.[70]

Having experienced repeated failures with the model based on the conceptual medical metaphor of war and the antibiotic arms race, the user is led by the game's "procedural rhetoric," or the processes and rule-based logic through which games make arguments,[71] toward the idea of coexistence. In medical-scientific terms, albeit simplistically, this follows a more ecological understanding of disease, one that looks to the broader context in which humans, pathogens, and their environments interact.[72] The most effective strategy for *Superbugs* that I was able to exercise, and which I later confirmed with the game's developers, is to balance bacterial growth with judicious applications of antibiotics and, even more carefully, with the activation of next-generation antibiotics. This means letting bacteria colonize to some degree and only using antibiotics to keep their growth manageable. The strategy applies even to resistant bacteria; an effective tactic is to let drug-resistant bacteria grow in insolation by not allowing new bacteria to make contact and become resistant through horizontal gene transfer, and to only activate a new antibiotic to eradicate the isolated colony when its size becomes unmanageable (fig. 8).

In practical terms, besides awareness and a more foundational understanding of AMR, the game inculcates a kind of lay antibiotic stewardship, or the "practices of daily behavior" it seeks to influence.[73] This stewardship involves using full courses of antibiotics, not sharing prescriptions, and not pressuring physicians to prescribe them when they are not necessary. More broadly, by highlighting the conceptual efficacy and sustainability of balance versus battle, it encourages more than specific practice; it changes the player's frame of mind.

And yet, despite this Socratic edutainment, *Superbugs*'s algorithms suggest that the antibiotic apocalypse is inexorable. Widely cited predications of the future of AMR are written into the game's procedural rhetoric. The NESTA team and PRELOADED, the game's designer, suggest that skilled players can make it to around fifty years, but after this point, the rate at which the bacteria grow and develop resistance makes it impossible to continue regardless of strategy. This number of years is not arbitrary; it corresponds to the "potential crisis point" described in the United Kingdom's recent "AMR Review."[74] As the game is targeted toward young teens, 2050 is a year that PRELOADED Studios suggests

---

[69] Alison Bashford and Claire Hooker, eds., *Contagion: Historical and Cultural Studies* (London, 2001), 3.

[70] On the early implementation of the socioeconomics of the "arms race," see Podolsky, *Antibiotic Era* (cit. n. 63), 146.

[71] Ian Bogost, *Persuasive Games: The Expressive Power of Videogames* (Cambridge, Mass., 2007), 30, 47.

[72] For a longer history of this ecological model, its supporters during the early to mid-twentieth century, and its relations to older holistic epistemes of disease (along the lines of the humoralism of Hippocrates and Sydenham), see Warwick Anderson, "Natural Histories of Infectious Disease: Ecological Vision in Twentieth-Century Biomedical Science," *Osiris* 19 (2004): 39–61.

[73] *Superbugs* (cit. n. 8).

[74] The UK Prime Minister commissioned the Review on Antimicrobial Resistance to address the growing global problem of drug-resistant infections. It was chaired by Jim O'Neill and supported by the

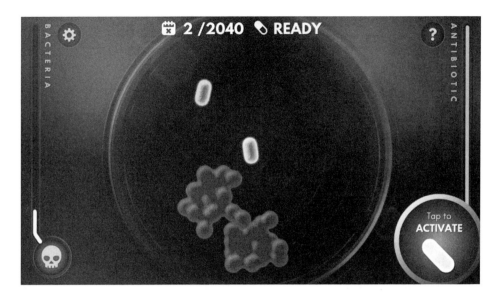

Figure 8. *Allowing for controlled growth and withholding activation of next-generation antibiotics. The two clusters at the bottom are drug-resistant bacteria.*

is "likely to fall within most of our players' lifetimes."[75] Although the data behind this predication are contested,[76] it continues to circulate in the popular media, contributing to SF-like speculations of the post-antibiotic apocalypse that sound like imaginative constructions of the future in the hyperbolized settings and tropes of SF texts.

The significance of trying to "game" the antibiotic apocalypse through *Superbugs* lies in the way this performs science fictionality, bringing together the history of science, SF, and game studies in a unique way to orient action. The video game medium provides the potential to demonstrate "one of the most valuable qualities about science fiction—its capacity to mobilize critical thinking about the world."[77] On the level of procedure, the game uses failure to prompt users to rethink the way they play. But with infinite retries limiting the cost of failure, video games allow us to spend more time and energy "thinking about *why we* failed, *what* we can do about it, and *how* it reflects on us personally."[78] That is, players can rethink their strategies, from the tim-

---

Wellcome Trust and the UK Government, but operated and spoke with full independence from both; "Antimicrobial Resistance: Tackling a Crisis for the Health and Wealth of Nations," 11 December 2014. A more recent (2016) review was also commissioned, and like the 2014 review on page 4, referenced the year 2050 (11); "Tackling Drug-Resistant Infections Globally: Final Report and Recommendations," 19 May 2016. A press release from the United Nations General Assembly on the problem of AMR in 2016 also warned that by the year 2050, 10 million people would die per year as a result of AMR infections, eclipsing cancer as the third leading cause of death; UN General Assembly, "World Health Leaders Agree on Action to Combat Antimicrobial Resistance, Warning of Nearly 10 Million Deaths Annually If Left Unchecked," *United Nations Coverage and Press Releases*, 21 September 2016, http://www.un.org /press/en/2016/ga11825.doc.htm.

[75] "Nesta, Longitude Prize, Superbugs," PRELOADED, http://preloaded.com/case-studies/superbugs/ (accessed 25 July 2017).

[76] Marlieke E. A. de Kraker, Andrew J. Stewardson, and Stephan Harbarth, "Will 10 Million People Die a Year Due to Antimicrobial Resistance by 2050?" *PLOS Medicine* 13 (2016): e1002184.

[77] Frelik, "Video Games" (cit. n. 15), 236.

[78] Juul, *Art of Failure* (cit. n. 65), 90 (emphasis in the original).

ing of screen presses to the general framework in which they approach the game's challenge. *Superbugs* facilitates this recognition by analogizing the history of a player's failures to the actual history of antibiotic failure.

If *Superbugs* confronts players with the model that underlies the system of its gameplay,[79] then it also unveils the epistemic model and cultural logics that have facilitated AMR. While the game, its sponsors, and much of the data contextualizing it is based in the United Kingdom, and the game's producers are transparent about this fact,[80] it is without question that AMR is a global, ecological problem. Evidenced in the vast expanse of the bacterial resistome,[81] AMR is less bound by state borders than those sensational viruses that travel among the interconnected circuits of capital in a globalized economy, as has been so often dramatized in outbreak narratives.[82] That is not to say that AMR does not have region-specific problematics; there is much work in medical and bioethics that questions the current relationship and obligations among first-world and developing nations with respect to AMR.[83] *Superbugs* doesn't instruct players in all of the nuanced histories and complexities of AMR; it is meant to spur awareness, interest, and action in present daily practices, and to encourage "citizen science" and possible future professional aspirations in children and young adults. This intent is evidenced by its use in science exhibits and classrooms.[84] And in that capacity, it has been successful. As I have shown, the way *Superbugs* achieves this is more than a matter of making scientific knowledge digestible and appealing to young audiences through interactive learning, however.[85] By using game studies and SF to think through the game's relation to broader discourse of the antibiotic apocalypse, we can see how *Superbugs* abstracts the grammar of AMR as history and future to scale it temporally, in effect, revealing a bellicose, anthropocentric, and unsustainable tenet of modernity—a naturalized war against bacteria, and a metaphor that has left such dire material consequences in its wake. By confronting its users with the logics that underlie the game's model of play, *Superbugs* makes them become complicit agents and authors of the same kind of history of drug development and utopian hopes that the antibiotic era passed through. Thus, if it is anything, *Superbugs* is surely a first step in mobilizing the understanding required to tackle AMR's com-

---

[79] Patrick Crogan, *Gameplay Mode: War, Simulation, and Technoculture* (Minneapolis, Minn., 2011), 143–8.

[80] The game's first version included the line, "Current version of the game is purposed for a UK audience," reflecting the fact that many of the statistics and information the players received were specific to the United Kingdom; *App Store Preview: Superbugs: The Game*, by NESTA, https://itunes.apple.com /us/app/superbugs-the-game/id1107524986?mt = 8 (accessed 1 September 2018).

[81] A resistome in this case is the mass of antibiotic-resistant genes spanning the genomes of both pathogenic and nonpathogenic bacteria; V. D'Costa, K. McGrann, D. Hughes, and G. D. Wright, "Sampling the Antibiotic Resistome," *Science* 311 5759: 374–7.

[82] Wald, *Contagious* (cit. n. 13), 25.

[83] Jasper Littmann and A. M. Viens, "The Ethical Significance of Antimicrobial Resistance," *Public Health Ethics* 8 (2015): 209–24. From an international relations perspective, see Stefaine R. Fishell, *The Microbial State: Global Thriving and the Body Politic* (Minneapolis, Minn., 2017).

[84] Iasha Iqbal, "What Does a Teacher Have to Say About Superbugs," *Longitude Prize*, 13 January 2017, http://longitudeprize.org/blog-post/what-does-teacher-have-say-about-superbugs; Nina Cromeyer Dieke, "Game on at the Museum," *Longitude Prize*, 27 July 2016, http://longitudeprize.org/blog-post/game -museum.

[85] "Superbugs takes the complex science behind the spread of antibiotic resistance in bacteria and models it as a casual mobile game that's not only great fun to play but scientifically sound." The game's developer, PRELOADED, goes on to suggest that "Games are a great tool for engaging people with science—they can make the complex seem simple, and by their very nature pull in players to construct their own learning experience." See "Nesta" (cit. n. 75).

plexities in terms of public health ethics, economics, and agrilogistics,[86] and in both its national insularities and broader global assemblages.

With games like *Superbugs* and SF-laden news media on the post-antibiotic apocalypse, it is not very hard to see how AMR is everyone's problem; what is less obvious is how AMR is "everyone's fault," as Dr. Brad Spellberg bluntly states: "It is a collective failure to respond to the problem."[87] Although we do not have infinite tries "in real life," what we do have is a critical tool, in the form of SF, to make the failure that began in the antibiotic era as productive as possible by understanding the relationship between the history, future, and present situation of AMR. Yet, because *Superbugs* works in the narrative frame of the apocalypse and prompts a sense of urgency, it also faces the pitfalls of using this extreme rhetoric to communicate science for the purpose of influencing the subtler practices of daily life. It might desensitize people, or make AMR seem as though it has unrealizable potential, much like the nuclear holocaust science fiction of the midcentury. However, the way the game and the discourse of the antibiotic apocalypse work science fictionally, not merely sensationally, suggests that this discourse, particularly when performed in the way *Superbugs* allows, might just permit the future of the antibiotic apocalypse to be less of a devastation and more of a revelation.

---

[86] Agrilogistics: "The specific logistics of agriculture that arose in the Fertile Crescent and that is still plowing ahead. Logistics, because it is a technical, planned, and perfectly logical approach to built space. Logistics, because it proceeds without stepping back and rethinking the logic. A viral logistics, eventually requiring steam engines and industry to feed its proliferation"; Timothy Morton, *Dark Ecology: For a Logic of Future Coexistence* (New York, N.Y., 2016), 42.

[87] Brad Spellberg, "Dr. Brad Spellberg: Antibiotic Resistance Is 'Everyone's Fault,'" interview by Sarah Childress, *Frontline*, 22 October 2013, PBS, http://www.pbs.org/wgbh/frontline/article/dr-brad-spellberg-antibiotic-resistance-is-everyones-fault.

Notes on Contributors

Peter J. Bowler is Emeritus Professor of the History of Science at Queen's University Belfast. He has published extensively on the history and impact of evolutionism, most recently with *Darwin Deleted* (Chicago, 2013). He now works on popular science and futurological speculation in the early twentieth century, and has published *A History of the Future* (Cambridge, UK, 2017). He is also working on a project to understand the changing structure of the idea of progress, using the impact of Darwinism as a model.

Lisa Garforth is Senior Lecturer in Sociology at Newcastle University, UK. Her recent monograph *Green Utopias: Environmental Hope Before and After Nature* (Cambridge, UK, 2017) brings together long-standing research on green utopianism, fiction, and environmental futures. From 2016 to 2018 she was coinvestigator on the Arts and Humanities Research Council project Unsettling Scientific Stories, which examined perceptions of socioecological crisis as they have been articulated in environmental discourse and science fiction, and explored what sociologies of readers and reading can reveal about popular understanding of science futures. She is currently working on how the science-fictional imagination can contribute to sociological epistemologies for the Anthropocene.

Nathanial Isaacson is Associate Professor of Modern Chinese Literature in the Department of Foreign Languages and Literature at North Carolina State University. His research interests include Chinese science fiction, Chinese cinema, cultural studies, and literary translation. Nathaniel has published articles in the *Oxford Handbook of Modern Chinese Literatures*, and journals including *Wenxue* and *Science Fiction Studies*, as well as translations of nonfiction, poetry, and fiction in the translation journals *Renditions*, *Pathlight*, and *Chinese Literature Today*. His book, *Celestial Empire: The Emergence of Chinese Science Fiction* (Middletown, Conn., 2017), examines the emergence of science fiction in late Qing China, and the relationship between science fiction and Orientalism.

David A. Kirby was a practicing evolutionary geneticist before leaving bench science to become Professor of Science Communication Studies at the University of Manchester, UK. His experiences as a member of the scientific community have informed his internationally recognized studies on the interactions between science and entertainment media. His book *Lab Coats in Hollywood: Science, Scientists, and Cinema* (Cambridge, Mass., 2011) examines the historic collaborations between scientists and the entertainment industry. He is currently writing a book titled *Indecent Science: Religion, Science, and Movie Censorship*, which will explore how movies served as a battleground over the role of science in influencing morality.

Nikolai Krementsov is Professor at the Institute for the History and Philosophy of Science and Technology of the University of Toronto. He is the author of several books and numerous articles on the history of biomedical sciences and medicine in twentieth-century Russia. His latest books include *A Martian Stranded on Earth: Alexander Bogdanov, Blood Transfusions, and Proletarian Science* (Chicago, 2011), *Revolutionary Experiments: The Quest for Immortality in Bolshevik Science and Fiction* (Oxford, 2014), and *With and Without Galton: Vasilii Florinskii and the Fate of Eugenics in Russia* (Cambridge, UK, 2018).

Erika Lorraine Milam is Professor of History and the History of Science at Princeton University, where she specializes in the history of the modern life sciences, especially evolutionary theory. She is author of *Looking for a Few Good Males: Female Choice in Evolutionary Biology* (Baltimore, 2010) and *Creatures of Cain: The Hunt for Human Nature in Cold War America* (Princeton, 2019). Together with Robert A. Nye she coedited *Scientific Masculinities, Osiris* vol. 30 (Chicago, 2015).

Colin Milburn is Gary Snyder Chair in Science and the Humanities and Professor of English, Science and Technology Studies, and Cinema and Digital Media at the University of California, Davis. His publications include *Nanovision: Engineering the Future* (Durham, N.C., 2008), *Mondo Nano: Fun and Games in the World of Digital Matter* (Durham, N. C., 2015), and *Respawn: Gamers, Hackers, and Technogenic Life* (Durham, N.C., 2018). At UC Davis, he directs the Program in Science and Technology Studies and the ModLab media laboratory.

Iwan Rhys Morus is currently Professor of History at Aberystwyth University, Wales. As a historian of science, he has a particular interest in Victorian science and culture and has published widely in the field. He is the coauthor with Peter Bowler of *Making Modern Science* (Chicago, 2005). He has also published a number of books aimed at a general audience, including *Michael Faraday and the Electrical Century* (London, 2004; new ed., 2017), and the *Oxford Illustrated History of Science* (2017). His latest book, *Nikola Tesla and the Electrical Future*, will be published later in 2019.

Projit Bihari Mukharji is Associate Professor in the History and Sociology of Science at the University of Pennsylvania. He studies the histories of science in modern South Asia, focusing primarily on how different traditions of knowledge interact. He is the author of *Nationalizing the Body: The Medical Market, Print and Daktari Medicine* (London, 2009) and *Doctoring Traditions: Ayurveda, Small Technologies, and Braided Sciences* (Chicago, 2016).

Joanna Radin is a historian of biomedical futures at Yale. She is the author of *Life on Ice: A History of New Uses for Cold Blood* (Chicago, 2017), coeditor with Emma Kowal of *Cryopolitics: Frozen Life in a Melting Age* (Cambridge, Mass., 2017), and is currently at work on a book about Michael Crichton, mass media, and the culture of biotechnology.

Lisa Raphals studies the cultures of early China and classical Greece, with interests in comparative philosophy, history of science, and science fiction studies. She is the author of *Knowing Words: Wisdom and Cunning in the Classical Traditions of China and Greece* (Ithaca, N.Y., 1992), *Sharing the Light: Representations of Women and Virtue in Early China* (Albany, N.Y., 1998), and *Divination and Prediction in Early China and Ancient Greece* (Cambridge, UK, 2013); coeditor of *Old Society, New Belief: Religious Transformation of China and Rome, ca. 1st–6th Centuries* (Oxford, 2017); and author of "Alterity and Alien Contact in Lao She's Martian Dystopia, *Cat Country*" (*Science Fiction Studies*, 2013).

Amanda Rees is Reader in Sociology at the University of York, UK. She works on the history of human-animal relationships, of field sciences, and of the future. Her earliest book was *The Infanticide Controversy: Primatology and the Art of Field Science* (Chicago, 2009), and her latest, *Human*, co-written with Charlotte Sleigh, will appear in Reaktion Books's "Animal" series. She has written for both the peer-reviewed and popular press on animal agency, the history of human prehistory, and the evolution of the history of the future, and is eagerly looking forward to eventually writing an entire book on John Wyndham.

Lorenzo Servitje is Assistant Professor of Literature and Medicine, a dual appointment in the Department of English and the Health, Medicine, and Society Program at Lehigh University. His current book project, *Medicine is War: The Martial Metaphor in Victorian Literature and Culture*, traces the metaphorical militarization of medicine in the nineteenth century. His articles have appeared in *Literature and Medicine*, *Journal of Medical Humanities*, and *Science Fiction Studies*, among other journals. He has coedited three collections: *The Walking Med: Zombies and the Medical Image*, with Sherryl Vint (University Park, Penn., 2016), *Endemic: Essays in Contagion Theory*, with Kari Nixon (London, 2016), and *Syphilis and Subjectivity: From the Victorians to the Present*, with Kari Nixon (London, 2017).

Charlotte Sleigh is Professor of Science Humanities at the University of Kent, UK. She has written a number of books in the history of natural history, history of science, and science and literature—and is slowly working her way toward a book on SF fandom in interwar Britain. She is currently editor of the *British Journal for the History of Science*.

Will Slocombe is Senior Lecturer in the Department of English at the University of Liverpool. He is the author of *Nihilism and the Sublime Postmodern* (New York, N.Y., 2006), and numerous articles on modern and contemporary literature and theory, including SF. He is interested broadly in the limits of narration and "states of mind," and the links between literature and technology. His current projects bring together these interests, and are engaged with the history of psychiatry and representations of artificial intelligence.

Alice White is a historian who works as a digital editor and Wikimedian at the Wellcome Collection. Her historical research explores scientific cultures in mid-twentieth-century Britain, and particularly how World War II shaped scientific networks and groups.

Martin Willis is Professor of English Literature at Cardiff University, Wales. His works include the award-winning *Vision, Science and Literature, 1870–1920* (Pittsburgh, Penn., 2011), as well as *Mesmerists, Monsters, and Machines* (Kent, Ohio, 2007). He contributed to *Repositioning Victorian Sciences* (London, 2007), and with Catherine Wynne coedited *Victorian Literary Mesmerism* (New York, NY, 2006). He has just completed the *Reader's Guide to Essential Criticism in Literature and Science* (London, 2014). Martin is also editor of the *Journal of Literature and Science* and Chair of the British Society for Literature and Science.

Index

340

SUGGESTIONS FOR CONTRIBUTORS TO OSIRIS

OSIRIS is devoted to thematic issues, conceived and compiled by guest editors who submit volume proposals for review by the OSIRIS Editorial Board in advance of the annual meeting of the History of Science Society in November. For information on proposal submission, please write to the Editors at pmccray@history.ucsb.edu and ss536@cornell.edu.

1. Manuscripts should be submitted electronically in Rich Text Format using Times New Roman font, 12 point, and double-spaced throughout, including quotations and notes. Notes should be in the form of footnotes, also in 12 point and double-spaced. The manuscript style should follow *The Chicago Manual of Style*, 16th ed.

2. Bibliographic information should be given in the footnotes (not parenthetically in the text), numbered using Arabic numerals. The footnote number should appear as superscript. "Pp." and "p." are not used for page references.

    a. References to books should include the author's full name; complete title of book in *italics*; place of publication; date of publication, including the original date when a reprint is being cited; and, if required, number of the particular page cited (if a direct quote is used, the word "on" should precede the page number). *Example*:

       [1] Mary Lindemann, *Medicine and Society in Early Modern Europe* (Cambridge, 1999), 119.

    b. References to articles in periodicals or edited volumes should include the author's name; title of article in quotes; title of periodical or volume in *italics*; volume number in Arabic numerals; year in parentheses; page numbers of article; and, if required, number of the particular page cited. Journal titles are abbreviated according to the journal abbreviations listed in *Isis Current Bibliography*. *Example*:

       [2] Lynn K. Nyhart, "Civic and Economic Zoology in Nineteenth-Century Germany: The 'Living Communities' of Karl Möbius," *Isis* 89 (1999): 605–30, on 611.

    c. All citations are given in full in the first reference. For succeeding citations, use an abbreviated version of the title with the author's last name. *Example*:

       [3] Nyhart, "Civic and Economic Zoology" (cit. n. 2), 612.

3. Special characters and mathematical and scientific symbols should be entered electronically.

4. A small number of illustrations, including graphs and tables, may be used in each volume. Hard copies should accompany electronic images. Images must meet the specifications of The University of Chicago Press "Artwork General Guidelines" available from the Editor.

5. Manuscripts are submitted to OSIRIS with the understanding that upon publication copyright will be transferred to the History of Science Society. That understanding precludes consideration of material that has been previously published or submitted or accepted for publication elsewhere, in whole or in part. OSIRIS is a journal of first publication.

OSIRIS (ISSN 0369-7827) is published once a year.

eBook edition eISBN: 978-0-226-47423-6

Single copies are $35.00.

Address subscriptions, single issue orders, claims for missing issues, and advertising inquiries to *Osiris*, The University of Chicago Press, Journals Division, 1427 E. 60th Street, Chicago, IL 60637-2902.

Postmaster: Send address changes to *Osiris*, The University of Chicago Press Subscription Fulfillment, 1427 E. 60th Street, Chicago, IL 60637-2902.

OSIRIS is indexed in major scientific and historical indexing services, including *Biological Abstracts, Current Contexts, Historical Abstracts*, and *America: History and Life*.

Paperback edition, ISBN: 978-0-226-68041-5

Osiris A RESEARCH JOURNAL DEVOTED
TO THE HISTORY OF SCIENCE
AND ITS CULTURAL INFLUENCES

A PUBLICATION OF THE
HISTORY OF SCIENCE SOCIETY

W. Patrick McCray
Co-Editor, Osiris
Department of History
University of California, Santa Barbara
Santa Barbara, CA 93106-9410 USA
pmccray@history.ucsb.edu

Suman Seth
Co-Editor, Osiris
Department of Science & Technology Studies
321 Morrill Hall
Cornell University
Ithaca, NY 14853 USA
ss536@cornell.edu